Analytical Supercritical Fluid Chromatography and Extraction

Analytical Supercritical Fluid Chromatography and Extraction

Edited by

MILTON L. LEE
Department of Chemistry
Brigham Young University

and

KARIN E. MARKIDES
Department of Analytical Chemistry
Uppsala University

Chromatography Conferences, Inc.
Provo, Utah

Contributing Authors

Marion R. Andersen Lee Scientific	*Supercritical Fluid Extraction*
William E. Artz University of Illinois	*Foods and Flavors, Lipids*
Keith D. Bartle University of Leeds	*FTIR, Multidimensional SFC*
Terry A. Berger Hewlett-Packard	*Mixed Mobile Phases, Flow Restriction*
Anthony J. Berry Polytechnic of Wales	*Mass Spectrometry*
Darryl J. Bornhop Linear Instruments	*UV Absorption Detection*
Claudio Borra Carlo Erba Strumentazione	*Mixed Mobile Phases, Programming*
H-C. Karen Chang Virginia Polytechnic Institute and State University	*Sulfur Chemiluminescence Detection, Electron Capture Detection*
Thomas L. Chester Procter & Gamble	*Relationship of SFC to GC and LC*
Mark S. Davis University of Massachusetts	*Photoionization Detection*
Paul A. David Amoco	*Sample Preparation*
Ilona L. Davies Dionex	*Multidimensional SFC*

Paul B. Farnsworth
Brigham Young University
Element-selective Plasma Detection

Steven M. Fields
Dow Chemical
Mobile Phases

Tyge Greibrokk
University of Oslo
Sample Introduction, Light Scattering Detection

Peter R. Griffiths
University of Idaho
FTIR Detection

Karl Grolimund
Ciba-Geigy
Industrial Applications

Steven B. Hawthorne
University of North Dakota
Supercritical Fluid Extraction

Jack D. Henion
Cornell University
Mass Spectrometry

Herbert H. Hill, Jr.
Washington State University
Fundamentals of Detection, Ion Mobility Detection

Frank Höfler
Dionex
Practice of SFC

Gunther Holzer
Georgia Institute of Technology
Natural Products

Eric C. Huang
Cornell University
Mass Spectrometry

Brian A. Jones
Lee Scientific
Open Tubular Column Technology

Jerry W. King
Northern Regional Research Center, Agricultural Research Service/USDA
Supercritical Fluid Extraction

Ernst G. Klesper
Technische Hochschule Aachen
History of SFC

Iina J. Koski
Brigham Young University
Foods and Flavors, Sulfur Chemiluminescence Detection

Joseph M. Levy
Suprex
Practical Aspects of Supercritical Fluid Extraction, Mobile Phase Selection

Dietger Leyendecker
BASF
Mobile Phases, Programming

Samuel B. Moore
Burlington Research
Textile Applications

Susan V. Olesik
Ohio State University
Reverse Micellar Mobile Phases, Flame Based Detectors, Mass Spectrometry

Kent M. Payne
Brigham Young University
Packed Column Technology

Paul A. Peaden
Unocal
Column Theory and Practice

J. David Pinkston
Procter & Gamble
Mass Spectrometry

Colin F. Poole
Wayne State University
Stationary Phases

Peter Pospisil
Chemical Waste Management
Environmental Pollutants

Douglas E. Raynie
Brigham Young University
Definition and History of SFC, Mobile Phase Mixing, Flow Restriction, Fluorescence Detection

Mark W. Raynor
Carlo Erba Strumentazione
FTIR Detection

Vernon N. Reinhold
Harvard School of Public Health
Carbohydrates, Biomolecules

Bruce E. Richter
Lee Scientific
Sample Introduction

Marja-Liisa Riekkola
University of Helsinki
Organometallic Compounds

Christine A. Rouse
Brigham Young University
Explosives

Robert M. Sauer, Jr.
University of Illinois
Foods and Flavors, Lipids

Franz P. Schmitz
Technische Hochschule Aachen and H.B. Fuller
Polymers, Programming

Wolfgang Schulz
Exxon
Petroleum, Hydrocarbons

Douglas M. Sheeley
Harvard School of Public Health
Carbohydrates, Biomolecules

Chung Hang Sin
Northern Illinois University
Supersonic Jet Spectroscopy

Sheldon R. Sumpter
Brigham Young University
Enantiomers

Larry T. Taylor
Virginia Polytechnic Institute and State University
Packed Column Characteristics and Selection

Peter C. Uden
University of Massachusetts
Photoionization Detection

Julia R. Wheeler
DuPont

Agrichemicals

Steven H. Wong
Johns Hopkins University
School of Medicine

Drugs and Pharmaceuticals

Bob W. Wright
Pacific Northwest
Laboratories, Battelle
Memorial Institute

Definition and Properties of Supercritical Fluids, Restrictors

Lily Q. Xie
Brigham Young University

Fraction Collection

Frank J. Yang
Dionex

Pumping Systems

Preface

A number of factors have brought about the recent revival of supercritical fluid chromatography (SFC). The desire to perform separations of high molecular mass compounds with higher efficiencies (per unit time) than can be performed using liquid chromatography, and the possibility of adopting both gas chromatographic and liquid chromatographic detectors for SFC to provide wide versatility in universal and selective detection have been among the most important factors. The rapid growth in SFC has occurred in only the last decade, and this trend continues to follow a steady growth pattern into the 1990s as new developments in instrumentation are made, and as analysts discover new methodologies and applications using this technique.

The closely related technique of supercritical fluid extraction (SFE) owes its current strong interest to efforts to simplify and increase the speed of sample preparation methods. For most analytical determinations today, sample preparation involves the most time-consuming and laborious steps in the analysis. The same properties that make supercritical fluids advantageous in chromatography, *i.e.*, high diffusivity and variable solvating power, provide similar advantages in sample preparation. High diffusivity leads to rapid sample penetration and extraction, while variable solvating power allows tuning of the extraction selectivity.

The literature relevant to SFC and SFE is widely scattered in various journals and reports. While the chromatographic literature is confined more to chromatographic journals, there is a wealth of related critical state information in the engineering and physical chemistry literature that has been extremely useful for both SFC and SFE. Understanding of critical state phenomena is important to properly and fully utilize supercritical fluids in analytical separations.

Until now, a comprehensive text on analytical SFC and SFE has not been available. Several useful compilations of selected topics, primarily as proceedings of conferences and symposia, have been published. While these publications provide much useful information, they do not provide complete and balanced treatment of the subjects. This book has been produced to provide a comprehensive reference source for those interested in the theoretical and technical aspects of analytical SFC and SFE, as well as for analysts concerned with details of the practice of the technique. Both the experienced chromatographer and the newcomer should find this book useful.

Chapters 2-4 include technical discussions of columns, mobile phases, and instrumentation for SFC, while Chapter 5 covers the technical aspects of SFE. Chapters 6 and 7 are more practical in nature, and they contain some rather unique sections. In Chapter 1, the relationship of SFC to LC and GC is treated rather extensively, and the unifying nature of SFC in the field of chromatography is clearly seen. Chapter 6 treats the details of the practice of SFC and includes a comprehensive trouble shooting section. This book has been designed to be a reference covering all aspects of analytical SFC and SFE.

This book was started approximately one year ago, and its timely completion is a credit to many of our scientist friends and associates. Particularly noteworthy are the efforts of the contributing authors who assisted us in planning, writing, and reviewing sections that comprise the text of the book. We appreciate the help of many who proofread the manuscript at various stages of completion. Special thanks and appreciation go to Douglas E. Raynie who spent many hours checking for consistency; LaVoy C. Hill who typed, retyped, and produced the camera-ready manuscript for reproduction; Sherri H. Aldous who prepared all of the figures for inclusion in the book; and finally our families who supported us in this undertaking.

Milton L. Lee
Karin E. Markides

Contents

3 MOBILE PHASES

4 INSTRUMENTATION

5 ANALYTICAL SUPERCRITICAL FLUID EXTRACTION

6 PRACTICE OF SFC

7 APPLICATIONS

List of Symbols

Arabic		Page
A	Fraction of h caused by eddy diffusion	24
A_b	Absorbance	251
A_c	Cross-sectional area of an extraction cell	337
A_p	Peak area	235
AU	Absorbance units	206
B	Fraction of h caused by longitudinal diffusion	22
C	Fraction of h caused by resistance to mass transfer	22
c	Concentration	21
cR	Chiral resolution	546
ΔC_p	Heat capacity	125
D_L	Detection limit	191
d_c	Open tubular column internal diameter	22
d_f	Stationary phase film thickness	42
d_p	Particle diameter	24
d_{pc}	Diameter of packed column	59
D	Solute binary diffusion coefficient	22
D_o	Sauter mean diameter	234
E_t	Electric field strength	240
F	Flow rate	150
ΔG	Free energy	29
h	Height equivalent to a theoretical plate	22
ΔH	Enthalpy	125
i	Index of linearity	190
K	Distribution coefficient	21
K_c	Mass transfer coefficient	323
K'_A	Modified mass transfer coefficient	322
K_o	Reduced mobility constant	240
k	Capacity factor	21
k_o	Ratio of particle pore volume to particle interstitial volume	47
L	Length	22
L	Sample volume loadability	59
LDR	Linear dynamic range	193

		Page
ℓ	Pathlength of optical detector flow cell	251
M	Mass	235
MDQ	Minimum detectable quantity	192
N	Detector noise (standard deviation of the baseline)	191
N'	Peak capacity	286
n	Number of theoretical plates	22
P	Pressure	12
R	Gas constant	29
R_d	Instantaneous detector response	190
R_e	Reynolds number	323
R_f	Response factor	235
RI	Refractive index	204
R_s	Resolution	28
ΔR_s	Fractional resolution loss	59
RMR	Relative molar response of a detector	192
RWR	Relative weight response of a detector	192
r	Radius	187
S	Detector sensitivity	190
S'	Detector sensitivity proportionality constant	190
ΔS	Entropy	128
Sc	Schmidt number	323
SEL	Detector selectivity	192
Sh	Sherwood number	323
S/N	Signal-to-noise ratio	160
s	Solute solubility	30
T	Temperature	12
t_r	Retention time (time required from moment of injection for elution of a chromatographic peak)	22
u	Linear velocity of the mobile phase	21
u_c	Charging linear velocity	169
V	Volume	21
v_i	Velocity of an ion in ion mobility spectrometry	240
w	Rate at which the density changes along the column	53
$w_{1/2}$	Width of a chromatographic peak measured at half its maximum height	22
W_o	Molar ratio of water-to-surfactant	111
X	Mole fraction	376
x	Amount of a sample component	21
Z	Fluid compressibility	320

Greek Page

α	Chromatographic selectivity	28
β	Phase ratio	21
δ	Solubility parameter	117
ϵ	Molar extinction coefficient	251
ϵ_e	Interstitial porosity	59
ϵ_i	Intraparticle porosity	59
ϵ_T	Total porosity	59
η	Viscosity	51
ϕ	Fugacity	126
π^*	Solvent polarizability (Solvatochromic polarity)	18
ψ	Solute-solvent interaction parameter	321
ρ	Density	14
σ	Standard deviation	22
σ^2	Variance	22
χ	Compressibility factor	149

Subscripts

Unless otherwise stated in the text, subscripts refer to the following:

A, B	System components
a	Analyte
c	Critical
d	Detector
g	Gas
i, j	Sample components
L	Limit, limiting
l, liq	Liquid
M	Mass
m	Mobile phase
min	Minimum
n	Oligomer number
o	Final, outlet conditions
obs	Observed
opt	Optimum
s	Stationary phase
V	Volume
v	Vapor

Superscripts

Unless otherwise stated in the text, superscripts refer to the following:

i	Linearity index
x	Coefficient of regression (slope) of response curve
∞	Infinite dilution

Bar above a symbol indicates molar quantity

Analytical Supercritical Fluid Chromatography and Extraction

While the similarities of gas and liquid chromatography are much more prominent than the differences, they can be stated much more succinctly; zone migration and separation occur by virtue of the same kinds of thermodynamic, flow, kinetic, and diffusion processes, and are thus subject to the same theoretical laws.

The differences between gas and liquid chromatography are naturally centered around the differences in gases and liquids. As far as chromatography is concerned, the most significant differences are related to some of the physical properties and their immense variation from liquids to gases.

J. Calvin Giddings
Dynamics of Chromatography
1965

CHAPTER 1

INTRODUCTION

1.1 Definition and Description of Supercritical Fluid Chromatography

Supercritical fluid chromatography (SFC) may be defined as a form of chromatography (*i.e.*, a physical separation method based on the interaction of an analyte in a mobile phase with a stationary phase) in which the mobile phase is subjected to pressures and temperatures near or above the critical point for the purpose of enhancing the mobile phase solvating power. Typically, one or both parameters (*i.e.*, pressure and temperature) extend into the critical region during a chromatographic run. This definition encompasses other less defined forms of chromatography such as dense gas chromatography, hyperpressure gas chromatography, and near (or sub-) critical fluid chromatography. As shown in Figure 1.1, the use of supercritical solvents for chemical separations, including both chromatography and extraction, represents a unique niche in high pressure research.

There is more to the definition of SFC than is implied from the strict definition of a supercritical fluid. In fact, helium is supercritical in gas chromatography (GC) when pressures in the column are greater than 2.26 atm. Since under these conditions the whole length of the column is not supercritical (the column outlet is usually at atmospheric pressure) and the mobile phase has essentially no solvating power, this cannot be considered SFC. Furthermore, if one investigates the practical operating conditions of SFC, it will be obvious that many of the chromatographic analyses are started at pressures below the critical pressure, and occasionally at subcritical temperatures. In general, three conditions must be met to truly define SFC: (a) the mobile phase must always be at temperatures and pressures near or above their critical point, (b) the mobile phase must

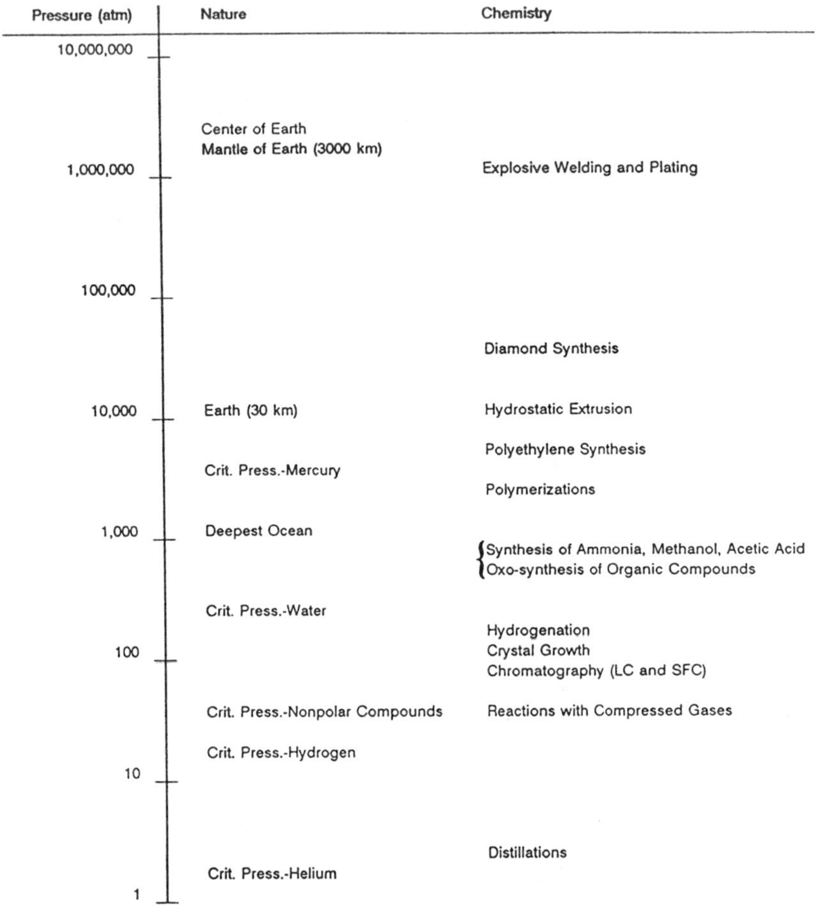

Figure 1.1. Pressure scale in nature and chemistry (according to similar compilations by Schneider[1] and Pilz[2]).

possess solvating power and, thus, be able to contribute to selectivity in the chromatographic process, and (c) the mobile phase must be subject to these conditions throughout the full length of the analytical column.

The nature of supercritical fluids is such that the mobile phase can easily be varied from gas-like to liquid-like. Based on judicious choice of mobile phase, this may lead to a variety of advantages for SFC over other separation techniques. Fluids with low critical temperatures permit operation at temperatures conducive to the analysis of thermally labile solutes and allow for greater stationary phase selectivity. The higher solvating ability of supercritical fluids permits the analysis of high molecular mass, nonvolatile compounds. SFC is compatible with essentially all detectors commonly used in both gas chromatography (GC) and liquid chromatography (LC). Both open tubular and packed columns, with the advantages and disadvantages associated with each column type, are widely used. SFC has a variety of parameters (*i.e.*, pressure, density, temperature, mobile phase composition, and stationary phase composition) which can be manipulated to effect solute retention. Finally, the coupling of extraction methods or multidimensional systems with SFC is possible, as in other forms of chromatography.

SFC can be conveniently divided into two categories based on column type: open tubular and packed. The choice of column type is due not only to the obvious chromatographic differences (*e.g.*, sample capacity, resolving power, *etc.*), but also to differences in column pressure drop and volumetric flow, which impose different constraints upon the system. Packed columns for SFC can be classified into three types, borrowing from developments in LC: (a) conventional packed columns that typically are of 2-4.6 mm internal diameter (i.d.), (b) 1-mm i.d. "microbore" columns, and (c) packed capillaries that have diameters of 0.2-0.8 mm. Packing materials of 3-10 μm diameter are similar to those used in LC. Peaden and Lee[3] have shown that to achieve resolution comparable to capillary GC with 200-μm i.d. columns, open tubular columns for SFC must have internal diameters smaller than 100 μm. Hence, 50-μm i.d. columns are commonly used in open tubular column SFC, although research on smaller diameter columns is in process.[4] The SFC equivalent to GC "megabore" columns (*i.e.*, >100-μm i.d. for SFC) has not come into common use. Column considerations are addressed in Chapter 2.

Of course, it is the nature of the mobile phase which is unique to SFC. The properties of supercritical fluids (*e.g.*, diffusivity, density, viscosity, *etc.*) are intermediate between those of gases and liquids, and can be varied by subtle changes in the operational parameters (*i.e.*, pressure and temperature). It is these properties that are exploited in SFC. Gases, supercritical fluids, and liquids have been compared as chromatographic mobile phases,[5-7] and criteria for selecting suitable mobile phases for SFC have been specified. These considerations include: (a) critical pressure, (b) critical temperature, (c) dipole moment, (d) chemical interactions with the stationary phase, (e) chemical interactions with the analyte, (f) compatibility with the detection system, (g) compatibility with seals, tubing, and pumps, (h) environmental and safety considerations, (i) cost, and (j) purity.

To enhance the solvating power of the supercritical mobile phase for polar or high molecular mass analytes, miscible organic dopants or entrainers, known as modifiers, are commonly added to SFC mobile phases. The modifier may interact with the analyte, mobile phase, or stationary phase to influence solute retention. Three advantages have been noted[8] as accompanying the use of modifiers: (a) improvement in the solubility of low-volatility and polar solutes, (b) modification of the pressure-volume-temperature (PVT) behavior of the supercritical fluid, and (c) improvement in selectivity. Various mechanisms have been proposed[9] to account for the modifier effect in SFC. These are: (a) interaction of polar molecules with stationary phase active sites, (b) solute-modifier interactions forming stable species which favor the mobile phase, (c) short-range clusters formed between the modifier and the mobile phase, and (d) a decrease in interfacial tension between the mobile and stationary phases. SFC mobile phase considerations are discussed in depth in Chapter 3.

As with any emerging technology, a variety of terminology has evolved, much of which is confusing or inconsistent. Through either developing trends, correct language usage, or convenience, the terminology and units suggested in this paragraph are recommended. The term "isopycnic" has been suggested to denote constant density at constant pressure and temperature.[10,11] Isopycnic is derived from the Greek word "pycnos," meaning dense, and therefore it denotes a body having equal density at different points; for example, pycnometers are density measuring instruments. Two isopycnic states may be attained at different temperatures merely by changing the pressure. Pressure units of atmospheres (atm) or bars are convenient and, consequently, programming rates of atm (or bar) min^{-1} follow. Density units of g mL^{-1} are advocated with density programming rates of g mL^{-1} min^{-1}. Since nearly all density (and pressure) programming is done isothermally, the operational temperature should be noted. To remain consistent with LC terminology, the term "gradient elution" should be restricted to mobile phase compositional gradients. Use of the term "simultaneous density-temperature programming" should only be used when density corrections for changing temperature have been made. Otherwise, the term "simultaneous pressure-temperature programming" would be preferred. Binary mobile phase systems are usually expressed in terms of concentration of modifier in the primary (or bulk) fluid. Mole percent, rather than weight or volume percent, is the preferred concentration unit since interactions occurring with modifiers are on a molecular basis and not on a mass or volume basis. Systems with more than two components should be specified as ternary, quaternary, *etc.*, mixtures.

One of the major advantages of SFC is the potential compatibility of the technique with the full array of detectors available for both GC and LC. As will be presented in Chapter 4, numerous different on-line detectors have been reported. The most popular SFC detection methods are the flame ionization detector (FID), ultraviolet absorption detector (UV), mass spectrometer (MS), and Fourier transform infrared spectrometer (FTIR).

A compilation of over 330 SFC chromatograms[12] was prepared for the 1988 Workshop on Supercritical Fluid Chromatography; a companion volume[13], with 350 chromatograms, was compiled for the 1989 meeting. These compilations represent the wide range of applications for which SFC is well-suited. Applications of SFC are wide-ranging, including drugs, chiral compounds, foods and natural products, fatty acids and derivatives, glycerides, steroids, biomolecules, explosives, pesticides and herbicides, fossil fuels and aromatic compounds, polymer additives, polymers, and miscellaneous chemicals and products. Analytical applications of SFC are extensively discussed in Chapter 7.

1.2 Historical Development of SFC

The phenomenon of the "critical state" has been known since 1822 when Cagnaird de la Tour noted[14] the lack of discontinuity (*i.e.*, disappearance of a meniscus) when passing between the gaseous and liquid states (see Table 1.1). Andrews' 1869 study[15] on CO_2 is considered to be the first systematic study of a gas-liquid critical point, although it has been claimed[16] that the general idea of the "critical state" was arrived at independently by Mendeleeff in 1861. Later, in 1879 and 1880, Hannay and Hogarth[17-19] published the first account of the enhanced solvating properties of supercritical fluids. Other investigators made similar observations at an early date.[20-22] In studying the solubility of cobalt chloride, ferric chloride, potassium bromide, and potassium iodide in supercritical carbonic acid[18], Hannay and Hogarth found a "perfect continuity of the liquid and gaseous states." Hannay summarized the findings by stating[19] that,

> The liquid condition of fluids has very little to do with their solvent power, but only indicates molecular closeness. Should this closeness be attained by external pressure instead of internal attraction, the result

Table 1.1. Chronological Listing of "Firsts" in the Development of SFC

Year	Event
1822	Phenomenon of "critical state" reported.
1869	First systematic study of critical region.
1879	Enhanced solubility in supercritical fluids reported.
1958	Use of supercritical fluids as chromatographic mobile phases suggested.
1962	Use of supercritical fluids as chromatographic mobile phase demonstrated.
1966	First published SFC chromatogram.
	Flame ionization detection in SFC reported.
1967	Ultraviolet absorbance detection in SFC reported.
1969	First report on use of mobile phase modifiers (*i.e.*, binary mobile phases).
	Differential refractometry detection in SFC reported.
1970	Initial use of pressure programming.
	Heat of adsorption detection in SFC reported.
1978	Mass spectrometric detection in SFC reported.
	First report on use of (negative) temperature programming.
	First report on use of simultaneous temperature-density programming.
1981	First report on use of capillary (open tubular) columns in SFC.
	Fluorometric detection in SFC reported.
	First report on use of mobile phase compositional gradient programming.
1982	Hewlett-Packard introduces SFC instrumentation at Pittsburgh Conference.
1983	Thermionic (nitrogen-phosphorus) and Fourier transform infrared detection in SFC reported.
1985	First report on multidimensional SFC.
	First report on coupling supercritical fluid extraction to SFC (on-line).
	First report on chiral separation by SFC.
1986	Capillary SFC instrumentation introduced at Pittsburgh Conference.
	Flame photometric and ion mobility detection in SFC reported.
	First report of solute derivatization in SFC analysis.
1987	Photoionization, light scattering, and inductively coupled plasma detection in SFC reported.
	Use of reverse micelles as SFC mobile phase demonstrated.
1988	First report on use of ion-pairing agents in SFC.
	Hydrogen atmosphere flame ionization, microwave induced plasma, thermal energy analysis, redox chemiluminescence, nuclear magnetic resonance, radiofrequency plasma, and electron capture detection in reported.
	Journal of Supercritical Fluids, first journal devoted entirely to supercritical measurements and applications, introduced.
1989	Sulfur chemiluminescence and supersonic jet spectroscopy detection in SFC reported.

> is that the same or even greater solvent power is obtained. . . . The gas must have a certain density before it will act as a solvent, and when its volume is increased more than twice its liquid volume, its solvent action is almost destroyed.

Considerable time passed, however, before this basic knowledge was utilized for extraction or for chromatography, although both of these processes occupied a central role in the separation of substances. Early reports of extractions with supercritical fluids[23,24] have eventually led to large scale extraction processes.

The first documented suggestion of what is now called SFC was made in 1958 in a notarized and witnessed private note by Lovelock, then at Yale University, who wrote down his idea about the extension of GC to nonvolatile ionic compounds using gases in their supercritical state as mobile phases, such as water, ammonia, sulfur dioxide, and hydrogen fluoride.[25,26]

In 1961, the idea of using dense gases for the transport of nonvolatile substances through a chromatographic column was independently conceived and reduced to practice by Klesper *et al.*[27] in a simple chromatographic apparatus using porphyrins as the eluates. The method was then called "high pressure gas chromatography" instead of SFC.

Before the first SFC instrument was built, it was ascertained that GC could not be applied to the porphyrins which were of interest to Corwin's group at Johns Hopkins University. Klesper designed an experimental apparatus which was built and put into operation in 1961. The design was very simple, but it was only intended to show the feasibility of the approach. A diagram of this apparatus is shown in Figure 1.2. No mechanical pump or elaborate detector was used. Instead, a heated pressure vessel provided a mobile phase at elevated pressures, and a spiral of copper tubing functioned as a final heat exchanger to reach the desired temperature. The column was contained in a glass high pressure gauge tube which allowed observation of the column and the colored bands of porphyrins which moved down the column. Chlorofluoroalkanes were used as mobile phases to move and separate bands of etioporphyrin II, Ni etioporphyrin II, and Ni mesoporphyrin IX dimethylester on the column. The chlorofluoroalkanes were used as mobile phases because of their low flammability and physiological inertness.

While the group of Corwin published their work on porphyrins, Giddings *et al.* were studying the general aspects of pressure induced equilibrium shifts in chromatography[28] and dense GC of nonvolatile species,[29,30] including size exclusion SFC.[31] Another group which contributed even earlier and very successfully to the development of

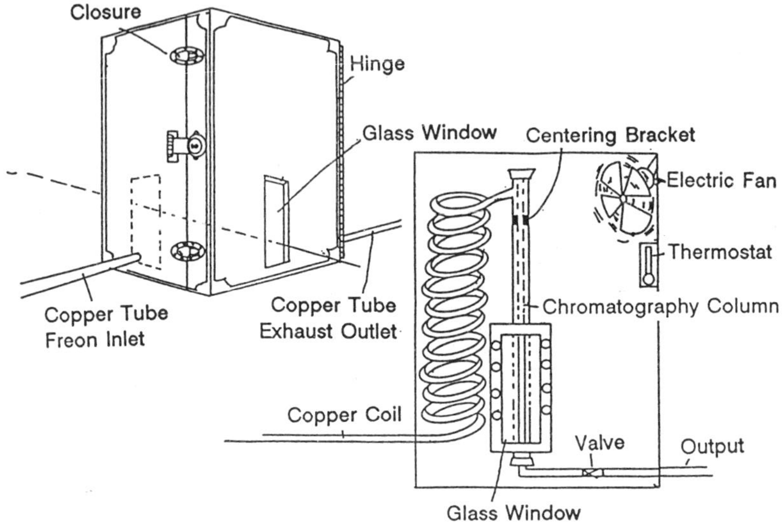

Figure 1.2. Original SFC apparatus used by Klesper.

SFC, was that of Sie and Rijnders at the Shell Laboratories in Amsterdam. This group performed in-depth studies on the effect of pressure on partition coefficients,[32] as well as the effect of mobile phase velocity and particle size on column plate height and permeability.[33] Moreover, they performed SFC both in the gas-liquid and gas-solid mode,[34,35] and investigated the use of porous polymeric stationary phases.[36]

In 1967, Sie and Rijnders[37] were the first to use the term "supercritical fluid chromatography" for this new chromatographic technique, and suggested the use of mobile phase pressure programming. Later, Jentoft and Gouw were the first to employ pressure programming[38] and mobile phase modifiers[39] (*i.e.*, methanol in *n*-pentane) to control retention in SFC. These workers published useful early reviews on SFC.[40,41] Figure 1.3 shows the temperature-pressure regions explored in these early SFC studies.

In 1971, Novotny *et al.* published an important paper concerning the effects of temperature, pressure, eluent composition, flow rate, and type of stationary phase on capacity factors.[42] Physicochemical measurements by SFC, particularly with respect to thermodynamics and mass transport, were reported by Bartmann and Schneider.[43] Schneider and coworkers continued in the years that followed to place emphasis on physicochemical aspects, with studies on the density dependence of capacity factors,[44] binary diffusion

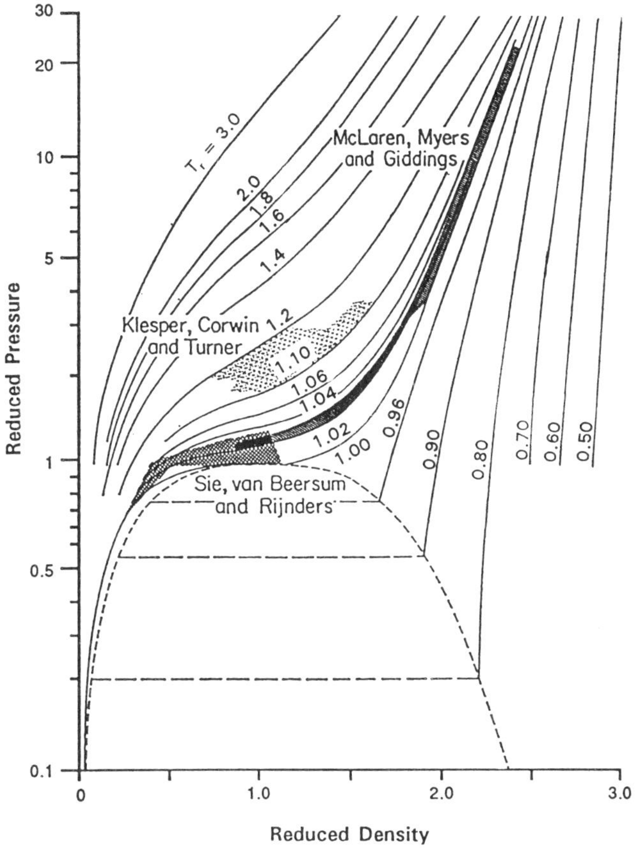

Figure 1.3. CO_2 pressure as a function of density at different temperatures, expressed in terms of reduced parameters. The area below the dotted line represents the two-phase gas/liquid equilibrium region. The indicated areas refer to conditions at which these researchers performed their experiments during the initial studies of SFC (reprinted with permission from Ref. 29).

coefficients,[45] and partial molar volumes.[46] More practical investigations were conducted by Rogers and coworkers, including the separation of oligomers by pressure programming, temperature programming, and modifiers, using both normal and reversed stationary phases.[47-49] Hybrid techniques made their first appearance in SFC when SFC was interfaced to MS by Randall and Wahrhaftig.[50]

The most rapid growth in SFC occurred during the 1980s. Packed column SFC became commercially available in 1982 with the introduction of a modified Hewlett-Packard liquid chromatograph.

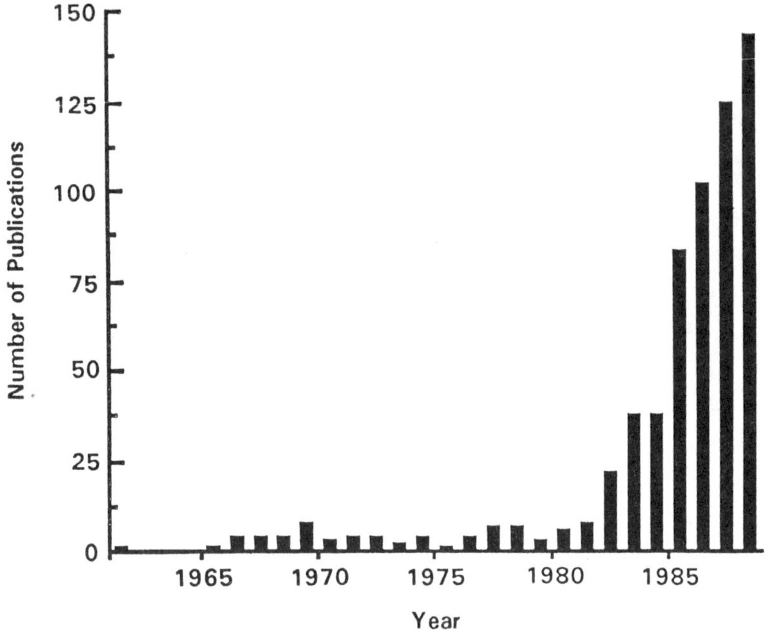

Figure 1.4. Frequency of SFC publications from 1962 to 1989.

The research groups of Novotny and Lee successfully introduced capillary SFC in 1981,[51] which increased the momentum for the development and general acceptance of the technique, and resulted in capillary instrumentation on the market by 1986. This growth is illustrated in Figure 1.4 which shows the number of papers on SFC published each year since its inception. Numerous detectors have been evaluated and modified for use with supercritical fluids, the column technologies for both packed and open tubular columns have advanced, multidimensional methods including on-line supercritical fluid extraction (SFE) have been reported, and the technique has been applied to a wide range of compound types. Detailed discussions of these developments are given in the following chapters.

1.3 History of Analytical SFE

SFE is a technique, like SFC, that employs a supercritical fluid phase to effect the solubilization and separation of solutes. It should be apparent that SFE is closely allied with SFC. In fact, SFE is embodied within SFC theory, since chromatographic plate theory[52] may be interpreted as a series of solute partitions or extractions between the mobile and stationary phases. This close correspondence between the two techniques was historically noted by early researchers in SFC, who found that chromatographic partition coefficients and capacity factors were dependent on the nature of the carrier gas[53] and the column operating pressure.[32] The possibility of enhancing the migration of high molecular mass solutes was noted by Giddings in 1964[54] and amply demonstrated by his research group in the late 1960s.[29,30] Nonetheless, the analytical potential of SFE lay dormant over the next decade and a half, until the mid-1980s.

Whereas the use of supercritical fluid media for performing analytical scale extractions is a recent development, the occurrence of supercritical fluid phenomena can be documented to the early 19th century.[14] Historically, as previously described, the solvent properties of supercritical fluids were initially reported by Hanney and Hogarth,[17-19] and an excellent review of these early studies is provided by Booth and Bidwell.[55] Unfortunately, the extraction potential of these fluids remained unrecognized for many years, although the phenomena of "retrograde condensation" of solutes from compressed gas solutions was recognized very early in the field of petroleum engineering.[56] The solvent properties of liquefied gases were also explored throughout the 1940s and 50s, and the pioneering research of Francis,[57] who compiled an extensive collection of ternary phase diagrams for liquid CO_2 with organic and inorganic compounds and estimated the solubilities of 261 compounds in near-critical CO_2, deserves particular mention. However, it was not until the appearance of a key patent by Zosel[23] that the potential of SFE as a processing technique became apparent. Since this time, there has been a steady growth in the number of applications of SFE in the chemical engineering field, leading recently to the construction of several plants designed to decaffeinate coffee and to process hops and spices. A recent book by McHugh and Krukonis[58] presents the chemical engineering perspective of SFE and is highly recommended as a complementary text.

By contrast, analytical SFE differs from preparative-scale SFE in that the levels of solutes dissolved in the supercritical fluid medium are quite low compared to levels normally encountered in preparative SFE. In addition, the apparatus used to perform the extraction is miniaturized compared to any preparative-scale extractors. This, in part, is due to the frequent coupling of analytical SFE with chromatographic instrumentation in what has become known as "on-line SFE." By contrast, SFE may also be used independently of any other analytical technique, and in this "off-line" mode, may approach a scale that is equivalent to bench-scale preparative units.[59]

In its simplest form, a supercritical fluid extractor consists of a source of pressurized extraction fluid, an extraction cell, and a collection device for isolating the extract. Traditional sources of supercritical fluids have consisted of tanks filled with pressurized gas or semiliquefied gas that is delivered to the extraction cell *via* pumps or compressors. The choice of the latter device is partly based on whether one desires to deliver the fluid in its liquid or gaseous state. There are many different configurations for extraction cells, and these options are discussed in Chapter 5. Collection of the extract is usually accomplished by reducing the pressure on the fluid stream as it exits the extraction cell; however, phase separation of the dissolved solute from the supercritical fluid may also be achieved by changing the temperature of the exiting fluid. In on-line SFE, ancillary devices, such as sorbent cartridges or retention gaps, may be used to focus the solutes after extraction in preparation for chromatographic separation.

From a historical perspective, it is somewhat difficult to precisely identify a specific study or individual that can be credited with inventing analytical-scale SFE. This is partly due to the simultaneous development of SFE in several technical disciplines, as well as the lack of definition as to what really constitutes "analytical SFE." However, there is little doubt that the efforts of Stahl and Schilz in 1976[60] to combine SFE with thin-layer chromatography demonstrated the considerable potential of the technique for analytical studies.

1.4 Definition of a Supercritical Fluid

The strict definition of a supercritical fluid can be visualized by reference to the pressure/temperature phase diagram shown in Figure 1.5. This phase diagram shows the relationship of the gas, liquid, and solid states of a substance as a function of temperature and pressure. The critical point is defined by a critical temperature (T_c) and a critical pressure (P_c) above which (as indicated by the dotted line in Figure 1.5) the substance is neither a gas nor a liquid, but possesses properties of both.

Supercritical fluids may be defined from a practical viewpoint as gases that are at temperatures usually above their critical temperatures and that are compressed to pressures (or densities) at which "liquid-like" interactions become significant. The combination of physical properties (*e.g.*, viscosity and diffusivity) with variable

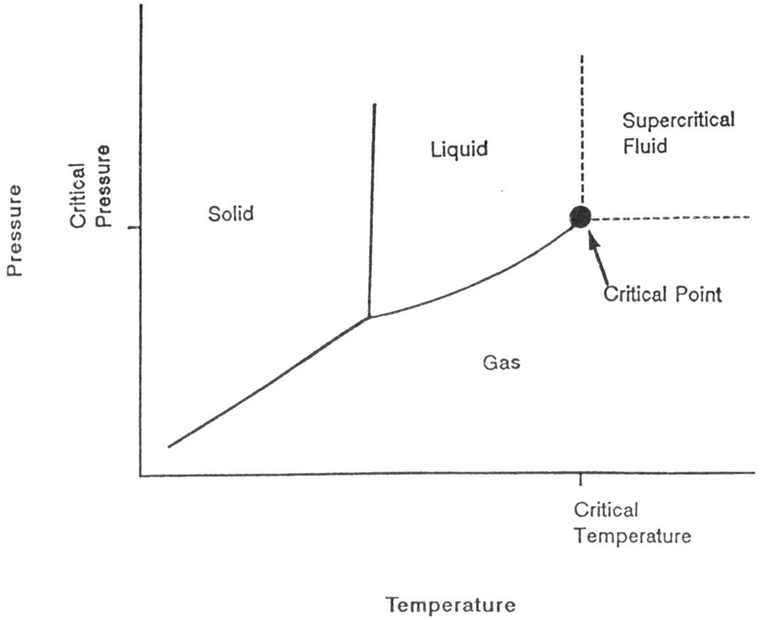

Figure 1.5. Pressure/temperature phase diagram of a substance.

solvating power provides the basis for the advantages of SFC and related analytical techniques. The physical properties of a supercritical fluid are variable between the limits of a normal gas and those of a liquid by control of pressure and temperature. Typically, supercritical fluids are used at densities ranging from 0.1 to 0.8 of their liquid density, and practical pressures for applications range from less than 50 atm to more than 500 atm. Under these conditions, the diffusion coefficients of supercritical fluids are substantially greater than those of liquids. For example, the diffusivity of supercritical CO_2 varies between 10^{-4} and 10^{-3} $cm^2\ s^{-1}$ over the range of conditions usually utilized, whereas liquids typically have diffusivities of less than 10^{-5} $cm^2\ s^{-1}$. Similarly, the viscosities of supercritical fluids mirror their diffusivities, and are typically 10 to 100 times lower than for liquids.[61] These more favorable physical properties provide the advantages of supercritical fluids in chromatography and extraction applications.

Figure 1.3 shows the pressure-density relationship for CO_2 in terms of reduced parameters (*e.g.*, pressure, temperature, or density divided by the appropriate critical parameter), including the two-phase vapor-liquid region. This relationship is generally valid for most single-component systems. The isotherms at several reduced temperatures show the variations in density that can be expected with changes in pressure. Thus, the density of a supercritical fluid will be typically 10^2 to 10^3 times greater than that of a gas at ambient temperatures. Consequently, molecular interactions increase due to shorter intermolecular distances. Table 1.2 gives the critical parameters for a number of common and potentially useful supercritical solvents.

For a neat supercritical fluid, and for very dilute supercritical fluid solutions relevant to chromatographic separations, the relationships between pressure, temperature, and density can be estimated with reasonable accuracy from equations of state, except near the critical point. However, it is often necessary to use a binary fluid mixture to obtain enhanced solvating power. The phase behavior of binary systems is highly varied and much more complex than in single-component systems, and has been well described in the literature for only selected model systems.[58] Six basic types of phase behavior have been identified, with the simplest type characterized

Table 1.2. Physical Parameters of Selected Supercritical Fluids

Fluid	Dipole moment (debyes)[a]	T_c (°C)[a]	P_c (atm)[a]	ρ_c (g mL^{-1})[a]	ρ_{400} (g mL^{-1})[b]	ρ_l (g mL^{-1})[a,c]
CO_2	0.00	31.3	72.9	0.47	0.96	0.71 (63.4 atm)
N_2O	0.17	36.5	72.5	0.45	0.94	0.91 (0°C)
						0.64 (59 atm)
NH_3	1.47	132.5	112.5	0.24	0.40	0.68 (-33.7°C)
						0.60 (10.5 atm)
n-C_5	0.00	196.6	33.3	0.23	0.51	0.75 (1 atm)
n-C_4	0.00	152.0	37.5	0.23	0.50	0.58 (20°C)
						0.57 (2.6 atm)
SF_6	0.00	45.5	37.1	0.74	1.61	1.91 (-50°C)
Xe	0.00	16.6	58.4	1.10	2.30	3.08 (111.75°C)
CCl_2F_2	0.17	111.8	40.7	0.56	1.12	1.53 (-45.6°C)
						1.30 (6.7 atm)
CHF_3	1.62	25.9	46.9	0.52	1.15	1.51 (-100°C)

[a]Data taken from Refs. 62 and 63.
[b]The density at 400 atm (ρ_{400}) and T_r = 1.03 was calculated from compressibility data.[64]
[c]Measurements were made under saturated conditions if no pressure is specified or were performed at 25°C if no temperature is specified.

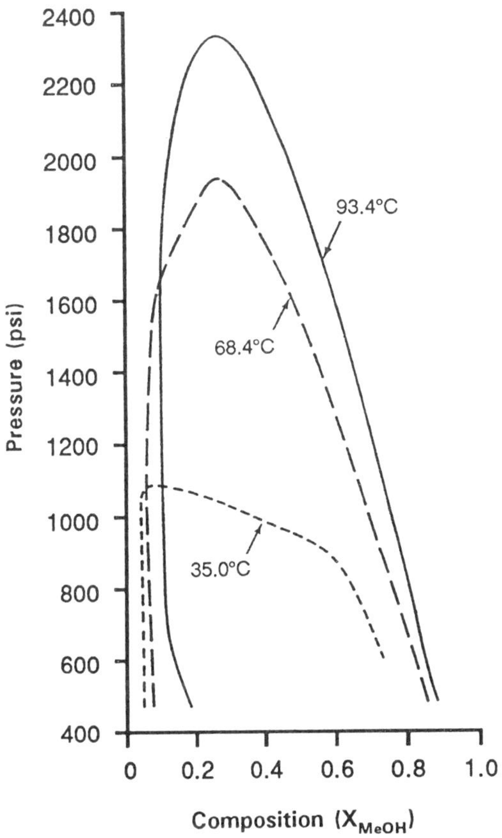

Figure 1.6. CO_2/methanol pressure composition vapor-liquid envelopes at various temperatures (reprinted with permission from Ref. 66).

by a continuous mixture curve for pressure and temperature conditions over the composition range of the two components. Examples of binary fluid systems of this type are CO_2 with isopropanol or methanol, and propane with isopropanol. A detailed discussion of the more complex types of binary fluid mixtures and the phase behavior of these systems can be found elsewhere.[65] The phase behavior of CO_2-methanol is shown in Figure 1.6, which gives the pressure-composition vapor-liquid envelopes for three temperatures.[66,67] The single-phase supercritical regions are above the envelopes, and the regions within the envelopes correspond to two-phase subcritical mixtures at their respective temperatures.

1.5 Characteristics of Supercritical Fluids Relevant to Separation Science

To exploit the chromatographic advantages of SFC, it is essential that the fluid mixtures used for mobile phases be selected such that they can be mixed and pumped as a single phase, preferably at ambient temperatures. Proper operating conditions must also be chosen to give a single-phase supercritical fluid, and care must be taken to avoid entering a two-phase region when operating over a range of pressures, as is typical in SFC. In the absence of actual phase equilibria data, simple mole fraction additivity methods used to obtain mixture critical parameters can result in considerable error and lead to inadvertent operation in the vapor-liquid region. More complex predictive methods utilizing equations of state[68,69] or surface fraction functions (Chueh and Prausnitz method)[70] generally provide more accurate estimates of the true critical parameters. These considerations are important when pressure programming methods are used, but are of lesser importance when relatively high isobaric pressures are used.

The chromatographic behavior observed for a fluid mixture can also be used to discern whether a single-phase or a two-phase system exists under a specified set of operating conditions. This is illustrated with the open tubular column SFC-MS chromatograms of a coal tar shown in Figure 1.7. These separations were obtained at three different temperatures, but otherwise under identical operating conditions, using a 10% (v/v) mixture of isopropanol in propane mobile phase. At 120°C (Figure 1.7A), the components were poorly resolved and eluted close together. When the temperature was raised to 130°C (Figure 1.7B), the components were much better resolved and eluted over a wider pressure range. At 150°C (Figure 1.7C), only slightly improved performance was obtained. This behavior can be explained by the existence of a subcritical liquid with strong solvating properties at the lower temperature that eluted the components close together. Under such conditions, pressure programming is of relatively little value because density does not change substantially. At the intermediate temperature of 130°C, the fluid mixture was probably in a single-phase supercritical condition, and the expected higher chromatographic efficiency and an increasing solvating power with increasing pressure were observed. The differences between the chromatograms obtained at 130°C and 150°C can be attributed to the different fluid properties (viscosity and density at a given pressure) and analyte volatilities at the two temperatures.

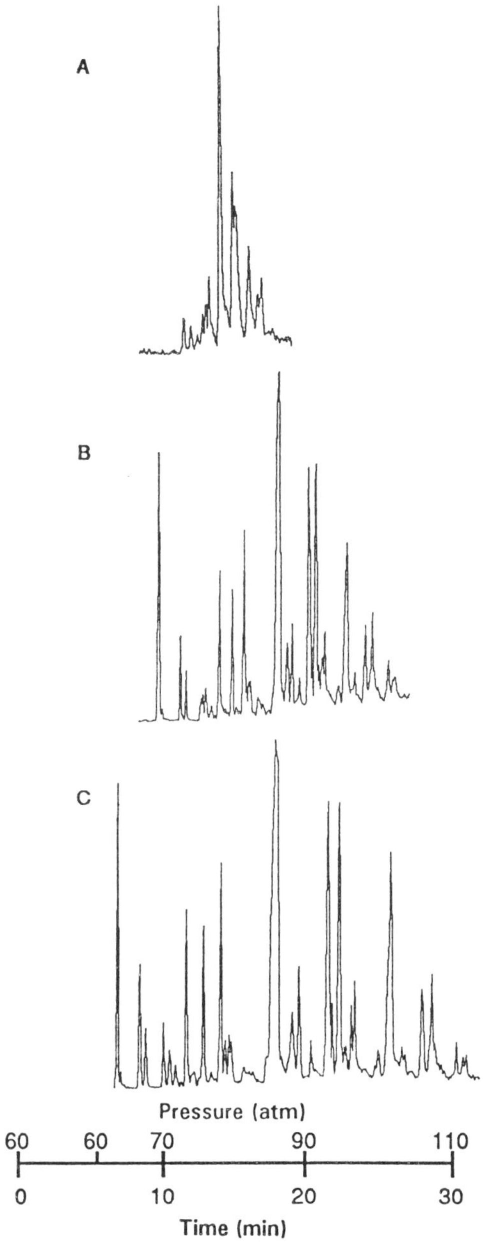

Figure 1.7. SFC/MS total-ion chromatograms of a coal tar extract obtained at (A) 120°C, (B) 130°C, and (C) 150°C. These chromatograms are aligned according to analysis time; apparent discrepancies are due to the delay in the start of data collection. Conditions: 10-m x 50-μm i.d. open tubular column, poly(phenyl)methylsiloxane stationary phase; 10% (v/v) isopropanol in propane; 60 atm for 5 min, 60 to 110 atm at 2 atm min^{-1} (reprinted with permission from Ref. 71).

The range of solvating power of practical supercritical fluids for SFC is of primary importance, and ultimately defines the limits of application. The solubility of analytes typically increases with density, and a maximum rate of increase in solubility with pressure is generally observed near the critical pressure, where the rate of increase of density with pressure is greatest.[72] There is often a linear relationship at constant temperature between log[solubility] and fluid density for dilute solutions of nonvolatile compounds (up to concentrations where solute-solute interactions become important). At constant pressure, when solute volatility is extremely low, and at densities less than or near the critical density, increasing temperature will typically decrease solubility.[73] However, solute entrainment in the fluid may increase at sufficiently high temperatures, where solute vapor pressure also becomes significant. Under conditions of constant density, solubility generally increases with temperature. Thus, while the highest supercritical fluid densities (at constant temperature) are obtained near the critical temperature, the greatest solubilities and lowest chromatographic retention will often be obtained at somewhat lower densities, but at higher temperatures.

As with liquids, polar solutes are most soluble in polar supercritical fluids, although nominally nonpolar fluids can be remarkably good solvents for many moderately polar compounds.[58] Carbon dioxide, for example, can exhibit solvating properties at higher pressures, intermediate between liquid *n*-pentane and dichloromethane. A comparison of the effective solvent polarity of seven fluids as a function of reduced density is shown in Figure 1.8.[74] Solvent polarity is defined in terms of solvent polarizability (π^*) which was developed by Kamlet *et al.*[75] to correlate different solvent-solute interactions based on the solvatochromic effect of the solvent on the π-π^* electronic transition of probe solutes. In this plot, π^* contains terms to account for solvent polarity (*i.e.*, dipolarity) and polarizability, but does not include effects from potential hydrogen bonding interactions.[76] At equal reduced densities, the various fluids have quite different π^* values, indicating that there are large differences in their effective polarities/polarizabilities. Ammonia has the largest π^* value, which supports the fact that it is the most polar solvent. The solvatochromic method also demonstrates the variable solvent properties of a supercritical fluid as a function of density.

Many polar solvents would offer highly specific solvating power but have excessively high critical temperatures, precluding practical operation with current stationary phases. The thermostability limits of the analytes themselves can also be exceeded. This has generated

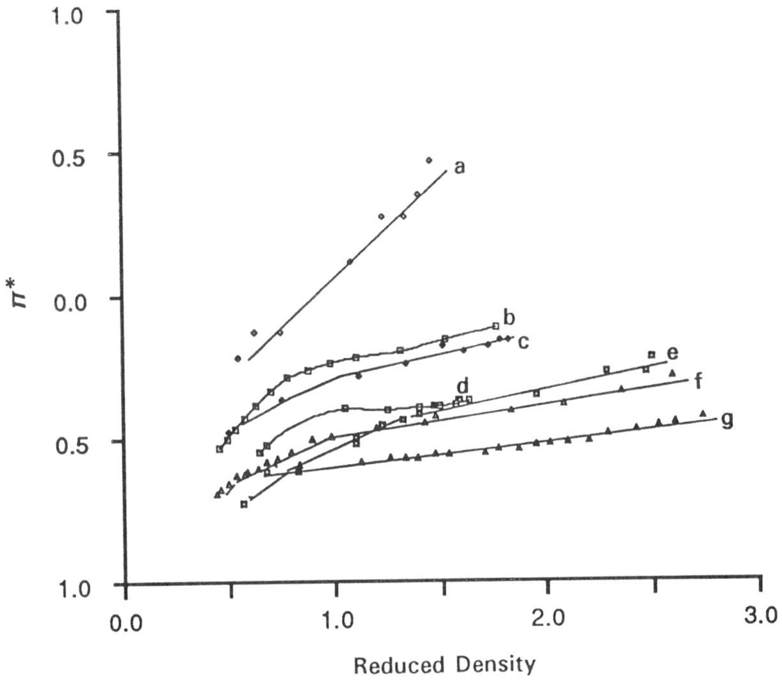

Figure 1.8. Solvent polarizability/polarity parameter (π*) for various supercritical fluids as a function of reduced density at a reduced temperature of 1.03. Supercritical fluids: (a) NH_3, (b) CO_2, (c) N_2O, (d) Xe, (e) CCl_2F_2, (f) C_2H_6, (g) SF_6 (reprinted with permission from Ref. 74).

interest in mixed or binary fluid mobile phases that can have enhanced solvating power at lower critical temperatures. Solvatochromic studies suggest that such fluid mixtures have a net enrichment of polar modifiers in the cybotactic region (nearest neighbor solvation sphere) of the analyte.[77]

1.6 Relationship of SFC to GC and LC from Theoretical Considerations

Since supercritical fluids seem to combine many characteristics of liquids and gases, it is not surprising to find that SFC lies between GC and LC in many respects. Randall described SFC as ". . . a chromatographic technique that is a combination of and complementary to GC and HPLC."[78] Whether SFC redundantly combines characteristics of GC and LC, or fills a void between GC and LC, is still a question for some (although opinions have been largely swayed toward the latter as more and more examples of problems uniquely solved by SFC have been published).

The link between SFC and GC is evident in the pioneering work of Sie, van Beersum, and Rijnders.[32] They proposed using a nonideal gas mobile phase, CO_2, at elevated pressures to increase interactions between the mobile phase and solutes. Their goal was to complement the solute volatility (and thus, the range of problems which could be addressed with GC) without raising the working temperature above the limits imposed by the stability of the solutes and stationary phase. Later, Sie and Rijnders measured volatility enhancement factors as high as 10^4 using supercritical *n*-pentane and isopropanol as mobile phase.[34] Also, Sie *et al.* noted that retention dropped rapidly as the mobile phase pressure was increased in the vicinity of the critical point, thus suggesting the possibility of eluting solutes too nonvolatile for ordinary GC.[32]

The bridge between SFC and LC was convincingly demonstrated by Lauer *et al.*[79] They showed that the logarithm of the capacity factor of solutes, measured at constant mobile phase density, changed linearly when plotted against reciprocal temperature, even as the critical isotherm is crossed (*i.e.*, in crossing from liquid to supercritical phase regions). Plots of the binary diffusion coefficients, measured at constant density, *vs.* reciprocal temperature, were also linear, even when crossing the isotherm.

The enhancement of solvation by the mobile phase can be quantitatively explained using an appropriate equation of state. The Peng-Robinson equation of state,[68] which works better than most for supercritical fluids, can be used to calculate the fugacity (activity) coefficient of the solute, which quantifies the solvating effect. The equation requires solute-mobile phase interaction parameters, which can be obtained from supercritical solubility measurements. Thus, for example, the maximum observed temperature dependence described earlier may be understood quantitatively using the Peng-Robinson equation.[80]

Martire and Boehm further linked GC, SFC, and LC with their "unified molecular theory of chromatography."[81] They developed a set of general equations applicable to GC, SFC, and LC which reduce to familiar retention expressions with the appropriate assumptions (see also Reference 82). Recently, it was demonstrated that an analysis using GC, SFC, and LC can be performed with each mode in series using a single column.[83] Ishii and Takeuchi have named this approach "unified fluid chromatography."[83] Both temperature and pressure were adjusted so that elution could be programmed and modes could be switched.

There are no distinct boundaries between SFC and the other two techniques, even though the equipment necessary to perform each technique has its unique features. In the remainder of this section, the theory necessary to consider the benefits of choosing one mobile phase over another, and how the mobile phase choice affects the resulting chromatography, will be considered.

Basic premises. All forms of elution chromatography basically work in the same way regardless of the nature of the mobile phase. Solutes are partitioned between the mobile and stationary phases according to a unique partition coefficient (K_i) for each solute (i) in the sample. The value of K_i is given by

$$K_i = \frac{c_{i,s}}{c_{i,m}} \tag{1.1}$$

where $c_{i,s}$ and $c_{i,m}$ represent the concentrations (as mass per volume) of component i in the stationary and mobile phases, respectively. The distribution of each solute between the stationary and mobile phase is described by the capacity factor (k_i)

$$k_i = \frac{x_{i,s}}{x_{i,m}} = \left(\frac{c_{i,s}}{c_{i,m}}\right)\left(\frac{V_s}{V_m}\right) = \frac{K_i}{\beta} \tag{1.2}$$

where x represents the masses of components i in each phase, V_s and V_m are the volumes of the stationary and mobile phases, respectively, and β is the phase ratio of the column, that is, V_m/V_s.

When the column is not overloaded, the velocity (u_i) of each solute in the column will be

$$u_i = \frac{u}{(1 + k_i)} \tag{1.3}$$

where u is the velocity of the mobile phase. For solutes with similar capacity factors, the distance (or time) separating the peaks increases in direct proportion to the distance traveled along the column, while the peak widths increase in proportion to the square root of the distance traveled. These peaks are resolved from each other at the column exit if the column, as used, has enough efficiency and/or selectivity.

Peak broadening in open tubular columns. Ideally, peaks are narrow and near Gaussian shaped at the detector. The width of a Gaussian peak can be described by its standard deviation (σ). If the peak is broadened by several "independent" mechanisms, then its width is described by

$$\sigma = (\sigma_a^2 + \sigma_b^2 + \ldots)^{1/2} \tag{1.4}$$

where the subscripts (a, b, . . .) denote the individual broadening mechanisms.

The efficiency of a column is described by the number of theoretical plates (n) that it can provide under a particular set of nonprogrammed (*i.e.*, isothermal, isocratic, isobaric, etc.) conditions

$$n = \left(\frac{t_r}{\sigma}\right)^2 = 5.54\left(\frac{t_r}{w_{1/2}}\right)^2 \tag{1.5}$$

for Gaussian-shaped peaks where t_r is the time from the moment of injection required to elute the peak, and $w_{1/2}$ is the width of the peak measured at half its maximum height. (Note, σ and $w_{1/2}$ must be described in the same units as t_r.)

To consider peak broadening, we need the concept of the height equivalent of a theoretical plate (h). Empirically, h is determined by finding n from the peak width, knowing the length of the column (L) and using the relation

$$h = \frac{L}{n} \tag{1.6}$$

The value of h for an open tubular column can also be predicted using the Golay equation[84]

$$h = \frac{2\,D_m}{u} + \frac{d_c^2\,(1 + 6\,k + 11\,k^2)u}{96\,D_m(1 + k)^2} + \frac{2\,k\,d_f^2\,u}{3(1 + k)^2\,D_s} \tag{1.7}$$

which in its simplified form is given by

$$h = \frac{B}{u} + Cu = \frac{B}{u} + (C_m + C_s)u \tag{1.8}$$

The B term describes the fraction of h caused by longitudinal diffusion, and is equal to 2 D_m, where D_m is the diffusion coefficient of the solute in the mobile phase. The C terms describe the fraction of h caused by resistance to mass transfer in the mobile and stationary phases (as denoted by the subscripts m and s). The C_m term results because of laminar flow through the open tube which gives rise to the characteristic parabolic velocity profile over the tube cross section. Solutes in the center of the flow move faster than solutes near the wall. Failure of the solute to rapidly diffuse in a radial direction (that is, resistance to mass transfer in the mobile phase) tends to keep the solute distributed on streamlines of differing velocity, thus broadening the peak. The C_s term describes the contribution to h caused by solute diffusion in the stationary phase. However, since this is usually negligible in open tubular columns (especially when thin coatings of stationary phase are used), and because it does not directly involve the effects of mobile phase choice, it can be ignored. Thus, C is essentially C_m. The value of C will be different for each solute, depending on the values of the individual capacity factors and diffusion coefficients (at least for those solutes with k values below about 10; the ratio of the two factors containing k approaches a constant value of 11 for increasing k, and can be considered practically constant for k values of 10 or greater). Thus, h and n for the column will be different for each solute. Factors which correct for mobile phase compressibility and peak broadening due to mobile phase decompression are unnecessary in this discussion. The inclusion of these factors would improve the accuracy of the equations, but would not change the final general conclusions.

By differentiation of Equation 1.8, it can be seen that the optimum mobile phase velocity (*i.e.*, the velocity that produces a minimum value for h and a maximum value for n) is

$$u_{opt} = \left(\frac{B}{C}\right)^{1/2} = \frac{4.2\,D_m}{d_c} \qquad (1.9)$$

for thin stationary phase films (where C_s is negligible) and values of k greater than 10 (indicating a strongly retained solute).

Substitution of u_{opt} into Equation 1.8 yields

$$h_{min} = 2(BC)^{1/2} = 0.9\,d_c \qquad (1.10)$$

under the same conditions.

Equations for packed columns. Similar equations can be derived for packed column SFC. The van Deemter equation for packed columns is given by

$$h = A + \frac{B}{u} + Cu \qquad (1.11)$$

where the A term represents eddy diffusion or flow nonuniformity and is equal to 2 λd_p, where λ is a structural factor describing the geometry of the packed bed, and d_p is the average particle diameter of the packing. For packed columns, the equation for h_{min} includes the A term.

$$h_{min} = A + 2(BC)^{1/2} \qquad (1.12)$$

1.7 Relationship of SFC to GC and LC from Practical Considerations

The choice of mobile phase obviously will have a great effect on mass transport in a chromatographic process. This will have a direct bearing on analysis time and efficiency. The role of solute volatility or solubility in the mobile phase will certainly depend on the mobile phase. Selectivity and solute stability may also change with the choice of mobile phase. Finally, various system characteristics, such as column inertness and detector compatibility, are affected by the choice of mobile phase.

Efficiency, speed of analysis, and resolution. Equations 1.10 and 1.12 indicate that the only variables affecting the best value of h that can be expected from the column is the column diameter (open tubular) or the particle size (packed). This is possible for any mobile phase, as long as the optimum velocity for that mobile phase is used, according to Equation 1.9. Thus, the optimum efficiency is a column characteristic (although the efficiency realized can be spoiled by inappropriate parameter choices, bad connections, extracolumn volumes, poor injections, etc.). Therefore, the major effect of mobile phase choice is not reflected in optimum column efficiency, but in analysis time (the optimum velocity depends directly on the diffusion coefficient of the solute in the mobile phase chosen).

In practice, different column types and dimensions are preferred for use in GC, SFC, and LC. Open tubular columns are generally preferred for GC and SFC, although packed columns are also popular for SFC. In LC, only packed columns are commonly used. Open tubular SFC columns usually have inside diameters between 25 and 100 μm, while typical GC columns range from 100 to 530 μm (although GC is occasionally performed using columns as small as 50-μm i.d.). The maximum efficiency of a 50-μm i.d. column in SFC is usually not realized in practice because this would require longer analysis times than are normally acceptable. Therefore, it is a common practice to operate SFC columns well above the optimum velocity to exchange efficiency for shorter analysis times.

Since GC is more rapid and more efficient than SFC and LC, it should be the first choice from among the three separation techniques whenever it is applicable to the problem at hand. From a speed and efficiency standpoint, SFC should be the second choice. Finally, when neither of these techniques are applicable, LC should be selected.

For open tubular column LC, column diameters would have to be between 5 and 10 μm in order to provide practical analysis times, and extracolumn effects would be even more difficult to overcome than in SFC. These difficulties make open tubular column LC an unpopular choice, despite the high efficiencies possible with such narrow-bore columns.

The primary factors affecting h in packed columns are the mean particle diameter (d_p) and the velocity. LC works well with small-diameter packings because of the low compressibility of the liquid mobile phase. High head pressures and large pressure drops over the column are of no great consequence. Thus, modern microparticle LC columns (packed with 3- to 10-μm diameter particles) can deliver reasonable efficiencies and reasonable analysis times. Pressure drops on these columns are much smaller with supercritical fluid mobile phases because of their low viscosities compared to liquids. Therefore, even though supercritical fluids are compressible, the relatively low pressure drops required to achieve optimum velocity make packed-column SFC with microparticle packings successful.

GC is not generally practical with microparticle columns, although the pressure drop would be near that of the SFC case at the same velocity (since viscosities are similar); this velocity is far too slow for a gaseous mobile phase. If operated at optimum velocity, the pressure drop would be very large. It is much more practical in packed column GC to use larger particles to reduce the pressure drop,

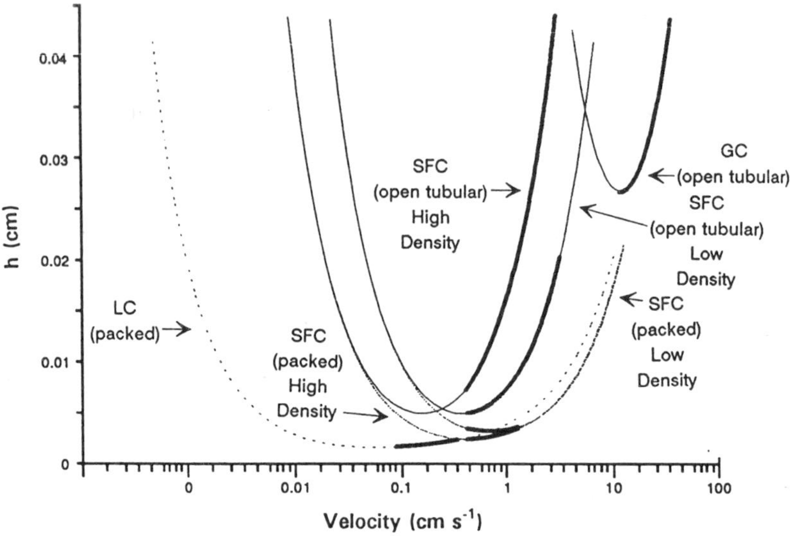

Figure 1.9. Efficiency as a function of mobile phase velocity. Commonly used efficiency-velocity ranges are highlighted. Packed column conditions: 5-μm particles, values calculated from the Knox equation with A = 1, B = 2, C = 0.05 (for LC), and C = 0.5 (for SFC). Open tubular column conditions: 50-μm i.d. (SFC), 300-μm i.d. (GC), values calculated from the Golay equation with k = 5. Values for D_m were 10^{-5} cm^2 s^{-1} for liquid, 2 x 10^{-4} cm^2 s^{-1} for high density supercritical fluid (100°C), 5 x 10^{-4} cm^2 s^{-1} for low density supercritical fluid (100°C), and 10^{-1} cm^2 s^{-1} for gas.

and then increase the column length to achieve the needed efficiency (or, better yet, just use an open tubular column).

Considering all pertinent dimensions and typical values of diffusion coefficients for the various mobile phases, van Deemter-type efficiency curves for the various combinations of parameters are plotted in Figure 1.9. The Golay equation was used for the open tubular columns, and the Knox equation[85] was used for the packed columns. The commonly used regions of these curves are highlighted.

The validity of the Knox equation under SFC conditions has not been experimentally well established, especially at high velocities. Hence, the packed column curves in Figure 1.9 were not extended much past the normal velocity ranges. The earlier approach of ignoring resistance to mass transfer in the stationary phase (*e.g.*, as in Reference 78) is also probably incorrect for microparticle packings with high surface areas and appreciable solute retention *via* adsorption. This was a more correct assumption in early SFC work in which GC packings were used. The importance of considering resistance to mass transfer in the stationary phase when using LC-type packings for SFC can be summarized as follows: if resistance to mass transfer in the stationary phase is important for a particular packing used under LC conditions, then its relative contribution to peak broadening will be even higher with a supercritical fluid mobile phase. If stationary phase mass transfer determines h at high velocities in both LC and SFC operation with a particular packing, then the efficiency curves for the LC and SFC modes will tend to merge at high velocities. This indicates that to achieve large benefits in analysis time with packed column SFC, the packings must operate as much as possible with a (fast) partition mechanism and that the (slower) adsorption/desorption retention mechanism must be minimized.

Figure 1.9 is potentially deceptive in terms of assigning merit to the different techniques. The value of h and its corresponding velocity are of fundamental importance, but are meaningless in a

Table 1.3. Efficiency and Analysis Time Ranges for Various Chromatographic Techniques

Technique[a]	Velocity range ($cm\ s^{-1}$)	Efficiency range (n)	Efficiency/-time range ($n\ s^{-1}$)	Elution time range[b] (min)	Practical analysis time range[c] (min)
LC (packed)	0.1 - 0.4	5,300 - 8,500	14 - 35	2.5 - 10	0.5 - 60
SFC (packed)					
Low density	0.5 - 1.5	3,300 - 3,700	31 - 79	0.7 - 2	0.3 - 30
High density	0.5 - 1.5	3,500 - 5,100	42 - 83	0.7 - 2	0.3 - 30
SFC (open tubular)					
Low density	0.5 - 4	50,000 - 221,000	18 - 33	25 - 200	5 - 90
High density	0.5 - 4	20,000 - 137,000	11 - 13	25 - 200	5 - 90
GC (open tubular)	15 - 50	64,000 - 112,000	93 - 180	6 - 20	1.5 - 60

[a]LC (packed): 10-cm column length with 5-μm packing.
SFC (packed): 10-cm column length with 5-μm packing.
SFC (open tubular): 10-m column length with 50-μm i.d.
GC (open tubular): 30-m column length with 300-μm i.d.
All except the last column are calculated for nonprogrammed elution with $k = 5$.
[b]Nonprogrammed conditions, for a solute with $k = 5$.
[c]Typical programmed conditions.

practical sense without consideration of the column length. The efficiency of the entire column is a more practical consideration. Table 1.3 lists the efficiency ranges of typical columns in each technique by combining the results of the calculations leading to Figure 1.9 with typical values of column lengths and the normal velocity range. These numbers still do not tell the whole story, because analysis time has not been considered. Using $k = 5$ and considering the same velocities and column lengths, the analysis time range for each technique was calculated and is also listed. Finally, the techniques can be compared on the basis of efficiency per unit time. This type of information is often used inappropriately to justify the selection of a particular technique, and must be interpreted with caution. Efficiency per unit time gives the power of a technique to generate theoretical plates, but says little about the ability of a column to separate peaks.

Resolution between two near equally retained peaks is simply the difference in elution times divided by the average peak widths measured at the baseline. In terms of chromatographic parameters, the resolution (R_s) is

$$R_s = \frac{n^{1/2}}{4}\left(\frac{\alpha - 1}{\alpha}\right)\left(\frac{k_2}{k_2 + 1}\right) \qquad (1.13)$$

where α is the column selectivity factor for the two peaks ($\alpha = k_2/k_1$ where the peak numbers are assigned such that $k_2 > k_1$). From this equation, it is clear that resolution depends on column efficiency, selectivity, and solute retention. If n or k approach zero, or if α approaches 1, the resolution will be lost, regardless of what the values of the other parameters might be.

The description so far has dealt only with nonprogrammed elution. Programming further complicates the situation. If the last peak of interest is eluted during a post-program hold period, then the analysis time with any chromatographic technique will be roughly the time of the program up to the final hold, plus $(1 + k_2)L/u$. Equation 1.13 still applies under programmed conditions, except that the constant k values must be replaced with dynamic values which reflect the effect of the program. The analysis time can be shortened and the peaks narrowed in any programmed chromatographic technique by using a faster program rate. However, this is analogous to eluting peaks with a lower average value of k, which, if reduced sufficiently, will lower the resolution of the technique and merge the peaks even though they are narrow.

Volatility or solubility. Among the partition chromatography methods to choose from in practical laboratory situations, GC is most rapid, delivers the highest efficiency, and has the greatest efficiency per unit time. However, GC is limited to thermally stable solutes with vapor pressures of a few torr, and to thermally stable stationary phases.

The partitioning of solutes into a supercritical fluid at a particular temperature is usually enhanced over what would be observed due to "ordinary" vapor pressure in GC. This enhancement is attributed to the solvating properties of supercritical fluids. A rigorous treatment of this phenomenon is beyond the scope of this book, but several excellent summaries are available (for example, References 74 and 86). For purposes of this discussion, it can be stated that the partition process in GC follows the van't Hoff equation

$$\ln K = \frac{\Delta G^{\theta}}{RT} \qquad (1.14)$$

or, upon application of Equation 1.2,

$$\ln k = -\frac{\Delta G^{\theta}}{RT} - \ln \beta \qquad (1.15)$$

where ΔG^{θ} is the free energy change associated with the partitioning of a solute between the stationary phase and an inert mobile phase, R is the gas constant, and T is the absolute temperature. In GC, the free energy change is essentially the free energy of solution of solute in the stationary phase. There is no significant interaction between the solute and the mobile phase. However, whenever the solute can be at least partially solvated by the mobile phase, the overall effect is a reduction in magnitude of the net free energy change of the partition process and a reduction in the values of K and k. The mobile phase now competes with the stationary phase for the solute, which lowers the retention. Thus, the first benefit of using a solubilizing supercritical fluid in place of an inert gas mobile phase is a reduction in solute retention at a given temperature. This, in turn, means that the molecular mass range can be substantially increased, or the temperature required for elution can be reduced. Thus, thermally labile materials may be successfully analyzed with SFC (if the fluid chosen has a low critical temperature) when GC would fail. These benefits could be combined by using a temperature intermediate between that required by GC and the critical temperature of the supercritical fluid.

The use of a liquid mobile phase can reduce retention even further by virtue of even stronger solvation. Of the three types of mobile phases considered, polar liquids are the only mobile phases suitable for dissolving common salts. Very high molecular mass solutes, such as proteins and large polymers, usually require liquid solvents and, thus, can only be analyzed by LC techniques. Most SFC work has been done, so far, for low-to-medium molecular mass and relatively nonpolar solutes. Polystyrenes with molecular masses up to 600,000 daltons have been successfully eluted from an SFC column.[87] However, the limits of SFC in terms of the permissible range of solute polarity and molecular mass are not clear.

In some circumstances,[88] there is a quantitative inverse relationship between the capacity factor and the solubility (s) of the solute in the mobile phase

$$s = \frac{C}{k} \tag{1.16}$$

where C is a constant for the solute, column, and temperature; it is independent of the mobile phase; and it has a temperature dependence of

$$C = A\, e^{B/T} \tag{1.17}$$

where A and B are constants. This relationship provides a route for the measurement of both liquid and supercritical fluid solubilities.

Selectivity. Selectivity in GC is determined exclusively by the stationary phase, since the mobile phase is only a carrier and has no interaction with the solutes. Since the selectivity factor is merely the ratio of the k values for two peaks, it follows from Equation 1.15 that the selectivity in GC is determined simply by the differences in heats of solution of the two solutes in the selected stationary phase (assuming the entropy change for partition of the two solutes is similar).

In LC, large adjustments in selectivity are possible by making compositional changes in the mobile phase, as well as by changing the stationary phase. Selectivity predictions are more empirical than in GC, because both phases interact with the solutes, and the retention process often has adsorption contributions.

In SFC, both the stationary and mobile phases can be varied as in LC. However, there are constraints on the mobile phase composition since the critical parameters are a function of the composition. The selectivities in SFC sometimes resemble GC and

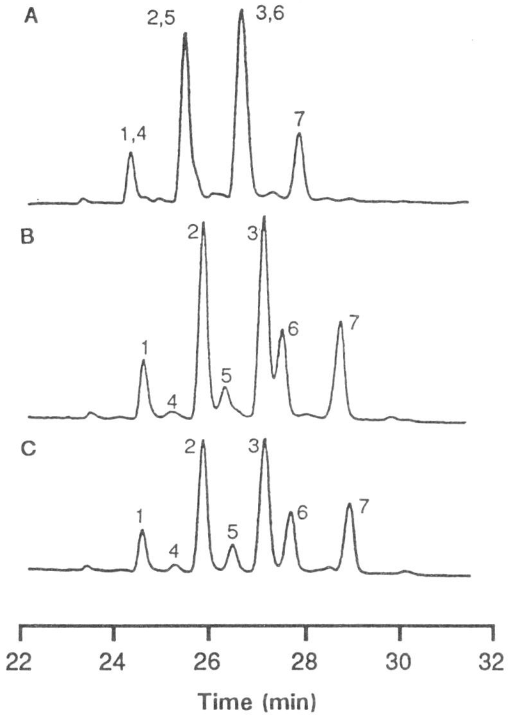

Figure 1.10. Selected segments of SFC chromatograms from a complicated sample, showing the effect of temperature on selectivity at (A) 140°C, (B) 160°C, and (C) 170°C. Two families of unknown peaks (1-3 and 4-7) overlap with some coelution at 140°C, but are resolved at 170°C. Conditions: 10-m x 50-μm i.d. open tubular column, poly(30% biphenyl)methylsiloxane stationary phase; CO_2; 5 atm min^{-1}, FID.

other times resemble LC. This is dependent on the chromatographic conditions, the mobile phase, and the solutes.

There is one noteworthy difference between selectivity adjustment in SFC and in GC and LC. Since mobile phase strength in SFC is a function of both temperature and density (for a fixed mobile phase composition), there are various combinations of temperature and density capable of eluting a particular solute with a desired value of k. However, if we manipulate temperature and density while keeping the k value of one solute fixed, there is no guarantee that the k values of other solutes will also stay fixed. In fact, shifts in relative retention are common, especially if the solutes contain different functional groups.[89] Thus, the selectivity can be changed to some extent by adjustment of the separation temperature.

This effect results from the unequal changes in solubilities of different solutes in the mobile phase as the temperature is varied. It is often possible to reverse the elution order of two close peaks or to split a merged pair. An example of how a relatively small temperature change can make a dramatic improvement in selectivity is shown in Figure 1.10.[90] This should usually be done for fine-tuning after selection of the best column/mobile phase combination for a particular separation.

Solute stability. SFC (with CO_2 mobile phase) can be performed at temperatures as low as about 35°C. Obviously, thermally labile solutes which cannot survive a GC separation at least have a chance of remaining intact after an SFC or LC separation, which may be at significantly lower temperatures. Chemical derivatives of solutes do not require the degree of thermal stability for SFC or LC analysis that is necessary for GC. Even compounds that may appear to be thermally stable (for example, when analyzed by thermogravimetric analysis) may still cause problems on certain columns. Reactions of solutes with active sites, or with impurities adsorbed on the column, are usually reduced at lower temperatures.

Column inertness. The silica surface in an open tubular column has no function other than to support the stationary phase. Thus, it is desirable to make the surface as inert as possible to minimize or eliminate undesirable interactions with the solutes. This is accomplished by extensive deactivation, as described in Section 2.2. Furthermore, the total surface area is rather small because the inside walls are smooth.

In contrast, many silica-based LC packings make use of interactions with the silica support in combination with the bonded stationary phase to achieve a particular degree of selectivity. Some adsorption/desorption equilibria occur with most bonded-phase, porous packings in addition to the solute interactions with the bonded phase. While this may be of benefit in LC, the slow adsorption/desorption kinetics cause problems in SFC because of the higher mobile phase velocities. The use of wide-pore silicas (with lower surface areas), polymer-coated silica packings, and polymer-based packings are all better alternatives for SFC.

Detection. From the detection viewpoint, SFC has an advantage over the other column chromatographic techniques. It can be interfaced to most GC and LC detectors, with only a few fairly obvious limitations. For example, doping of inorganic mobile phases with organic solvents cannot be done when it is intended to detect organic solutes with a flame ionization detector. An electron capture detector cannot be used with electron-capturing mobile phases (*e.g.*, fluorinated hydrocarbons) or when electron-capturing impurities are present in the mobile phase. Also, strongly absorbing mobile phases cannot be used with spectroscopic detectors. The commonly used supercritical fluid mobile phases, particularly CO_2, have a tremendous range of detector compatibility. Specific information on the use of various detectors is given in Chapter 4.

1.8 General Utility of SFE Off-Line and On-Line with Chromatography

Extractions using supercritical fluids, whether off-line or on-line, are attractive for a number of reasons when compared to conventional liquid extractions. Their lower viscosities and higher solute diffusivities improve mass transfer from solid or liquid matrices and, thus, decrease the overall time needed for extraction. By increasing the density, the solvent strength of the supercritical fluid usually increases. Therefore, conditions can be optimized for the extraction of a specific solute or class of solutes from a complex matrix by changing the extraction pressure or temperature. Temperature or pressure changes, when near the critical point of the supercritical fluid, can change solute solubilities by as much as a factor of 100, or even 1,000. By using different supercritical fluids, selective extractions can be achieved for different solutes. Moreover, the use of fluids that have low critical temperatures (*i.e.*, CO_2, N_2O, and SF_6) allow extractions under thermally mild conditions, protecting thermally labile components.

Using certain supercritical fluids that are gases at room temperature, off-line component collection or concentration is greatly simplified. Because these supercritical fluids undergo expansive (*i.e.*, Joule-Thompson) cooling during decompression, even volatile components can be quantitatively and efficiently collected off-line in solvents. It is also possible to directly interface SFE with analytical chromatography. Recent reports have demonstrated the potential of using SFE as an alternative to time consuming, less efficient, and less quantitative liquid solvent extraction techniques. Specific solutes, ranging from environmental priority pollutants to spices and

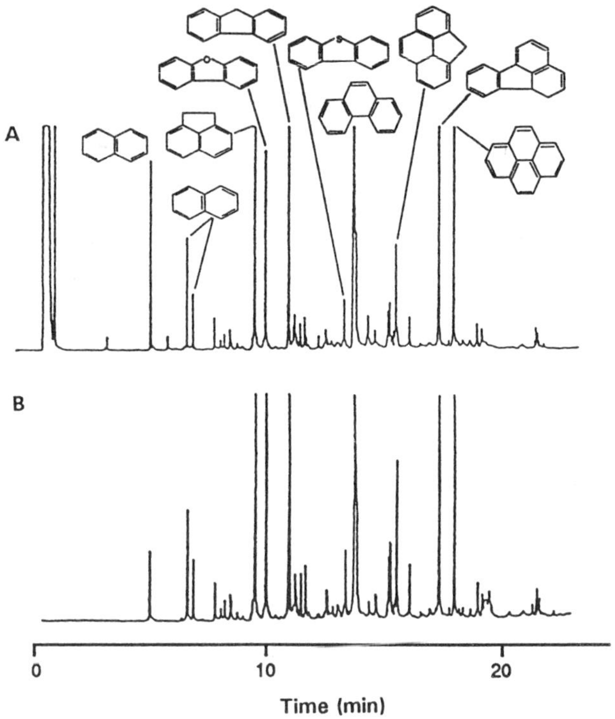

Figure 1.11. GC chromatograms of spiked river sediment extracts analyzed by (A) on-column GC injection and (B) on-line SFE/GC analysis. SFE conditions: 10-mg sample; N_2O; 300 atm; 45°C; 10 min. GC conditions: 30-m x 0.32-mm i.d., poly(5% phenyl)methylsiloxane stationary phase (d_f = 1 μm); 100°C to 320°C at 8°C min^{-1}; FID (reprinted with permission from Ref. 91).

fragrance components, have been qualitatively and quantitatively extracted.

The advantage of off-line SFE is that extraction effluents can be transferred to appropriate collection solvents and analyzed using a variety of analytical techniques (*e.g.*, GC, SFC, LC, MS, NMR, and FTIR). For many samples, collection solvents are not necessary and, as the supercritical fluid effluent vaporizes, the extracted solutes can be quantitatively deposited onto a variety of surfaces (*e.g.*, KBr disc, MS direct probe, NMR tube, or empty autosampler vial). Because of the ability of SFE to rapidly and efficiently remove target solutes from a variety of matrices, off-line SFE can be very effective as a sample cleanup and concentration tool. By the addition of appropriate standards to the sample matrix or to the collecting solvent, quantitative results of recoveries and concentrations can be obtained.

The features of directly coupled SFE to GC or SFC are that no sample handling is required between the extraction and the GC or SFC separation stages, and that extraction effluents can be quantitatively and reproducibly transferred for on-the-fly analyses. When employing an FID, there is no detector response when using extremely pure inorganic supercritical fluids for extractions. This permits the determination of solutes of interest which are often masked by liquid solvents when using conventional extraction techniques. Moreover, when modifiers are used to augment the solubilizing power of primary supercritical fluids, additional components can be analyzed. While SFE/GC is limited to volatile organic compounds, SFE/SFC can extend the analysis to higher molecular mass materials.

The ability of directly coupled SFE/GC to yield good quality chromatograms with minimum extraction and analysis times is demonstrated in Figure 1.11. As can be seen, comparable chromatographic efficiency is obtained with SFE sample introduction as with conventional syringe injection. Since the entire sample can be transferred from the extraction vessel to the GC column, only very small samples are needed to achieve low detection levels.

The use of SFE for the selective extraction of specific solutes or classes of solutes from complex matrices can be seen in Figure 1.12 by the SFE/SFC analysis of coal tar pitch with supercritical CO_2 at various pressures.[92] As the pressure was increased, higher molecular mass components were extracted. Temperature can also be used to alter extraction selectivities.

Figure 1.13 demonstrates the selectivities obtained when using two different fluids, SF_6 and CO_2, for the SFE/GC analysis (actually the conditions were subcritical) of an aromatic and alkane test mixture. With SF_6, only alkanes (*i.e.*, decane, dodecane, and pentadecane) were selectively extracted, regardless of pressures, temperatures, or extraction durations. With CO_2, under identical extraction conditions, both alkanes and aromatics through chrysene were extracted.

Using modifiers, the solvent strengths of supercritical fluids can also be greatly enhanced for the extraction and chromatography of polar or high molecular mass solutes. Most modifiers give responses in the FID, requiring the use of other detectors. Using a UV detector, 1 ppm of quinclorac (3,7-dichloro-8-quinoline carboxylic acid) was selectively determined in soil with a total sample preparation and analysis time of 30 min.[94] The conventional sample preparation method involved a time-consuming reflux with aqueous base, solvent extraction, and methylation before GC/MS or GC/ECD analysis.

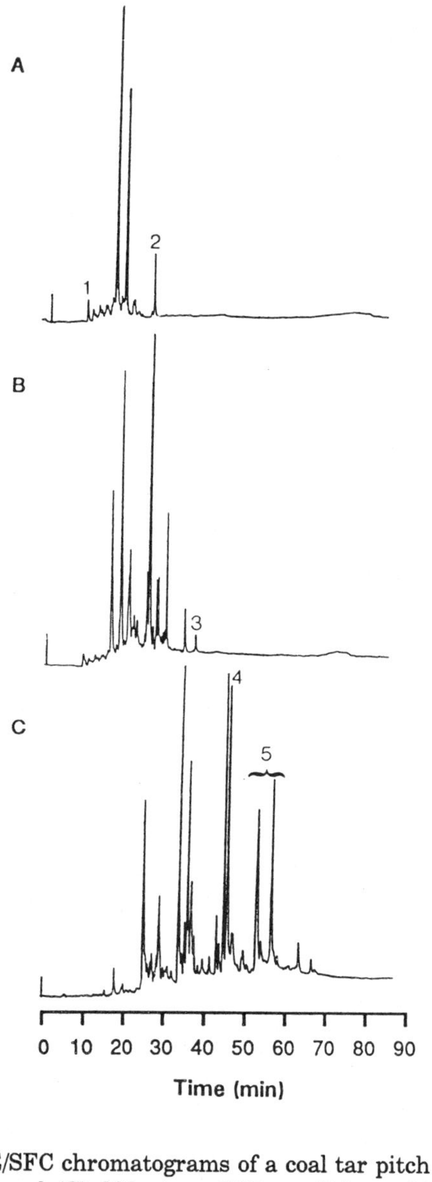

Figure 1.12. SFE/SFC chromatograms of a coal tar pitch extracted at (A) 70 atm, (B) 100 atm, and (C) 200 atm. SFE conditions: CO_2; 43°C; 1 h. SFC conditions: 10-m x 50-μm i.d. open tubular column, poly(30% biphenyl)-methylsiloxane stationary phase (d_f = 0.25 μm); CO_2; 110°C; density program from 0.25 g mL^{-1} to 0.74 g mL^{-1} at 0.006 g mL^{-1} min^{-1} after an initial 20-min isopycnic period; FID. Peak identifications: (1) naphthalene, (2) phenanthrene, (3) pyrene, (4) chrysene, (5) benzopyrenes and benzofluoranthenes (reprinted with permission from Ref. 92).

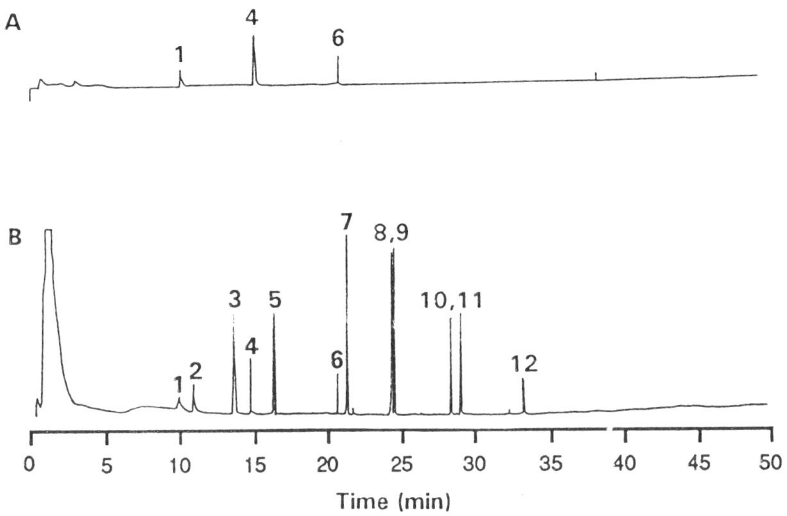

Figure 1.13. SFE/GC chromatograms of aliphatic and aromatic hydrocarbon test mixture obtained by (A) SF_6 extraction and (B) CO_2 extraction. SFE conditions: sample adsorbed on alumina; 375 atm; 25°C; 30 min. GC conditions: 50-m x 200-μm i.d. open tubular column, polymethylsiloxane stationary phase (d_f = 0.5-μm); 30°C for 8 min, then to 310°C at 7°C min^{-1}; FID. Peak identifications: (1) *n*-decane, (2) butylbenzene, (3) naphthalene, (4) *n*-dodecane, (5) 1-methyl naphthalene, (6) *n*-pentadecane, (7) fluorene, (8) phenanthrene, (9) anthracene, (10) fluoranthene, (11) pyrene, (12) chrysene (reprinted with permission from Ref. 93).

The current major limitation to on-line SFE/chromatography is the unavailability of autosampling devices. At present, samples must be analyzed individually, requiring the constant attention of the analyst. The tremendous potential of this technique will only be realized when reliable autosamplers become available.

References

1. G.M. Schneider, *Pure & Appl. Chem.* **47**, 277 (1976).
2. V. Pilz, *Verfahrenstech* **9**, 280 (1975).
3. P.A. Peaden and M.L. Lee, *J. Chromatogr.* **259**, 1 (1983).
4. E.C. Huang, R.J. Skelton, Jr., S.R. Sumpter, K.E. Markides, and M.L. Lee, *Proc. Ninth Int. Symp. Cap. Chromatogr.*, P. Sandra, Ed. (Huethig, Heidelberg, 1988), p. 251.
5. P.J. Schoenmakers and F.C.C.J.G. Verhoeven, *Trends Anal. Chem.* **6**, 10 (1987).
6. L. Blomberg, *Chimicaoggi* **1**, 17 (1988).

7. R.B. Denszyn, Scott Specialty Gases Application Note, Plumsteadville, PA, 1986.
8. W.J. Schmitt and R.C. Reid, *Fluid Phase Equilib.* **32**, 77 (1986).
9. J.H. Phillips and R.J. Robey, *J. Chromatogr.* **465**, 177 (1989).
10. S.M. Fields, Ph.D. Dissertation, Brigham Young University, 1987.
11. D. Leyendecker, D. Leyendecker, B. Lorenschat, F.P. Schmitz, and E. Klesper, *J. Chromatogr.* **398**, 89 (197).
12. *SFC Applications: 1988 Workshop on Supercritical Fluid Chromatography*, K.E. Markides and M.L. Lee, Comps. (Brigham Young University Press, Provo, UT, 1988).
13. *SFC Applications: 1989 Symposium/Workshop on Supercritical Fluid Chromatography*, K.E. Markides and M.L. Lee, Comps. (Brigham Young University Press, Provo, UT, 1989).
14. C. de la Tour, *Ann. Chim. Phys.* **21**, 127 (1822).
15. T. Andrews, *Phil. Trans. Roy. Soc.* **159**, 575 (1869).
16. S. Glasstone, *Textbook of Physical Chemistry*, Vol. 1 (Van Nostrand, New York, 1940), p. 429.
17. J.B. Hannay and J. Hogarth, *Proc. Roy. Soc. (London)* **29**, 324 (1879).
18. J.B. Hannay and J. Hogarth, *Proc. Roy. Soc. (London)* **30**, 178 (1880).
19. J.B. Hannay, *Proc. Roy. Soc. (London)* **30**, 484 (1880).
20. P. Villard, *Z. Physik. Chemie* **23**, 373 (1897).
21. M. Centnerszwer, *Z. Physik. Chemie*, **46**, 427 (1903).
22. A. Smits, *Z. Elektrochem.* **9**, 663 (1903).
23. K. Zosel, US Patent 3969 196 (13.7.1976); Belgium Patent 646 641 (16.10.1964), *Chem. Abstr.* **63**, 11045b (1965).
24. T.P. Zhuse, *Vestn. Akad. Naukk SSSR* **29**, 47 (1959); *Petroleum* (London) **23**, 298 (1960).
25. W. Bertsch, Ph.D. Dissertation, University of Houston, 1973.
26. C.M. White, in *Modern Supercritical Fluid Chromatography*, C.M. White, Ed. (Hüthig, Heidelberg, 1988), p. v.
27. E. Klesper, A.H. Corwin, D.A. Turner, *J. Org. Chem.* **27**, 700 (1962).
28. J.C. Giddings, *Sep. Science* **1**, 73 (1966).
29. J.C. Giddings, M.N. Myers, L. McLaren, and R.A. Keller, *Science* **162**, 67 (1968).
30. J.C. Giddings, M.N. Myers, and J.W. King, *J. Chromatogr. Sci.* **7**, 276 (1969).
31. J.C. Giddings, L.M. Bowman, Jr., and M.N. Myers, *Anal. Chem.* **49**, 243 (1977).
32. S.T. Sie, W. van Beersum, and G.W.A. Rijnders, *Sep. Sci.* **1**, 459 (1966).
33. S.T. Sie and G.W.A. Rijnders, *Sep. Sci.* **2**, 699 (1967).
34. S.T. Sie and G.W.A. Rijnders, *Sep. Sci.* **2**, 729 (1967).
35. S.T. Sie and G.W.A. Rijnders, *Sep. Sci.* **2**, 755 (1967).
36. S.T. Sie, J.P.A. Bleumer, and G.W.A. Rijnders, in *Gas Chromatography-1968*, C.L.A. Harbourn, Ed. (The Institute of Petroleum, London, 1969), p. 235.
37. S.T. Sie and G.W.A. Rijnders, *Anal. Chim. Acta* **38**, 31 (1967).

38. R.E. Jentoft and T.H. Gouw, *J. Chromatogr. Sci.* **8**, 138 (1970).
39. R.E. Jentoft and T.H. Gouw, *J. Polym. Sci. Polym. Lett.* **7**, 811 (1969).
40. T.H. Gouw and R.E. Jentoft, *J. Chromatogr.* **68**, 303 (1972).
41. T.H. Gouw and R.E. Jentoft, *Adv. Chromatogr.* **13**, 1 (1975).
42. M. Novotny, W. Bertsch, and A. Zlatkis, *J. Chromatogr.* **61**, 17 (1971).
43. D. Bartmann and G.M. Schneider, *J. Chromatogr.* **83**, 135 (1973).
44. U. van Wasen and G.M. Schneider, *Chromatographia* **8**, 274 (1975).
45. I. Swaid and G.M. Schneider, *Ber. Bunsenges. Phys. Chem.* **83**, 969 (1979).
46. U. van Wasen and G.M. Schneider, *J. Phys. Chem.* **84**, 229 (1980).
47. J.A. Nieman and L.B. Rogers, *Sep. Sci.* **10**, 517 (1975).
48. F.J. van Lenten, J.E. Conaway, and L.B. Rogers, *Sep. Sci.* **12**, 1 (1977).
49. J.E. Conaway, J.A. Graham, and L.B. Rogers, *J. Chromatogr. Sci.* **16**, 102 (1978).
50. L.G. Randall and A.L. Wahrhaftig, *Anal. Chem.* **50**, 1703 (1978).
51. M. Novotny, S.R. Springston, P.A. Peaden, J.C. Fjeldsted, and M.L. Lee, *Anal. Chem.* **53**, 407A (1981).
52. H. Purnell, *Gas Chromatography* (John Wiley & Sons, New York, 1962), p. 86.
53. D.H. Desty, A. Goldup, G.R. Luckhurst, and W.T. Swanton, in *Gas Chromatography-1962*, M. van Swaay, Ed. (Butterworths, Washington, D.C., 1962), p. 67.
54. J.C. Giddings, in *Gas Chromatography-1964*, A. Goldup, Ed. (Elsevier, Amsterdam, 1965), p. 22.
55. H.S. Booth and R.M. Bidwell, *Chem. Rev.* **44**, 477 (1949).
56. D.L. Katz and F. Kurata, *Ind. Eng. Chem.* **32**, 817 (1940).
57. A.W. Francis, *J. Phys. Chem.* **58**, 1099 (1954).
58. M.A. McHugh and V.J. Krukonis, *Supercritical Fluid Extraction* (Butterworths, Boston, 1986).
59. J.W. King, J.H. Johnson, and J.P. Friedrich, *J. Agric. Food Chem.* **37**, 951 (1989).
60. E. Stahl and W. Schilz, *Z. Anal. Chem.* **280**, 99 (1976).
61. K. Stephan and K. Lucas, *Viscosity of Dense Fluids* (Plenum Press, New York, 1979).
62. *Matheson Gas Data Book, 6th Ed.*, W. Braker and A.L. Mossman, Eds. (Matheson, Secaucus, NJ, 1980).
63. *CRC Handbook of Chemistry and Physics, 65th Ed.*, R.C. Weast, Ed. (Chemical Rubber Company, Boca Raton, FL, 1984).
64. G.N. Lewis and M. Randall, *Thermodynamics, 2nd Ed.*, K.S. Pitzer and L. Brewer, Rev. (McGraw-Hill, NY, 1961).
65. W.B. Street, in *Chemical Engineering at Supercritical Fluid Conditions*, M.E. Paulaitis, J.M.L. Penninger, R.D. Gray, and P. Davidson, Eds. (Ann Arbor Science, Ann Arbor, MI, 1983).
66. R.D. Smith, H.T. Kalinoski, and H.R. Udseth, *Mass Spec. Reviews* **6**, 445 (1989).
67. E. Brunner, *J. Chem. Therm.* **17**, 671 (1985).
68. D. Peng and D.B. Robinson, *AIChE. J.* **23**, 137 (1977).

69. R.A. Heidemann and A.H. Khalil, *AIChE. J.* **26**, 769 (1980).
70. P.L. Chueh and J.M. Prausnitz, *AIChE. J.* **13**, 1099 (1967).
71. R.D. Smith, B.W. Wright, and H.T. Kalinoski, in *Progress in HPLC*, Vol. 4, M. Yoshioka, S. Parvez, T. Miyazaki, and H. Parvez, Eds. (VSP, Utrecht, 1989), p. 99.
72. M. Gitterman and I. Procaccia, *J. Chem. Phys.* **78**, 2648 (1983).
73. R.D. Smith and H.R. Udseth, *Anal. Chem.* **55**, 2266 (1983).
74. C.R. Yonker and R.D. Smith, in *Supercritical Fluid Extraction and Chromatography: Techniques and Applications, ACS Symposium Series 366*, B.A. Charpentier and M.R. Sevenants, Eds. (American Chemical Society, Washington, DC, 1988), p. 161.
75. M.J. Kamlet, J.L. Abboud, and R.W. Taft, *J. Am. Chem. Soc.* **99**, 6027 (1977).
76. S.L. Frye, C.R. Yonker, D.R. Kalkwarf, and R.D. Smith, in *Supercritical Fluids, Chemical and Engineering Principles and Applications, ACS Symposium Series 329*, T.G. Squires and M.E. Paulaitis, Eds. (American Chemical Society, Washington, DC, 1987), p. 29.
77. C.R. Yonker and R.D. Smith, *J. Phys. Chem.* **92**, 2374 (1988).
78. L.G. Randall, in *Ultra High Resolution Chromatography, ACS Symposium Series 250*, S. Ahuja, Ed. (American Chemical Society, Washington, DC, 1984), p. 135.
79. H.H. Lauer, D. McMannigill, and R.D. Board, *Anal. Chem.* **55**, 1370 (1983).
80. K.D. Bartle, A.A. Clifford, J.P. Kithinji, and G.F. Shilstone, *J. Chem. Soc. Faraday Trans. 1* **84**, 4487 (1988).
81. D.E. Martire and R.E. Boehm, *J. Phys. Chem.* **91**, 2433 (1987).
82. D.E. Martire, *J. Liq. Chromatogr.* **10**, 1569 (1987).
83. D. Ishii and T. Takeuchi, *J. Chromatogr. Sci.* **27**, 71 (1988).
84. M.J.E. Golay, in *Gas Chromatography*, D.H. Desty, Ed. (Butterworths, London, 1958), p. 35.
85. J.H. Knox, *J. Chromatogr. Sci.* **15**, 352 (1977).
86. P.J. Schoenmakers, *J. Chromatogr.* **315**, 1 (1984).
87. F.P. Schmitz and E. Klesper, *Polymer Commun.* **24**, 142 (1983).
88. K.D. Bartle, A.A. Clifford, and S.A. Jafar, *J. Chem. Soc. Faraday Trans.*, in press.
89. T.L. Chester, *J. Chromatogr.* **299**, 424 (1984).
90. T.L. Chester and D.P. Innis, Procter and Gamble Company, unpublished results.
91. S.B. Hawthorne, D.J. Miller, and M.S. Krieger, *Fres. Z. Anal. Chem.* **330**, 211 (1988).
92. M.W. Raynor, I.L. Davies, K.D. Bartle, A.A. Clifford, A. Williams, J.M. Chalmers, and B.W. Cook, *J. High Resolut. Chromatogr. Chromatogr. Commun.* **11**, 766 (1988).
93. J.M. Levy, R.A. Cavalier, T.N. Bosch, A.F. Rynaski, and W.E. Huhak, *J. Chromatogr. Sci.* **27**, 341 (1989).
94. J.M. Levy, A. Rosselli, D. Boyer, K. Cross, B. Ravey, P. Zimmerman, and J. Kane, *J. High Resolut. Chromatogr.*, in press.

CHAPTER 2

COLUMNS

2.1 Introduction

The column is the heart of SFC, as it is in all forms of column chromatography. The instrumentation and analytical processes are designed around the column, with the hope of being able to take full advantage of its intricately powerful separating ability. The goal is to eliminate all extracolumn contributions to the chromatographic process, so that retention and resolution are only dependent on interactions between the mobile phase, stationary phase, and solute molecules.

The column considerations can be divided into two categories: configurational and chemical. The first includes parameters such as packed or open tubular, internal diameter and length, particle size and porosity, and stationary phase film thickness. The second category includes column deactivation and stability, stationary phase structure and selectivity (*i.e.*, solute/stationary phase interactions), and solute diffusivity in the stationary phase.

The following sections in this chapter cover both configurational and chemical column considerations. Both theoretical calculations and practical measurements are included. Additional practical information is given in Chapter 6. Considerable emphasis is given in this chapter to the column technology essential in preparing high performance columns.

2.2 Column Efficiency

Open tubular columns. The simplified form of the Golay equation was given in Chapter 1 (Equation 1.7). This equation gives the plate height (h) as a function of column dimensions, solute diffusivity, and various other chromatographic parameters. The expanded Golay equation is given by

$$h = \frac{2\ D_m}{u} + \frac{d_c^2\ (1 + 6k + 11k^2)\ u}{96\ (1 + k)^2\ D_m} + \frac{2\ k\ d_f^2\ u}{3(1\ +\ k)^2\ D_s} \qquad (2.1)$$

where u is the average mobile phase linear velocity along the column, d_c is the column internal diameter, k is the capacity factor, d_f is the stationary phase film thickness, and D_m and D_s are the solute diffusion coefficients in the mobile and stationary phases, respectively.

Plots of plate height *vs* linear velocity (van Deemter plots) can be determined experimentally by measuring the number of theoretical plates that a column delivers (Equation 1.5) as a function of the linear velocity of the mobile phase. The linear velocity is determined by using an unretained compound. For efficiency measurements, the solutes should have k values greater than 2, and preferably greater than 5. In any case, the k value should be given for each efficiency measurement reported. The values of h can be calculated by dividing the length of the column by the measured values of n. These values can then be plotted against linear velocity.

Theoretical van Deemter curves can be calculated from Equation 2.1. All of the variables are easily obtained, except for the solute diffusion coefficients, D_m and D_s. Published values of applicable diffusion coefficients are very sparse, and assumptions often must be made. For example, in a study of the effect of column diameter on efficiency in open tubular column SFC, no available D_m or D_s values were found in the literature for C_{10} to C_{12} normal hydrocarbons which matched the operating conditions of the study.[1] However, literature values indicated that for *n*-C_{10} to *n*-C_{12} in supercritical CO_2 at densities up to 0.7 g mL^{-1}, D_m values[2-4] were of the order of 10^{-4} to 10^{-3} $cm^2\ s^{-1}$, and D_s values[5-8] were of the order of 10^{-7} to 10^{-5} $cm^2\ s^{-1}$. Extrapolation of published D_m values[9,10] for several aromatic hydrocarbons (benzene, alkylated benzenes, and naphthalene) in CO_2 at 40°C and densities greater than 0.25 g mL^{-1} predicted that these solutes would have D_m values of 2.5×10^{-4} to 4×10^{-4} $cm^2\ s^{-1}$.

Figure 2.1 shows calculated van Deemter curves for open tubular columns with internal diameters from 25 to 100 μm[1]. Since both pressure and stationary phase dimensions were held constant, the k values increased in Figure 2.1 with decreasing column diameter. The D_m (CO_2 mobile phase at 40°C and 72 atm) and D_s values were assumed to be 2×10^{-4} $cm^2\ s^{-1}$ and 1×10^{-6} $cm^2\ s^{-1}$, respectively.

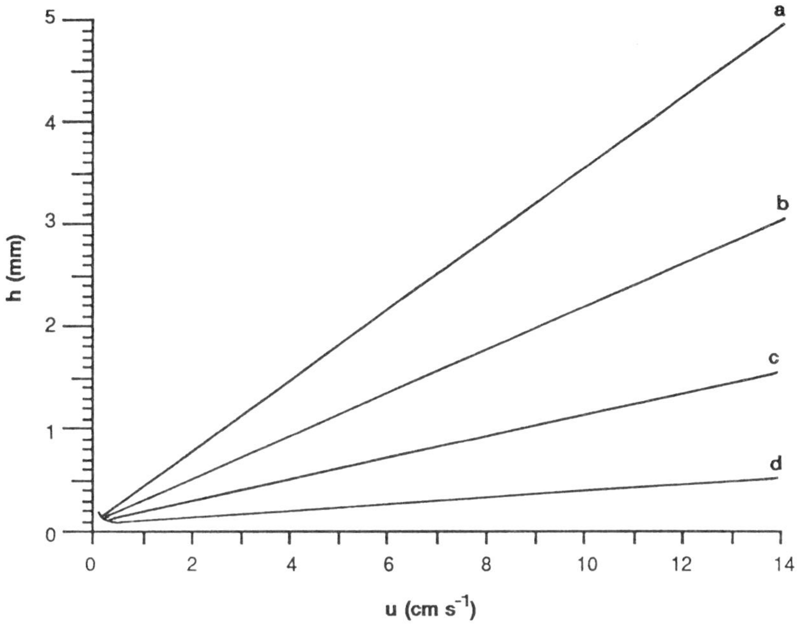

Figure 2.1. SFC van Deemter plots for n-C_{12} on (a) 100-μm i.d. (k = 2.24), (b) 75-μm i.d. (k = 2.72), (c) 50-μm i.d. (k = 3.90), and (d) 25-μm i.d. (k = 11.36) open tubular columns. Conditions: CO_2; 40°C; 72 atm (reprinted with permission from Ref. 1).

These conditions give a mobile phase density of 0.22 g mL^{-1}. Table 2.1 lists the calculated u_{opt} and h_{min} values for the 25 to 100 μm i.d. columns at a k value of 3. In addition, since analysis times are generally too long when operating at u_{opt}, the h values at 10 u_{opt} are also listed.

Table 2.1. Theoretical Values of u_{opt}, h_{min}, and h at 10 u_{opt} for Different Open Tubular Column Diameters[a]

d_c (μm)	u_{opt} (cm s^{-1})	h_{min} (mm)	h at 10 u_{opt} (mm)
100	0.102	0.078	0.396
75	0.136	0.059	0.298
50	0.203	0.039	0.198
25	0.402	0.020	0.101

[a]Conditions are given in the text.

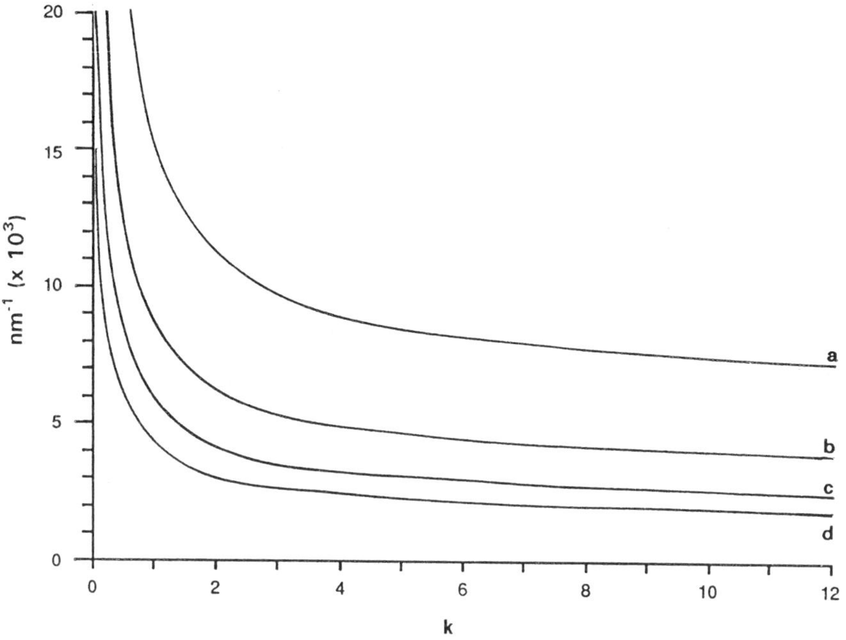

Figure 2.2. Plates per meter as a function of k at 10 u_{opt} for (a) 25-μm i.d., (b) 50-μm i.d., (c) 75-μm i.d., and (d) 100-μm i.d. open tubular columns. Conditions: CO_2; 40°C; 72 atm (reprinted with permission from Ref. 1).

At u greater than about 0.5 cm s^{-1}, the contribution to plate height from longitudinal diffusion in the mobile phase becomes negligible, and h becomes a linear function of the resistance to mass transfer in the mobile and stationary phases. In practice, u is usually greater than 1 cm s^{-1}. The linear velocity at 10 u_{opt} for the 50-μm i.d. column in Figure 2.1 was approximately 2 cm s^{-1}.

The practical efficiencies obtainable are better illustrated in plots of plates per meter as a function of k. Figure 2.2 shows such plots for the four column diameters under discussion at their 10 u_{opt} linear velocities[1]. The 50-μm i.d. column should generally produce greater than 4,000 plates per meter under these conditions. Table 2.2 lists the experimental performance data for the four different diameter columns studied[1] at a k = 3. From the point of view of analysis time, it is easy to see why 50- and 25-μm i.d. columns, and columns less than 15 m in length, are preferred.

So far, the discussion of column efficiency has been limited to low density supercritical fluid conditions where diffusion coefficients are largest and efficiencies are highest. At higher densities, the results are not as favorable.

There are two factors that must be considered when evaluating column efficiencies at increasing densities. The first is the effect of density alone; as the density increases, D_m decreases and h increases at u larger than u_{opt}. The second factor is the inherently lower D_m values characteristic of larger solute molecules that are eluted at the higher densities. Both of these factors were considered in the calculation of the van Deemter curves shown in Figure 2.3. The D_m values were taken from the literature[10] for three different compounds in CO_2 at 40°C, with the assumption that the three compounds would elute at 0.28, 0.45, and 0.79 g mL^{-1}, respectively.

If one were to control the linear velocity at 2 cm s^{-1} during density programming, efficiencies would decrease by nearly 75% from low to high density. For very large compounds, diffusivities would be even lower; for a D_m of 3×10^{-5} cm^2 s^{-1}, the column efficiency would drop to about 700-800 plates m^{-1}. From these theoretical predictions, it is obvious that significant losses in efficiency could occur at high density in open tubular column SFC. This arises from the fact that the slope of the van Deemter curve becomes very steep at high density, and the practical operating linear velocities become $\geq 10\ u_{opt}$ with density programming.

There are two possible solutions to mitigate the loss of efficiency at higher density: (a) decrease the column diameter and (b) increase the operating temperature. While excellent results have been obtained using 25-μm i.d. columns, more immediate results have been

Table 2.2. Practical Open Tubular Column Efficien-cies at 10 u_{opt} for Different Column Diameters at k = 3[a]

d_c (μm)	L (m)	10 u_{opt} (cm s^{-1})	h (mm)	n	n m^{-1}	t_R (min)	n min^{-1}
100	24	1.1	0.44	5×10^4	2300	145	370
75	24	1.4	0.30	8×10^4	3300	114	700
50	23	2.0	0.22	1×10^5	4400	77	1300
25	7[b]	4.3	0.18	4×10^4	5600	11	3500

[a]Conditions are given in the text.
[b]Length was shorter because of pressure drop.

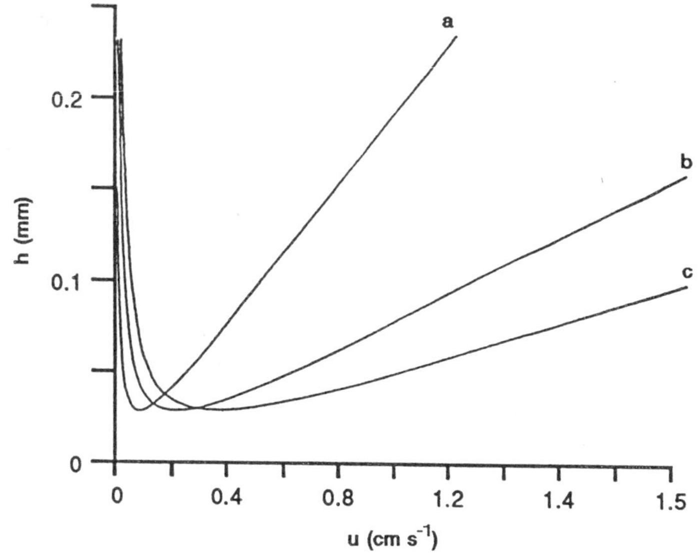

Figure 2.3. SFC van Deemter plots for three compounds with D_m values of (a) 0.79 g mL^{-1}, (b) 0.45 g mL^{-1}, and (c) 0.28 g mL^{-1}. Conditions: 50-μm i.d. open tubular column; CO_2; 40°C (data taken from Ref. 11).

obtained by increasing the temperature.[11] Figure 2.4 shows van Deemter curves for naphthalene at 60, 90, and 120°C. The dramatic decrease in the slope of the curve by going to higher temperatures is obvious. At constant density, an increase in temperature can result in three favorable effects: first, an increase in solute diffusion coefficients; second, an increase in solubility; and third, an increase in solute volatility, with the last two effects leading to a corresponding decrease in retention (*i.e.*, solutes elute at lower densities). It should be pointed out here that most SFC separations performed today using CO_2 are carried out at temperatures near or at 100°C.

One final consideration is that mobile phase linear velocity usually changes during density programming. When using a fixed restrictor, which is the usual case in open tubular column SFC, the linear velocity can more than double when programming from low to high density. When a 7-m x 50-μm i.d. open tubular column was programmed from 88 atm (0.47 g mL^{-1}) to 400 atm (0.96 g mL^{-1}) at 40°C, the linear velocity changed[12] from 1.3 to 10.2 cm s^{-1}. Here, much greater than 10 times the optimum velocity was produced, and the mass transfer term clearly dominated column efficiency. It should be noted that the linear velocity increased only 2.5 times for

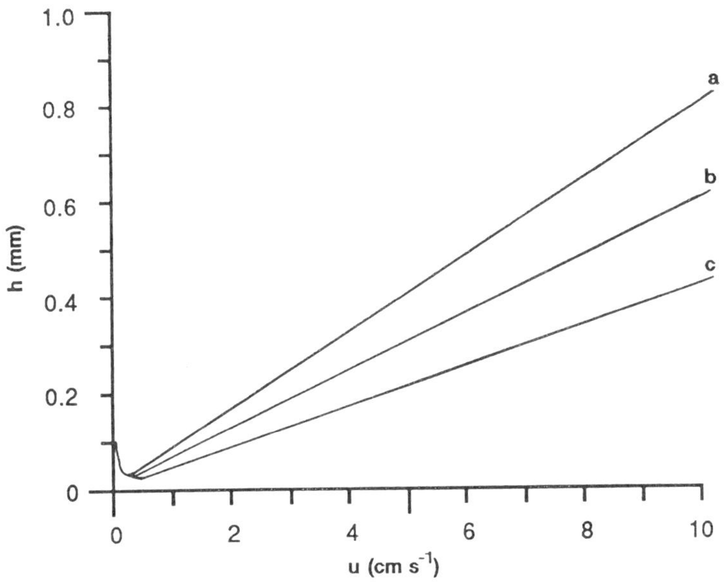

Figure 2.4. SFC van Deemter plots for naphthalene at (a) 60°C, (b) 90°C, and (c) 120°C. Conditions: 50-μm i.d. open tubular column; CO_2; 0.6 g mL^{-1} (data taken from Ref. 11).

the same pressure range at the more practical temperature of 100°C. This demonstrates a situation, not seen in LC or GC, where column efficiency drastically changes because the magnitude of the mass transfer term changes considerably during the course of a pressure or density programmed run.

Packed columns. The relationship between h and u for a packed column is more complex,[13] and several expressions have been reported. Schwartz *et al.*[14] selected the Horvath-Lin equation[15] because it incorporates the resistance to mass transfer in the interior of the porous stationary phase particles. A simplification of the Horvath-Lin equation is given by

$$h = 1.5\, d_p + \frac{1.4\, D_m}{u} + \frac{2\,(k_o + k + k_o k)^2\, d_p^2\, u}{15\,(1 + k_o)^2\,(1 + k)^2\, D_m} \qquad (2.2)$$

where d_p is the particle diameter and k_o is the ratio of particle pore volume to particle interstitial volume in the column. The other

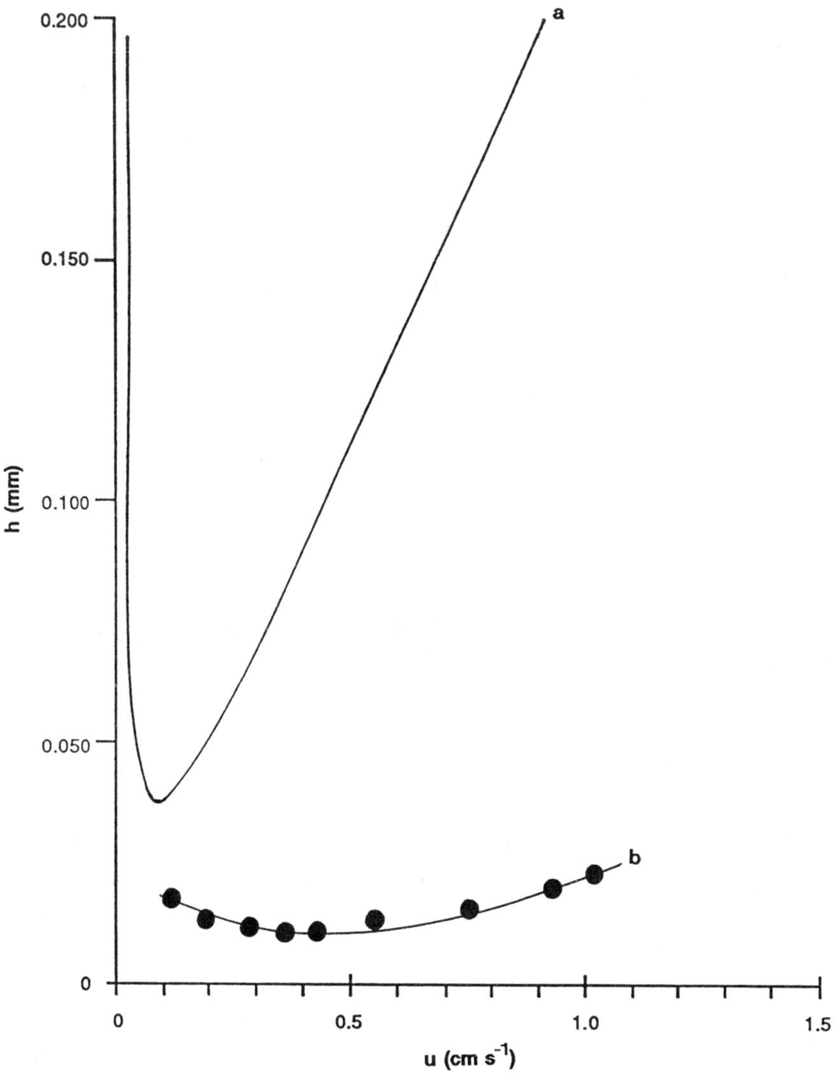

Figure 2.5. SFC van Deemter plots for pyrene using (a) open tubular and (b) packed columns. Curve **a** was calculated from the Golay equation, and curve **b** was determined experimentally. Conditions: (a) d_c = 50-μm, k = 2.30; CO_2; 40°C; 0.8 g mL^{-1}; (b) 10-cm x 4.6-mm i.d. packed column, C_{18} bonded silica particles (Hypersil), 5-μm, k = 2.30; CO_2; 40°C; 0.8 g mL^{-1} (reprinted with permission from Ref. 16).

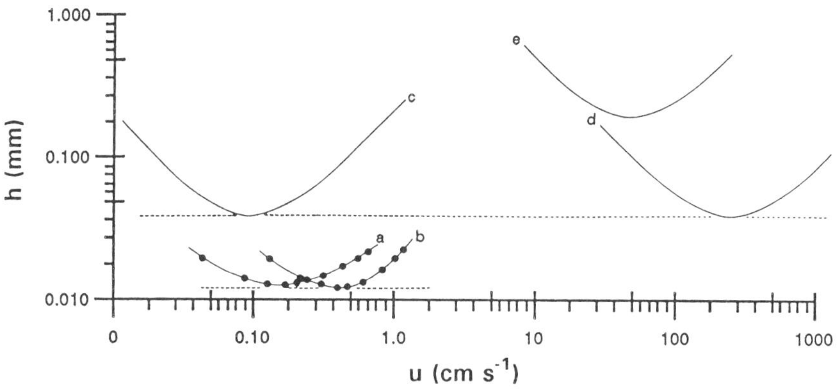

Figure 2.6. Van Deemter curves for various column chromatographic techniques: (a) packed column LC, (b) packed column SFC, (c) open tubular column SFC, (d,e) open tubular column GC. Curves **a** and **b** are from experimental data, while curves **c**-**e** are calculated from the Golay equation. Conditions: (a) 10-cm x 4.6-mm i.d. packed column, C_{18} bonded silica particles (Hypersil), 5-μm, k = 2.85; 30% H_2O in CH_3CN; 40°C; (b) same as for **a** except CO_2 at 0.8 g mL^{-1}, k = 2.30; (c) 50-μm i.d. open tubular column, k = 2.30, CO_2; 40°C; 0.8 g mL^{-1}, D_m = 8 x 10^{-5} cm^2 s^{-1}; (d) 50-μm i.d. open tubular column, k = 2.30, D_m = 0.2 cm^2 s^{-1}; (e) same as for **d** except 250-μm i.d. open tubular column (reprinted with permission from Ref. 16).

parameters are as defined for Equation 2.1. The parameter k_o typically has a value around 0.5 for most column packings.

Randall[16] experimentally determined van Deemter curves for pyrene using a packed column with 5 μm C_{18} bonded silica (Hypersil) under LC (CH_3CN/H_2O) and high density SFC (CO_2) conditions. From these curves, D_m for pyrene in supercritical CO_2 (40°C and 0.8 g mL^{-1}) was calculated to be 8 x 10^{-5} cm^2 s^{-1}. This value was substituted into the Golay equation (Equation 2.1) to generate a van Deemter curve for open tubular column SFC (50-μm i.d. column) at the same high density conditions. Figure 2.5 shows both curves plotted in the same figure for comparison. Obviously, there is a much more significant loss in efficiency at higher than optimum linear velocities for the open tubular column than for the packed column. Furthermore, the theoretical h_{min} for the packed column is approximately 1/3 of that for the open tubular column. It should be pointed out that this is a worse case situation for an open tubular column. Pyrene elutes from an open tubular column at a much lower density (below 0.6 g mL^{-1}), and since higher temperatures are normally used, the slope of the van Deemter curve is considerably lower.

Figure 2.6 gives a comparison of the efficiencies of the different column chromatographic methods at high density conditions. Plate heights are the smallest for the packed column methods (Curves a and b for LC and SFC, respectively), and highest for typical GC conditions (Curve e). GC can be operated at the highest linear velocities, and, therefore, it is the most rapid analysis technique. When comparing the same column types, SFC is more rapid than LC.

Two ways to look at column efficiency are from the viewpoint of plate height (h) or total number of theoretical plates (n). The smallest plate heights are obtained by introducing packing material to form packed columns. This minimizes the contribution to the plate height from the resistance to mass transfer in the mobile phase. However, the packing also reduces the permeability of the column and creates a resistance to flow, such that the length of the column

Table 2.3. Calculated[a] Total Theoretical Plates for Packed and Open Tubular Columns Under Practical Conditions

			n		u (cm s^{-1})	
Column type	d_p,d_c (μm)	Length (m)	Low density[b]	High density[c]	Low density[b]	High density[c]
Packed	5	0.1	5,200	9,100	0.6 (0.2 u_{opt})	2.1 (1.3 u_{opt})
Open tubular	50	10	102,000	19,000	2.5 (5.0 u_{opt})	5.8 (2.8 u_{opt})

[a]Plate heights were calculated from Equations 2.1 and 2.2 for open tubular and packed columns, respectively. The following conditions were assumed:

Open tubular column	Packed column
$D_s = 1 \times 10^{-6}$ cm^2 s^{-1}	$k_o = 0.5$
$d_f = 0.25$ μm	$d_p = 5$ μm
$k = 2$	$k = 2$

[b]Low density conditions were assumed for CO_2 mobile phase at 100°C, 100 atm, 0.19 g mL^{-1}, and $D_m = 5 \times 10^{-4}$ cm^2 s^{-1}.

[c]High density conditions were assumed for CO_2 mobile phase at 100°C, 400 atm, 0.76 g mL^{-1}, and $D_m = 2 \times 10^{-4}$ cm^2 s^{-1}.

must usually be limited to less than 25 cm. Table 2.3 gives a comparison of total plates calculated for both packed and open tubular columns under practical conditions at 100°C. The efficiencies at low density were calculated for practical initial linear velocities: ~5 u_{opt} (u_{opt} = 0.5 cm s^{-1}) for the open tubular column, and 0.2 u_{opt} (u_{opt} = 3.0 cm s^{-1}) for the packed column, which is close to the practical u_{opt} when 5-μm particles are used.[17] Practical u_{opt} values are known to differ from theoretical u_{opt} for packed columns, especially for small particle sizes. As density increases in the packed column during density programming, the linear velocity should move to a more favorable value of 1.3 u_{opt} (u_{opt} = 1.6 cm s^{-1}) with a concommitant increase in efficiency; however, these efficiencies are never realized in practice. The open tubular column, on the other hand, starts out with extremely high efficiency at a practical velocity of 5 times the theoretical u_{opt}, but the dramatic increase in linear velocity that is achieved at the final density corresponds to 28 u_{opt} (u_{opt} = 0.2 cm s^{-1}), giving a 5.4 times reduction in efficiency.

2.3 Pressure Drop Effects on Resolution

So far in this chapter, little reference has been made to the pressure drop along the length of the column, and how this pressure drop affects resolution. Pressure, itself, has a direct effect on retention, diffusion, and selectivity, which all affect resolution.

The pressure drop is important in SFC because density and, hence, solvating power is a direct result of the pressure. If there is a pressure drop, there is also a density drop, and the solubilities of solutes decrease along the length of the column. This is especially important in considering the type of column to use; typical open tubular columns have minimal pressure drops, while significant drops are usually associated with packed columns. In most cases, pressure drop effects will be insignificant for open tubular columns.[18]

Pressure drop effects on column efficiency are reduced by operating at linear velocities higher than u_{opt} and under constant mass flow conditions.[19] This is because the product $\eta \rho D_m$ (where η is the viscosity and ρ is the density of the mobile phase) is reasonably constant and, under constant mass flow, the mobile phase velocity is inversely proportional to its density. However, changes in D_m due to viscosity changes do cause losses in efficiency even under constant mass flow conditions. The change in mobile phase viscosity with density causes diffusion rates to change faster than corresponding shifts in the linear velocity. Unfortunately, this means that the optimum mass flow rate varies as the mobile phase density changes along the length of a column. The net effect is an apparent increase in the measured mass transfer term, a decrease in the optimum mass flow rate, and an increase in the minimum plate height achievable. For packed columns, this seems to be a limiting factor in the total efficiency possible.[17,20,21] As higher pressures and densities are used in packed columns, the magnitude of this pressure drop effect decreases because the relative density change along the column becomes smaller. It is because of the effects of pressure drop that most packed column applications are usually done at higher pressures and densities where the relative change in density with pressure is smaller.

To help understand the effect of pressure drop on column efficiency, reference can be made to Figure 2.7. If the optimum velocity is achieved for a compound at some point in the middle of the column, the condition at that point in the column is represented by the section labeled "b". This means that earlier in the column, k will be lower and diffusion rates slower than the optimum. Point "a" in Figure 2.7 identifies this region. Further in the column, the conditions shown by the region labeled "c" are reached. Diffusion rates have now increased significantly compared to the linear flow rate, and the mobile phase moves slower than the optimum velocity at this point. Averaging these effects over the whole column leads to a net larger plate height. These consequences can be summarized as: (a) the measured optimum linear flow rate is higher than that of an ideal "zero density drop" column, (b) the measured minimum plate height is greater than the "zero density drop" column, and (c) sections a and c of the van Deemter plot in Figure 2.7 both rise more sharply.

These changes in diffusion rate with changes in viscosity and density create similar consequences for pressure programming. Although not easily accomplished, maintaining a constant mass flow during pressure programming would help to compensate for these effects. Under these conditions, the mobile phase linear velocity gradually decreases as its density increases, helping to compensate for the lower diffusion rates. The overall mass flow must decrease to totally compensate for the lower diffusion rates.

In reality, constant mass flow is difficult to achieve because fixed restrictors are normally used. Flow rates generally increase during density programming as discussed in the previous section. Schoenmakers and Uunk[17] observed decreases in packed column efficiency when operating with very low and high pressure drops.

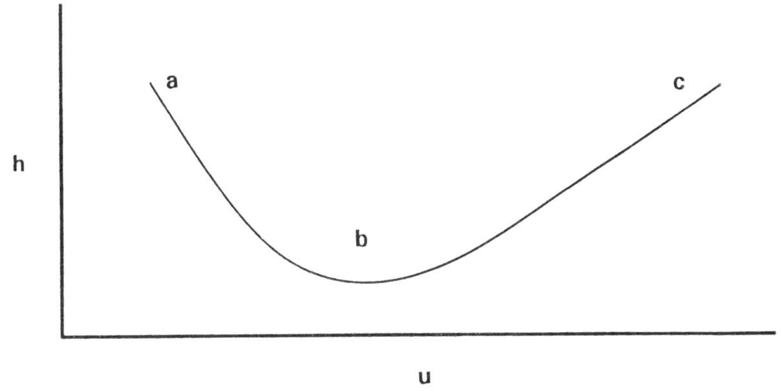

Figure 2.7. Van Deemter curve showing regions **a**, **b**, and **c**.

The effect at low pressure drop was presumably due to the resultant lower than optimum linear velocity and, hence, large contribution to h from longitudinal diffusion. In this study, the pressure drop was established by keeping the inlet pressure constant while changing the outlet pressure with a variable restrictor. Obviously in practice, a minimum outlet pressure must be maintained in order to achieve solute transfer out of the column.

One obvious effect of pressure drop on efficiency is that higher column inlet pressures are needed to reach high enough density at the column outlet to elute samples from columns with greater pressure drops. This leads to higher elution densities and lower efficiencies. It is almost always observed that compounds are eluted at lower densities on open tubular columns than on packed columns. As stated, with typical open tubular columns in SFC, pressure drop is never a concern. Figure 2.8 shows plots of column diameter as a function of column length at 10 u_{opt} which generate a 3% pressure drop at k values of 1, 3, and 5 for CO_2 at 40°C and 72 atm. For a 50-μm i.d. column, lengths greater than 32 m would be required to produce greater than 3% pressure drop. Typically, shorter columns are used.

The effect of pressure drop on selectivity is less significant than its effect on efficiency. The observed values of the capacity factor (k_{obs}) and the selectivity (α_{obs}) are as follows[19]

$$k_{obs} = \frac{e^{(a-b\rho)}(e^{bwL}-1)}{bwL} \qquad (2.3)$$

$$\alpha_{obs} = \frac{e^{(f-q\rho)}(e^{qwL} - 1)}{qwL} \quad (2.4)$$

where ρ is the density of the mobile phase at the head of the column, w is the rate at which the density changes along the column, L is the total length of the column, and a, b, f, and q are constants that depend on the compound types being separated, the nature of the mobile phase, the nature of the stationary phase, and the temperature of the column. In both equations, wL is simply the density change (Δρ) along the full length of the column. From these equations and an evaluation using published practical data, a number of conclusions were drawn:[19]

(a) The observed retention and selectivity are dependent on the total density drop and not on the rate at which the density changes along the column.

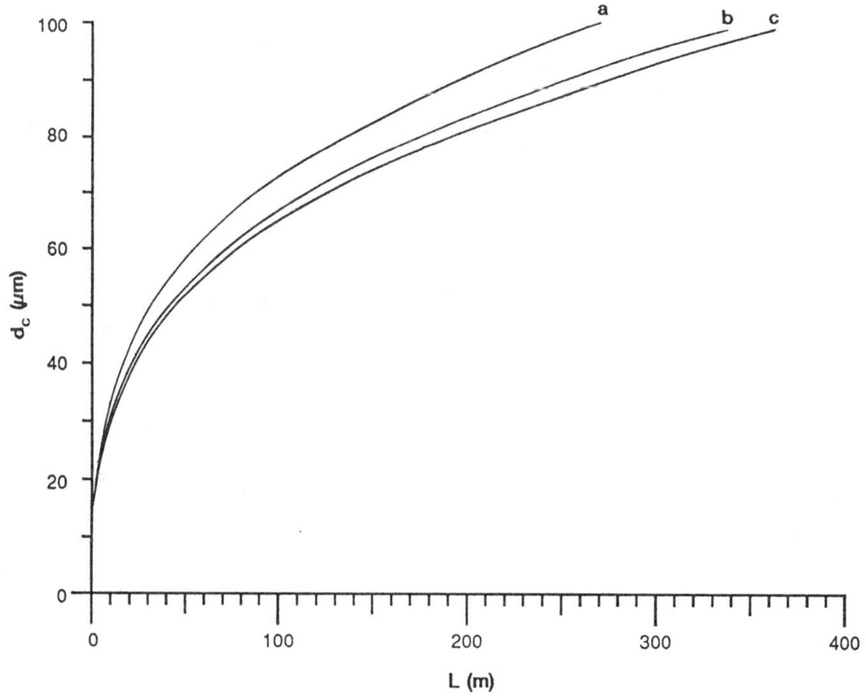

Figure 2.8. Column diameter as a function of column length at 10 u_{opt} for open tubular columns which generate a 3% pressure drop at k values of (a) 1, (b) 3, and (c) 5. Conditions: CO_2; 40°C; 72 atm. Calculations were made using the Poiseuille Equation (reprinted with permission from Ref. 1).

(b) In order to maintain constant retention or selectivity as the density drop is increased, the density at the head of the column must be increased.

(c) If k_{obs} is kept constant, selectivity will always decrease, and if selectivity is kept constant, k_{obs} will always increase, as the density drop increases.

(d) The sensitivity of α_{obs} to density drop increases at lower k values.

(e) Selectivity at constant k_{obs} decreases more rapidly as density drops become large.

(f) Resolution losses can occur because of losses in selectivity.

Because lower densities cause increases in selectivity, it could be argued that density drops in columns should cause selectivity improvements. This is not true if comparisons are made at a constant observed k, where in one case there is a significant density drop and the other case there is not.[19] In order to keep a constant k, the column head pressure must be increased as the density drop is increased. The fundamental cause of selectivity losses with pressure drops is that the higher mobile phase density in the front end of the column causes a loss in selectivity which is not possible to compensate for without a net increase in retention. Yet another way of saying this is that the exit end of the column cannot provide resolution (selectivity) to compensate for poor performance at the front end of the column. Selectivity losses are probably small in most cases and not noticeable except in cases where k and α are strong functions of the mobile phase density.

The effect of pressure drop on resolution limits will be greatest at conditions near the critical point where the density change with pressure is the largest. For compounds with an α value of 1.01 under zero pressure drop conditions, 80% resolution loss would result from only a 10% pressure drop.[19] Based on pressure drops at mobile phase linear velocities of 1 cm s^{-1} experienced on 10-cm long columns packed with 10-, 5-, and 3-μm particles, losses in selectivity in the region of maximum change in density with pressure are given in Table 2.4.

It should be mentioned that the significance of this resolution loss depends on the application of interest. If the α values are large enough, the loss in selectivity with pressure drop is of no consequence. Furthermore, if higher pressures are used, which is common practice in packed column SFC, the density change is small along the length of the column and selectivity losses become negligible. High pressure drops can be tolerated if one does not work in the pressure regions where density changes rapidly with pressure.

Table 2.4. Particle Size Effects on Selectivity

Particle diameter (μm)	$\Delta\alpha_{obs}$ range expected
10	0.01-0.07
5	0.03-0.22
3	0.09-0.29

2.4 Speed of Analysis

The analysis time in chromatography for a given solute can be calculated from

$$t_R = \frac{L}{u}(1 + k) \qquad (2.5)$$

where the terms are as previously defined. When fast analysis times are sought, higher than optimum mobile phase linear velocities are used. As the value of u increases, the ratio h/u approaches the value of $h_{min}/2\ u_{opt}$, which is given by the values of h and u at the minimum in the van Deemter curve.

If we assume that k = 2, the value of u_{opt} for an open tubular column can be calculated from Equation 2.1

$$u_{opt} = \frac{5.5\ D_m}{d_c} \qquad (2.6)$$

and

$$h_{min} = 0.7\ d_c \qquad (2.7)$$

With the same assumption of k = 2 for a packed column, u_{opt} and h_{min} can be calculated from Equation 2.2

$$u_{opt} = \frac{4.2\ D_m}{d_p} \qquad (2.8)$$

$$h_{min} = 2.2\ d_p \tag{2.9}$$

Equal speeds of analysis for packed and open tubular columns occur when the values of h_{min}/u_{opt} for the two columns are equal. This occurs when $d_c \sim 2\ d_p$, as determined from Equations 2.6-2.9. Therefore, the open tubular column internal diameter would have to be reduced to approximately 10 μm in order to generate theoretical plates as fast as a packed column containing 5-μm particles.

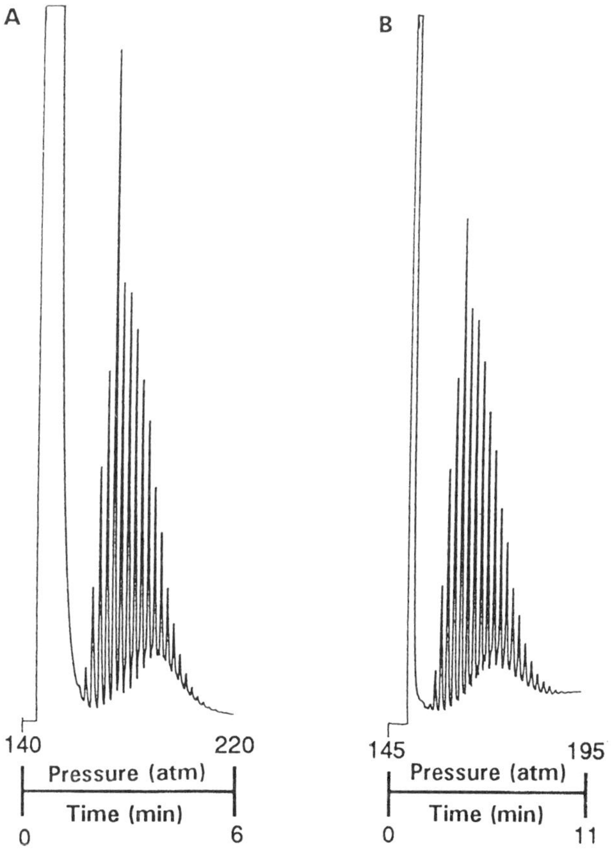

Figure 2.9. SFC chromatograms of paraffin wax on (A) packed capillary column and (B) open tubular column. Conditions: (A) 12.5-cm x 250-μm i.d.; C_8-bonded silica particles (Nucleosil, 5-μm); CO_2; 100°C; pressure program from 140 atm to 220 atm at 20 atm min^{-1}, (B) 3-m x 50-μm i.d. open tubular column; poly(50% octyl)methylsiloxane stationary phase; CO_2; 100°C; pressure program from 145 atm to 195 atm at 5 atm min^{-1} (reprinted with permission from Ref. 12).

In making a direct comparison of the speed of analysis for packed and open tubular columns, it must be assumed that both columns can produce a sufficient number of total theoretical plates to accomplish the desired separation. Speed is of no value if resolution is not satisfied. If n is set equal for both columns at 5,000 total plates, the analysis times can be calculated for both low and high densities from Equation 2.5. For the conditions described in Table 2.3, it was found that a 1-m x 50-μm i.d. open tubular column could give an analysis time of 2.0 min compared to 0.8 min using a 10-cm packed column with 5-μm particles at low density. While the open tubular column can approach the speed of the packed column at lower densities, the differences are more pronounced at higher densities.

The analysis times for open tubular columns can be reduced by reducing the column internal diameter, shortening the column, increasing the linear velocity, increasing the density program ramp rate, or increasing the temperature. There are, however, practical limitations to changing these parameters. As a practical example for comparison, Figure 2.9 shows chromatograms of a paraffin wax sample analyzed using open tubular and packed capillary columns. The open tubular column was pushed to a linear velocity of 4 cm s^{-1}. The dead time of the packed capillary column was adjusted to match the open tubular column (corresponding to u = 0.8 cm s^{-1}). The pressure program ramp rate was then adjusted for the packed capillary column until the resolution between the n-C_{18} and n-C_{20} peaks was close (R_s = 0.94) to that provided by the open tubular column (R_s = 1.08). Even though the open tubular column conditions were pushed to a relatively fast analysis time, the packed capillary column was still 1.4 times faster.

2.5 Sample Capacity and Loadability

Unlike in open tubular column chromatography, the packed column internal diameter does not, in general, affect the separation efficiency, and it can be increased to increase sample capacity and loadability. The sample capacity refers to the maximum amount of sample solute that can be introduced on the column without causing significant fronting of the chromatographic peak. On the other hand, sample loadability, as defined here, refers to the maximum volume of sample (solutes plus solvent) that can be introduced on the column without causing significant degradation of column efficiency.

For open tubular columns, an equation has been derived[19] to estimate allowable injection volumes based on column diameter and length, plate height, capacity factor, and acceptable resolution loss

$$V_i = 0.866\ \pi d_c^2 (Lh)^{1/2} \left[\frac{1}{(1 - \Delta R_s)^2} - 1 \right]^{1/2} (1 + k) \quad (2.10)$$

where V_i is the injection volume, ΔR_s is the fractional resolution loss, and the other variables are as previously defined. For a 10-m x 50-μm i.d. column under practical operating conditions ($u = 10\ u_{opt}$ and $k = 2$), and for a 1% allowable resolution loss, the maximum sample volume that can be injected is 96 nL. Obviously in practice, methods for sample volume reduction and solute focussing must be used for sample introduction into open tubular columns, especially when smaller diameter columns are used. These methods are described in Chapter 4.

The sample volume loadability (L) of a sample component on a chromatographic column has been defined as

$$L = c_{m,max} V \quad (2.11)$$

where $c_{m,max}$ is the maximum concentration of a particular component that can be accommodated by the mobile phase without causing more than 5% deviation from the linear distribution isotherm of the phase system for a particular injection volume (V).[14] Assuming equal distribution coefficients (K) an equation relating the maximum loadabilities for packed and open tubular columns can be derived

$$\frac{L_p}{L_c} = 1.24 \left(\frac{d_{pc}}{d_c} \right)^2 \left[\frac{1 + K\, \epsilon_i (1 - \epsilon_e)/\epsilon_T}{1 + 4\, K\, d_f/d_c} \right] \left(\frac{L_p}{L_c} \right)^{1/2} \left(\frac{d_p}{d_c} \right)^{1/2} \quad (2.12)$$

where L_p and L_c are the loadabilities of the packed and open tubular columns, respectively; d_{pc} is the diameter of the packed column; ϵ_i, ϵ_e, and ϵ_T are the intraparticle, interstitial, and total porosities, respectively; d_f is the stationary phase film thickness in the open tubular column; L_p and L_c are the lengths of the packed and open tubular columns, respectively; and the other variables are as previously defined.

Table 2.5. Sample Loadability Ratios for Different Distribution Coefficients

K	L_p/L_c
1	31
10	101
100	373

Packed column: L = 20 cm, d_p = 5 μm, d_{pc} = 1 mm, ϵ_i = 0.5, ϵ_e = 0.4, ϵ_T = 0.7.
Open tubular column: L = 10 m, d_c = 50 μm, d_f = 0.2 μm
(Taken from Ref. 14)

Equation 2.12 shows that the maximum sample loadability strongly depends on column diameter. Table 2.5 lists the sample loadability ratios for different distribution coefficients and typical packed and open tubular columns. As expected, sample loadabilities are much greater on packed columns.

As has been shown, the particle size in packed columns has the same effect on efficiency and speed of analysis as column diameter in open tubular columns. This leaves the total surface area of packed columns as a variable to increase sample capacity. Both specific surface area and packed column diameter can be varied to change the sample capacity. However, since most packed columns in use contain porous particles with maximized surface area (100-300 m^2 g^{-1}), the column diameter becomes the most easily alterable variable.

One approach to increase the sample capacity of open tubular columns is to increase the stationary phase film thickness. However, this leads to a reduction in column efficiency. Figure 2.10 shows that film thicknesses up to 1 μm produce only small losses in efficiency.[22] At high linear velocity ($u >> u_{opt}$), the fraction of the total plate height due to the stationary phase, h_s, can be given by

$$h_s = \frac{64\ kA}{(1 + 6k + 11\ k^2) + 64\ kA} \qquad (2.13)$$

where

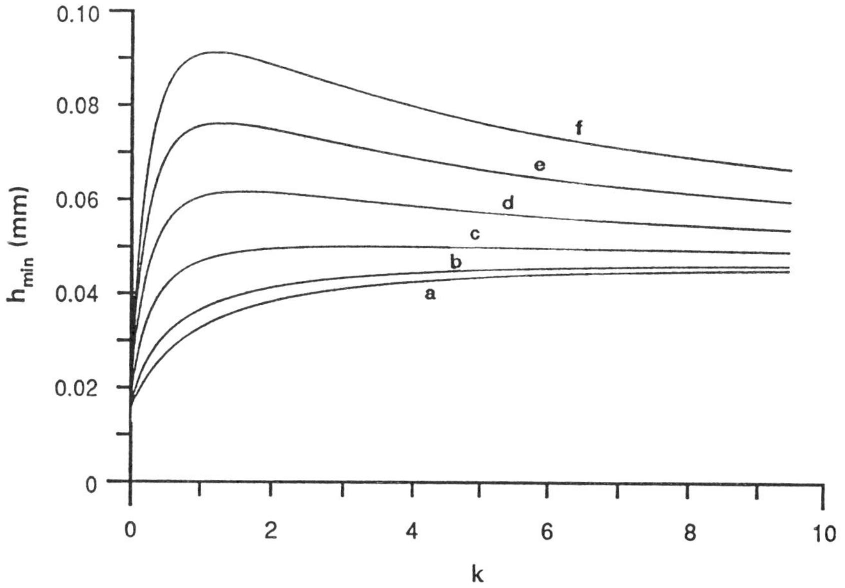

Figure 2.10. Plot of h_{min} *vs* k for film thicknesses of (a) 0.25 μm (β = 50), (b) 1.0 μm (β = 12.5), (c) 2.0 μm (β = 6.75), (d) 3.0 μm (β = 4.2), (e) 4.0 μm (β = 3.1), and (f) 5.0 μm (β = 2.5). Conditions: 50-μm i.d. open tubular column; $D_m = 2 \times 10^{-4}$ cm^2 s^{-1}; $D_s = 1 \times 10^{-6}$ cm^2 s^{-1}, and u_{opt} for each d_f and k (reprinted with permission from Ref. 22).

$$A = \left(\frac{D_m}{D_s}\right)\left(\frac{d_f}{d_c}\right)^2 \tag{2.14}$$

These equations can be used to determine the maximum stationary phase film thickness for any application, assuming that the compounds of interest can still be eluted from the column at the greater k values. For constant L, k, and α, the fractional loss in resolution from the increased total plate height due to the stationary phase is given by

$$-\frac{\Delta R_s}{R_s} = 1 - (1 - h_s)^{1/2} \tag{2.15}$$

From plots of $-\Delta R_s/R_s$ *vs* percent d_f/d_c, it was evident that at film thicknesses larger than 2% of the column diameter (*i.e.*, $\beta<15$), the resolution loss became very significant.[22]

From the previous considerations, stationary phase film thicknesses of up to 1 μm can be used in 50-μm i.d. open tubular columns for SFC with only minimal losses in chromatographic efficiency. Even greater film thicknesses can be used, depending on the resolution losses that can be tolerated. Based on a definition of sample capacity as the amount injected for which it takes twice the length of time for the leading edge of the chromatographic peak as for the trailing edge, practical measurements have led to a sample capacity of 100 ng (alkanes) on a 10-m x 50-μm i.d. column coated with a 0.25-μm film of a polymethylsiloxane stationary phase. A 1-μm film should increase this capacity to approximately 400 ng with less than a 10% resolution loss.

2.6 General Requirements for Column Technology

Theoretical and practical considerations of SFC column configurations have been discussed in the previous sections. The effects of column type and dimensions, including stationary phase film thickness, on the important chromatographic performance measurements of efficiency, selectivity, resolution, speed of analysis, sample capacity, and sample loadability were considered. In the remainder of this chapter, the important aspects of preparing high performance columns for use in SFC are treated. Furthermore, some discussion of stationary phase selectivity is included.

The column dimensions used in SFC are tabulated in Table 2.6. The traditional names that have been adopted for the different

Table 2.6. Typical Column Dimensions

Column type	Internal diameter (mm)	Length (m)
Conventional packed	2 - 4.6	0.03 - 0.25
Microbore packed	0.5 - 2	0.03 - 0.25
Packed capillary	0.1 - 0.5	0.05 - 0.5
Open tubular	0.025 - 0.1	1 - 35

packed column types are borrowed from LC. The conventional 4.6-mm i.d. packed column is not the most common dimension of packed columns in SFC as it is in LC. Limitations in volumetric pumping speeds and/or volumetric flow rates for the detectors of choice (see Table 2.7) have helped to make packed columns of micro sizes increasingly common in SFC. The packed capillary column has the advantages of lower volumetric flow rates, making them more compatible with mass flow sensitive detectors, smaller elution peak volumes which provide greater sensitivity in concentration sensitive detectors, easier sample transfer in multidimensional systems, and higher permeability, which allows the use of longer columns for better resolution. Open tubular columns for SFC are typically 3 to 10-m x 50-μm i.d., which represents a practical compromise in efficiency and speed.

The requirements that must be satisfied by the column technology include (a) well deactivated surfaces and stationary phases, (b) uniform stationary phase films, and (c) well immobilized stationary phase films. There continues to be a demand for more inert columns, and this demand has resulted in innovations in deactivation procedures for column and support materials. The most deactivated columns have been open tubular columns. This is a result of their low surface area and much lower porosity when compared to silica or alumina packing materials. In addition to causing adsorption of polar analytes, active sites lead to poorly controlled and nonreproducible retention. Active sites generally necessitate the use of polar modifiers.

Table 2.7. Expected Volumetric Flow Rates for Various Column Sizes

	Open tubular column i.d. (mm)		Packed column i.d. (mm)		
	0.01	0.05	0.25	1.0	4.6
Supercritical fluid flow[a] (mL min^{-1})	0.001	0.01	0.1	1	10
Gas flow[a] (mL min^{-1})	0.1	1	10	100	1,000

[a]Within an order of magnitude

The coating methods for open tubular columns or packed columns must provide stationary phase films which are extremely even in thickness and which completely cover the surface. A number of advancements have been made over the last 10 years in the coating of both open tubes and packed column support material. In the packed column, the stationary phase is normally near monomolecular thickness, polymerized, and chemically bonded to the support. Immobilization (generally crosslinking of the polymeric phase) is an essential ingredient in the preparation of open tubular columns. It must be performed to resist dissolution, but without lowering solute diffusion within the phase.

2.7 Packed Column Technology

Packing materials. Since the early days of SFC, packed column SFC technology has depended on materials available from the current state of the art LC technology. It is not surprising that LC packing materials perform well under SFC conditions, since both techniques depend on the ability of the mobile phase to solvate analyte molecules.

In their pioneering work in SFC, Klesper *et al.*[23] mentioned having tried both silica gel and Woelm alumina, although the latter did not perform adequately for their particular experiment. Through the remainder of that decade, these two materials, as well as glass beads and diatomaceous earth, were mentioned in different SFC publications. However, spherical porous silica particles are now available that provide (a) more uniformly shaped, (b) higher surface area, and (c) higher sample capacity materials that pack more efficiently and are applicable to a wide range of applications. Currently, interest is growing in alternative packing materials, especially polymer beads. These materials promise more inert surfaces and chemical stability compared to silica packings.

Particle sizes referred to in publications normally vary from 3 to 10 µm in diameter with pore sizes ranging from 100 to 300 Å (corresponding to a surface area of ~100 to 300 $m^2\ g^{-1}$). Of these, the most commonly used particle size is 5-µm diameter. The popularity of this particular size can be attributed to the fact that it is small enough to give relatively small plate heights, while being commercially available in sufficient uniformity and narrow distribution to allow efficient packing to be accomplished. Smaller particles provide smaller plate heights; however, they also reduce permeability and increase the pressure drop across the column. The feasibility of working with small diameter particles in SFC has been discussed by several groups.[20,24-27]

Chemical Treatment. The surface activity of silica packing materials is a serious limitation for packed column SFC when mobile phases of low polarity (*e.g.*, CO_2) are used to elute polar analytes. The silica surface tends to exhibit reversible adsorption of polar compounds, resulting in tailing peak shapes. Polar organic modifiers are often used to alleviate this problem. However, the interfering detector response of the modifier often limits detector capabilities. Therefore, much attention has been focused on improvement of deactivation methods in order to allow the elution of more polar compounds when using CO_2 as the mobile phase.

Various methods have been used to modify silica particles for use in chromatography. Due to the ability of a supercritical mobile phase to solvate large molecules, liquid stationary phases can be stripped away. Stationary phases which are chemically bonded to, or otherwise immobilized on, the support surface usually perform better. According to a recent review of bonded phases for LC,[28] the majority of commercial alkyl phases are monomeric in nature (*i.e.*, monolayer coverage of the solid support as a result of chemically bonded low molecular mass monofunctional reagents). Monofunctional reagents, such as chloro- or alkoxysilanes, react with silanol groups on the silica material to produce monomeric stationary phases. This is usually accomplished by refluxing the reagent with the particles in an organic solvent, such as toluene. Di- and trichlorosilanes are more chemically reactive than their monofunctional counterparts. However, surface attachment occurs with only one or two siloxane bonds from the same reagent molecule, due to steric constraints imposed by bond lengths. The unreacted functional groups can be hydrolyzed to silanols in the presence of trace amounts of water. Subsequently, polymerization of the polyfunctional reagent occurs. In order to ensure monomeric coverage by the phase, care must be taken to exclude water from the reaction mixture. Studies[29] have shown that even though the polyfunctional reagents may be more reactive and may modify a greater portion of the active surface sites, they actually increase the number of silanols available as a result of the hydrolysis of unreacted moieties left behind. In the past, these additional silanols have often been used advantageously to increase the selectivity of the packing material for closely related solutes.

Due to differences in reactivity of the individual silanol groups on the particle surface and possible steric hindrance associated with the reagent molecules themselves, chemical modification of all of the silanol groups is unlikely. In an attempt to provide a more complete coverage, several research groups[30-32] have taken advantage of the tendency of the chlorosilane reagents to polymerize in the presence of water. By the introduction of controlled amounts of water, thicker films of polymeric phases result, possessing greater sample capacities and unique shape selectivities. However, because additional silanol groups are created during polymerization, further endcapping is required to remove these somewhat less reactive silanol sites. A new approach to preparing polymeric phases in LC has recently been introduced by Figge *et al.*[33] This method is based on free radical initiated crosslinking procedures used to immobilize stationary phases in capillary GC. This type of polymeric phase does not directly involve chemical modification of the silica surface. Silanols can be endcapped prior to coating of the phase, which improves deactivation. This crosslinked phase creates a shield that hinders the solutes from interacting with the residual surface silanol groups. These crosslinked phases are claimed to be more uniformly distributed than conventional "monomeric" bonded phases.

Recently[34], a deactivation procedure was developed specifically for SFC, in which hydrosiloxane polymers are chemically bonded to porous silica. This procedure simultaneously deactivates the surface and provides a homogeneous stationary phase. These materials have often been called polymer encapsulated phases, and they have been shown to produce improved peak shapes for polar analytes when compared to conventional packing materials. In addition, enhanced separation (*i.e.*, selectivity) was demonstrated for nonpolar, as well as polar, analytes.

Packing Methods. The advantages of a variety of methods for slurry packing LC columns of the dimensions commonly used in SFC have been discussed elsewhere.[35,36] There are many methods available for efficiently packing LC columns, but no one best method for all packing sizes or column types. However, for LC columns of analytical dimensions commonly used in SFC (0.10-4.6 mm i.d. using particles <20 μm diameter), the packing method of choice includes the use of a high pressure slurry (see Table 2.8). The efficiencies of columns produced by slurry packing are attributed to the high pressures used. The particles are packed by filtration and not sedimentation. Large volumetric flow rates are required which favor the use of lower viscosity solvents. This allows higher impact speeds to be attained by the particles, avoiding differential movement according to particle size and resulting in more stable columns. As

Table 2.8. General Parameters for Column Packing

Parameter	Conventional (2-4.6-mm i.d.)	Microbore (0.5-2-mm i.d.)	Packed capillary (0.1-0.5-mm i.d.)
Slurry solvent	Viscous solvents (MeOH/glycol)	Viscous solvents (MeOH/glycol)	Nonviscous solvents (acetonitrile)
Slurry concentration	5-30%	1-5%	20%
Packing pressure	300-1000 atm	600-1800 atm	300-500 atm
Flow direction (down or up)	Both	Down	Down
Column	As is or polished metal	Polished or glass-lined metal	As is or inner wall coated fused silica

in LC, particles that are spherically shaped and have smaller size distributions are more efficiently packed. These factors become more important as particle sizes become smaller. Each particular combination of particle and column size has its own optimum set of conditions.

Additionally, a dry packing method has been demonstrated[37] for the preparation of 0.25-mm i.d. fused silica columns using 5-μm particles. Static charging and particle aggregation problems, which normally restrict the use of this technique to larger particles, were addressed. Recently,[38] a study was conducted in which several packing techniques were examined, including a new method for packing columns >0.5-mm i.d. using supercritical CO_2 as the packing solvent. The study concluded that this technique provided equal efficiency and better permeability than slurry packing. This is the only method exclusively designed for SFC to date.

Since most packed columns are originally intended for LC use, the column is usually received with some residual solvent still present. Therefore, conditioning of the column at supercritical conditions before use is recommended.

2.8 Open Tubular Column Technology

Open tubular columns for SFC must possess the usual qualities of high efficiency, inertness, and lasting stability which are characteristic of open tubular columns for gas chromatography.[39] The requisite use of small i.d. columns (<100 μm) for efficient separations in SFC has led to modifications in the procedures established for preparing GC columns. Traditional manufacturing processes developed for the larger bore GC columns often fail when applied to small capillaries. Attempts at using this technology often result in plugging, unacceptable activity, or columns of low efficiency.

Small diameter open tubular columns require high levels of inertness.[40] Active compounds must be able to be analyzed at low levels with no adsorption. This requires the use of columns with the best levels of deactivation attainable. Because the surface area to internal volume ratio of open tubes varies inversely with diameter, wall effects and their inherent contributions to activity become more important as column size is reduced.

Contributions of surface polarity to overall column polarity become more significant in narrow bore capillaries. If SFC columns of reproducible selectivity and polarity are to be obtained, it is imperative that the deactivation process effectively blocks active sites on the support surface and provides a layer with the same polarity as the stationary phase.

The restrictive nature of narrow bore capillaries requires the use of high pressures (up to 1,000 psi) to force solvents and typically viscous coating solutions through at reasonable rates. Steel or brass enclosings of the glass pressure chambers used in GC column manufacture allow convenient manipulation of solutions through small i.d. columns.

Pretreatment. While glass has been shown to be a suitable substrate for capillary SFC columns,[41] most work has been performed with fused silica as the base material because of its ease of handling and inherent inertness. Because of its popularity, only column technology for fused silica will be discussed here.

The fused silica drawing process produces a surface with variable layers of water hydrogen-bonded to surface silanol groups. In addition, 300-700 ppm levels of chloride are present, the result of incomplete hydrolysis of silicon chloride precursors in the tube preforms from which the capillary tubing is drawn. This chloride is in the form of silicon chlorides and hydrogen chloride. Most manufacturing processes include the use of a hydrothermal pretreatment or preconditioning step to bring levels of chloride, silanol, and water to regulated low levels. This is important since deactivation methods give optimum performance only when these variables are controlled. Wet tubing often reacts with deactivation reagents to produce thick irregular layers that plug small i.d. capillaries. Deactivation layers of changing film thickness have led to variations in surface energies, and the production of regions not wettable by stationary phases coated later.

Deactivation. Deactivation reagents are usually introduced as dilute solutions in volatile solvents, and are dynamically coated on the capillaries to produce thin, evenly distributed films. Many of the same classes of reagents used successfully with larger bore GC columns have given good performance with small i.d. SFC columns.[40] These include disilazanes as well as cyclic and linear siloxanes which are thermally cured at ~400°C, and the hydrosilane and silanol terminated reagents which bond at temperatures between 250-350°C. Excess reagent is removed by rinsing the columns with methylene chloride or other suitable solvents.

Wettability rules for stationary phases on deactivated surfaces that are standard in GC column technology are even more rigorous with narrow bore capillaries.[42] Small inhomogeneous stationary phase film regions that may go unnoticed in larger columns inevitably result in plugging with SFC columns. Deactivation layers with higher surface energies than those used for GC columns with a given stationary phase must be used with narrow bore capillaries to ensure uniform film deposition. Attempts at coating narrow bore columns with stationary phases that are close to the limit with respect to critical surface tension requirements result in high failure rates and generally low efficiencies.

Coating. The static method of coating is used exclusively in the commercial production of narrow bore open tubular columns for SFC. The film thickness deposited is controlled by the concentration of stationary phase in the volatile coating solvent.

Coating should be performed as quickly and at as low a temperature as possible in order to minimize efficiency losses and plugging tendencies which result from Rayleigh instability,[42,43] a phenomenon affecting the stability of an annular coating of liquids on the inside of cylindrical tubes. Once coated, the stationary phase film begins to rearrange through formation of standing waves that grow in amplitude until droplets are formed. The rate of degradation of the homogeneous film is directly proportional to the cube of the film thickness, while it is inversely proportional to the fourth power of the internal diameter of the tube and to the film viscosity. From a practical manufacturing standpoint, this means that stationary phases of very high viscosity must be used,[44] and that efficient narrow bore columns with thick films are difficult to obtain.[22]

Because film rearrangement is time dependent, fast coating is desirable. One way of speeding the coating process is to raise the temperature,[45] but gains are offset by the increased thermal motion of the polymer chains which reduce film viscosity and enhance the rate of droplet formation. Viscosity is also reduced by residual solvent remaining in the freshly deposited stationary phase film. Coating of narrow bore capillaries is most commonly performed at 25-50°C with volatile solvents such as Freon 11 or 21, *n*-pentane, or methylene chloride.[46] Mixtures of these solvents have also been used with success.[47]

Immobilization. To stabilize the stationary phase film against rearrangement and rinse-out when in contact with supercritical fluids, crosslinking is employed. The strong solvating power of supercritical fluid mobile phases necessitates efficient crosslinking of SFC stationary phases. Processes developed for GC columns may need to be enhanced to give films with the stability required for use in SFC. Free radical processes using azo-*t*-butane[39] and dicumyl peroxide[48] are commonly used along with specialized alkoxy silane condensation of silanol terminated stationary phases.[41] Thermal treatment under hydrogen gas flow can immobilize stationary phases containing low levels of silicon hydride.[49]

Extensive crosslinking provides necessary stability, but reduces the ability of solutes to diffuse freely, with resulting efficiency loss.[50] Less interchain crosslinking is required if higher molecular mass linear polymers are used as stationary phases. Efficiency is not lost by increasing the molecular mass of the linear stationary phase polymer, but by interfering with chain mobility through crosslinking into less flexible three-dimensional networks.

Extensive crosslinking of the stationary phase film is required for mobile phases, such as ammonia, which react with siloxane bonds to some degree.[50,51] A column lifetime of several months was experienced with a highly crosslinked poly(octyl)methylsiloxane stationary phase using ammonia as mobile phase.[51] Even so, limited column lifetimes are to be expected under such rigorous conditions, and more suitable phases should be developed for this application.

To prevent excessive column bleed during analysis, SFC columns should be well rinsed with solvents which have a strong affinity for the respective stationary phase employed. For example, nonpolar polymethylsiloxane phases may be rinsed with pentane, but polar poly(cyanopropyl)methylsiloxane phases require stronger solvents, such as methylene chloride. The ideal rinsing solvent is the intended mobile phase, but at higher density. Careful column testing procedures (see Section 6.6) should be applied periodically to evaluate the stability of the column with time and use.

2.9 Stationary Phases for SFC

The stationary phase plays an important role in achieving high performance in SFC. Many stationary phases developed for either LC or GC can be adopted for use in SFC. This includes phases exhibiting all types of solute-stationary phase interactions and selectivities, such as adsorption, dispersion, dipole-induced dipole, dipole-dipole, and size and shape, as well as combinations of these interactions.

Before describing the specific stationary phases for packed and open tubular columns, it is important to consider the general criteria for a phase to be useful in SFC. A high degree of thermostability is usually not an essential property for a stationary phase in SFC, whereas resistance to dissolution and chemical reaction is critical. Thus, when choosing the proper stationary phase/mobile phase combination, it is of utmost importance to consider the solvating strength of the mobile phase and its inherent possibility for chemical reaction with the stationary phase. For example, a high molecular mass polar Carbowax stationary phase is not soluble in CO_2. Therefore, it may be used even without immobilization. On the other hand, if ammonia is the mobile phase, the Carbowax phase would rapidly deteriorate. Even though these extreme examples are not commonly encountered, they stress the important fact that the stationary and mobile phases should not be chosen independently.

A more common consideration is that the stationary phase in SFC should be covalently bonded by crosslinking or immobilization on support particles or on the column wall. Supercritical mobile phases, can dissociate physically or ionically bound stationary phases, and if the stationary phase then becomes soluble in the mobile phase, it will rinse out of the column. Macaudière *et al.*[52] reported that commercial LC phases with amide linkages contain about 25% ionic bonds, and methanal/CO_2 extraction was found to be a good tool for measuring the percentage of ionic bonds that hold the phases in place.

Temperature is one of the most useful variables in SFC, but it is also a parameter which strongly influences the properties of the stationary phase. Therefore, the temperature for an analysis must be chosen with the stationary phase in mind. The selectivity of the stationary phase is usually reduced with increasing temperature, while the efficiency, on the other hand, may decrease at lower temperature because of lower rates of diffusion. It is important to strive to attain stationary phases that possess good diffusion at low temperatures and strong interactions at higher temperatures, making the phases versatile over wide temperature ranges. The limit of diffusion in the stationary phase, or the glass transition point, defines the lower operating temperature limit, below which the phase cannot be utilized to its full resolving potential.

Stationary phases for packed columns. The packed columns used today in SFC are usually columns developed for LC. Up to an order of magnitude greater resolution per unit time is achieved by simply changing from a liquid to a supercritical mobile phase. Limited effort to date has been dedicated to developing packed column stationary phases specifically for SFC. The development of new inert packing materials and the fast growing acceptance of packed capillary columns are examples of important trends that most certainly will increase the demand for packed column stationary phases in the future.

There is a fine line between the effects of the stationary phase support and the stationary phase in packed columns, since they both usually contribute to the retention mechanism. Furthermore, the mobile phase and/or mobile phase modifiers interact with the stationary phase to form a modified surface. This final surface should be considered as the real stationary phase.

Adsorbents, such as silica and alumina, have been extensively used as stationary phases in the past. These phases are useful for nonpolar compounds, however, they lead to both reversible and irreversible adsorption of polar solutes in SFC, especially when neat CO_2 is used as the mobile phase. The limited success experienced to date in achieving a high level of deactivation of these materials suggest their rather limited future potential.

Modification of the typical small particle size silicas and aluminas with bonded stationary phases such as octyl, octadecyl, cyanoalkyl, aminoalkyl, and diolalkyl provide less adsorptive packing materials and a wide range of polarities for dipole-dipole and dipole-induced dipole interactions. In most cases, except for the most nonpolar molecules, polar organic modifiers are required for elution of analytes from these materials.

Most commercial phases are monomeric in nature because they produce a monolayer coverage of phase on the solid support. Excess silanol groups in this monolayer may be either endcapped, used to induce polymerization within this monolayer, or they may be left to take part in selective interactions as part of the stationary phase.

Polymeric phases are claimed to be more homogeneous than monomeric phases, but they have only been used to a limited extent in SFC. By the introduction of controlled amounts of water during chemical reaction of the stationary phase with the support as discussed earlier, thicker polymeric phases result which possess greater sample capacities and unique shape selectivities. Polymerization of the stationary phase may also be induced by reaction by free radical initiation[33,49] or with a dehydrocondensation reaction.[34]

The dehydrocondensation reaction produces surfaces more inert than any produced by the other polymerization methods. This method either endcaps or covers surface active groups on the silica or alumina with a highly crosslinked resin surface. While octyl, octadecyl, cyano, and phenyl groups have been incorporated in this type of phase to date, other functional groups could be incorporated in the same manner. SFC chromatograms illustrating the gain in inertness for a poly(octadecyl)hydrosiloxane treated silica compared to a commercial Nucleosil C_8 are shown in Figure 2.11.

These so-called polymer encapsulated phases currently commercially available include phases with hydrophobic functionalities (octyl, methyl, and phenyl) and a polar phase with a strong dipole (cyano). A phase which possesses a weak dipole and the ability to hydrogen bond is the recently introduced polyethylene glycol bonded silica (PEG). Figure 2.12 compares the elution of a five component polarity test mixture under identical chromatographic conditions on octyl, cyano, and PEG polymer encapsulated packed columns.[53] A better separation was obtained on the PEG column for the more acidic components (*e.g.*, palmitic acid and naphthol). Clearly, the understanding of packed column selectivity as it relates to SFC is in its infancy.

Yang[54] recently reported on the development of a new line of inert packing materials for use in SFC. These are based on macroporous divinylbenzene polymeric materials having tightly controlled particle size and special dispersions. The polymerization

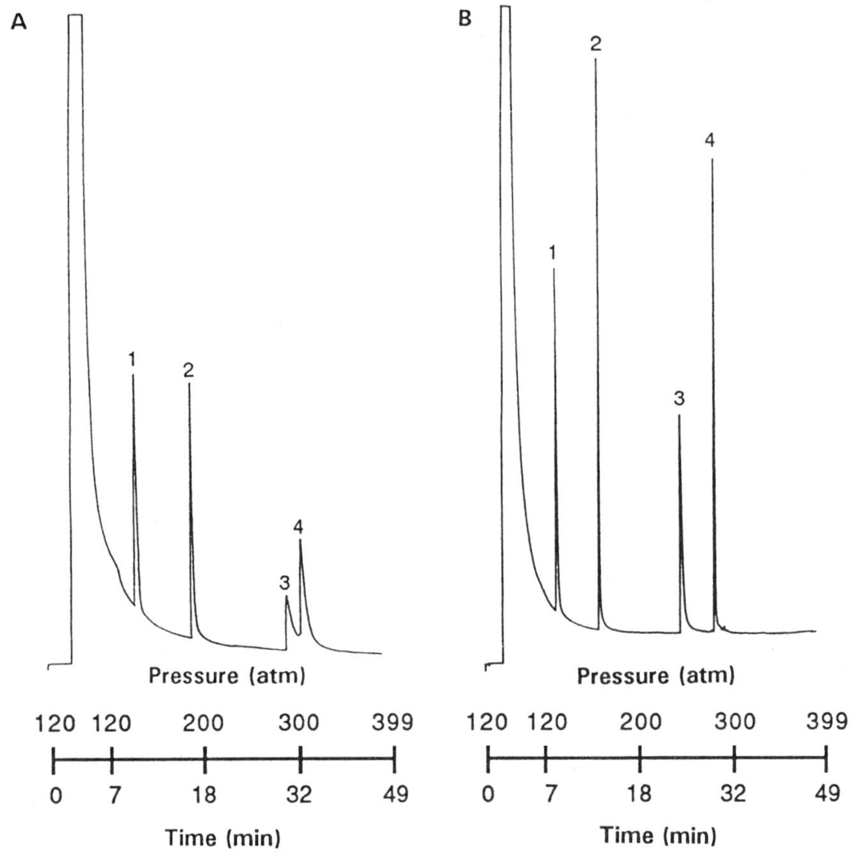

Figure 2.11. SFC chromatograms of a polarity test mixture separated on (A) C_8 bonded silica (Nucleosil) and (B) poly(25% octadecyl)methylsiloxane bonded silica (Vydac). Conditions: 30-cm x 250-μm i.d. packed capillary column; 5-μm particles; CO_2; 100°C; pressure program from 120 atm to 400 atm at 7 atm min^{-1} after an initial 7-min isobaric period. Peak identifications: (1) acetophenone, (2) naphthol, (3) benzo[*f*]quinoline, (4) cholesterol (reprinted with permission from Ref. 34).

process is regulated to create a rigid inert surface with porosities which can be adjusted between 100 and 2,000 Å. These packing materials can withstand pressures to 10,000 psi, they are compatible with most solvents, and they are stable at pH values from 0 to 14. An example of the use of one of these polymeric packed columns is given in Figure 2.13 in which free fatty acids were chromatographed using neat CO_2 mobile phase. Limited fundamental studies have also been conducted using polystyrene polymer packings.[55]

Considerable work has been reported with packed columns for the separation of enantiomers using chiral stationary phases. Mourier

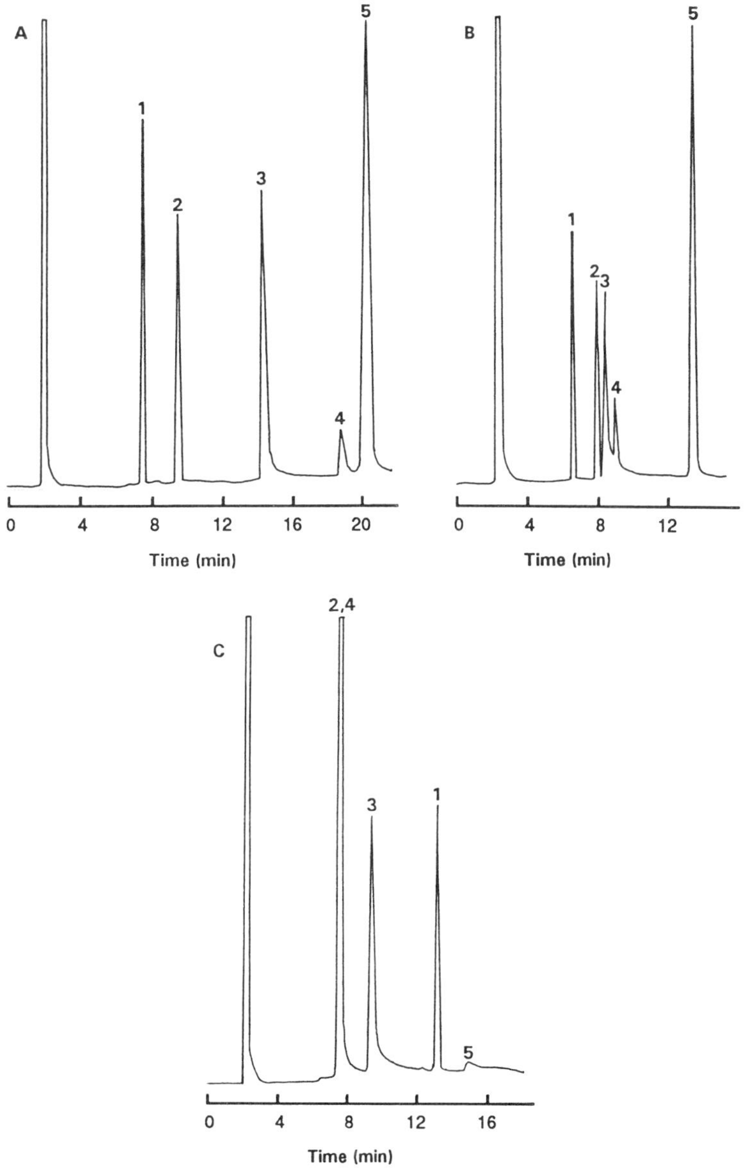

Figure 2.12. SFC chromatograms of a polarity test mixture on (A) polyethylene glycol, (B) cyanopropyl, and (C) *n*-octyl polymer bonded (Deltabond) columns. Conditions: 100-mm x 1-mm i.d. packed column, 5-μm particles; CO_2; 60°C; pressure program from 100 atm to 400 atm at 8.5 atm min^{-1}; FID. Peak identifications: (1) *n*-C_{28}, (2) diphenylamine, (3) palmitic acid, (4) 1-naphthol, (5) caffeine (reprinted with permission from Ref. 53).

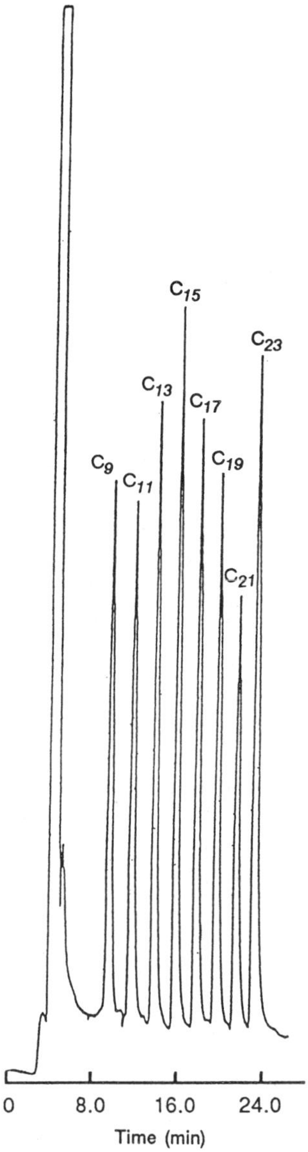

Figure 2.13. SFC chromatogram of C_9 - C_{23} free fatty acids. Conditions: 15-cm x 0.75-mm i.d. packed column, 6-μm divinylbenzene (μ-PRN-300); CO_2; 170°C; pressure program from 200 atm to 415 atm at 8 atm min^{-1} after a 3-min isobaric period (reprinted with permission from Ref. 54).

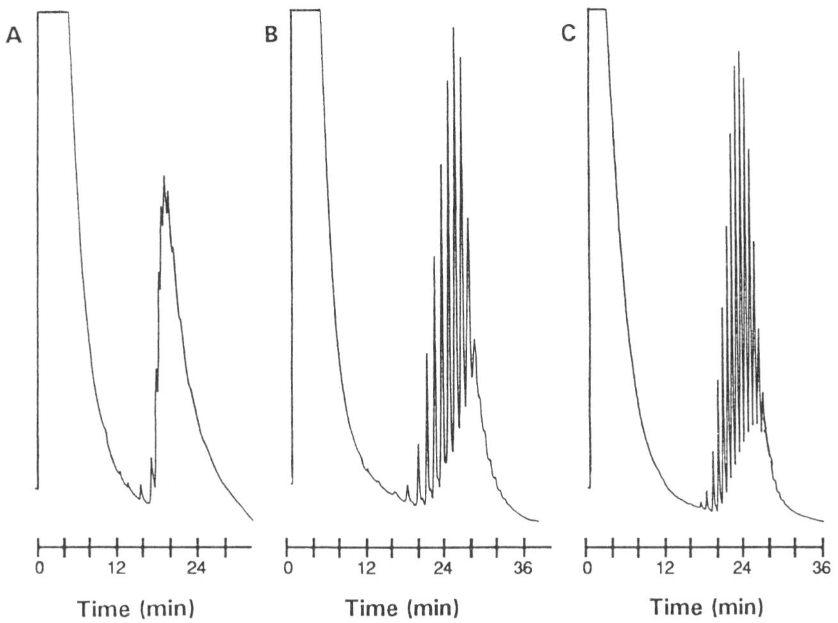

Figure 2.14. SFC chromatograms of Triton X-100 showing the effect of a liquid organic salt stationary phase. Conditions: (A) 15-cm x 1-mm i.d. packed column, C_{18} bonded silica particles (Alltech); (B) 15-cm x 1-mm i.d. packed column; C_{18} bonded silica particles (Alltech), 0.4% (w/w) Ethoquad 18/25 coating; and (C) 7-cm x 1-mm i.d. packed column, C_{18} bonded silica particles (Vydac), 3.6% (w/w) Ethoquad 18/25 coating, d_p = 10 μm, 300-Å pore size, CO_2; 80°C; pressure program from 100 atm to 400 atm at 10 atm min^{-1} after a 10-min isobaric period (reprinted with permission from Ref. 38).

et al.[56] and Macaudière *et al.*[52] have evaluated a large number of commercial stationary phases including bonded Pirkle-type phases and a cyclodextran phase. Dobashi *et al.*[57] synthesized a chiral diamide stationary phase that was bonded to silica particles through a long hydrocarbon spacer to make the diamide accessible for the solutes in the mobile phase.

Liquid organic salts have also been used as stationary phases in SFC.[38] Ethoquad 18/25 (stearylmethyldipolyoxyethylammonium chloride), which has an average molecular weight of 994 was found to meet the requirements of nonextractability by the mobile phase over the density range used for analysis. In Figure 2.14, the separations of Triton X-100 are shown on a pellicular octadecyl silanized packing with and without coating with 0.4% (w/w)

Ethoquad 18/25, and on a wide pore octadecyl silanized silica packing coated with 3.6% (w/w) Ethoquad. This sample elutes fairly easily from the low surface area pellicular packing in contrast with the difficulty associated with elution from the high surface area uncoated support.

Stationary phases for open tubular columns. Polysiloxanes are extensively used as polymeric backbones in stationary phases for open tubular columns. The chemical and physical stabilities of the polysiloxanes, along with the desirable flexibility of the Si-O bond which leads to good diffusion of sample analytes, make them ideal as stationary phases. Polysiloxanes have been substituted with a wide range of chemical groups for selective interactions with different types of samples.

Dispersion interactions are commonly used in open tubular column SFC. The great inertness and efficiency of columns coated with polymethylsiloxanes are utilized in SFC, but enhanced partitioning was demonstrated using *n*-octyl substituted polysiloxanes compared to methyl substituted phases.[58] This *n*-octyl phase also has a sufficient density of C-C bonds such that these columns could be used for a limited time with neat NH_3 as the mobile phase.[51]

The biphenyl phase with 30 mole% substitution is usually preferred over the 50% phenyl phase because the larger, more polarizable biphenyl group provides greater interaction with the analytes.[59] In addition, the biphenyl phase contains a higher percentage of methyl groups than the corresponding 50% phenyl phase and is therefore easier to immobilize on the column wall. Analytes containing either electron donating or electron withdrawing groups can induce polarity in the biphenyl stationary phase. The lack of polar interactions makes this phase ideal for the separation of closely related polar solutes without excessive retention.

The most widely used polar stationary phases in open tubular column SFC are the cyanopropyl polysiloxanes.[60] With CO_2 as mobile phase, these stationary phases have been particularly useful for the analysis of compounds containing carboxylic acid functional groups.

Rokushika *et al.*[61] found that a liquid crystalline stationary phase in a 150-μm i.d. open tubular column gave equivalent efficiency and improved peak shape in SFC compared to GC, but with an overall improved resolution. The phase was composed of liquid crystalline moieties substituted on a hydrocarbon backbone. The solubility of this phase in supercritical CO_2 was so low that immobilization was not considered necessary.

A highly ordered liquid crystalline polysiloxane stationary phase was reported by Chang *et al.*[62,63] for use in SFC. A dramatic enhancement in resolution over GC was demonstrated for selected geometrical isomers. The SFC elution was performed at 120°C where the stationary phase was more ordered than at the 230°C elution temperature in GC.

Chiral separations in SFC to date have been primarily explored using packed column technology developed for LC analysis. In a preliminary study by Röder *et al.*[49] using a chiral polymeric phase of the Pirkle type (amide) in an open tubular column, it was clearly demonstrated how strongly the enantioselective separation depends on the temperature. A thermally stable chiral amide phase developed for GC[64] was found to give higher resolution in SFC than in GC for some derivatized amino acids. The gain in selectivity at the lower elution temperature more than compensated for the loss in efficiency from the lower diffusion in the supercritical fluid. (See Section 7.16 for a more detailed discussion of chiral separations.)

2.10 Other Stationary Phase Considerations

Phase structure. It is well known that diffusion is about 1,000 times faster in GC than in LC, and that diffusion in SFC is intermediate between the two and varies over the range of the pressure/density program. It is also known, as a result of this fact, that the active substituents in the stationary phase are usually very short in GC (*e.g.*, methyl is the most common stationary phase) but must be considerably longer in LC (*e.g.*, octadecyl is the most common stationary phase). From the same reasoning, it can be expected that substituents of intermediate lengths are favorable in SFC, making them available for interaction with the analytes as they move down the column. This accessibility of the selective sites for solutes in the mobile phase can also be accomplished by incorporation of a long spacer between the particle surface or polymeric backbone and the interaction site.

Modification of the stationary phase by the mobile phase. A surface excess adsorption of carbon dioxide on an ODS stationary phase was reported by Strubinger and Parcher.[65] The adsorbed layer was 1-3 molecules thick and had a negative dependence of thickness on pressure and a positive dependence on temperature. It is clear

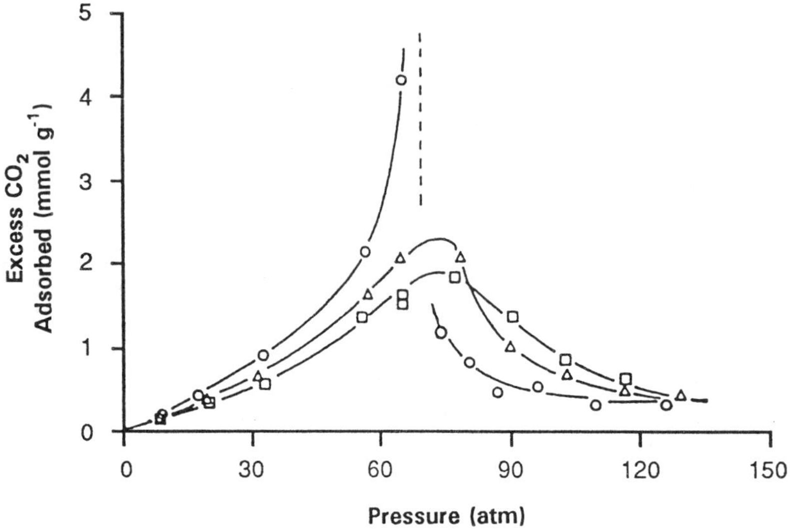

Figure 2.15. Gibbs adsorption isotherms of CO_2 on C_{18} bonded silica particles: (○) 30°C, (△) 40°C, and (□) 50°C. The dotted line represents the vapor pressure of pure CO_2 at 30°C (reprinted with permission from Ref. 65).

that the mobile phase and its additives play an important role in modifying the stationary phase.

Adsorption and uptake of the mobile phase by the stationary phase has been investigated using mass spectrometric tracer pulse chromatography (MSTPC). MSTPC provides a method by which physicochemical measurements can be made under the high temperature and pressure conditions found in SFC. For MSTPC, a supercritical fluid chromatograph is interfaced with a mass spectrometer to allow an isotopically labeled component of the mobile phase to be detected. An accurately determined dead time (usually with a homologous series or an inert gas) is used to calculate the adjusted retention time for the isotopically labeled tracer pulse. The association of the tracer pulse (*i.e.*, absorbed, partitioned, or dissolved mobile phase) with the stationary phase is a function of the adjusted retention time and the volumetric or mass flow rate of the system. The resulting value is a measure of the association of the mobile phase component with the support or stationary phase.

Most of the MSTPC studies show similar trends. In Figure 2.15, the absolute adsorption isotherm for carbon dioxide on an octadecyl silica (ODS) based packed column material is shown.[65] These isotherms were measured at temperatures and pressures both below and above the critical points. Below the critical point, an increase in the adsorption of CO_2 is observed with increasing pressure, but once supercritical conditions are attained, the reverse is true, *i.e.*, CO_2 adsorption decreases with continued increase in pressure. At supercritical pressures, the excess adsorbed CO_2 increases with temperature at constant pressure. This again is the reverse of what is observed for subcritical pressures. These data indicate that there is at least a monolayer of CO_2 mobile phase associated with the support under normal operating conditions, and that at conditions near the critical point, the coverage is in excess of a monolayer. It is still not clear in what way this behavior affects the retention process, if at all.

MSTPC has been used to study the type of interactions that take place between a supercritical fluid mobile phase, *n*-pentane, and the polymeric phases used in open tubular columns.[66] The "association" plots are very similar in shape to those for CO_2 on the ODS silica under supercritical conditions. The word "association" is used for this description because the interactions indicate more than simple adsorption phenomena. Even though the curves have similar shapes, they vary in overall magnitude, depending on the composition and quantity (*i.e.*, film thickness) of the stationary phase. *n*-Pentane associates or partitions to a greater extent with a nonpolar phase (*e.g.*, SE-30) than with a more polar phase (*e.g.*, SE-54). For both phases, an increase in the film thickness results in a greater association of *n*-pentane, providing further evidence for a partition mechanism.

In other MSTPC studies, the interactions between mobile phases containing solvent modifiers and crosslinked polymeric phases in open tubular columns have been investigated. The results observed for the addition of methanol to *n*-pentane indicate that the modifier has a more pronounced effect on the stationary phase than the mobile phase.[67] In Figure 2.16, the effect of methanol on the partition isotherms of *n*-pentane, adsorbed or dissolved, was reduced slightly. This reduction may be explained by some of the polar sites being occupied by methanol, thereby producing a binary and subsequently more polar stationary phase. With the addition of 1% methanol, the association of *n*-pentane is maximized, possibly due to a complete coverage of active sites by the methanol, thereby rendering the stationary phase as nonpolar as possible. When 5% methanol is added, the association of pentane is once again reduced, possibly due to direct competition with the methanol present in the mobile phase for adsorption sites in partitioning.

In another study with MSTPC, the interaction of 2-propanol with a crosslinked SE-54 stationary phase was investigated.[68] The results

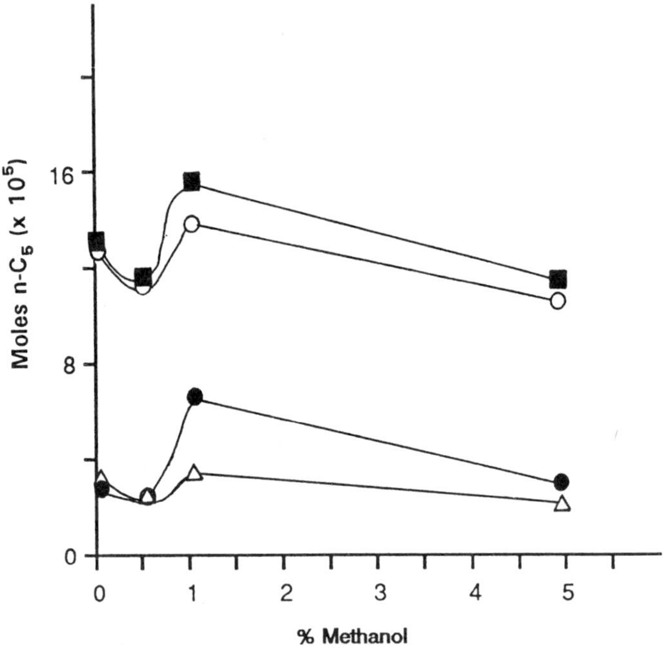

Figure 2.16. Effect of methanol concentration on the solubility/adsorption of supercritical *n*-pentane: (■) 515 psi, (○) 555 psi, (●) 750 psi, and (△) 800 psi. Conditions: 22-m x 0.2-mm i.d. open tubular column, poly(5% phenyl)methylsiloxane stationary phase (d_f = 0.11 μm), immobilized with azo-bis-*t*-butane (reprinted with permission from Ref. 67).

of this study revealed a definite partitioning of the modifier (2-propanol) into the stationary phase. This was revealed by calculation of the amount of 2-propanol necessary to provide a monolayer, and comparing this value to the actual amount of 2-propanol associated with the stationary phase. In all cases, the amount of 2-propanol associated with the stationary phase was reported to be at least two orders of magnitude greater than that required for a monolayer coverage. It is unlikely that simple adsorption would result in an adsorbed layer exceeding a monolayer under the temperature conditions (>110°C) used in this study.

Results similar to those above have also been reported using conventional methods for determining association isotherms.[69,70] Adsorption isotherms were determined for ethyl acetate, methanol, and hexanol on silica at various temperatures and densities by the peak maximum method. Trends similar to the above were observed at various temperatures and densities. Saturation coverage was found to occur when the concentration of modifier in the mobile phase reached 1%. Methanol was found to have the highest molar stationary phase coverage, but hexanol, based on its larger mean molecular area, was demonstrated to provide the best surface coverage and the greatest degree of deactivation.

Swelling of polymeric stationary phases. The results of the previous studies raise several interesting questions with regard to the swelling of polymer-based stationary phases. Polymeric stationary phase shrinking or swelling has been demonstrated by comparing the linear velocity of an unretained solute in an uncoated column to that in one that has been coated with polymeric phase.[71,72] The swelling factor is defined as the relative expansion of the stationary phase volume under supercritical conditions. The swelling factor was determined for several polymeric stationary phases in the presence of supercritical carbon dioxide and butane. Swelling factors for SE-30 and SE-54 phases were less for CO_2 than for butane, possibly because of the lower critical temperature and lower solubilities of CO_2 in these phases. The swelling factors for *n*-butane in methyl-, phenyl-, and cyanopropyl-containing polysiloxane phases decreased from the least polar to the most polar in accordance with the similarity in polarity between the mobile and stationary phases. The addition of 5% methanol caused a complete reversal of the observed relative swelling factors, again in agreement with polarity considerations. A more complicated relationship was found for a polar cyanopropyl stationary phase that responded parabolically to mobile phase uptake of a polar modifier in butane as a function of density. The initial swelling by methanol uptake in the stationary phase collapsed at densities over 0.25 g mL^{-1}, indicating that butane then had gained enough solvating power to force methanol out of the stationary phase. For an increase in temperature at constant density, an increase in the swelling factor was observed. This suggests that the stationary phase may undergo a physical change with changes in temperature. Swelling factor values are also observed to increase with density at constant temperature until a plateau is reached as the density nears 0.40 g mL^{-1}.

The observed increase in swelling factor with temperature tends to be somewhat contradictory to what is observed from the MSTPC association isotherms. It has been suggested that perhaps there are at least two separate processes that occur when polymeric phases are exposed to supercritical fluids.[68] The first is a kinetically slow process, characterized by a conditioning of the polymer over several hours. This slow process is indicated by the slow (several hours) decrease in background modifier concentration after switching to pure mobile phase. The second is a kinetically fast process, which would be amenable to MSTPC. The slow process could account for the swelling of the polymeric phase, while the fast temperature and pressure dependent process may limit the amount of modifier that can exchange through the partition interface.

It has also been suggested that the swelling phenomena may not be associated with the uptake of mobile phase, but rather the result of thermal expansion or relaxation.[67] This expansion or relaxation is accompanied by greater mobile phase permeability and faster mass transfer, resulting in a reduction of retention measured by MSTPC.

Swelling of polymeric supports has also been observed for certain polystyrene divinylbenzene packing materials.[38] These materials were observed to extrude from the end of a packed column some time after exposure to supercritical CO_2. They were also reported to undergo a significant loss in efficiency at higher densities, possibly due to changes in shape of the packing with increasing pressure. Other observations have indicated that the polymeric supports are fractured by exposure to supercritical fluid mobile phases.[73]

Chemical modification of silica after exposure to supercritical fluids. Chemical reactions between mobile phase components and active sites on the support surface, particularly for silica based column packings, have been demonstrated. Some of the earliest bonded phases in GC were formed by the reaction of alcohols and isocyanates with silanol groups on the surface of a silica based support.[74] As indicated earlier in this chapter, the addition of a polar modifier, often an alcohol, to a nonpolar supercritical fluid mobile phase significantly reduces the retention of polar analytes. The addition of ethanol to *n*-hexane mobile phase was shown to have a significant effect on the retention of polystyrene oligomers on a column packed with silica.[75,76] When 0.1% ethanol was added to the mobile phase, the retention of styrene oligomers was greatly reduced. The retention times of the oligomers continued to decrease until 1% ethanol in *n*-hexane had been reached. For concentrations of ethanol greater than 1%, the retention of the styrene oligomers began to increase. A reduction in retention at concentrations less than 1% can be attributed to deactivation of the silica support by adsorption of ethanol. Beyond a modifier concentration of 1%, no further deactivation of the silica surface can take place, but rather a decrease in the solubilizing strength of the binary mobile phase causes an increase in the retention of the polystyrene oligomers. When this

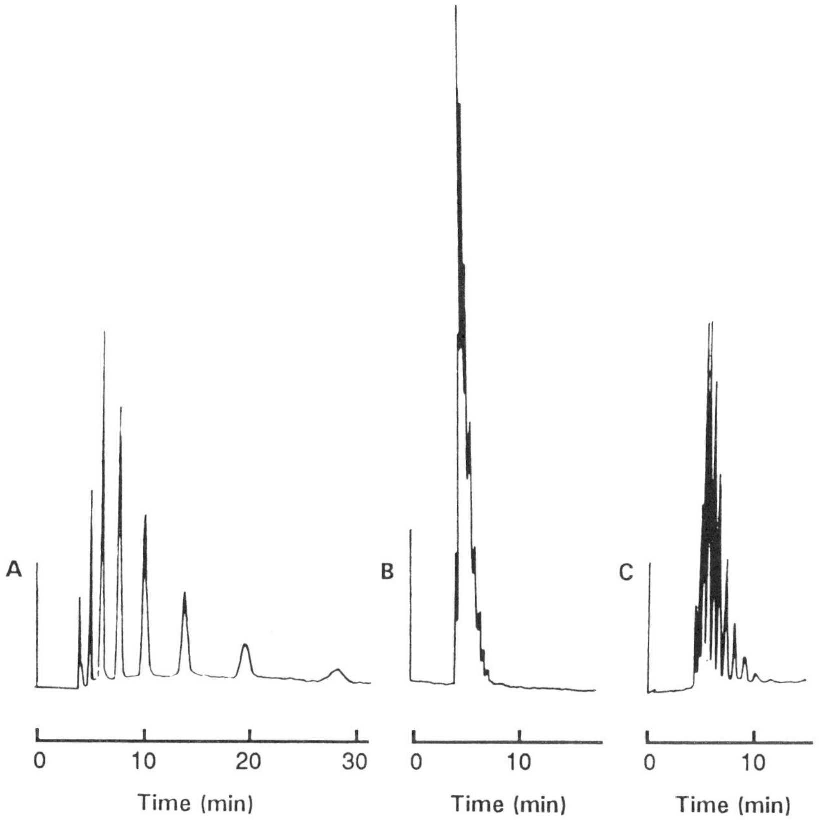

Figure 2.17. SFC chromatograms of polystyrene 580 on (A) a column only used with *n*-hexane mobile phase, (B) after conditioning with 20% ethanol in *n*-hexane, and (C) as for B after washing with 50% acetonitrile in water. Conditions: 40-60-cm x 0.2-mm i.d. packed column, methyl bonded silica (Develosil-10), 10-μm particles; *n*-hexane; 260°C; 50-atm inlet pressure (reprinted with permission from Ref. 75).

same study was applied to a bonded phase silica, there was no reduction in the retention of the styrene oligomers, but instead a steady increase in retention was observed as the mobile phase solubilizing strength was reduced with the addition of ethanol. These results indicate that the ethanol modifier has a very definite effect on the retention properties of the silica support. In Figure 2.17A, the separation of a styrene oligomer sample is shown on a silica column with supercritical *n*-hexane as the mobile phase. The same oligomer mixture is chromatographed after treatment of the column with 20% ethanol at 200°C (Figure 2.17B) and after exhaustive flushing of the column with 50:50 acetonitrile/water

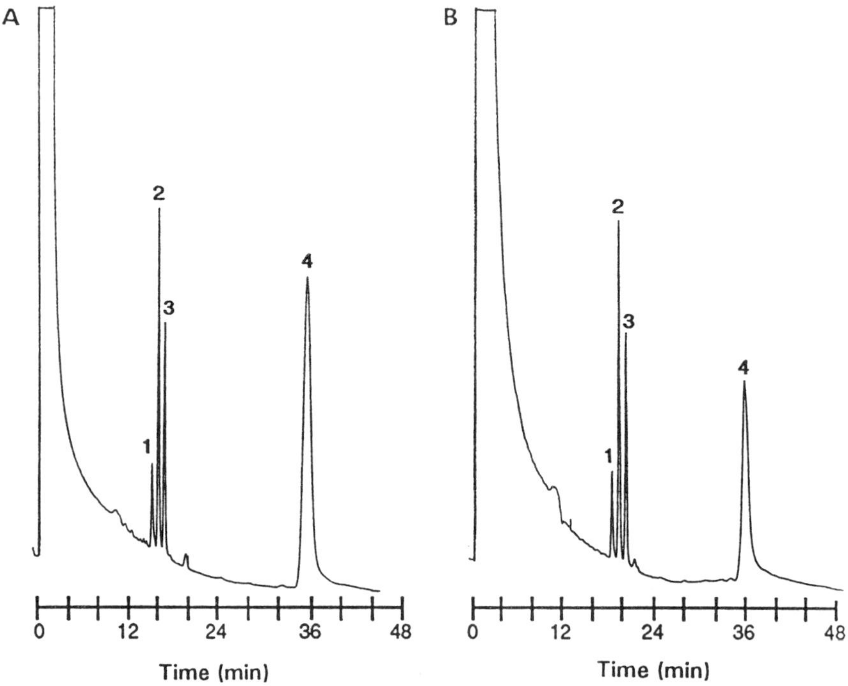

Figure 2.18. SFC chromatograms of cholestane hydrocarbons and cholesterol at (A) 80°C and (B) 100°C. Conditions: 10-cm x 1-mm i.d. packed column, silica particles (Corasil); CO_2; pressure program from 80 atm to 300 atm at 7.3 atm min^{-1} after a 10-min isobaric period. Peak identifications: (1) cholestane, (2) 5-cholestene, (3) 3,5-cholestadiene, (4) cholesterol (reprinted with permission from Ref. 38).

(Figure 2.17C). Since the original activity of the silica cannot be recovered, it is suggested that some of the ethanol may be left bonded to the silica through an ether linkage.

In a similar study, 1,4-dioxane was shown to irreversibly modify the surface of a silica support.[77] After high temperature (250°C) treatment of the column with 1,4-dioxane, the peak shapes of an aromatic test mixture were observed to improve. Peak efficiencies increased by as much as an order of magnitude, and retention was significantly reduced. To further characterize the reaction, the carbon content of the silica was determined before and after treatment with 1,4-dioxane. In all cases, the carbon content of the treated silicas was increased by 1-2% by weight. The proposed reaction mechanism is simple cleavage of the cyclic ether by the acidic silanol groups at high temperatures, and subsequent formation of the silyl ether. The final form of the modified silica surface may consist of single bonded chains with alcohol termination, bridged chains bonded at both ends, and polymerized chains.

In Figure 2.18, the reaction of an injected component is demonstrated on a silica support with CO_2 as the mobile phase.[38] Adsorption interactions are useful for the separation of the cholestane hydrocarbons which are difficult to resolve on a bonded phase, but the later eluting cholesterol peak is broadened and has a poor peak shape. An increase in analysis temperature from 80 to 100°C results in a substantial loss of peak area for cholesterol. Further increases in temperature result in complete loss of the cholesterol peak. This reaction has been observed on the most deactivated supports that are commercially available. The cholesterol loss is delayed to a higher temperature when supports of lower surface area or higher deactivation are used.

References

1. S.M. Fields, R.C. Kong, J.C. Fjeldsted, M.L. Lee, and P.A. Peaden, *J. High Resolut. Chromatogr. Chromatogr. Commun.* **7**, 312 (1984).
2. U. van Wasen, I. Swaid, and G.M. Schneider, *Angew. Chem. Int. Ed. Engl.* **19**, 575 (1980).
3. S.T. Sie and G.W.A. Rijnders, *Sep. Sci.* **2**, 699 (1967).
4. H.H. Lauer, D. McManigill, and R.D. Board, *Anal. Chem.* **55**, 1370 (1983).
5. J.M. Kong and S.J. Hawkes, *J. Chromatogr. Sci.* **14**, 279 (1976).
6. W. Millen and S.J. Hawkes, *J. Chromatogr. Sci.* **15**, 148 (1977).
7. W. Millen and S.J. Hawkes, *Macromolecules* **10**, 1250 (1977).
8. C.A. Cramers, C.E. van Tilburg, C.P.M. Schutjes, J.A. Rijks, G.A. Rutten, and R. DeNijs, *Proc. Fifth Int. Symp. on Cap. Chromatogr.*, P. Sandra, Ed. (Huethig, Heidelberg, 1983), p. 76.
9. I. Swaid and G.M. Schneider, *Ber. Bunsenges. Phys. Chem.* **83**, 969 (1979).
10. R. Feist and G.M. Schneider, *Sep. Sci. Tech.* **17**, 261 (1982).
11. S.M. Fields and M.L. Lee, *J. Chromatogr.* **349**, 305 (1985).
12. K.M. Payne, K.E. Markides, and M.L. Lee, Brigham Young University, unpublished results.
13. E. Katz, K.L. Ogan, and R.P.W. Scott, *J. Chromatogr.* **250**, 51 (1982).
14. H.E. Schwartz, P.J. Barthel, S.E. Moring, and H.H. Lauer, *LC/GC* **5**, 490 (1987).
15. C. Horváth and H.-J. Lin, *J. Chromatogr.* **149**, 43 (1978).
16. L.G. Randall, in *Ultrahigh Resolution Chromatography, ACS Symp. Ser. 250*, S. Ahuja, Ed. (American Chemical Society, Washington, DC, 1984), p. 135.
17. P.J. Schoenmakers and L.G.M. Uunk, *Chromatographia* **24**, 51 (1987).

18. M. Roth and A. Ansorgova, *J. Chromatogr.* **465**, 169 (1989).
19. P.A. Peaden and M.L. Lee, *J. Chromatogr.* **259**, 1 (1983).
20. J.A. Graham and L.B. Rogers, *J. Chromatogr. Sci.* **18**, 75 (1980).
21. D.E. Martire, *J. Chromatogr.* **461**, 165 (1989).
22. S.M. Fields, R.C. Kong, M.L. Lee, and P.A. Peaden, *J. High Resolut. Chromatogr. Chromatogr. Commun.* **7**, 423 (1984).
23. E. Klesper, A.H. Corwin, and D.A. Turner, *J. Org. Chem.* **27**, 700 (1962).
24. D.R. Gere, *Anal. Chem.* **54**, 736 (1982).
25. M. Novotny, W. Bertsch, and A. Zlatkis, *J. Chromatogr.* **61**, 17 (1971).
26. S.T. Sie and G.W.A. Rijnders, *Anal. Chim. Acta* **38**, 31 (1967).
27. J.C. Giddings and M.N. Meyers, *Sep. Sci.* **1**, 761 (1966).
28. L.C. Sanders and S.A. Wise, *Crit. Rev. Anal. Chem.* **18**, 299 (1987).
29. P. Roumeliotis and K.K. Unger, *J. Chromatogr.* **125**, 115 (1976).
30. L.C. Sander and S.A. Wise, *Anal. Chem.* **56**, 504 (1984).
31. R.E. Majors and M.J. Hopper, *J. Chromatogr. Sci.* **12**, 767 (1986).
32. M. Verzele and P.J. Mussche, *J. Chromatogr.* **254**, 117 (1983).
33. H. Figge, J. Kohler, and G. Schomburg, *J. Chromatogr.* **351**, 393 (1986).
34. K.M. Payne, B.J. Tarbet, J.S. Bradshaw, K.E. Markides, and M.L. Lee, *Anal. Chem.* **62**, 1379 (1990).
35. F.J. Yang, in *Microbore Column Chromatography: Chromatographic Science Series* **45**, F.J. Yang, Ed. (Marcel Dekker, New York, 1988), p. 13.
36. L.R. Snyder and J.J. Kirkland, *Modern Liquid Chromatography* (John Wiley, New York, 1974), p. 187.
37. G. Crescentini, F. Bruner, F. Mangani, and A.R. Mastrogiacomo, *Proc. Ninth Int. Symp. Cap. Chromatogr.*, P. Sandra, Ed.(Huethig, Heidelberg, 1988), p. 385.
38. T.A. Dean and C.F. Poole, *J. Chromatogr.* **468**, 127 (1989).
39. R.C. Kong, S.M. Fields, W.P. Jackson, and M.L. Lee, *J. Chromatogr.* **289**, 105 (1984).
40. R.C. Kong, C.L. Wooley, S.M. Fields, and M.L. Lee, *Chromatographia* **18**, 362 (1984).
41. W. Blum, K. Grolimund, P.E. Jordi, and P. Ramstein, *J. High Resolut. Chromatogr. Chromatogr. Commun.* **11**, 441 (1988).
42. C.L. Woolley, B.J. Tarbet, K.E. Markides, J.S. Bradshaw, and M.L. Lee, *J. High Resolut. Chromatogr. Chromatogr. Commun.* **11**, 113 (1988).
43. K.D. Bartle, C.L. Woolley, K.E. Markides, M.L. Lee, and R.S. Hansen, *J. High Resolut. Chromatogr. Chromatogr. Commun.* **10**, 128 (1987).
44. B.W. Wright, P.A. Peaden, and M.L. Lee, *J. High Resolut. Chromatogr. Chromatogr. Commun.* **5**, 413 (1982).
45. R.C. Kong and M.L. Lee, *J. High Resolut. Chromatogr. Chromatogr. Commun.* **6**, 319 (1983).
46. S.R. Sumpter, C.L. Woolley, K.M. Markides, and M.L. Lee, *J. Chromatogr.*, in press.
47. R.C. Kong and M.L. Lee, *Chromatographia* **17**, 451 (1983).
48. K. Markides, L. Blomberg, J. Buijten, and T. Wännman, *J. Chromatogr.* **267**, 29 (1983).

49. W. Röder, F.-J. Ruffing, G. Schomburg, and W.H. Pirkle, *J. High Resolut. Chromatogr. Chromatogr. Commun.* **10**, 665 (1987).
50. M.L. Lee and K.E. Markides, *J. High Resolut. Chromatogr. Chromatogr. Commun.* **9**, 652 (1986).
51. J.C. Kuei, K.E. Markides, and M.L. Lee, *J. High Resolut. Chromatogr. Chromatogr. Commun.* **10**, 257 (1987).
52. P. Macaudière, M. Caude, R. Rosset, and A. Tambuté, *J. Chromatogr. Sci.* **27**, 383 (1989).
53. R.A. Henry, Keystone Scientific, unpublished results.
54. F.J. Yang, *1989 Symposium/Workshop on Supercritical Fluid Chromatography*, Snowbird, UT, June 13-15, 1989.
55. B. Gemmel, B. Lorenschat, and F.P. Schmitz, *Chromatographia* **27**, 605 (1989).
56. P.A. Mourier, E. Eliot, M.H. Caude, R.H. Rosset, and A.G. Tambute, *Anal. Chem.* **57**, 2819 (1985).
57. A. Dobashi, Y. Dobashi, T. Ono, S. Hara, M. Saito, S. Higashidate, and Y. Yamauchi, *J. Chromatogr.* **461**, 121 (1989).
58. J.C. Kuei, B.J. Tarbet, W.P. Jackson, J.S. Bradshaw, K.E. Markides, and M.L. Lee, *Chromatographia* **20**, 25 (1985).
59. J.C. Kuei, J.I. Shelton, L.W. Castle, R.C. Kong, B.E. Richter, J.S. Bradshaw, and M.L. Lee, *J. High Resolut. Chromatogr. Chromatogr. Commun.* **7**, 13 (1984).
60. K.E. Markides, S.M. Fields, and M.L. Lee, *J. Chromatogr. Sci.* **24**, 254 (1986).
61. S. Rokushika, K.P. Naikwadi, A.L. Jadhav, and H. Hatano, *J. High Resolut. Chromatogr. Chromatogr. Commun.* **8**, 480 (1985).
62. H-C.K. Chang, K.E. Markides, J.S. Bradshaw, and M.L. Lee, *J. Chromatogr. Sci.* **26**, 280 (1988).
63. H-C.K. Chang, K.E. Markides, J.S. Bradshaw, and M.L. Lee, *J. Microcol. Sep.* **1**, 131 (1989).
64. J.S. Bradshaw, S.K. Aggarwal, C.A. Rouse, B.J. Tarbet, K.E. Markides, and M.L. Lee, *J. Chromatogr.* **405**, 169 (1989).
65. J.R. Strubinger and J.F. Parcher, *Anal. Chem.* **61**, 951 (1989).
66. M.I. Selim and J.R. Strubinger, *Fres. Z. Anal. Chem.* **330**, 246 (1988).
67. J.R. Strubinger and M.I. Selim, *J. Chromatogr. Sci.* **26**, 579 (1988).
68. C.R. Yonker and R.D. Smith, *Anal. Chem.* **61**, 1348 (1989).
69. C.H. Lochmüller and L.P. Mink, *J. Chromatogr.* **409**, 55 (1987).
70. C.H. Lochmüller and L.P. Mink, *J. Chromatogr.* **471**, 357 (1989).
71. S.R. Springston, P. David, J. Steger, and M.V. Novotny, *Anal. Chem.* **58**, 997 (1986).
72. M.V. Novotny and P. David, *J. High Resolut. Chromatogr. Chromatogr. Commun.* **9**, 647 (1986).
73. J.W. King, A.L. Eissler, and J.P. Friedrich, in *Supercritical Fluid Extraction and Chromatography: Techniques and Applications, ACS Symp. Ser. 366*, B.A. Charpentier and M. Severants, Eds. (American Chemical Society, Washington, DC, 1988), p. 63.

74. C.F. Poole and S.A. Schuette, *Contemporary Practice of Chromatography* (Elsevier, Amsterdam, 1984), p. 65.
75. Y. Hirata, *J. Chromatogr.* **315**, 31 (1984).
76. Y. Hirata, *J. Chromatogr.* **315**, 39 (1984).
77. F.P. Schmitz, D. Leyendecker, D. Leyendecker, *J. Chromatogr.* **389**, 245 (1987).

CHAPTER 3

MOBILE PHASES

3.1 Introduction

The nature of the mobile phase sets SFC apart from the other types of chromatography. As was explained in Chapter 1, the mobile phase must exhibit solvating properties that can be varied by changing its density. Unfortunately, the number of fluids that have critical parameters within the realm of practicality, and which also have solvating power, is somewhat limited. On the other hand, it is fortunate that temperature and mobile phase composition can be varied, in addition to pressure, to change the solvating power of the mobile phase and, hence, the chromatographic selectivity. Combinations of any two (or potentially all three) of the programming methods offer greater possibilities for adjusting selectivity and optimizing elution than any other form of chromatography.

Retention in SFC is controlled by the relative distribution of solutes between the mobile and stationary phases. This relative distribution between these two phases is controlled by the temperature and the density of the mobile phase. Much of the early work, as well as recent studies, in SFC involved the relationship between pressure, temperature, and retention.[1-6] Changes in retention with pressure are not linear, especially when working close to the critical point. This is because it is the density of the mobile phase and not its pressure that controls solubility. The relationship of pressure to density is not linear, but is better described by cubic and other more complex equations. The relationship of density to retention, expressed as the logarithm of the capacity factor (k), is much simpler and can usually be described as linear for short regions; but over a wider range, a quadratic function gives a better fit.[1,5,7]

Phase diagrams have been used to explain linear and quadratic solubility behaviors at different temperatures as pressures are changed.[1] Another approach uses the solubility parameter (δ) to explain and make qualitative predictions of the relative abilities of different fluids to dissolve solutes at different densities (see Section 3.6). This parameter works well for van der Waals type interactions, but it becomes less reliable when dipole and hydrogen bonding interactions are present. Solutes having solubility parameters which are much higher or lower than that of the supercritical fluid will not dissolve well. Good mobile phases have solubility parameters in the same range as the solutes of interest.

Poor solubility does not mean that a substance cannot be successfully chromatographed; however, it will limit the sample capacity of a system.

The pressures required to achieve significant solubilities can be quite high, especially if the solubility parameter of the solute is considerably larger than that of the mobile phase. In order to overcome this difficulty, mixed mobile phases can be used, but care must be taken to assure miscibility under all pressure and temperature combinations. Density programming with single component mobile phase systems is often described in terms of reduced parameters.[8] This is also the case for mixed mobile phases when density calculations are based on pseudocritical constants.[7,9,10]

Temperature effects are also easier to predict when density is used as the primary variable. This is because the logarithm of the capacity factor changes in a linear manner with respect to 1/T at constant density (cf., Equation 3.7).[2,4] Besides solvation by the mobile phase, solute volatility can also play a significant role in retention.[11-13] The effects of these factors are not limited to the supercritical fluid region, but represent a continuum between the liquid and gas states of the mobile phases. "Unified" theories have been proposed to predict retention behavior across the gas, supercritical, and liquid regions.[14,15]

In this chapter, the fundamentals of mobile phase selection and utilization are described, including the potential use of more exotic fluids such as reverse micellar mobile phases. Furthermore, the principles behind density, temperature, and eluent composition programming of the mobile phase are presented. The more practical details of these programming methods are given in Chapter 6.

3.2 Single Component Mobile Phases

Critical parameters. The critical temperature is the major physical property that determines whether or not a substance is suitable as a supercritical fluid mobile phase for a given application. As shown in Table 3.1, a wide range of critical temperatures is encompassed by fluids of practical interest. Only a few neat fluids can be used in the analysis of thermally labile samples (*i.e.*, T_c <100°C). When temperature stability is not a limiting factor, the analyst has a much larger selection of fluids from which to choose. The use of a particular fluid is then determined by other factors, such as instrumental considerations or reactivity with solutes. Of course, toxicity, flammability, or instability would preclude the use of certain fluids.

The critical pressures of all but two of the fluids in Table 3.1 fall within the range of 30-80 atm. Water and ammonia are these exceptions, but even their critical pressures are well below the pressure ratings of most supercritical equipment. However, the chemical incompatibilities of water and ammonia with system components and analytes pose other significant problems.

Excellent sources of critical and other relevant data are *The Properties of Gases and Liquids*,[16], the *Matheson Gas Data Book*,[17] and the *Gas Encyclopedia*.[18] These references contain critical data and graphs of vapor-liquid equilibria for a number of compounds of interest.

Density ranges. The solvent power of a fluid, to a first approximation, is proportional to its density. It is important to understand the range of densities available in supercritical fluid applications to fully utilize the technique, as well as to establish whether limits have been reached in a system. The selective solvating properties of supercritical fluids can be modified by changing their densities, an advantageous property unique to supercritical fluids.

A comparison of densities for selected fluids, at their critical points, at 400 atm and a reduced temperature of 1.03, and as liquids is given in Table 1.2. In most cases, the density of the supercritical fluid at high pressure approaches that of the liquid. For CO_2, the most commonly used fluid, the density can be varied by nearly an order of magnitude, by changing the pressure and temperature (Table 3.2). Higher densities than those listed for 400 atm in Tables 1.3 and 3.2 can be generated by using equipment with pressure ratings

Table 3.1. Critical Parameters of Selected Fluids[a]

Fluid	T_c (°C)	P_c (atm)	ρ_c (g mL^{-1})	V_c (mL mol^{-1})
CO_2	31.1	72.8	0.468	94
N_2O	36.4	71.5	0.452	97
SF_6	45.5	37.0	0.738	198
SO_2	158	78	0.525	122
CS_2	279	78	0.448	170
Xe	16.6	57.6	1.113	118
H_2O	374.1	217.6	0.322	56
Methanol	239.4	79.9	0.272	118
Ethanol	243.0	63.0	0.276	167
Propanol	263.5	51.0	0.275	218
2-Propanol	235.1	47.0	0.273	220
Dimethyl ether	127	52.6	0.259	178
Methyl ethyl ether	164.7	43.4	0.272	221
Diethyl ether	193.6	36.3	0.267	280
Acetonitrile	274.7	47.7	0.237	173
NH_3	132.3	111.3	0.235	72
Methylamine	156.9	40.2	0.222	148
Dimethylamine	164.6	52.4	0.241	187
Trimethylamine	160.1	40.2	0.233	254
Triethylamine	258.9	30	0.259	390
Ethane	32.4	48.3	0.203	140
Ethylene	10.0	51.2	0.227	124
Propane	96.8	42.0	0.220	203
n-Butane	152.0	37.5	0.228	255
n-Pentane	196.6	41.7	0.554	304
Benzene	288.9	48.3	0.302	259
Trifluoromethane	25.9	47.7	0.516	136
Tetrafluoromethane	-45.7	41.4	0.628	140
Chlorotrifluoromethane	28.8	38.7	0.58	180
Chlorodifluoromethane	96.0	49.1	0.524	165
Dichlorodifluoromethane	111.7	39.4	0.557	217

[a]Data taken from Ref. 16.

up to 650 atm or more, although the rate of density change with pressure decreases at these higher pressures.

When comparisons of fluids are made, careful consideration must be given to the selection of conditions. Identical temperature and

Table 3.2. Density Ranges of CO_2 at Various Temperatures[a]

Temp (°C)	Density (g mL^{-1})	
	72 atm	400 atm
40	0.22	0.96
62	0.17	0.89
100	0.13	0.76
150	0.11	0.62

[a]Calculated using compressibility and acentric factor data.[19]

density conditions for two different fluids are not comparable. For example, the number of CO_2 molecules per unit volume is 3.3 times that of SF_6 at the same temperature and density. A mobile phase comparison based on free volume (*i.e.*, unoccupied intermolecular volume) removes the molecular weight contribution.[20] The inverse of the free volume, the molecular density (mol mL^{-1}), is analogous to mass density, but it also removes the molecular weight dependency.

Reduced parameters have often been used to normalize conditions. (A reduced value is the operating condition divided by the critical condition). According to the theory of corresponding states,[8] different substances will exhibit equal compressibility at the same reduced pressure and temperature. Reduced parameters yield a relative measure of actual pressure, temperature, and/or density to the respective critical condition.

Polarity or "solvent power" ranges. In the present practice of SFC, there is a gap in mobile phase polarity. In the low polarity region are CO_2 and the alkanes. The alternative very polar fluids, such as ammonia, are seldom used because of many instrumental problems. The gap between these extremes is being filled to some degree with mixed fluid systems (see Section 3.3 for further discussion), including ternary and quaternary systems. It appears unlikely that any single neat fluid will fill this gap.

When discussing polarity or solvent power scales, the natural frame of reference has been the liquid state, since gases generally exhibit little solvating ability. Supercritical fluid systems form a continuum between classical gas and liquid conditions, with solvating power becoming more liquid-like as density increases. Solubility on a molecular scale involves molecular clustering and local density gradients,[21,22] analogous to hydration spheres of ions in aqueous solution. Specific interactions can contribute significantly to the loading of solute in the fluid phase.

Simplified concepts of polarity generally hold for supercritical fluids; higher polarity solutes are usually those containing a greater number of functional groups, such as hydroxyl, amino, or acidic. Such solutes generally are more soluble, or elute at lower density, in more polar fluids. Similarly, a higher density is often needed to transport higher polarity solutes.

Giddings proposed a scale of polarity based on the solubility parameter, or cohesive energy density.[23] The solubility parameter was obtained from the Van der Waals equation of state, which is applicable to gases. If the derivation had been started with the Berthelot equation, then the solubility parameter would have incorporated a temperature term. The solubility parameter provides a reasonable comparison of supercritical fluids to liquids, to a first approximation.

Research into fundamental fluid properties has yielded valuable and interesting data. Solvatochromic measurements[24-27] indicated two distinct regions of solute/fluid interaction for a wide variety of fluids (see Figure 1.8). A physical description of this phenomenon postulates regions of gas-phase-like and liquid-phase-like interactions.[24] The point at which the transition in interaction character takes place correlates with a rapid decrease in solute partial molar volume (immediately before a minimum), and to a "threshold" density at which solubility increases rapidly with increasing density.[24] Under higher density conditions, the cybotactic region around the solute contains more ordered solvent molecules.

A large number of chromatographic comparisons of the solvating powers of various fluids have been published, a few of which are mentioned here. The criterion for a "better" fluid has usually been faster elution in chromatographic systems. N_2O and CO_2 were compared for alkaloids,[28] low polarity and nonpolar compounds,[29] petroleum fractions, and polycyclic aromatic compounds (PAC).[30] N_2O has generally been shown to be similar to CO_2, but may have a significant advantage for the analysis of basic analytes. The solubility parameters of CO_2, N_2O, $CHClF_3$, and NH_3 were correlated with retention data;[26] the resulting increasing solvent strengths were in the order listed. Carbon dioxide and pentane were compared at equal reduced temperatures for PAC analytes[31]; pentane was found to be superior.

Polystyrene standards of molecular masses of 110,000 and 2,950,000 were separated with supercritical dichloromethane.[32] This is the highest molecular mass substance reported that has been transported with a supercritical fluid. The highly polar supercritical ammonia has been shown to transport a number of polar compounds.[23,33,34]

Solute chemical compatibilities. The question of whether a solute will react in a particular supercritical fluid system is important. If it is assumed that all surfaces within the system are nonreactive and noncatalytic, then several possible reactions include (a) reaction with the fluid, (b) reaction with impurities in the fluid, and (c) reaction with solvents, or impurities in the solvents, used for sample introduction.

The possible reaction of solute and fluid can often be clearly inferred from simple chemistry. Amines with at least one available hydrogen (primary or secondary) can react with CO_2. However, symmetrical chromatographic peaks have been obtained for certain aliphatic amines with CO_2 as mobile phase.[35] The amount of water present in the CO_2 may be the most important factor. Tertiary aliphatic amines, although they are stronger bases in aqueous solution than NH_3, have been shown to elute with no problems in CO_2.[36,37] Similarly, the reaction of acidic solutes with basic fluids can also be postulated. Reaction of supercritical ammonia with caffeine and theophylline has been observed.[38] Some unspecified amino acids were also noted to react.

The nature and quantity of impurities in the fluid can lead to irreproducibility. Care should always be taken to remove impurities in a consistent and thorough manner.

Instrumental considerations. A fluid delivery system must withstand continued exposure to different fluids under varying temperature and pressure conditions. Typically, the pumping system is at room temperature or slightly cooled, but the separation system is at supercritical temperatures. High quality stainless steel, such as Nitronic 60, is compatible with almost any fluid in common use. Sealing materials in pumps, valves, and column connections can create major problems if not chosen correctly. Cooling may harden pump seals and lead to fluid leakage. Manufacturers should be consulted for available data, and material exposure experiments should be performed if no data are available. While Teflon may swell when in contact with liquid or supercritical CO_2, it has proven to be adequate in most applications, although it can deform with sustained use under high pressure. The upper temperature limit for Teflon in seals and valves is usually <175°C (at near atmospheric pressure). Polyimide-based materials will dissolve in basic fluids, as will nearly all polysiloxane stationary phases.[33] Compatibility data for a number of fluids with various polymeric materials can be found elsewhere.[18]

Detector compatibility. Nearly all GC and LC detectors have been adapted for use with supercritical fluids (see Chapter 4). Detection may be either universal or selective. Table 3.3 lists several important detectors with their compatibilities with various fluids. The choice of detector in a particular application may be guided more by selectivity needs than by compatibility.

The combination of CO_2 with the FID provides a universal detection system for nearly all organic compounds. The purity of CO_2 can greatly affect the FID performance. Many chromatograms using CO_2 with an FID have shown rising baselines during pressure or density programming. This increase in signal arises from the increasing mass flow rate through the column with pressure, which delivers increasing levels of impurities to the FID. These impurities can be extracted from valves or other system components as well as being present in the original CO_2. In packed column systems, flow rates of CO_2 may be too high to sustain combustion, so splitting at the detector may be needed.

Table 3.3. Detector/fluid Compatibilities[a]

	Detector					
Fluid	FID	NPD	SCD	UV/Fluor	IR	MS
CO_2	+	+	+	+	0	+
N_2O	+[b]	+[b]	+	+	0	+
NH_3	–	–	–	+	0	+
SF_6	+[c]	–	–	+	0	+
Xe	+	+	+	+	+	+
Hydrocarbons	–	–	+	+	0	+

[a](+) denotes no signal from the fluid. (0) denotes small interferences from the fluid. (-) denotes complete incompatibility.
[b]Demonstrated but may have some interference problems.
[c]Gold and nickel plated collectors are necessary to prevent corrosion by HF.

Baseline compensation or dual detection has been required for N_2O introduction into the FID during pressure or density programming; the baseline changes are usually much more drastic than with CO_2. Recent results suggest that the impurities in the fluid may be the cause of the flow-dependent background signal.[39] Although the GC literature indicates a response for NH_3 in an FID, it has proven impractical to use in supercritical fluid applications. The high levels of NH_3, when it is used as a mobile phase, give rise to sufficient ions to give an excessive background.[33] SF_6 is compatible with the FID,[40,41] provided some precautions are taken. The collector assembly must be coated to protect it from the corrosive action of HF formed in the flame. Nickel- and gold-plated collectors have been successfully used. The exhaust gas must also be vented through a scrubbing system.

Xenon has slowly received increasing attention as a supercritical fluid mobile phase, especially for use in IR detection.[42] Cost has been an obvious obstacle. Recent research suggests that xenon may be nearly as useful a fluid as CO_2 in SFC.[42]

Any hydrocarbon fluid, including halocarbons, is primarily limited to spectroscopic detectors. Furthermore, the addition of a significant amount of an organic modifier to any FID-compatible fluid will preclude the use of the FID.

UV detectors can also produce drifting signals during pressure or density programming. This drift does not arise from an increase in mass flow, but rather from refractive index changes in the cell. The more compressible a fluid is, the greater will be the changes in density and refractive index.[43,44] Cooling of cells, and baseline compensation, have been solutions to this problem. A study of refractive index changes in a cylindrical cell (capillary tubing) has produced a system that virtually eliminates such signal changes.[43]

3.3 Binary Fluid Mobile Phases

The main limitation of the most often used mobile phase, CO_2, is its limited ability to dissolve polar molecules, even at very high densities. Other possible neat fluids for such purposes often are reactive, flammable, or toxic.

Alternatively, the characteristics of the supercritical fluid mobile phase can be varied by the addition of miscible compounds to the primary fluid. The term "modifier" is generally used for the second component of the mobile phase. Substances that are liquids at ambient conditions have been extensively employed as modifiers. A second supercritical fluid or a solid compound can also be used as a mobile phase modifier. Adding further components to supercritical fluid solvents can additionally alter their solvating properties.

Capacity factors and separation ratios are not only functions of modifier properties but also of modifier concentration. Therefore, the use of binary phase systems offers great flexibility since the modifier identity and concentration can be easily and widely altered, thereby allowing the adjustment of the mobile phase for optimum chromatographic conditions. In Table 3.4 are listed the most widely used modifiers and their physico-chemical properties. One of the main drawbacks in utilizing modifiers is their significant response in the FID, precluding the use of this rugged and universal detector that has good sensitivity for most organic compounds.

Table 3.4. Frequently Used Modifiers in SFC

Modifier	T_c (°C)	P_c (atm)	Molecular mass	Dielectric constant[a] at 20°C	Polarity index[a]
Methanol	239.4	79.9	32.04	32.70	5.1
Ethanol	243.0	63.0	46.07	24.3	4.3
1-Propanol	263.5	51.0	60.10	20.33	4.0
2-Propanol	235.1	47.0	60.10	18.3	3.9
1-Hexanol	336.8	40.0	102.18	13.3	3.5
2-Methoxy ethanol	302	52.2	76.10	16.93	5.5
Tetrahydrofuran	267.0	51.2	72.11	7.58	4.0
1,4-Dioxane	314	51.4	88.11	2.25	4.8
Acetonitrile	275	47.7	41.05	37.5	5.8
Dichloromethane	237	60.0	84.93	8.93[b]	3.1
Chloroform	263.2	54.2	119.38	4.81	4.1
Propylene carbonate	352.0		102.09	69.0	6.1
N,N-Dimethylacetamide	384		87.12	37.78[b]	6.5
Dimethyl sulfoxide	465.0		78.13	46.68	7.2
Formic acid	307		46.02	58.5[c]	
Water	374.1	217.6	18.01	80.1	10.2
Carbon disulfide	279	78.0	76.13	2.64[c]	

[a]Data from Burdick and Jackson Solvent Guide, 1984.
[b]25°C.
[c]Data from *Eastman Organic Chemical Bull.* **47**, 1975, pp. 1-12.

Thus far, only formic acid (at low levels),[45] water,[46] and SF_6[47] have been shown to be both useful as modifiers in CO_2 and compatible with the flame ionization detector. However, the detector must be gold-plated when using SF_6 because of the formation of HF in the flame. Preliminary results indicated that a properly designed nitrogen thermionic detector gives negligible loss of sensitivity with mixed mobile phases containing up to 10 mol% of methanol or 2-propanol.[48] Mixed mobile phases are often employed in conjunction with UV or fluorescence detectors. The effect of fluid composition on UV detection baseline drift during pressure gradients was investigated.[44] The rate of baseline drop as pressure increased was inversely proportional to the mobile phase critical temperature.

Mass spectrometers can also be employed as universal detectors when mixed mobile phases are used.[29] It has been shown that, depending on the conditions in the ionization chamber of the mass spectrometer, the modifier can also act as reactant gas for chemical ionization of the separated compounds.[49] The light-scattering detector (LSD), a detector with limited structure-related response, has been successfully employed in conjunction with mixed mobile phase SFC.[50,51] Preliminary results using an inductively coupled plasma (ICP) detector[52] show some promise for use of this detector for element selective response, even when modifiers are added.

Binary fluid properties. The nature of mixed fluid solvent systems and their role in various supercritical fluid applications is complicated by the increased complexity of their phase behavior. Several different types of phase behavior have been identified,[53] some of them involving multiphase equilibria.

Many investigations have been carried out using mixtures of CO_2 and methanol. These are of interest because they combine the low critical temperature and pressure of CO_2 with the affect of a polar, less volatile modifier. This system exhibits relatively simple phase behavior, and many research groups have studied the critical points of these mixtures. While the critical temperatures of different mixture compositions lie between those of the two neat components, the critical pressure shows a maximum of approximately 160 atm at intermediate composition, a value higher than that of either pure CO_2 or pure methanol. The addition of methanol to CO_2 increases the supercritical mobile phase polarity, but together with other modifiers, its effects vary depending on the column used (as discussed later). Solvatochromic studies[54] have also suggested that at high methanol concentrations (over approximately 10 mol%), the region surrounding a solute molecule may have a higher modifier content than the bulk solution. Similar studies have also been employed to establish an empirical modifier polarity scale under supercritical conditions.[55]

The critical parameters of potential binary fluids must be carefully considered to ensure that a single mobile phase is employed during the separation, and that a two-phase vapor/liquid carrier fluid is avoided. The methods for calculating PVT properties of mixed fluids necessarily include the additional variable of composition. For applying the corresponding states methods to mixtures, rules must be devised for relating composition to pseudocritical mixture constants. Many of these rules have been suggested.[16,56] These rules give surprisingly accurate results in calculations of pseudocritical parameters for determining mixture properties when the two components are similar or they do not have specific interactions. Less satisfactory results are found for mixtures of dissimilar components, especially if one or more of the components is polar or shows tendencies to associate into dimers, to form clusters, *etc.* For these mixtures, binary constants are generally approximated by back-calculations from experimental data, and very few of them are reported in the literature. Therefore, since critical parameters of polar fluid mixtures are subject to significant errors, density estimates are subject to the same errors. Moreover, critical parameters and densities change with mole fraction. Consequently, it is most practical (because of limited reliable phase equilibria data) to compare mixed fluid phase results on the basis of absolute temperature and pressure.

Generation of mixed mobile phases. Supercritical fluid mobile phases can be premixed, or mixtures can be generated using two high pressure pumps. Most of the suppliers of pure fluids for SFC also offer combinations of common supercritical fluids and solvents. This approach is easy, but the limited solubilities of many organic solvents in CO_2 and in other common fluids at ambient temperature and pressures of a few tens of atmospheres (the typical pressure found in commercial cylinders) reduces the choices of commercially available mixtures. For example, 12 mol% is the highest concentration of methanol which will maintain a homogeneous mixture of modifier and CO_2 below 90 atm and between -5°C and 29°C.[57] Under similar conditions, chloroform and acetonitrile are soluble in CO_2 only up to about 1.5 mol%. It is quite obvious that complete miscibility of the two fluid components is a prerequisite for reproducible analysis. The composition of the fluid stored in the bottle should be constant for typical room temperatures and for a reasonable range of pressures, because in the reservoir, fluid pressure gradually decreases with fluid consumption. When employing syringe type pumps, the miscibilities at temperatures at which the pumps are maintained for efficient solvent delivery, must be taken into account. There is a risk that a layer of organic modifier may separate and settle on the cylinder bottom, thus changing the mobile phase composition. Recently, methods have been reported[58-60] for in-house mixing of organic solvents and supercritical fluids. These methods allow the preparation of mixtures which are not commercially available; however, they have the same drawbacks as described before. Moreover, the necessary equilibration time for fluids under pressure inside a bottle is relatively long, *i.e.*, several hours.

The use of a single pump permits easy use of pressure programming, regardless of the fluid composition. The premixed fluid approach, however, has the disadvantage of limited flexibility. In fact, if, after an analytical run, the mobile phase must be varied, it is necessary to stop the flow, replace the tank, and purge the pump.

The second way of generating mixed supercritical fluid mobile phases involves the use of two pumps: one for the primary fluid, the other for the modifier or a modifier mixture. The second pump can, alternatively, deliver a premixed binary fluid with a high modifier content. This approach is similar to that routinely used in LC,[61] but requires more sophisticated equipment due to the relatively high compressibilities of supercritical fluids. By itself, compressibility is not of concern, but it becomes important when two fluids with different compressibilities, such as a supercritical fluid and a liquid, are simultaneously delivered to generate a mixed mobile phase. For example, if a pressure gradient at constant composition is required, then as the pressure increases, the relative speed of a syringe pump delivering CO_2 decreases compared to the speed of a syringe pump delivering a less compressible organic modifier.[66]

The use of two pumps instead of a single one is a much more flexible approach for mixing mobile phases. Besides pressure gradients, it allows the generation of composition gradients similar to those employed in LC, and simultaneous pressure/composition gradients.

As is also the case for neat fluids, reciprocating or piston pumps are usually utilized in mixed mobile phase SFC using conventional packed columns (4.6-mm i.d. or similar), while either reciprocating pumps followed by a splitting system and a back-pressure regulator,[63] or syringe-type pumps,[62] have been employed for small-bore columns that require flow rates in the range from a few μL min^{-1} to hundreds of μL min^{-1}.

Packed column SFC. All reports regarding the use of mixed mobile phases in packed column SFC note that even a very small amount (less than 1%) of a polar modifier added to CO_2 greatly influences retention and reduces peak tailing when using silica-based stationary phases. This is a clear effect of the modifier altering the packed column stationary phase. The large number of active sites, generally free silanols, present on a typical packing material are dynamically deactivated by direct interaction with polar modifiers.[64] It has become clear that for certain compounds the use of better deactivated packing materials results in better peak shapes and more rapid elution, often without the need for a polar modifier. Deactivated packings are also less prone to modifier memory effects.[65] The effects of modifiers are clearly shown in Figure 3.1 in which azo-dyes were analyzed with pure CO_2, 6 mol% dimethylacetamide, and 6 mol% propylene carbonate.[66]

Many groups have investigated packed column SFC using several different binary mobile phases. Klesper and coworkers have studied the elution of polymers and polar compounds employing CO_2 or *n*-alkanes, from ethane to *n*-pentane, mixed with ethers as mobile phases.[67-72] The use of an *n*-pentane/1,4-dioxane mixed fluid permitted[73] the elution of polystyrene samples with molecular weights up to 600,000, as shown in Figure 3.2. As the molecular weights became larger, elution required higher dioxane content and higher pressure. Levy and Ritchey studied the effects of a large variety of modifiers added to CO_2 with different stationary phases.[45,74] McNally *et al.*[75] have shown that different trends in retention can be related to solute polarity and functionality, in addition to modifier polarity and stationary phase interactions.

The retention time and peak area reproducibility found in SFC on packed columns with binary mobile phases have been reported to be of the order of 0.5-1% RSD and 1-4% RSD, respectively.[50,52,62]

Open tubular column SFC. In contrast to packed column SFC, small amounts (*i.e.*, <2%) of polar modifiers added to CO_2 do not drastically change capacity factors and peak shapes in open tubular column SFC.[76] This difference in behavior can be attributed to the highly inert surfaces and polymeric stationary phases employed in open tubular columns compared to the large number of active sites

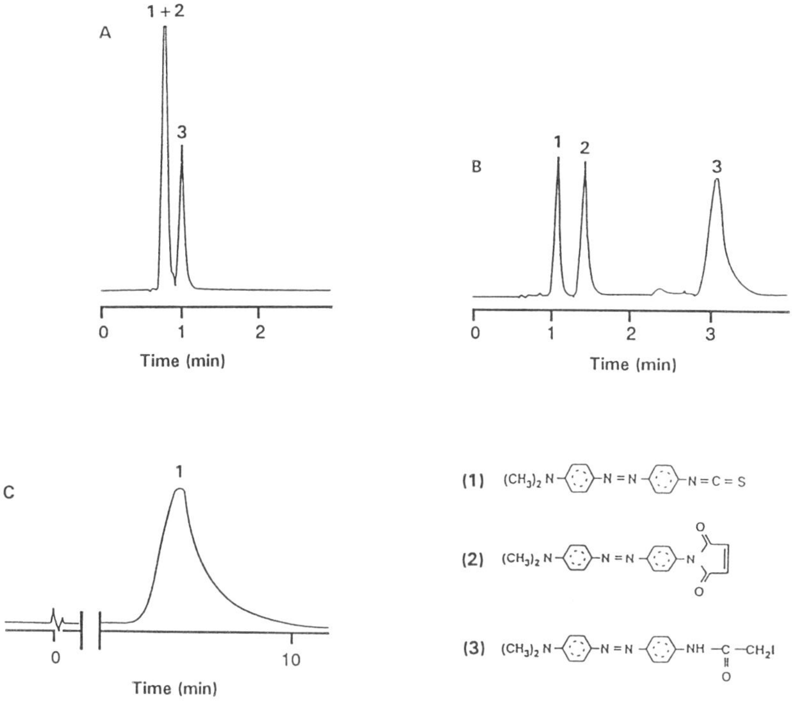

Figure 3.1. SFC chromatograms of azo-dyes showing the effect of modifier in CO_2 with (A) 6 mol% propylene carbonate, (B) 6 mol% dimethylacetamide, compared to (C) without modifier. Conditions: 25-cm x 4.6-mm i.d. packed column, diol modified silica, 5-μm particles; CO_2; 60°C; 370 atm; UV (420 nm) (reprinted with permission from Ref. 66).

present on silica based packing materials. On the other hand, if modifier percentages are increased to 5-20%, the changes in retention are more significant.[60] These results support the hypothesis that the observed decreases in retention are due to modifiers changing the mobile phase solvation power.

Studies have been carried out[77,78] to determine the amount of modifier absorbed by the bonded polymeric stationary phases. At constant temperature, the amount of modifier found in the stationary phase increased as the pressure increased. At constant pressure, the modifier amount associated with the stationary phase decreased with increasing temperature. It was also demonstrated that 2-propanol partitioned into the polymeric phase rather than adsorbing onto the polymer surface.[78] It follows that solute retention in mixed fluid

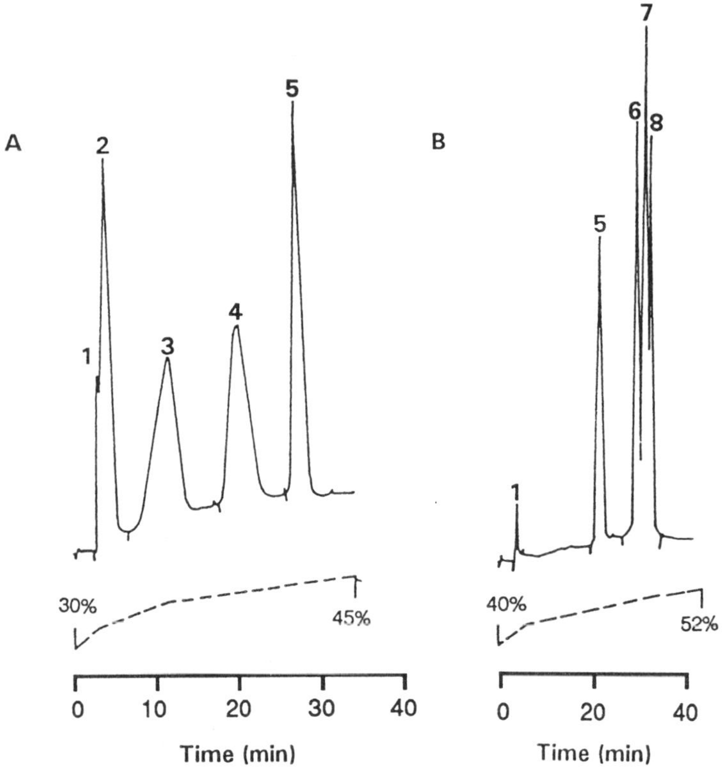

Figure 3.2. SFC chromatograms of polystyrenes with (A) 30 to 45% modifier gradient, and (B) 40 to 52% modifier gradient. Conditions: 25-cm x 4.6-mm i.d. packed column, silica, 10-μm particles; 1,4-dioxane modifier in n-pentane; 250°C; 166 atm; UV (254 nm). Peak identifications: (1) solvent; polystyrenes with molecular mass of (2) 3,600, (3) 20,400, (4) 53,700, (5) 93,000, (6) 254,000, (7) 392,000, and (8) 900,000, (reprinted with permission from Ref. 73).

open tubular column SFC can also be affected by changes in stationary phase characteristics, including polarity.

Another difference between the behavior of packed and open tubular columns is that in open tubular column SFC, equilibration times after changing pressure and/or temperature are negligible,[79] allowing fast and reproducible pressure and temperature programming.

3.4 Ternary Mixed Fluids

Modifiers that are miscible with CO_2, such as methanol, are of limited polarity, and more polar modifiers tend to separate or build up on the column. The polarities of common modifiers are often inadequate to compete with very polar solutes for active sites in the column. Low concentrations of highly polar compounds added to the modifier offer an alternative means of covering active sites. The more polar the additive, the more successfully it will compete with solutes for these sites, eliminating one retention mechanism and preventing tailing. Very polar additives, thus, can provide a dynamic form of end capping.

There have only been a few examples in the literature[80-83] of the use of binary fluid additives to make a ternary fluid in SFC, and no systematic studies on the role played by such additives have been reported. Preliminary measurements of the effects of additive polarity and concentration on the chromatography of some very polar solutes using packed columns are presented in the next few sections.

Effect of additive polarity. The relative polarities of the additive, the solutes, and the stationary phase determine the effectiveness of the additive in improving peak shape and affecting retention. Acetic, chloroacetic, dichloroacetic, trichloroacetic, and trifluoroacetic acids comprise a series of increasingly polar additives, which allow the effect of additive polarity to be easily studied.[84] Benzoic, phthalic, and trimesic acids were chosen as representative of very polar solutes. Benzoic acid usually can be eluted from open tubular or packed columns using neat CO_2 as the mobile phase. The other two solutes could not be eluted from packed columns even when a large concentration (*i.e.*, more than 10%) of methanol was added to the mobile phase. The addition of even the least polar of the additives (*i.e.*, acetic acid) brought about the elution of all of the solutes.

Effect of stationary phase polarity. The role played by the additive appears to be related to the polarity of the bonded stationary phase. On less polar phases, such as *n*-octyl, phenyl, or cyanopropyl, additives dramatically improve peak shapes, but do not significantly affect retention times.

On more polar columns, such as a diol column, additives appear to affect both retention and peak shape.[84] With no additive, only benzoic acid eluted of the series of acids discussed in the previous section. Low polarity additives caused all three acids to elute, but trimesic acid eluted before phthalic acid. The peak order reversed when the additive was changed from dichloroacetic acid to trichloroacetic acid. Since all of these chromatograms were obtained using similar silica-based packing materials, it is unlikely that the observed differences in peak shape and retention were due to differences in the number or type of active sites on the columns.

The effect of additive concentration. The effect of increasing concentrations of citric acid (added to 6% methanol in CO_2) on the retention of some very polar aromatic acids[84] is presented in Figure 3.3. These acids are too polar to be analyzed by reversed phase liquid chromatography and are, instead, usually analyzed by ion chromatography. The solutes include benzoic acid; 2-, 3- and 4-hydroxybenzoic acids, and 1,2-, 1,3-, and 1,4-benzene-dicarboxylic acids. Only benzoic acid eluted with pure CO_2. Several of the other acids eluted when methanol was added, but the remaining acids did not elute until an additive was introduced into the mobile phase. With the addition of additives, the retention of some solutes changed more dramatically than others. Those analytes that changed the most have substitutions in the 1,2-positions (peaks 2 and 7) and are probably less polar because of intermolecular hydrogen bonding.

Observations related to the ternary additive mechanism. Very polar additives tend to adsorb strongly onto the packed stationary phase. This is probably due to interactions with active sites and perhaps also with the bonded phase. When an additive is first introduced into a column, dramatic changes in peak shape and sometimes retention are observed. If the additive is subsequently removed from the mobile phase, the column performance generally does not rapidly return to the performance originally observed. Instead, peak shapes tend to retain much (or at least some) of the improvement over long periods.

The presence of adsorbed trifluoroacetic acid (TFA) was confirmed on a packed glass diol column which had been exposed to the additive, but had subsequently been washed with more than 50 column void volumes using a mobile phase of methanol in CO_2.[84] The modifier was then changed in several stages to hexane. Injections of a solvatochromic dye, Nile Red, dissolved in hexane, produced a moving band of an intense blue color, characteristic of Nile Red in trifluoroacetic acid (the dye is a pale yellow-orange color

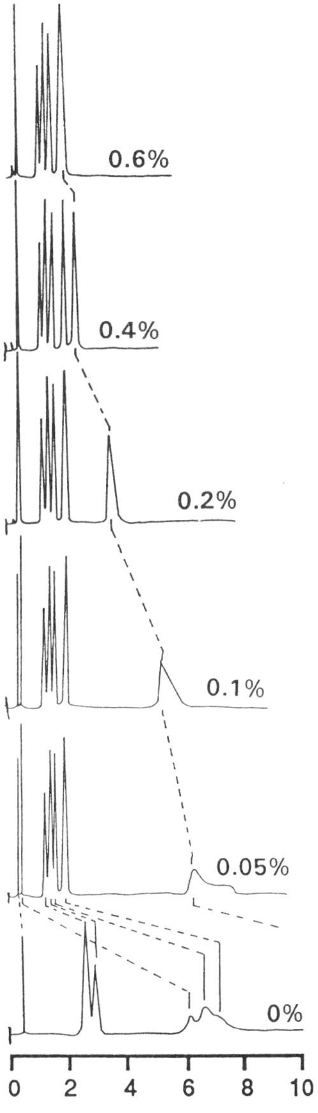

Figure 3.3. SFC chromatograms showing the effect of a third component additive on peak shape and retention. Conditions: 10-cm x 2-mm i.d. packed column, diol stationary phase, 7-μm particles (Nucleosil); 6 v/v% methanol + citric acid in 94% CO_2 (actual citric acid concentration in methanol is listed beside each chromatogram); 40°C; 207 atm outlet pressure; 3 mL min^{-1} flow rate. Components of mixture: benzoic acid, 2-hydroxybenzoic aicd, 1,4-benzenedicarboxylic acid, 1,3-benzene dicarboxylic acid, 3-hydroxybenzoic acid, 4-hydroxybenzoic acid, and 1,2-benzenedicarboxylic acid (reprinted with permission from Ref. 84).

in hexane). A column which had never been exposed to trifluoroacetic acid produced a pale pink to purple color characteristic of a solvent in the polarity range slightly more polar than methanol. The fact that the residual adsorbed trifluoroacetic acid could still change the color of the dye indicates that the adsorbed additive can interact with solutes and does not act as merely an inert cap on the active sites.

Solvent strengths of solutions containing additives. Solvatochromic dye studies indicate a major increase in apparent solvent strength when trichloroacetic acid is added to acetonitrile in CO_2.[84] Such measurements suggest that modifiers and additives do much more than simply cover active sites on the column surface.

The use of third-component additives in SFC is new, but they offer very exciting possibilities. The role of the additive appears to be multifunctional, especially in packed columns: blocking of active sites, dynamic stationary phase modification, and possibly increasing the local solvent strength of the mobile phase inside the modifier clusters. Additives may produce mobile phases with solvent strengths approaching that of water, but retaining the low viscosity and high diffusivity of CO_2.

3.5 Reverse Micelle Organized Media as Mobile Phases

Fundamental properties of reverse micelles in supercritical fluids. Reverse micelles are thermodynamically stable clusters that are typically tens of nanometers in diameter. They consist of a polar core of water molecules with amphiphilic, surfactant molecules sheathed around the polar core. The polar "head" groups of the surfactant molecules dip into the water core, while the hydrophobic tail groups extend into the nonpolar solvent.

Dorsey and coworkers[85] first considered using reverse micelle mobile phases in LC, in the hope that reverse micelle LC would provide an alternative to normal-phase LC for the separation of polar compounds. Normal-phase LC suffers from irreproducible solute retention, while this is much less of a problem in reverse micelle LC. Above a water content of approximately 0.1%, the retention of various test compounds was independent of the water-to-surfactant concentration ratio. However, the efficiencies of the reverse micellar separations were approximately half those of the nonmicellar separations.

Partitioning of polar compounds between the water core of the reverse micelles and the stationary phase in SFC could potentially bring about the efficient separation of polar compounds without the need for highly polar supercritical fluids. It is, therefore, hoped that reverse micellar SFC can extend the range of molecules that can be separated by SFC.

AOT, *bis*-2-ethylhexyl sulfosuccinate, is the surfactant used in all reverse micelle SFC and LC studies to date. The two primary reasons for its use are that its properties have been thoroughly characterized, and it forms large thermodynamically stable micelles without the addition of a cosurfactant.

Phase behavior of reverse micelles in supercritical fluids. Reverse micelles in conventional liquid solvents, such as isooctane, are practically unaffected by substantial changes in pressure. For example, in liquid isooctane at 25°C, the maximum molar ratio of water-to-surfactant ($W_{o,max}$) is either invariant or it decreases as pressure increases.[86] Conversely, the phase behavior of reverse micelles in supercritical fluids is strongly dependent on pressure.[87,88] Figure 3.4 shows the variation of $W_{o,max}$ with pressure in supercritical and liquid ethane, as well as in liquid propane.[86] The maximum amount of water in supercritical ethane that the micellar system can dissolve is very low; $W_{o,max}$ ≤5, for pressures ≤450 atm. When the pressure is increased substantially to *ca.* 800 atm, then $W_{o,max}$ increases sharply to 40, which is similar to that found for conventional liquids, such as isooctane at atmospheric pressure. $W_{o,max}$ also varies considerably with pressure in liquid propane. The value of $W_{o,max}$ = 40 was attained at lower pressures, *ca.* 200 atm for propane compared to *ca.* 800 atm for ethane.[86]

The critical micelle concentration (CMC) was recently determined[89] for both supercritical propane and supercritical ethane with W_o = 1. The apparent CMC for AOT in supercritical propane at 199°C and 267 atm was 3.3 x 10^{-3} M. The approximate CMC for ethane at 39°C and 274 atm was 24 x 10^{-3} M. These values are approximately the same as those found in liquid reverse micelle systems.

Transport properties of reverse micelles in supercritical fluids. Figure 3.5 shows a comparison of the diffusion coefficients measured by dynamic light scattering for reverse micelles in liquid hydrocarbons and supercritical ethane for W_o = 5.[86] The diffusion

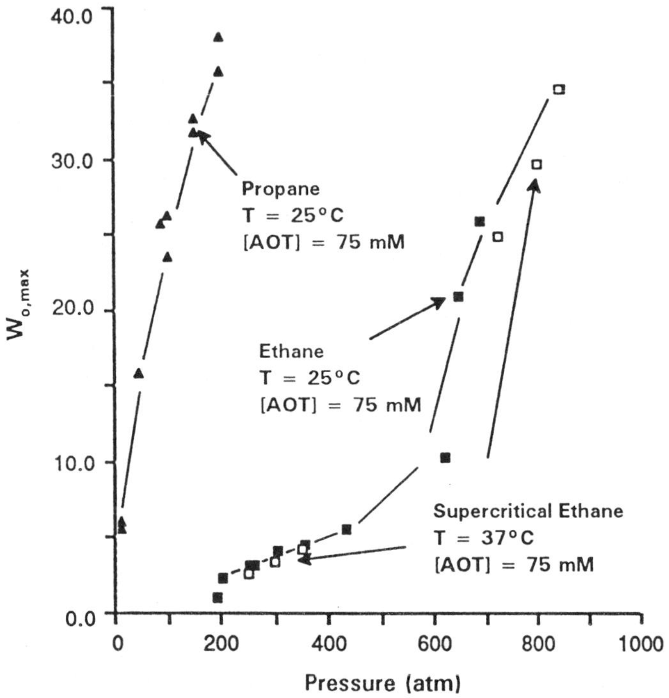

Figure 3.4. Variation of maximum molar water to AOT surfactant ratio ($W_{o,max}$) as a function of pressure for ethane at 25 and 37°C, and propane at 25°C (reprinted with permission from Ref. 86).

coefficients of the reverse micelles in liquids decreased by approximately 10-15% when the pressure was increased from 1 to 350 atm. The diffusion coefficient of the reverse micelles in supercritical ethane at 37°C increased as the pressure was increased over the range of 400-600 atm.[86] The different diffusion behavior of reverse micelles in liquids and supercritical fluids has been attributed to strong micelle-micelle interactions in the supercritical fluids.[90]

Preparation of reverse micelle supercritical fluid mobile phases. Similar problems arise in the preparation of reverse micelle mobile phases as in the preparation of modifier systems. In each case, the questions that must be addressed are: (a) is the surfactant that was added completely soluble in both the liquid and supercritical solvent; (b) does the phase behavior of the mixture cause precipitation of the surfactant as the pressure is increased; and (c) is the solution that is delivered to the analytical column homogeneous? As with modifiers, the solubility of the surfactant and the variation

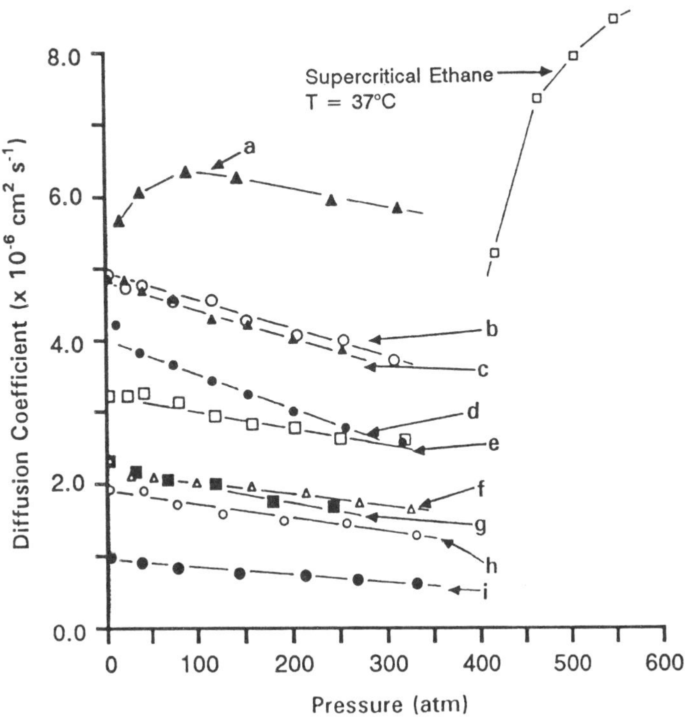

Figure 3.5. Diffusion coefficients for AOT/water reversed micelles at W = 5 in supercritical ethane and liquid alkanes (25°C) as a function of pressure. Liquid alkane: (a) propane, (b) *n*-pentane, (c) *iso*-pentane, (d) *neo*-pentane, (e) *n*-hexane, (f) *n*-heptane, (g) *iso*-octane, (h) *n*-octane, (i) *n*-decane (reprinted with permission from Ref. 86).

of the solubility with pressure must be determined before use in SFC. The homogeniety of the solution is typically ensured by mechanically mixing the fluid over an extended period of time, *e.g.*, overnight, before use in SFC.

Reverse micelle SFC separations. Gale *et al.*[91] were the first to separate polar compounds using reverse micelle SFC. The separation of phenol, 2-naphthol, and resorcinol on a microbore column packed with 5-μm particles (Figure 3.6) using AOT reverse micelle/supercritical propane mobile phase was compared to that found when using neat supercritical propane. In addition, these data were compared to a complementary separation in neat liquid hexane and AOT reverse micelle/liquid hexane systems. In each case, the reverse micelle system provided a decrease in the capacity factor of each solute. When liquid reverse micelle systems were used, the

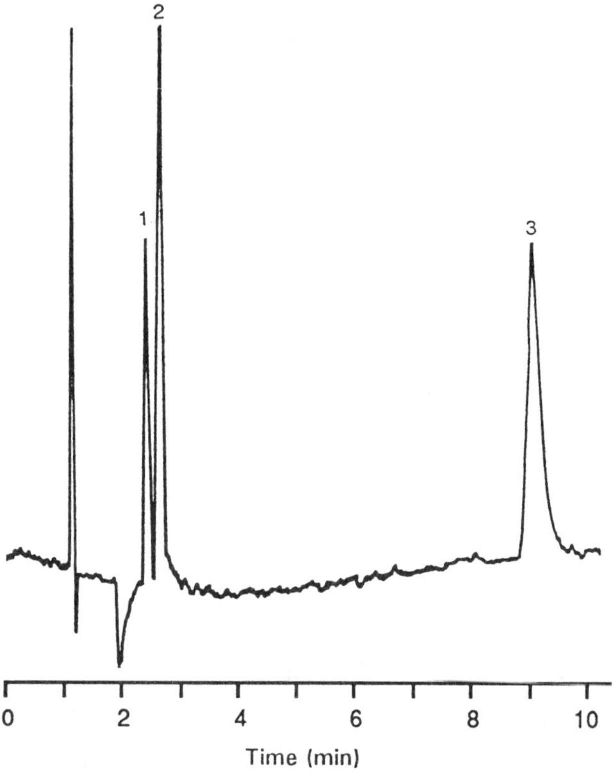

Figure 3.6. SFC chromatogram of phenolic standard compounds using normal-phase reverse micelle modifier. Conditions: 25-cm x 1-mm i.d., silica (Spheri-5) stationary phase, 5-μm particles; 5 x 10^{-2} M AOT (W_o = 5) in propane; 110°C; 250 atm; UV (254 nm). Peak identifications: (1) phenol, (2) 2-naphthol, (3) resorcinol (reprinted with permission from Ref. 91).

efficiency of the separation was half that of the separation with the neat liquid. However, when the supercritical fluid reverse micelle mobile phase was used, the efficiency was similar to that in the neat supercritical solvent.

Gradient elution methods. At least two additional types of gradient methods are possible in reverse micellar chromatography as compared to the typical polarity gradient method. These methods involve: (a) changing the water-to-surfactant concentration ratio, W_0, or (b) changing the pH, ionic strength, or chemical composition of the water core. The first method is expected to be used most in the near future. Before the second method can be used routinely, further studies on reverse micelle phase behavior in supercritical fluids are required.

When W_0 is increased, the water pool in each reverse micelle enlarges, which causes the solvent strength of the expanding water pool to increase. A solvent strength gradient can, therefore, be created by varying W_0, and the separation factor can be changed substantially by using W_0 programming.

Detection methods. The use of reverse micelle mobile phases limits detection options substantially. The strong response of the surfactant molecules prohibits the use of flame-based detection in SFC. The range of mass spectral ionization methods that can be used is also seriously limited. Electrospray ionization may be the only applicable mass spectral ionization method. Due to the absorbance of surfactant and the water pool, the use of FTIR is also highly restricted.

Optical spectroscopies such as conventional UV and fluorescence should be unaffected, or even improved, by the presence of reverse micelle assemblies in the supercritical fluid. For example, room temperature phosphorescence can be stabilized by placing the molecule inside, or at interfacial regions of, micelles.[92] New methods of detection that utilize the unique environment of reverse micelles can be anticipated.

Conclusions. The use of reverse micelle SFC for the separation of polar molecules has only received preliminary attention. Significant advancements are clearly needed in the understanding and use of reverse micelles in supercritical fluids. Advances will probably come more quickly in the area of supercritical fluid extraction than in supercritical fluid chromatography due to limitations such as the pressure dependence of physical properties and reduced detector compatibility. However, the chromatographic efficiencies demonstrated in the reverse micelle supercritical fluid chromatograms clearly show the potential of the technique.

3.6 Density Programming

The density of the mobile phase is the key parameter governing retention in SFC, comparable to temperature in GC or eluent composition in LC. Due to the high densities of supercritical fluids, interactions between molecules of the fluid mobile phase and the

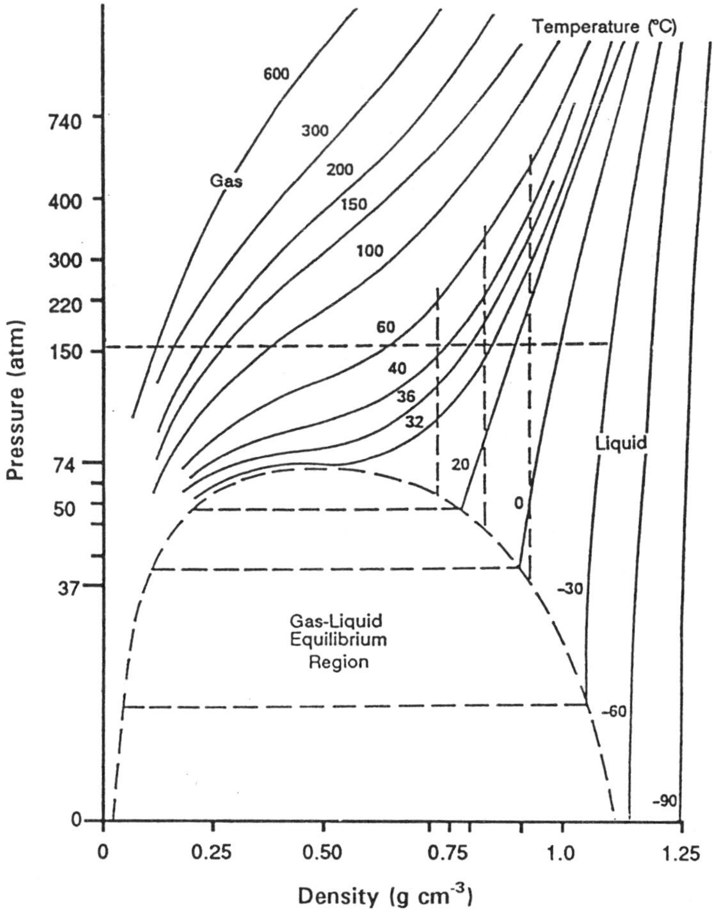

Figure 3.7. Carbon dioxide pressure as a function of density at different temperatures.

sample are the most important forces affecting transport of the solute along the separation column. The retention time of a solute in the chromatographic column depends on both its affinity for the stationary phase and its solubility in the mobile phase. Sample volatility may play a role at elevated analysis temperatures in SFC, but is of less importance than in GC because the analytes are solubilized by the mobile phase. SFC is able to analyze substances of much higher molecular mass than can be done using GC.

The density of a supercritical fluid is a variable which is related to both the temperature and the pressure. The characteristic dependence of density or volume on temperature and pressure is given by equations of state. The Redlich-Kwong equation[93] or the Peng-Robinson equation[94] describe the ρPT relationships. Since supercritical fluids are very compressible near the critical point, compressibility must be taken into account for the calculation of pressure/temperature/density relationships. Commercially available SFC instruments use computers to calculate density values from pressure and temperature inputs using the theory of corresponding states and generalized compressibility factor tables.

Before commercial SFC instruments were available, pressure programming was the most favorable way to perform SFC analyses. As Figure 3.7 illustrates for CO_2, P-ρ isotherms are nonlinear at temperatures near the critical temperature of the eluent.[95] In the liquid state, as well as in the gas-like region at high temperatures, the isotherms become more linear. A linear pressure program at temperatures near T_c (greater than or equal to T_c) yields a nonlinear density program. The pressure *vs* density isotherm for $T = T_c$ shows a horizontal tangent at the critical pressure. Working at the critical temperature, or slightly above, small changes in pressure result in huge density variations, which are accompanied by variations in solubility. While varying the analysis pressure is the most popular way to perform density programmed analyses,[51,65,96-112] temperature is the other important parameter which can effect density changes.[108,111,113-117] An increase in density results either from increasing the pressure or by decreasing the temperature. The influence of analysis temperature will be discussed in detail in Section 3.7.

Qualitatively, the solvent strength of an eluent may be described by the solubility parameter, δ, introduced by Hildebrand and Scott.[118] They stated that the solubility parameters of the solvent and the solute should be similar to achieve dissolution of the solute. The solvent becomes stronger as its δ-value increases. Giddings *et al.*[23] expressed the solubility parameter as a function of the density (ρ) of a supercritical fluid

$$\delta = 1.25\, P_c^{1/2} \left(\frac{\rho}{\rho_{liq}} \right) \tag{3.1}$$

where P_c is the critical pressure of the eluent, and ρ_{liq} is the density of the eluent in the liquid reference state.

In SFC, the solubility parameter varies from 0 up to liquid-like values of 10 at high densities. Near-linear plots of δ *vs* ρ resulted[119] from a more detailed equation based on a thermodynamic approach derived by Tijssen *et al.*[120] and on experimental results.[121] The solubility parameter of CO_2 increases by an order of magnitude within the typical density range of SFC. The influence of temperature on δ is much less significant.

The concentration (c_i) of a solute i in a supercritical phase was found to depend logarithmically on solvent density[122]

$$\ln c_i = a_i \ln \rho + b_i \tag{3.2}$$

where a_i and b_i are solute dependent constants. In a given chromatographic system, the capacity ratio of a solute i (k_i) is a measure of the solubility of the solute in the mobile phase. Van Wasen *et al.*[5] developed an equation describing the pressure dependence of k_i:

$$\left(\frac{\partial \ln k_i}{\partial P}\right)_T = \frac{1}{RT}(\overline{V}^{\infty}_{i,m} - \overline{V}^{\infty}_{i,s}) + \frac{1}{\overline{V}_m}\left[\frac{\partial \overline{V}_m}{\partial P}\right]_T \tag{3.3}$$

where $\overline{V}^{\infty}_{i,m}$ is the partial molar volume of solute i at infinite dilution in the mobile phase, $\overline{V}^{\infty}_{i,s}$ is the partial molar volume of solute i at infinite dilution in the stationary phase, and $\overline{V}_m$ is the molar volume of the mobile phase. Yonker *et al.*[123] showed that the results calculated from Equation 3.3 agreed well with experiment. While plots of log k *vs* pressure show sigmoidal shapes at temperatures slightly above T_c (Figure 3.8A), log k *vs* density plots are more linear (Figure 3.8B).[1] According to Equation 3.3, the linearity of log k *vs* log ρ plots is nearly perfect (Figure 3.8C). Due to the decrease in log k when increasing either P or ρ, analyses become faster at higher pressures or densities. However, increasing the density to extremely high values may not further increase solute solubility. Liphard *et al.*[124] reported an increase of ten orders of magnitude in the solubility of squalane in CO_2 when raising the pressure from ambient up to 400 atm, but a decrease in solubility at higher pressures. On the other hand, there may exist a "threshold density" or a "threshold pressure" which must be exceeded to dissolve the solutes.[125]

In Figure 3.8C, the log k *vs* log ρ isotherms for two different solutes converge at higher densities. Therefore, the selectivity between two solutes, as defined by the ratio of the capacity ratios of the solutes, decreases at the same time. This is one reason why SFC should be carried out at the lowest possible density to achieve good separations. Besides selectivity, the separation efficiency also

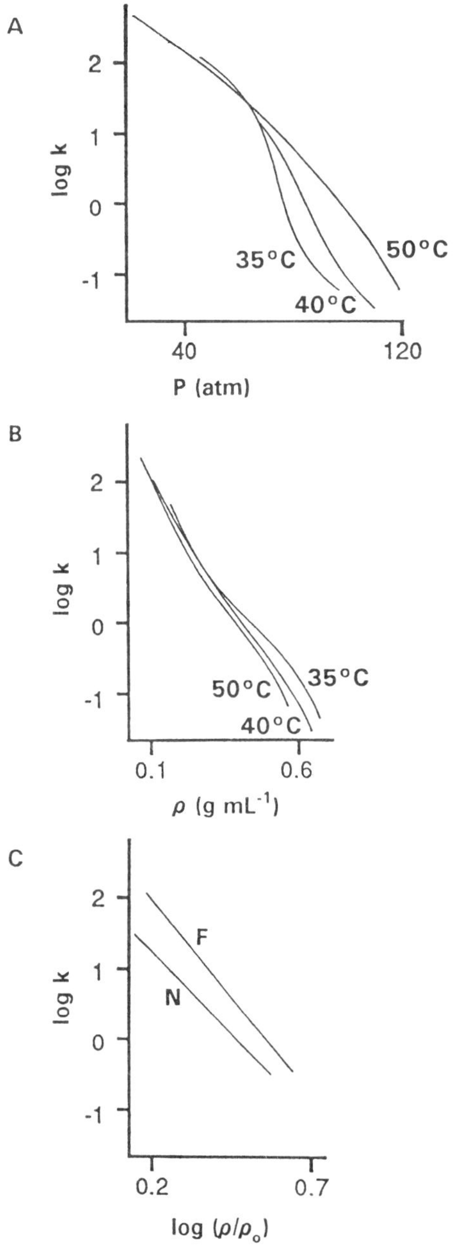

Figure 3.8. Capacity factor as a function of (A) pressure, (B) density, and (C) reduced density; (A) log k of napthalene as a function of pressure at 35, 40 and 50°C, (B) log k of naphthalene as a function of density at 35, 40, and 50°C, and (C) log k of naphthalene (N) and fluorene (F) as a function of log (ρ/ρ_o) at 40°C. Conditions: 30.5-cm x 3-mm i.d. packed columns, octyl stationary phase, 30-40 μm particles (Perisorb); CO_2 (reprinted with permission from Ref. 1).

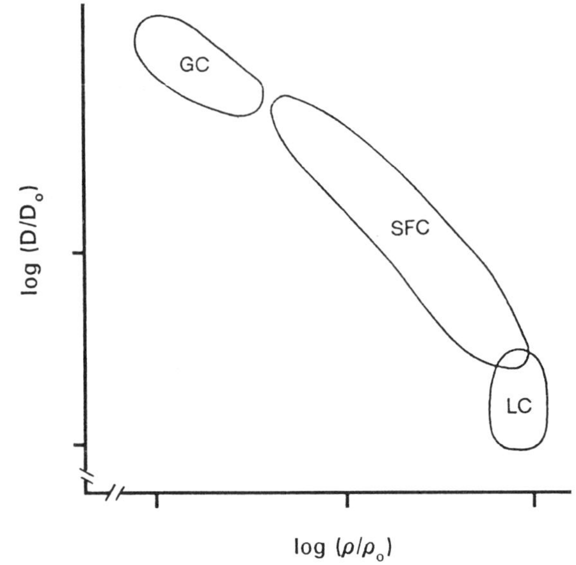

Figure 3.9. Schematic diagram illustrating the relationship between the mobile phase density and the diffusion coefficients of solutes, indicating areas of gas (GC), liquid (LC), and supercritical fluid (SFC) mobile phases (reprinted with permission from Ref. 127).

decreases at higher densities. At higher densities, there is a decrease in self-diffusion coefficients of the mobile phase, as well as in diffusion coefficients of the analytes,[126] resulting in peak broadening. At the same time, the viscosity of the mobile phase increases. Assuming constant retention behavior (k = const.), the increase in plate height at linear velocities greater than the minimum in the van Deemter plot is less steep at lower densities than at high densities (see Figure 2.3). Although the plate height at the minimum is the same for all densities, the minimum is shifted to higher linear velocities as the density decreases. Since open tubular column SFC separations usually are performed at linear velocities 5-10 times higher than the optimum linear velocity, the loss in efficiency at higher densities is obvious.

Summarizing the factors mentioned above, there are two regions of SFC separations depending on eluent density: a "gas-like" region where density is low; and a "liquid-like" region at higher densities. In the "gas-like" region, the solvent strength of the mobile phase is low and the sample molecules are transferred into the mobile phase mostly by volatilization. The good chromatographic properties of the mobile phase, such as low viscosity and high diffusion coefficients, result in efficient separations with high selectivities. On the other hand, at high densities, the mobile phase assumes more liquid-like properties and the solutes dissolve in the eluent. The disadvantages at high densities are low selectivities and low separation efficiencies due to low diffusion coefficients. In Figure 3.9, the typical operating regions for GC, SFC, and LC are indicated in a plot of diffusion coefficient *vs* density.[127] From this point of view, as well as from a consideration of the viscosity and density of the mobile phase, supercritical fluids bridge the gap between liquids and gases. The logical means of taking advantage of both the "liquid-like" region and the "gas-like" region is by density programming, *i.e.*, ramping the density up from low values. Many samples, *e.g.*, oligomer mixtures, contain both volatile and nonvolatile components. SFC using density programming is the ideal technique to achieve both efficient separations of the volatile components and elution of the nonvolatile components.

3.7 Temperature Programming

There is an increasing number of applications utilizing temperature programming in addition to density or pressure programming in SFC.[70,111,116,118,120-122,128] Since density is related to both pressure and temperature, density changes may be achieved by variations of both, either singularly or combined.

The influence of temperature on SFC separations should be considered for both constant pressure and constant density conditions. At constant analysis pressure, temperature variations result in density changes, whereas the pressure must be adjusted to compensate for temperature variations at constant density. At constant pressure, solute retention depends on temperature in a very characteristic manner. Experiments involving different mobile and stationary phases, as well as different solutes, have resulted in plots[6,13,69,72,129-139] of the capacity ratio (k) *vs* analysis temperature (T) similar to those shown in Figure 3.10 for four aromatic hydrocarbons and CO_2 as eluent.[140] Increasing the temperature from ambient, *i.e.*, in the liquid state for most eluents, k values decrease and reach a minimum, thus reflecting an increase in solute solubility at higher temperatures. After passing the critical temperature of the eluent (indicated by a vertical line in Figure 3.10), the k values increase significantly. This increase is more pronounced as the analysis pressure is decreased. At the critical temperature, the

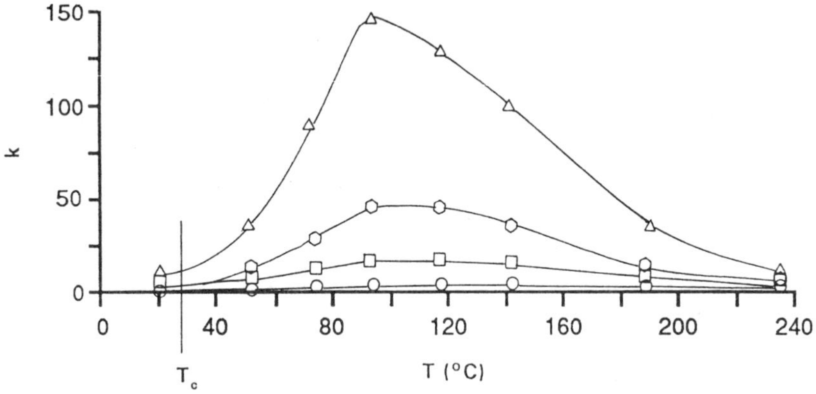

Figure 3.10. Capacity ratio as a function of temperature for increasing molecular mass polycyclic aromatic hydrocarbons. Conditions: 25-cm x 4.6-μm i.d. packed column, silica stationary phase, 10-μm silica particles; CO_2; 150 atm. Analytes: (o) napthalene, (□) anthracene, (⬡) pyrene, (△) chrysene. The vertical line indicates the critical temperature (reprinted with permission from Ref. 140).

density of the eluent decreases as it moves from liquid to supercritical. As was outlined in Section 3.6, analyte solubility is low at low densities, which is reflected by higher k values. Counteracting this, the vapor pressure of the analyte increases at higher temperatures, leading to an increase in solubility. The second effect overcompensates for the first at a temperature somewhat above the critical temperature and at moderate pressures, resulting in a maximum in capacity ratio. At high pressures, the density decrease of the eluent with increasing temperature is too small for the first effect to be significant and, therefore, the k values do not reach a maximum. The combined influences of pressure and temperature on retention behavior can be read from a three-dimensional plot of k *vs* P and T in Figure 3.11,[141] showing distinct k maxima at low pressures, but flat curves at high pressures.

The dependence of retention on temperature at constant pressure has been explained by three thermodynamic approaches. Yonker *et al.*[4] used Equation 3.4

$$\left(\frac{\partial \ln k_i}{\partial T}\right)_p = \frac{1}{RT^2}\left[\Delta\overline{H}_i^{\infty}(T_o) + \int_{T_o}^{T} \Delta\overline{C}_{p,i\infty} dT\right] + \frac{1}{\overline{V}_m}\left(\frac{\partial \overline{V}_m}{\partial T}\right)_p \qquad (3.4)$$

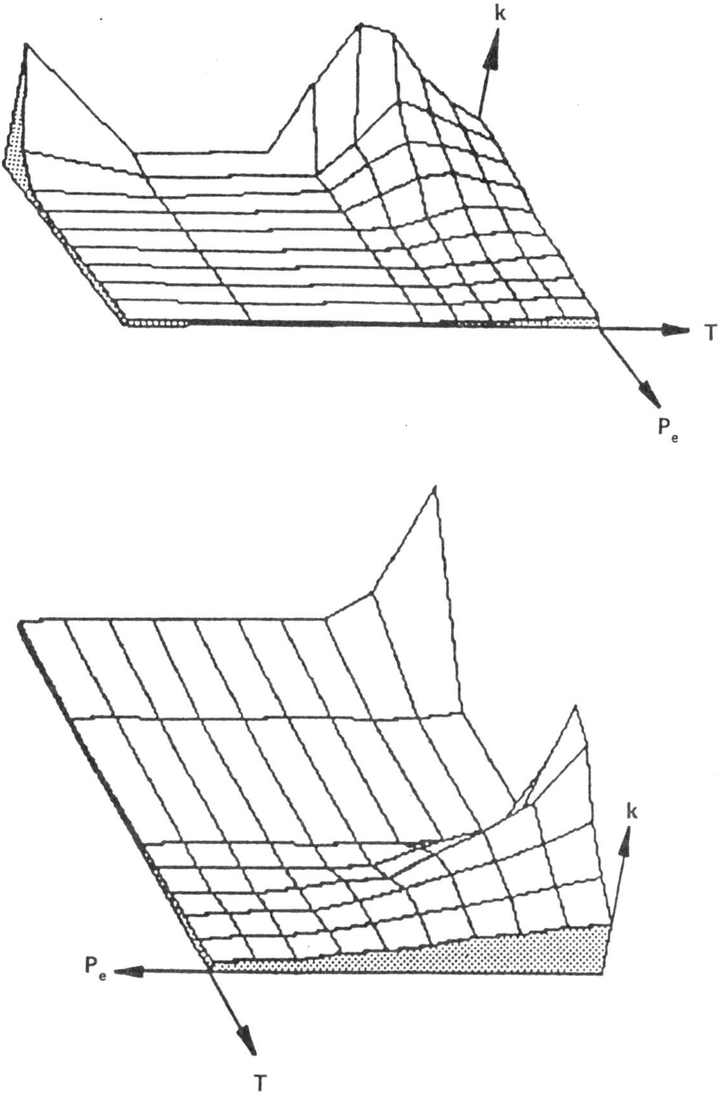

Figure 3.11. Capacity ratios of pyrene as a funciton of column end pressure (P_e) and temperature (T). Three dimensional network shown in two perspectives. Conditions: 25-cm x 4.6-mm i.d. packed column, 10-μm silica particles; *n*-pentane (reprinted with permission from Ref. 141).

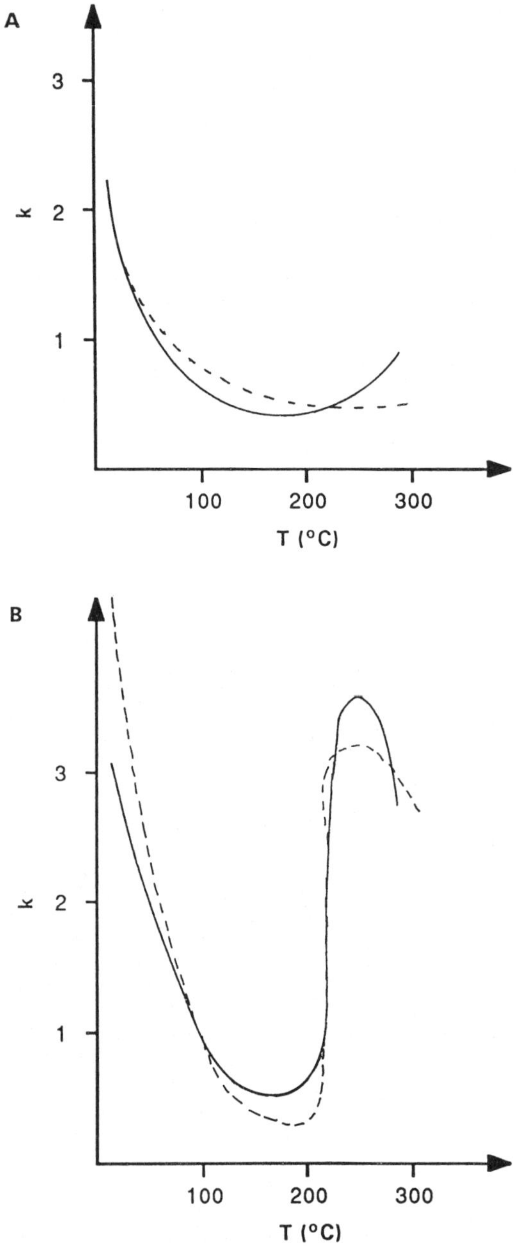

Figure 3.12. Capacity ratio of chrysene as a function of temperature at (A) 36 atm and (B) 70 atm showing experimental data[142] (—) and calculated data[4] (---). Conditions: 25-cm x 4.6-mm i.d. packed column, 10-μm silica particles; *n*-pentane (reprinted with permission from Ref. 143).

where $\Delta\overline{H}_i^{\infty}$ is the transfer enthalpy of substrate i between the mobile and stationary phases at infinite dilution, $\Delta\overline{C}_{p,i}^{\infty}$ is the molar heat capacity of substrate i at infinite dilution, and $\overline{V}_m$ is the molar volume of the mobile phase. Since chromatographic separations usually satisfy the conditions of infinite dilution, Equation 3.4 may be assumed applicable. Figure 3.12 demonstrates for chrysene as the solute, *n*-pentane as the mobile phase, and silica as the stationary phase, that this thermodynamic model describes the real chromatographic observations quite well.[142] The k maximum at low pressure was predetermined, as well as the continuous decrease in k at higher pressure. This model may be used for liquid eluents as well as for supercritical and gaseous eluents.

Another thermodynamic approach, by Chester *et al.*,[13] separated the k (temperature) dependencies at constant pressure into GC-like and LC-like regions using Equation 3.5

$$\log k = \frac{-0.43\ \Delta H_s}{RT} - \log \beta + \frac{0.43\ \Delta H_m}{RT} \qquad (3.5)$$

where ΔH_s is the partial molar heat of solution of the solute i in the stationary phase, ΔH_m is the partial molar heat of solution of the solute i in the mobile phase, and β is the column phase ratio. In Figure 3.13, the GC-like behavior is demonstrated[13] by the dashed straight line when plotting log k *vs* T^{-1}. Since ΔH_m is a function of the mobile phase density, which varies with temperature in a nonlinear way, the right hand term in Equation 3.5 ($0.43\ \Delta H_m/RT$) causes a deviation from linear behavior. At low temperatures, *i.e.*, high 1/T values, mobile phase solvation is the dominant parameter determining retention behavior. Ideally, the intercept would be the negative logarithm of the phase ratio, but any entropy change occurring in the separation process would also affect the intercept.

Bartle *et al.*[144] were able to account rigorously for the temperature variation of retention at constant pressure in SFC by means of the expression

$$\ln k - \ln \phi_m + \ln\left(\frac{\rho}{c^{\theta}}\right) = C + \frac{\Delta S_s^{\theta}}{R} - \frac{PV + \Delta H_s^{\theta}}{RT} \qquad (3.6)$$

where ϕ_m is the fugacity coefficient of the solute in the mobile phase, ρ is the density of the supercritical fluid, c^{θ} is the standard concentration, and C is a constant. The left-hand side of equation 3.6 was obtained as follows: experimental values of k; values of ρ/c^{θ} from the IUPAC equation of state for CO_2; and values of ϕ_m were calculated from the Peng-Robinson equation of state. Figure 3.14 shows the individual contributions and total of the left-hand side of Equation 7 for fluorene. It can be seen how the addition of the steepening curves of both ln (p/c^{θ}) and ln ϕ_m to the curve of ln k, with its maximum, gives a total which is approximately a straight line when plotted against 1/T. Graphs of this total for four polycyclic aromatic hydrocarbons are shown in Figure 3.15.

In addition to absolute retention at constant pressure, the relative retention between two solutes, indicated by the selectivity (α) depends on temperature.[145] As was seen in Figure 3.10, the relative retentions of the different solutes vary with temperature. For chemically different solutes, even a reversal in retention may be observed.

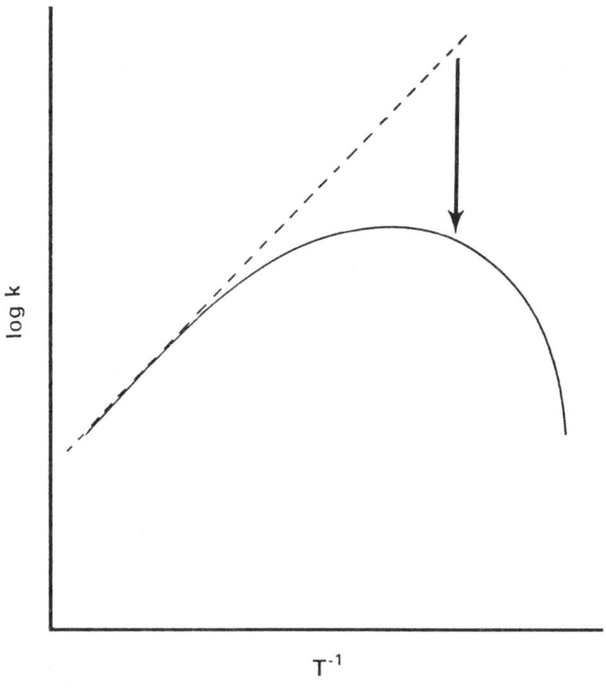

Figure 3.13. Model of observed SFC retention behavior (reprinted with permission from Ref. 13).

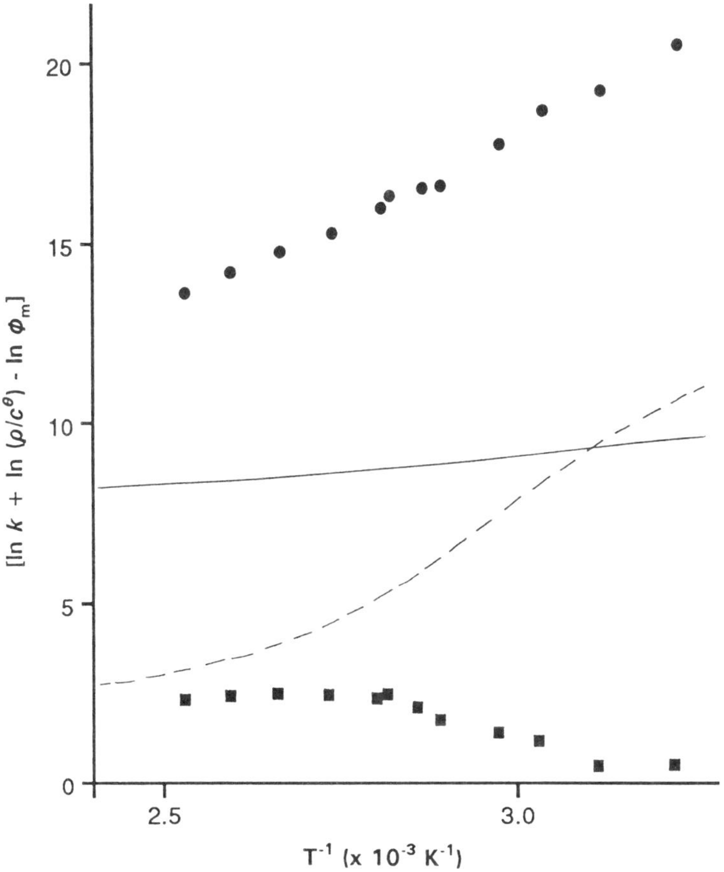

Figure 3.14. Contributions to [ln k + ln (p/c^{θ}) - ln ϕ_m] for fluorene as a function of reciprocal temperature: (■) ln k, (—) ln (p/c^{θ}), (---) ln ϕ_m, (●) [ln k + ln (p/c^{θ}) - ln ϕ_m] (reprinted with permission from Ref. 144).

Since the resolution between different solutes is a function of retention, selectivity, and efficiency, resolution exhibits a distinct maximum at a certain temperature in the supercritical region (Figure 3.16[141]). Again, the maximum is more pronounced at lower analysis pressures.

At constant density, the correlation between retention and temperature is much more uniform. As the "liquid-like" contribution to retention, *i.e.*, solvent strength of the mobile phase, expressed by its density, remains constant, only the "gas-like" factor remains important. From basic thermodynamic principles, it can be derived

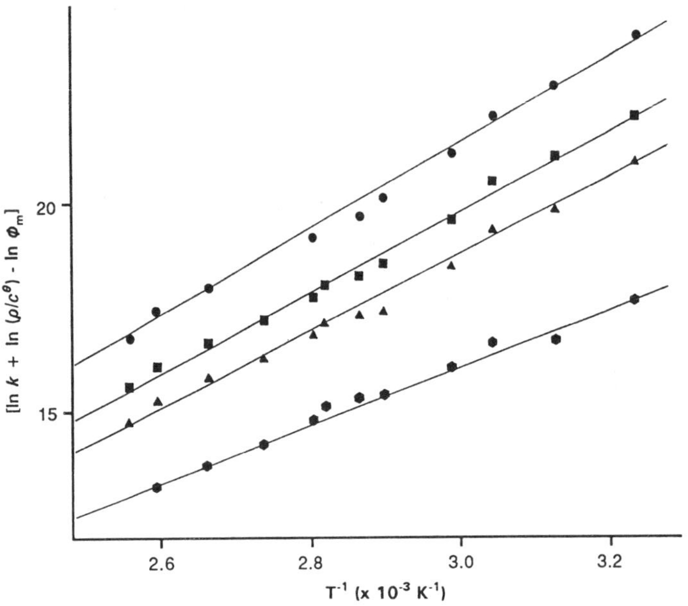

Figure 3.15. Plots of [ln k + ln (p/c^θ) - ln ϕ_m] *vs.* reciprocal temperature: (●) pyrene, (■) phenanthrene, (▲) fluorene, (⬢) naphthalene (reprinted with permission from Ref. 144).

that ln k is a function of the reciprocal temperature at constant density[146]

$$\ln k = -\frac{\Delta\overline{H}_i^\infty}{RT} + \frac{\Delta\overline{S}_i^\infty}{R} - \ln\beta \tag{3.7}$$

where $\Delta\overline{H}_i^\infty$ is the transfer enthalpy of solute i between mobile and stationary phase at infinite dilution, $\Delta\overline{S}_i^\infty$ is the transfer entropy of solute i between mobile and stationary phase at infinite dilution, and β is the column phase ratio. Assuming that $\Delta\overline{S}_i^\infty$ and β are independent of temperature, differentiation of Equation 3.6 results in a linear dependence of ln k on T^{-1}

$$\left(\frac{\partial \ln k}{\partial(1/T)}\right)_\rho = -\frac{\Delta\overline{H}_i^\infty}{R} \tag{3.8}$$

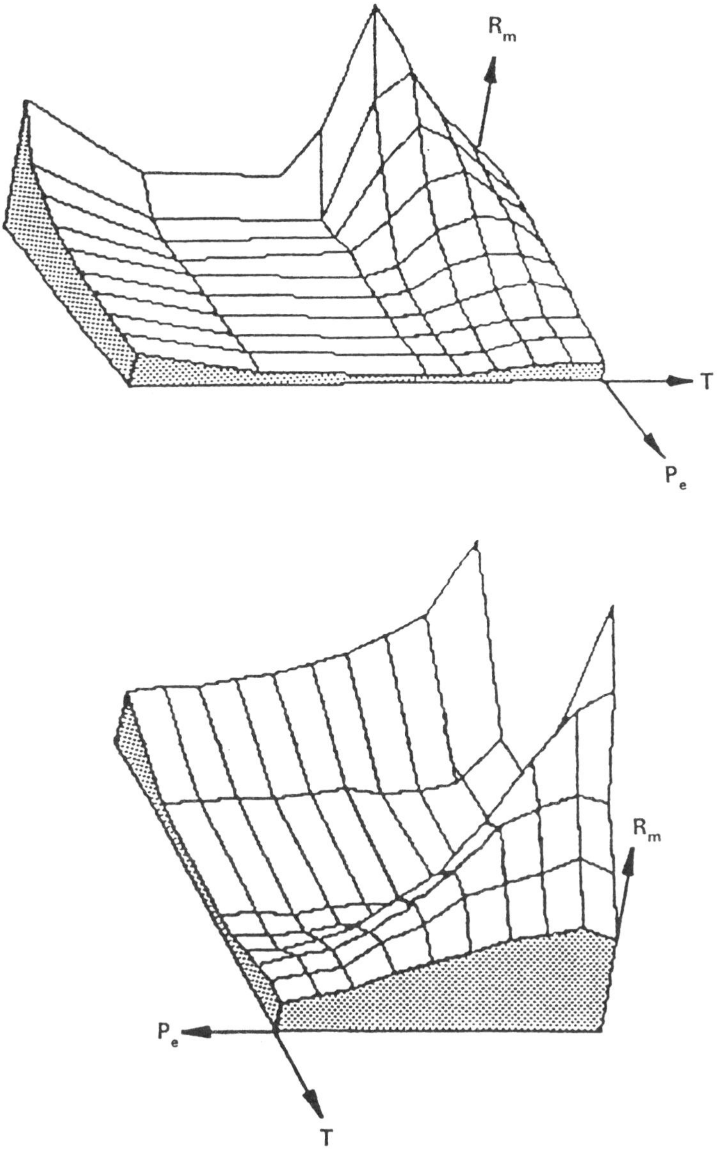

Figure 3.16. Mean resolution (R_m) between naphthalene, anthracene, pyrene, and chrysene, as a function of column end pressure (P_e) and temperature (T). Three dimensional network shown in two perspectives. Conditions: 25-cm x 4.6-mm i.d. packed column, 10-μm silica particles; *n*-pentane; 0.288 g mL^{-1} (reprinted with permission from Ref. 141).

Plots showing this linear relationship are called van't Hoff plots (Figure 3.17).[140] Increasing the analysis temperature at constant density always results in a continuous decrease in capacity ratio. Again, chemically different solutes may show a reversal in retention order for increasing temperature.

At higher temperatures, the chromatographic system, including the mobile phase, the stationary phase, and the solutes, contains more thermal energy. Therefore, diffusion of the solute molecules is accelerated, resulting in higher separation efficiencies. The increase in plate height with higher linear velocities after the typical minimum in the van Deemter plot, is less steep at higher temperatures (Figure 2.4[147]).

In conclusion, the analysis temperature has the following effects in SFC:

(a) At constant pressure, the retention time of a solute, as well as selectivity and resolution, may either be increased or decreased by increasing the temperature. This is dependent on the combined

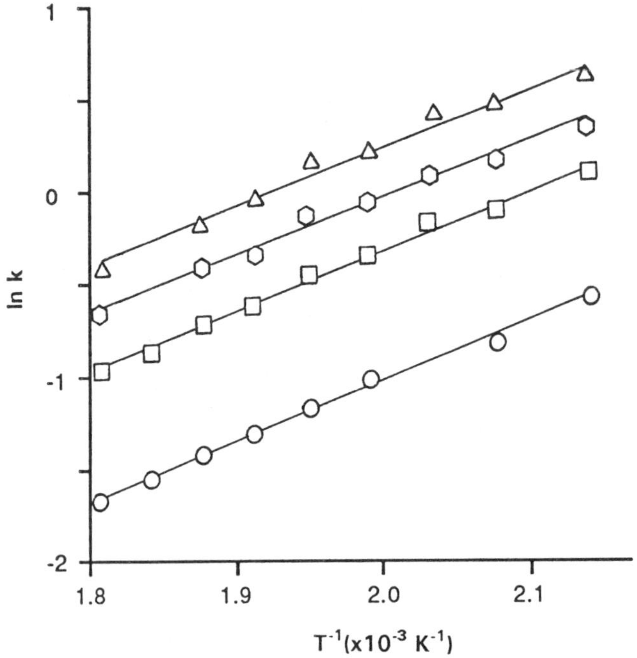

Figure 3.17. Logarithm of capacity ratio as a function of reciprocal temperature (van't Hoff plot). Conditions: 25-cm x 4.6-mm i.d. packed column, 10-μm silica particles, *n*-pentane; 0.288 g mL^{-1}. Analytes: (○) naphthalene, (□) anthracene, (⬡) pyrene, (△) chrysene (reprinted with permission from Ref. 140).

influences of solubility in the stationary and mobile phases and on vapor pressure. Therefore, the chromatographic performance may be tuned by adjusting the analysis temperature.

(b) At constant density, retention is always decreased at higher analysis temperatures. At the same time, separation efficiencies are increased.

(c) The retention order of two chemically different solutes may be changed by temperature; selectivity is not only a product of the mobile and the stationary phases, but also of temperature.

(d) A positive density program may not only be obtained by a positive pressure program, but also by a negative temperature program. Examples will be shown in Section 6.10.

3.8 Eluent Composition Programming

Eluent composition programming is very popular in LC. In fact, increasing the amount of a better solvating component in a mixed mobile phase is the only programming technique which is generally applied in LC, and has been termed "gradient elution." Eluent composition programming was introduced in SFC in 1981,[148,149] and the use of this technique in SFC has been recently reviewed.[150-152] While this technique has had limited practical use in SFC to date, an understanding of the principles and problems with its use can help in making it more applicable.

Principles. Due to their solvating nature, the chemical compositions of supercritical mobile phases have a strong influence on the retention of analytes, as is the case for the use of liquid mobile phases. When a composition program is applied, retention is influenced by changes in the composition of the mobile phase, mainly because of changes in solvent properties; stationary phase effects can often be neglected (*cf.*, Sections 3.3 and 3.4).

During a composition gradient, the chemical nature of the mobile phase is altered, *e.g.*, a mobile phase composed of CO_2 and methanol will show an increase of polarity with increasing methanol content. Thus, the increase in solvent strength for a given sample, which is attainable with a composition gradient, may well exceed the solvent strength which can be reached with a single-component mobile phase at high pressures/densities. This has been observed experimentally for the separation of 2-vinylnaphthalene oligomers which are readily eluted with a gradient of increasing content of 1,4-dioxane in CO_2, whereas elution was successful only for *ca.* decamers when a density gradient was applied with a neat CO_2 mobile phase.[153,154] For other combinations of sample and mobile phase, a comparison of pressure/density programming and composition gradients may yield different results.[155]

Procedures. Different procedures, depending on the instrumental design, can be used to effect composition programming. The overall pressure and/or mobile phase flow rate may vary or be held constant. This depends on changes of viscosity and density when a fixed restriction is used, or on their control with back-pressure type valves, or when the valve is part of a feed-back loop. Thus, composition gradients can be performed at: (a) constant overall flow and constant pressure; (b) constant overall flow and varying pressure; (c) varying overall flow and constant pressure; and (d) varying overall flow and varying pressure.

The variation in flow can result from either an increase in the flow rate of the secondary mobile phase component (modifier) at constant flow of the primary component, or a decrease in the flow rate of the primary component at a constant modifier flow.

For maintaining a constant overall flow rate, an increase in the modifier flow must be compensated for by a corresponding decrease in the flow rate of the primary component.

Pressure variations under constant flow conditions are mainly due to differences in the viscosities of the mobile phase components; in most cases, an increase in pressure will be observed as a result of the higher viscosity of the compounds used as modifiers. Increases in flow rate and, especially, pressure contribute to the reduction in analysis time obtained when using a composition gradient. However, they may adversely affect the separation efficiency; an increase in the flow rate leads to greater deviation from the minimum in the van Deemter curve, and a higher pressure yields higher density, and hence, smaller solute diffusion coefficients. Effects of the change of composition itself on the efficiency of the separation have not yet been elucidated.

Furthermore, the pressure drop over the length of the column and the linear velocity of the mobile phase in the column may vary with the changing composition of the mobile phase. An increase in the viscosity will cause an increase in the pressure drop (especially with packed columns), while the linear velocity becomes altered as a result of changes in density. The influence of these parameters on the efficiency of composition gradient separations has not yet been studied; experimental viscosities and densities for supercritical mobile phase mixtures are not yet available.

Comparison with LC. Compared to LC, the main advantage of composition programming in SFC is the higher diffusivity of the mobile phase. This results in faster column equilibration between analyses and when changing mobile phase components,[156] which makes the SFC technique more suitable for routine analysis. Additionally, the separation itself can be carried out faster and with better resolution.[157,158] As an example, Figure 3.18 shows chromatograms of an oligomer sample, obtained with the same mobile phase under different conditions. Chromatogram B was obtained with a liquid mobile phase, whereas Chromatogram A was begun at supercritical conditions. Even though a transition to the subcritical (liquid) state can be assumed for the second half of Chromatogram A (as estimated from the critical data of CO_2 and acetonitrile), the SFC chromatogram clearly demonstrates more advantageous conditions. Frequently, it has been observed that a transition to the subcritical liquid state does not have a significant detrimental effect on the separation, if the major part of the separation is carried out under supercritical conditions.

Choice of pressure and temperature. Generally, there are significant changes in the critical parameters of the mobile phase as the composition changes. The critical parameters of the mixture determine the values of pressure and temperature for starting the chromatogram. The initial values of temperature and pressure are also determined by the solvating properties of the mobile phase components for a given sample. This is demonstrated by Figures 3.19 and 3.20.

Figure 3.19 shows separations of styrene oligomers with supercritical propane/1,4-dioxane and hexane/1,4-dioxane. Both combinations are appropriate for this separation, but the conditions for achieving the same analysis time are quite different. Moreover, due to the higher solvating ability of hexane as compared to propane, the first part of the separation is compressed, although the initial dioxane content is much lower. A similar observation is made for the comparison between butane/1,4-dioxane and pentane/1,4-dioxane (Figure 3.20). A comparison between the propane-based and butane-based separations (Figures 3.19A and 3.20A) shows that with the same gradient, the analysis time is shorter for the butane-based separation, although the column pressure is lower. Finally, in

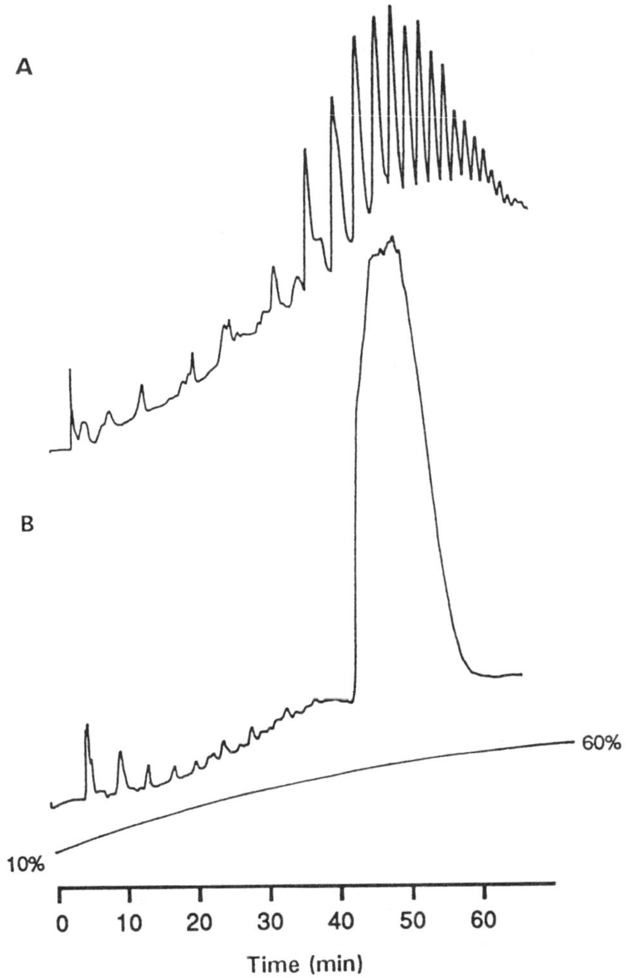

Figure 3.18. Separation of caprolactone diol oligomers at (A) SFC conditions and (B) LC conditions. Conditions: 25-cm x 4.6-mm i.d. packed column, 10-μm silica particles (LiChrosorb); acetonitrile modifier in CO_2 (v/v% gradient is shown in the figure); 135°C in (A) and 25°C in (B); 232 atm column exit pressure at start of chromatogram; 1 mL min^{-1} flow rate (liquid); UV (220 nm) (reprinted with permission from Ref. 158).

Figure 3.21, a separation is shown which was obtained with a CO_2/1,4-dioxane mobile phase. Comparison with, *e.g.*, the pentane-based separation in Figure 3.20B, the pressure had to be much higher, and more dioxane was necessary in order to yield the same analysis time; on the other hand, a much lower temperature could be selected for the CO_2-based mobile phase.

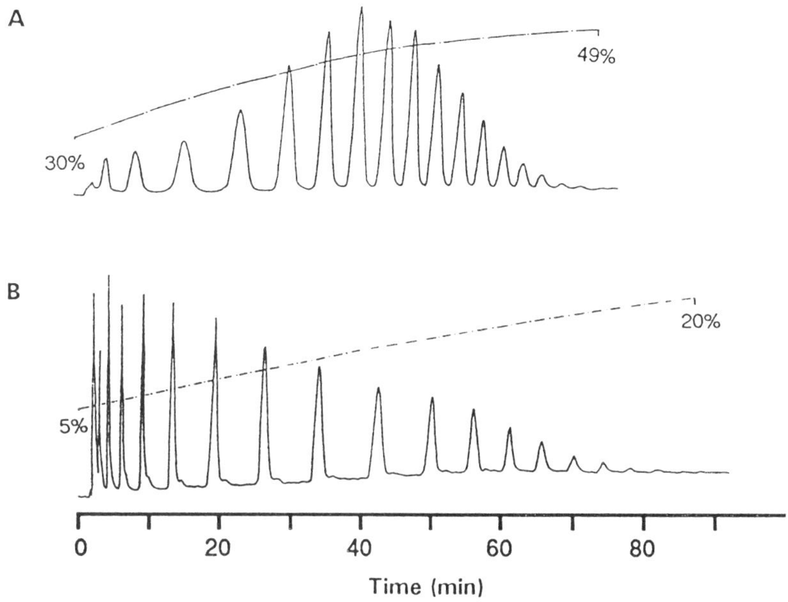

Figure 3.19. SFC chromatograms of styrene oligomers with (A) 30 to 49% 1,4-dioxane in propane and (B) 5 to 20% 1,4-dioxane in hexane. Conditions: Column and flow rate as in Figure 3.18; 240°C in (A) and 275°C in (B), 80 atm column exit pressure at start of chromatogram in (A) and 41 atm in (B), UV (278 nm) (reprinted with permission from Ref. 159).

Generally, the levels of pressure and temperature must be tuned to the mobile phase composition and to the selected gradient shape and slope in order to attain optimum performance. Apart from their dependence on the critical parameters of the mixture, the minimum temperature and pressure which can be applied depend also on the miscibility of the mobile phase components.

Detection and columns. When composition gradients are used, there are limitations with respect to detection, which in general also apply to isocratic work with mixed mobile phases. Flame-based detection, as well as on-line infrared detection, are not applicable. For UV detection, the detection range depends strongly on the nature of the mobile phase components. While some mixtures do not allow detection at low wavelengths, mixtures of CO_2 and acetonitrile allow detection even at 190 nm during composition gradient elution.[161] Other detectors can be expected to be less sensitive to

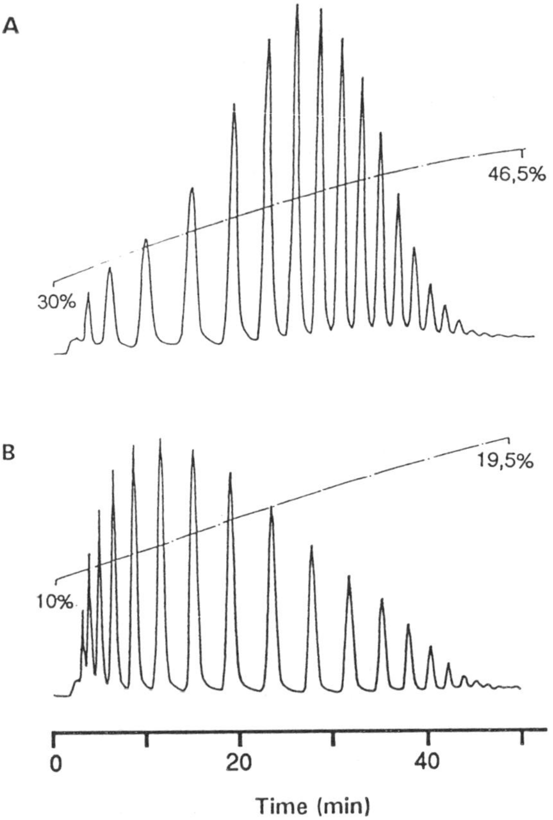

Figure 3.20. SFC chromatograms of styrene oligomers with (A) 30 to 46.5% 1,4-dioxane in butane and (B) 10 to 19.5% 1,4-dioxane in pentane. Conditions: column and flow rate as in Figure 3.18; 240°C; 69 atm column exit pressure at start of chromatogram in (A) and 50 atm in (B); UV (278 nm) (reprinted with permission from Ref. 159).

changes in mobile phase composition (*e.g.*, mass spectrometry and evaporative light scattering).

Packed columns have been employed for nearly all of the work done with composition gradients. Recent studies, however, have been reported which demonstrate that the method is also useful when using open tubular columns.[162]

Multiple gradients. Methanol doped with tetramethylammonium hydroxide (TMAOH) has been used as one component in composition programming,[81,82] yielding a composition gradient separation with a ternary mobile phase (composed of CO_2, methanol, and TMAOH in this case). However, separate pumping of more than

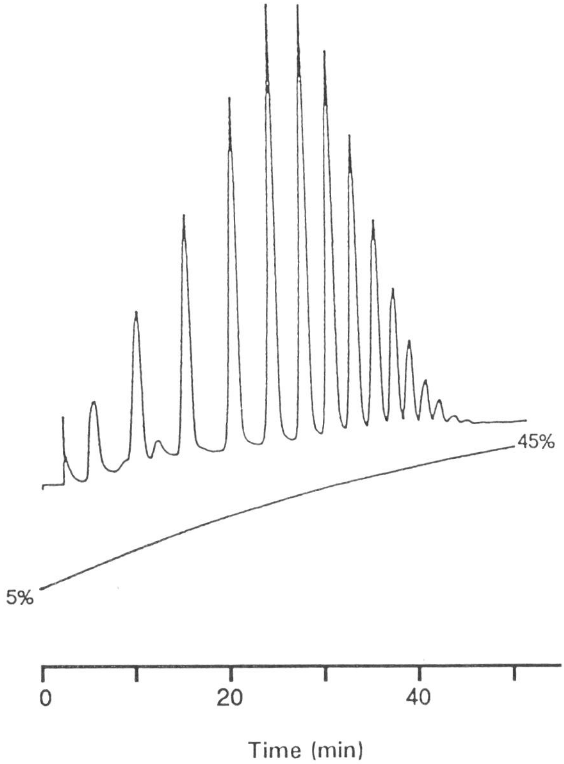

Figure 3.21. SFC chromatogram of styrene oligomers with 5 to 45% 1,4-dioxane in CO_2. Conditions: column and flow rate as in Figure 3.18; 145°C; 250 atm column exit pressure at start of chromatogram; UV (278 nm) (reprinted with permission from Ref. 160).

two components has not yet been reported with composition gradients in SFC; in principle, ternary, *etc.*, gradients are as applicable as in LC.

As has already been mentioned above, composition gradients may be coupled with other programming techniques. Thus, coupling has been done with pressure[102,161-163] or flow[181,182,164,165] gradients which enhances the effect of the composition gradient on elution. Alternatively, a negative pressure gradient (decreasing pressure) can be coupled to a composition gradient in order to improve the separation efficiency.[163] Such an improvement can also be achieved by coupling composition and temperature gradients (increasing temperature).[68,153,158] Finally, composition gradients may be included as part of a consecutive gradient sequence[166]; such consecutive procedures may be applicable for samples with components that differ very widely in their dissolution behavior.

References

1. U. van Wasen, I. Swaid, and G.M. Schneider, *Angew. Chem. Int. Ed. Engl.* **19**, 575 (1980).
2. H.H. Lauer, D. McManigill, and R.D. Board, *Anal. Chem.* **55**, 1370 (1983).
3. J.C. Giddings, L.M. Bowman, Jr., and M.N. Myers, *Anal. Chem.* **49**, 243 (1977).
4. C.R. Yonker, B.W. Wright, R.C. Petersen, and R.D. Smith, *J. Phys. Chem.* **89**, 5526 (1985).
5. U. van Wasen and G.M. Schneider, *Chromatographia* **8**, 274 (1975).
6. S.T. Sie, W. van Beersum, and G.W.A. Rijnders, *Sep. Sci.* **1**, 459 (1966).
7. D.E. Martire, *J. Liq. Chromatogr.* **10**, 1569 (1987).
8. K.S. Pitzer, D.Z. Lippmann, R.F. Curl, Jr., C.M. Huggins, and D.E. Petersen, *J. Am. Chem. Soc.* **77**, 3433 (1955).
9. K.S. Pitzer and G.O. Hultgren, *J. Am. Chem. Soc.* **80**, 4793 (1958).
10. A. Kreglewski and W.B. Kay, *J. Phys. Chem.* **73**, 3359 (1969).
11. B.P. Semonian and L.B. Rogers, *J. Chromatogr. Sci.* **16**, 49 (1978).
12. T.H. Gouw and R.E. Jentoft, *Adv. Chromatogr.* **13**, 1 (1975).
13. T.L. Chester and D.P. Innis, *J. High Resolut. Chromatogr. Chromatogr. Commun.* **8**, 561 (1985).
14. D.E. Martire, *J. Chromatogr.* **452**, 17 (1988).
15. D.E. Martire and R.E. Boehm, *J. Phys. Chem.* **91**, 2433 (1987).
16. R.C. Reid, J.M. Prausnitz, and B.E. Poling, *The Properties of Gases and Liquids*, 4th Ed. (McGraw-Hill, New York, 1987).
17. *Matheson Gas Data Book*, W. Braker and A.L. Mossman, Eds. (Matheson, Secausus, New Jersey, 1980).
18. *Gas Encyclopedia* (Elsevier, Amsterdam, 1976).
19. J.C. Fjeldsted, Ph.D. Dissertation, Brigham Young University, 1986.
20. F.P. Schmitz, *J. High Resolut. Chromatogr. Chromatogr. Commun.* **10**, 650 (1987).
21. P.G. Debenedetti, *Chem. Eng. Sci.* **42**, 2203 (1987).
22. W.J. Schmitt and R.C. Reid, *Process Technol. Proc.* **3**, 123 (1985).
23. J.C. Giddings, M.N. Myers, L. McLaren, and R.A. Keller, *Science* **162**, 67 (1968).
24. C.R. Yonker and R.D. Smith, *J. Phys. Chem.* **92**, 235 (1988).
25. C.R. Yonker and R.D. Smith, *J. Phys. Chem.* **93**, 1261 (1989).
26. C.R. Yonker, S.L. Frye, D.R. Kalkwarf, and R.D. Smith, *J. Phys. Chem.* **90**, 3022 (1986).

27. C.R. Yonker and R.D. Smith, in *Supercritical Fluid Extraction and Chromatography: Techniques and Applications, ACS Symposium Series 366*, B.A. Charpentier and M.R. Sevenants, Eds. (American Chemical Society, Washington, DC, 1988), p. 162.
28. E. Stahl, W. Schiltz, E. Schuetz, and E. Willing, *Angew. Chem. Int. Ed. Engl.* **17**, 731 (1978).
29. B.W. Wright, H.T. Kalinoski, and R.D. Smith, *Anal. Chem.* **57**, 2823 (1985).
30. E. Lundanes, B. Iversen, and T. Greibrokk, *J. Chromatogr.* **366**, 391 (1986).
31. E. Klesper, D. Leyendecker, and F.P. Schmitz, *J. Chromatogr.* **366**, 235 (1986).
32. C. Fujimoto, T. Watanabe, and K. Jinno, *J. Chromatogr. Sci.* **27**, 325 (1989).
33. J.C. Kuei, K.E. Markides, and M.L. Lee, *J. High Resolut. Chromatogr. Chromatogr. Commun.* **10**, 257 (1987).
34. K. Grolimund, W.P. Jackson, M. Joppich, W. Nussbaum, K. Anton, and H.M. Widmer, in *Proc. Seventh Int. Symp. Cap. Chromatogr.*, D. Ishii, K. Jinno, and P. Sandra, Eds. (Huethig, Heidelberg, 1986), p. 625.
35. O. Gyllenhaal and J. Vessman, *1989 Symposium/Workshop on Supercritical Fluid Chromatography*, Snowbird, UT, June 13-15, 1989.
36. S.M. Fields and K. Grolimund, *J. High Resolut. Chromatogr. Chromatogr. Commun.* **11**, 727 (1988).
37. P. Sandra and F. David, *J. High Resolut. Chromatogr. Chromatogr. Commun.* **11**, 897 (1988).
38. C.R. Yonker and R.D. Smith, *Fluid Phase Equil.* **22**, 175 (1985).
39. S. Olesik, Ohio State University, unpublished results.
40. S.M. Fields and K. Grolimund, *J. Chromatogr.* **472**, 197 (1989).
41. H.E. Schwartz and R.G. Brownlee, *J. Chromatogr.* **353**, 77 (1986).
42. M.W. Raynor, G.F. Shilstone, K.D. Bartle, A.A. Clifford, M. Cleary, and B.W. Cooks, *J. High Resolut. Chromatogr.* **12**, 300 (1989).
43. A.E. Bruno, E. Gassmann, N. Pericles, K. Anton, *Anal. Chem.* **61**, 876 (1989).
44. S.M. Fields, K.E. Markides, and M.L. Lee, *Anal. Chem.* **60**, 802 (1988).
45. J.M. Levy and W.M. Ritchey, *J. Chromatogr. Sci.* **24**, 242 (1986).
46. F.O. Geiser, S.G. Yocklovich, S.M. Lurcott, J.W. Guthrie, and E.J. Levy, *J. Chromatogr.* **459**, 173 (1988)
47. R.M. Campbell, N.M. Djordjevic, K.E. Markides, and M.L. Lee, *Anal. Chem.* **60**, 356 (1988).
48. P.A. David and M. Novotny, *J. Chromatogr.* **452**, 623 (1988).
49. H.T. Kalinoski and R.D. Smith, *Anal. Chem.* **60**, 529 (1988).
50. S. Hoffmann and T. Greibrokk, *J. Microcol. Sep.* **1**, 35 (1989).
51. D. Nizery, D. Thiebaut, M. Caude, R. Rosset, M. Lafosse, and M. Dreux, *J. Chromatogr.* **467**, 49 (1989).
52. C. Fujimoto, H. Yoshida, and K. Jinno, *J. Microcol. Sep.* **1**, 19 (1989).

53. P.H. Konynenberg and R.L. Scott, *Phil. Trans. Roy. Soc.* **298**, 495 (1980).
54. S.L. Frye, C.R. Yonker, D.R. Kalkwarf, and R.D. Smith, in *Supercritical Fluids, ACS Symposium Series 329*, T.G. Squires and M.E. Paulaitis, Eds. (American Chemical Society, Washington, DC, 1987), p. 29.
55. J.M. Levy and W.M. Ritchey, *J. High Resolut. Chromatogr. Chromatogr. Commun.* **10**, 493 (1987).
56. R.C. Reid, J.M. Prausnitz, and T.K. Sherwood, *The Properties of Gases and Fluids*, 3rd Ed. (McGraw-Hill, New York, 1977), p. 73.
57. K.L. Maguire and R.B. Denyszyn, in *Modern Supercritical Fluid Chromatography*, C.M. White, Ed. (Huethig, Heidelberg, 1988), p. 45.
58. C.R. Yonker and R.D. Smith, *J. Chromatogr.* **361**, 25 (1986).
59. P. Lamarche, S. McClintock, and M. Di Maso, *LC/GC* **6**, 154 (1988).
60. D.E. Raynie, S.M. Fields, N.M. Djordevic, K.E. Markides, and M.L. Lee, *J. High Resolut. Chromatogr.* **12**, 51 (1989).
61. D.R. Gere, R. Board, and D. McManigill, *Anal. Chem.* **54**, 736 (1984).
62. C. Borra, G. Mapelli, F. Andreolini, A. Trisciani, and S. Trestianu, in *Proc. Tenth Int. Symp. Cap. Chromatogr.*, P. Sandra and G. Redant, Eds. (Huethig, Heidelberg, 1989), p. 1205.
63. K. Anton, N. Pericles, S.M. Fields, and H.M. Widmer, *Chromatographia* **26**, 224 (1988).
64. A.L. Blilie and T. Greibrokk, *Anal. Chem.* **57**, 2239 (1985).
65. M. Ashraf-Khorassani, L.T. Taylor, and R.A. Henry, *Anal. Chem.* **60**, 1529 (1988).
66. J.M. Levy and J.P. Guzowski, *J. Chromatogr. Sci.* **26**, 194 (1988).
67. D. Leyendecker, D. Leyendecker, F.P. Schmitz, and E. Klesper, *J. Chromatogr.* **371**, 93 (1986).
68. D. Leyendecker, D. Leyendecker, F.P. Schmitz, and E. Klesper, *J. High Resolut. Chromatogr. Chromatogr. Commun.* **9**, 525, 1986.
69. D. Leyendecker, F.P. Schmitz, and E. Klesper, *Chromatographia* **23**, 171 (1987).
70. D. Leyendecker, D. Leyendecker, F.P. Schmitz, and E. Klesper, *J. Chromatogr.* **392**, 101 (1987).
71. D. Leyendecker, D. Leyendecker, F.P. Schmitz, B. Lorenschat, and E. Klesper, *J. Chromatogr.* **398**, 105 (1987).
72. D. Leyendecker, F.P. Schmitz, D. Leyendecker, and E. Klesper, *J. Chromatogr.* **393**, 155 (1987).
73. F.P. Schmitz and E. Klesper, *Polym. Commun.* **24**, 142 (1983).
74. J.M. Levy and W.M. Ritchey, *J. High Resolut. Chromatogr. Chromatogr. Commun.* **8**, 503 (1985).
75. M.E. McNally, J.R. Wheeler, and W.R. Melander, *LC/GC* **6**, 816 (1988).
76. B.W. Wright and R.D. Smith, *J. Chromatogr.* **355**, 367 (1986).
77. J.R. Strubinger and M.I. Selim, *J. Chromatogr. Sci.* **26**, 579 (1988).
78. C.R. Yonker and R.D. Smith, *Anal. Chem.* **61**, 1348 (1989).
79. S.M. Fields, K.E. Markides, and M.L. Lee, *J. High Resolut. Chromatogr. Chromatogr. Commun.* **11**, 25 (1988).

80. W. Steuer, M. Schindler, G. Schill, and F. Erni, *J. Chromatogr.* **477**, 287 (1988).
81. M. Ashraf-Khorassani, M.G. Fessahaie, L.T. Taylor, T.A. Berger, and J.F. Deye, *J. High Resolut. Chromatogr. Chromatogr. Commun.* **11**, 352 (1988).
82. T.A. Berger, J.F. Deye, M. Ashraf-Khorassani, and L.T. Taylor, *J. Chromatogr. Sci.* **27**, 105 (1989).
83. A. Giorgetti, N. Pericles, K. Anton, H.M. Widmer, and P. Dätwyler, *J. Chromatogr. Sci.* **27**, 318 (1989).
84. T.A. Berger, Hewlett-Packard, unpublished results.
85. M.A. Hernandez-Torres, J.S. Landy, and J.G. Dorsey, *Anal. Chem.* **58**, 744 (1986).
86. R.D. Smith, J.L. Fulton, H.K. Jones, R.W. Gale, and B.W. Wright, *J. Chromatogr. Sci.* **27**, 309 (1989).
87. J.P. Blitz, J.L. Fulton, and R.D. Smith, *J. Phys. Chem.* **92**, 2702 (1988).
88. J.L. Fulton and R.D. Smith, *J. Phys. Chem.* **92**, 2903 (1988).
89. S.V. Olesik and C. Miller, *Langmuir* **5**, 1 (1990).
90. J.L. Fulton, J.P. Blitz, J.M. Tingey, and R.D. Smith, *J. Phys. Chem.* **93**, 4198 (1989).
91. R.W. Gale, J.L. Fulton, and R.D. Smith, *Anal. Chem.* **59**, 1977 (1987).
92. L.J. Cline-Love, M. Skrilec, and J.G. Habarta, *Anal. Chem.* **52**, 754 (1980).
93. O. Redlich and J.N.S. Kwong, *Chem. Rev.* **44**, 233 (1949).
94. D.Y. Peng and D.B. Robinson, *Ind. Eng. Chem. Fundam.* **15**, 59 (1976).
95. D. Leyendecker, in *Supercritical Fluid Chromatography*, R.M. Smith, Ed. (The Royal Society of Chemistry, London, 1988), p. 53.
96. R.E. Jentoft and T.H. Gouw, *J. Chromatogr. Sci.* **8**, 138 (1970).
97. J.A. Graham and L.B. Rogers, *J. Chromatogr. Sci.* **18**, 75 (1980).
98. L.M. Bowman, Jr., M.N. Myers, and J.C. Giddings, *Sep. Sci. Technol.* **17**, 271 (1982).
99. E.W. Albaugh, D. Borst, and P. Talarico, *Org. Coat. Appl. Polym. Sci. Proc.* **48**, 622 (1983).
100. J.C. Fjeldsted, W.P. Jackson, P.A. Peaden, and M.L. Lee, *J. Chromatogr. Sci.* **21**, 222 (1983).
101. P. Mourier, M. Caude, and R. Rosset, *Analusis* **13**, 299 (1985).
102. A.L. Blilie and T. Greibrokk, *J. Chromatogr.* **349**, 317 (1985).
103. R.D. Smith, E.G. Chapman, and B.W. Wright, *Anal. Chem.* **57**, 2829 (1985).
104. A. Wilsch and G.M. Schneider, *J. Chromatogr.* **357**, 239 (1986).
105. K. Matsumoto, S. Tsuge, and Y. Hirata, *Chromatographia* **21**, 617 (1986).
106. K.H. Linnemann, A. Wilsch, and G.M. Schneider, *J. Chromatogr.* **369**, 39 (1986).
107. W. Ecknig, H.-J. Polster, and U. Lohse, *Fres. Z. Anal. Chem.* **327**, 39 (1987).

108. W. Gmür, J.O. Bosset, and E. Plattner, *Mitt. Geb. Lebensm. Hyg.* **78**, 21 (1987).
109. K.R. Jahn and B.W. Wenclawiak, *Anal. Chem.* **59**, 382 (1987).
110. A. Nomura, J. Yamada, and K.-I. Tsunoda, *J. Chromatogr.* **488**, 87 (1988).
111. G. Schomburg and W. Roeder, *Proc. Ninth Int. Symp. Cap. Chromatogr.*, P. Sandra and G. Redant, Eds. (Huethig, Heidelberg, 1988), p. 276.
112. A. Giorgetti, N. Pericles, H.M. Widmer, K. Anton, and P. Dätwyler, *J. Chromatogr. Sci.* **27**, 318 (1989).
113. B. Wenclawiak, *Fres. Z. Anal. Chem.* **323**, 492 (1986).
114. Y. Hirata, F. Nakata, and S. Murata, *Chromatographia* **23**, 663 (1987).
115. T. Takeuchi, K. Ohta, and D. Ishii, *Chromatographia* **25**, 125 (1988).
116. B.W. Wenclawiak, *Fres. Z. Anal. Chem.* **330**, 218 (1988).
117. B. Gemmel, F.P. Schmitz, and E. Klesper, *J. High Resolut. Chromatogr. Chromatogr. Commun.* **11**, 921 (1988).
118. J.H. Hildebrand and R.L. Scott, *The Solubility of Nonelectrolytes* (Reinhold, New York, 1950).
119. P.J. Schoenmakers, *J. Chromatogr.* **315**, 1 (1984).
120. R. Tijssen, H.A.H. Billet, and P.J. Schoenmakers, *J. Chromatogr.* **122**, 185 (1976).
121. E.B. Bayley, T.P. Nelson, S.-A. Chen, and J.W. Barlow, *Ind. Eng. Chem. Fundam.* **10**, 955 (1971).
122. E. Stahl, and W. Schilz, *Chem.-Ing. Tech.* **50**, 535 (1978).
123. C.R. Yonker, S.L. Frye, H.R. Udseth, B.W. Wright, and R.D. Smith, *Am. Chem. Soc., Div. Fuel Chem. Prepr.* **30**(3), 183 (1985).
124. K.G. Liphard and G.M. Schneider, *J. Chem. Thermo.* **7**, 805 (1975).
125. J.C. Giddings, M.N. Meyers, and J.W. King, *J. Chromatogr. Sci.* **7**, 276 (1969).
126. G.M. Schneider, *Angew. Chem.* **90**, 762 (1978).
127. P.J. Schoenmakers, in *Supercritical Fluid Chromatography*, R.M. Smith, Ed. (The Royal Society of Chemistry, London, 1988), p. 102.
128. D.W. Later, E.R. Campbell, and B.E. Richter, *J. High Resolut. Chromatogr. Chromatogr. Commun.* **11**, 65 (1988).
129. S.T. Sie and G.W.A. Rijnders, *Sep. Sci.* **2**, 729 (1967).
130. S.T. Sie and G.W.A. Rijnders, *Sep. Sci.* **2**, 755 (1967).
131. S.T. Sie and G.W.A. Rijnders, *Sep. Sci.* **2**, 699 (1967).
132. U. van Wasen, I. Swaid, and G.M. Schneider, *Angew. Chem. Int. Ed. Engl.* **19**, 575 (1980).
133. D.R. Gere, R. Board, and D. McManigill, *Publication No. 43-5953-1647* (Hewlett Packard Co., Avondale, 1982), p. 7.
134. D. Leyendecker, F.P. Schmitz, and E. Klesper, *J. Chromatogr.* **315**, 19 (1984).
135. F.P. Schmitz, H. Hilgers, D. Leyendecker, B. Lorenschat, U. Setzer, and E. Klesper, *J. High Resolut. Chromatogr. Chromatogr. Commun.* **7**, 590 (1984).
136. F. Bickmann and B. Wenclawiak, *Fres. Z. Anal. Chem.* **320**, 261 (1985).

137. D. Leyendecker, F.P. Schmitz, D. Leyendecker, and E. Klesper, *J. Chromatogr.* **321**, 273 (1985).
138. M. Novotny, W. Bertsch, and A. Zlatkis, *J. Chromatogr.* **61**, 17 (1971).
139. Y. Hirata, *J. Chromatogr.* **315**, 31 (1984).
140. D. Leyendecker, PhD Thesis, RWTH Aachen, 1986.
141. D. Leyendecker, D. Leyendecker, F.P. Schmitz, and E. Klesper, *J. High Resolut. Chromatogr. Chromatogr. Commun.* **9**, 566 (1986).
142. F.P. Schmitz, D. Leyendecker, and E. Klesper, *Ber. Bunsenges. Phys. Chem.* **88**, 912 (1984).
143. F.P. Schmitz, *Chromatographie mit überkritischen verdichteten Gasen zur Trennung von Oligomeren und Polymeren* (Profil Verlag, München, F.R.G., 1988).
144. K.D. Bartle, A.A. Clifford, J.P. Kithinji, and G.F. Shilstone, *J. Chem. Soc. Faraday Trans.* **84**, 4487 (1988).
145. D. Leyendecker, D. Leyendecker, F.P. Schmitz, and E. Klesper, *J. High Resolut. Chromatogr. Chromatogr. Commun.* **10**, 141 (1987).
146. C.R. Yonker and R.D. Smith, *J. Chromatogr.* **351**, 211 (1986).
147. S.M. Fields and M.L. Lee, *J. Chromatogr.* **349**, 305 (1985).
148. F.P. Schmitz and E. Klesper, *Polym. Bull.* **5**, 603 (1981).
149. F.P. Schmitz and E. Klesper, *Makromol. Chem., Rapid Commun.* **2**, 735 (1981).
150. F.P. Schmitz and E. Klesper, *J. Chromatogr.* **388**, 3 (1987).
151. E. Klesper and F.P. Schmitz, in *Modern Supercritical Fluid Chromatography*, C.M. White, Ed. (Huethig, Heidelberg, 1988), p. 1.
152. E. Klesper and F.P. Schmitz, *J. Supercrit. Fluids* **1**, 45 (1988).
153. F.P. Schmitz, D. Leyendecker, D. Leyendecker, and B. Gemmel, *J. Chromatogr.* **395**, 111 (1987).
154. F.P. Schmitz, B. Gemmel, D. Leyendecker, and D. Leyendecker, *J. High Resolut. Chromatogr. Chromatogr. Commun.* **11**, 339 (1988).
155. A. Giorgetti, N. Pericles, H.M. Widmer, K. Anton, and P. Dätwyler, *J. Chromatogr. Sci.* **27**, 318 (1989).
156. W. Steuer, M. Schindler, and F. Erni, *J. Chromatogr.* **454**, 253 (1988).
157. F.P. Schmitz, H. Hilgers, B. Lorenschat, and E. Klesper, *J. Chromatogr.* **346**, 69 (1985).
158. F.P. Schmitz and B. Gemmel, in *Progress in HPLC*, Vol. 4, M. Yoshioka, S. Parvez, T. Miyazaki, and H. Parvez, Eds. (VSP, Utrecht, 1989), p. 73.
159. F.P. Schmitz, H. Hilgers, and E. Klesper, *J. Chromatogr.* **267**, 267 (1983).
160. B. Gemmel and F.P. Schmitz, Univ. of Bochum, unpublished results.
161. B. Gemmel, B. Lorenschat, and F.P. Schmitz, *Chromatographia* **27**, 605 (1989).
162. C.R. Yonker and R.D. Smith, *Anal. Chem.* **59**, 727 (1987).
163. S. Küppers, B. Lorenschat, F.P. Schmitz, and E. Klesper, *J. Chromatogr.* **475**, 85 (1989).
164. A.J. Berry, D.E. Games, and J.R. Perkins, *J. Chromatogr.* **363**, 147 (1986).

165. R.C. Simpson, J.R. Gant, and P.R. Brown, *J. Chromatogr.* **371**, 109 (1986).
166. B. Gemmel, F.P. Schmitz, and E. Klesper, *J. Chromatogr.* **455**, 17 (1988).

CHAPTER 4

INSTRUMENTATION

4.1 Introduction

Instrumentation for SFC has similarities to both GC and LC instrumentation, as well as having some unique aspects itself. The most significant among these are the pressure-controlled pumping system which is essential for pressure or density programming, and the flow or pressure restrictor at the end of the column which is required to maintain supercritical pressures inside the column. Figure 4.1 shows a schematic diagram of an SFC instrument. The major components are (a) high pressure pumping system, (b) sample introduction system, (c) precision temperature controlled oven, (d) detector, and (e) microprocessor-based control system.

This chapter includes the requirements and descriptions of the various components that comprise the SFC system, except for the special case of supercritical fluid extraction. This topic is covered in detail in Chapter 5.

4.2 Pumping Systems

Progress in the development of high pressure LC pumping systems in the last 20 years has provided three distinctly different pump designs for contemporary SFC, *i.e.*, pneumatic amplifier, syringe, and reciprocating piston. The pneumatic amplifier pump was utilized by Giddings *et al.*[1-4] in the early 1970s to reach SFC fluid pressures up to 2,000 atm. Schmitz *et al.*,[5,6] Gere *et al.*,[7,8] Greibrokk *et al.*,[9] and Lauer *et al.*[10] modified commercially available reciprocating piston pumps for packed column SFC in the early 1980s. Novotny *et al.*[11] adapted a high pressure syringe pump for open tubular column SFC applications. Syringe pumps are now the

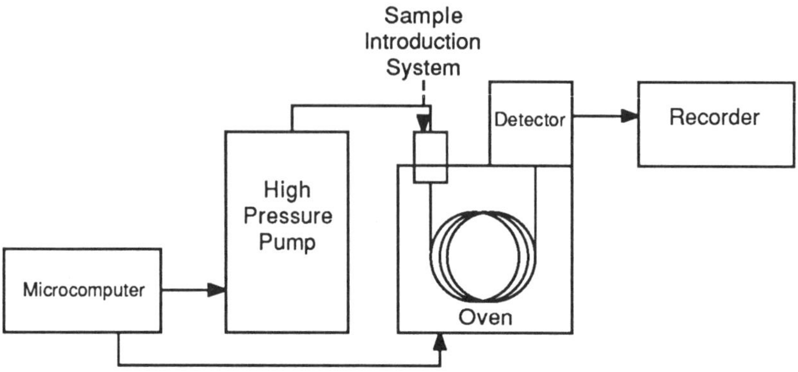

Figure 4.1. Schematic diagram of an SFC instrument.

most widely utilized SFC fluid delivery system, and they are available from several SFC instrument manufacturers. The popularity of the syringe pump design for SFC applications is mainly due to the ease of obtaining pulseless flow and pressure control. The recent interest in composition gradient elution, and in SFE continuous flow applications, has increased efforts in developing reciprocating piston pump designs for SFC. The following sections address the benefits, constraints, and applications of the above three pump designs.

Pneumatic amplifier pump. As shown in Figure 4.2, a pneumatic amplifier pump is composed of two pump cylinders that are different in piston cross-sectional area. The piston cross-sectional area ratio between the two cylinders equals the pressure amplification factor from the low pressure cylinder to the high pressure cylinder, and also equals the flow rate attenuation factor from the high flow rate cylinder to the low flow rate (high pressure) cylinder. In practice, an area ratio of 5 to 10 is recommended for reasons such as safety, reliability of ultra-high pressure seals and connectors, fluid compressibility, and high pressure cylinder volume.

Two methods for providing the low pressure source are (a) gas pressure which is regulated from a high pressure gas cylinder or air pump and (b) water pressure delivered by a high pressure LC pump. The advantages of using high pressure gas are fast response time to pressure change, low cost, and ease of operation. However, a high pressure gas amplifier pump is limited in its use to isobaric operation. Pressure programming operations essential to SFC applications cannot be implemented easily. Pneumatic amplifier pump operation is also constrained by its limited high pressure cylinder volume, and

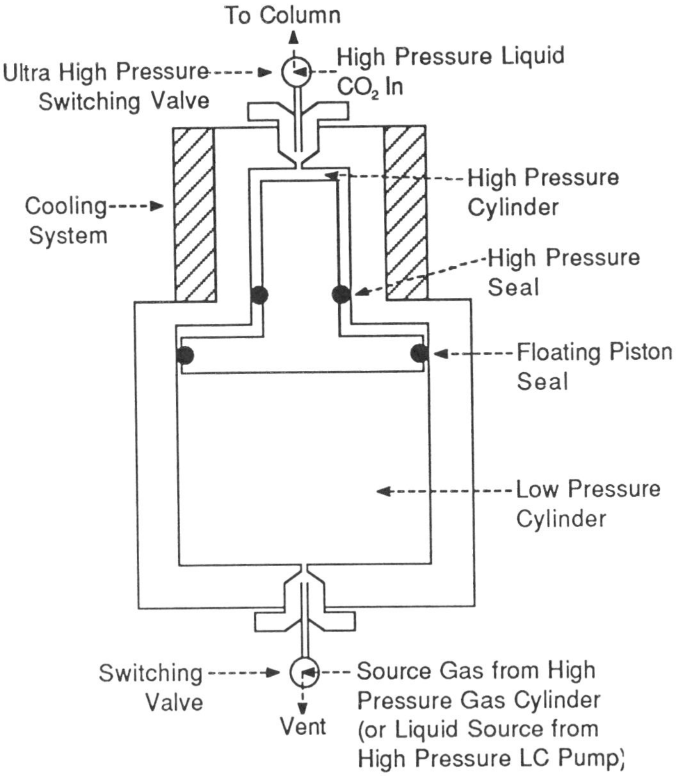

Figure 4.2. Schematic diagram of a pneumatic amplifier pump.

thus requires frequent refill. As a result, pump refill pulsation and noise often are problems in practice. The large pressure pulsation of a small volume pneumatic amplifier pump, however, can be very beneficial to extraction processes where fast pressurization and depressurization steps can enhance the efficiency of the extraction.

The use of a high pressure LC pump to deliver water to the source cylinder (low pressure side) allows either pressure or flow control and time programming of the SFC fluid in the amplifier's high pressure cylinder. A conventional LC reciprocating piston pump which provides a 400 atm pressure range and 1 μL min^{-1} to 10 mL min^{-1} flow range, with an amplification of 10 times, could produce an SFC fluid pressure as high as 4,000 atm and a flow rate range from 0.1 μL min^{-1} to 1 mL min^{-1}. Such a pressure and flow rate could match the requirements for open tubular column and microbore packed column SFC in isobaric or pressure (density) programming operations. The use of a reciprocating piston LC pump as a pressure

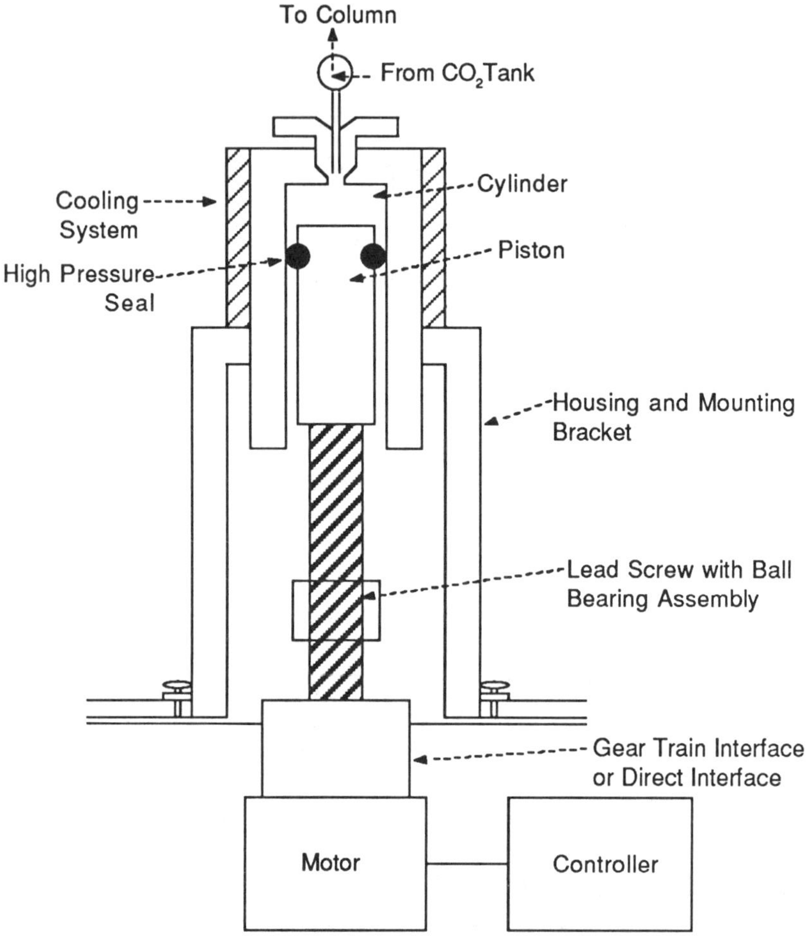

Figure 4.3. Schematic diagram of a syringe pump.

source for the amplifier, however, is severely limited by the speed of response to pressure change and the rate of pressure ramp-up and ramp-down.

The pneumatic amplifier pump is a good choice for ultra-high pressure SFC applications. High pressure seals and connector technology have been well developed for ensuring pump reliability. The major drawbacks of the pneumatic amplifier pump are its high cost in employing two amplifier pumps for composition gradient SFC elution and its poor gradient performance due to fluid compressibility effects as discussed in the following section.

Syringe pump. The syringe pump is widely employed today for both open tubular column and packed column SFC applications. Microprocessor control allows reproducible SFC fluid pressure or density programming. The syringe pump design offers several advantages for SFC applications, including pulseless flow delivery, micro-flow rate control, ease of pressure or density programming, and rapid pressure ramp-up and ramp-down operations. Figure 4.3 shows a schematic diagram of a typical syringe pumping system. The major variation in pump design is the plunger lead-screw interface to the pump driver motor. The lead-screw can be driven directly by a motor or by interfacing through a gear train assembly. The direct drive mechanism allows rapid pressure ramp-up and ramp-down and is more reliable with less wear on parts and components. However, it requires a more powerful motor (more expensive) and motor driver power supply. The gear-train driver mechanism requires a less expensive motor and power supply. However, there are more parts for service, and the cost can be substantially higher than a direct motor driver syringe pump.

The major drawbacks of a large volume syringe pump are (a) high cost, (b) long refill time, and (c) poor composition gradient performance. The first two can be minimized by improving the design. The last drawback, poor composition gradient performance, is difficult to eliminate due to the effect of fluid compressibility on gradient linearity and flow rate stability.[12-17]

For a compressible fluid, the compressibility factor can be calculated by

$$\chi = \frac{\ln\left(\frac{V}{V_o}\right)}{P - P_o} \tag{4.1}$$

where V and V_o are the volumes of the fluid in the pump cylinder at the initial pump pressure (P) and at the end pressure (P_o). The compressibility factors measured for CO_2 and acetonitrile at 25°C are 7.3×10^{-4} mL atm^{-1} and 4.5×10^{-5} mL atm^{-1}, respectively. For CO_2, the compressibility factor can be reduced significantly by reducing the pump cylinder temperature, *e.g.*, $\chi = 4.5 \times 10^{-4}$ mL atm^{-1} at 10°C. Because of the large compressibility of CO_2, it is necessary to compress CO_2 to a smaller volume than that for acetonitrile to reach the same pressure. As shown in Equation 4.2, the mixing of two fluids with different compressibility factors will affect the total gradient flow rate (F)

$$F = F_o - (\chi_A V_A + \chi_B V_B)\frac{dP}{dt} \tag{4.2}$$

where F_o is the total flow rate delivered by a gradient dual syringe pump, χ_A and χ_B are compressibility factors for solvent A (*e.g.*, CO_2) and solvent modifier B (*e.g.*, acetonitrile or methanol), V_A and V_B are the initial volumes of solvent A and solvent modifier B in the pump cylinder, and dP/dt is the pressure change with time due to the viscosity change as a function of solvent mixture composition during composition gradient programming.

As shown in Figure 4.4, the gradient flow rate obtained using a dual syringe pump is not constant in a gradient run. In addition, because F is also a function of V_A and V_B, the gradient flow rates are also dependent on the initial solvent volumes, V_A and V_B, in pump cylinders A and B.

An additional difficulty in using a dual syringe pump for gradient SFC elution is that, due to the large difference in CO_2 and organic solvent modifier (*e.g.*, acetonitrile or methanol) compressibilities, high pressure mixing of pump A and pump B solvents can cause the least compressible solvent (*e.g.*, acetonitrile, $\chi = 4.5 \times 10^{-5}$ mL atm^{-1}) to flow past the mixing chamber into the more compressible CO_2 (7.3×10^{-4} mL atm^{-1}) pump delivery tubing (*i.e.*, ringing effect). Hence, the composition reproducibility and accuracy of a solvent gradient may be affected. Problems such as cross-contamination of the syringe pump cylinders due to this ringing effect because of poor pump design must be taken into account.

Reciprocating piston pump. The reciprocating piston pump is a continuous flow pump that is the choice of pump design for both isocratic and gradient elution in LC. Today, reciprocating pumps probably account for over 85% of the pumps used in LC. The reasons for this overwhelming acceptance of the reciprocating piston pump for LC applications are the ease of operation, rapid solvent change, continuous flow operation, reproducible flow rate control, ease of service, and superior gradient performance.

In SFC, the utility of the reciprocating piston pump has been demonstrated by Schmitz *et al.*,[5,6] Gere *et al.*[7,8] and Greibrokk *et al.*;[9] however, it has not yet become popular. The emergent interest in SFE for sample preparation may promote the development of reciprocating pumps for both isocratic and gradient SFC and SFE applications.

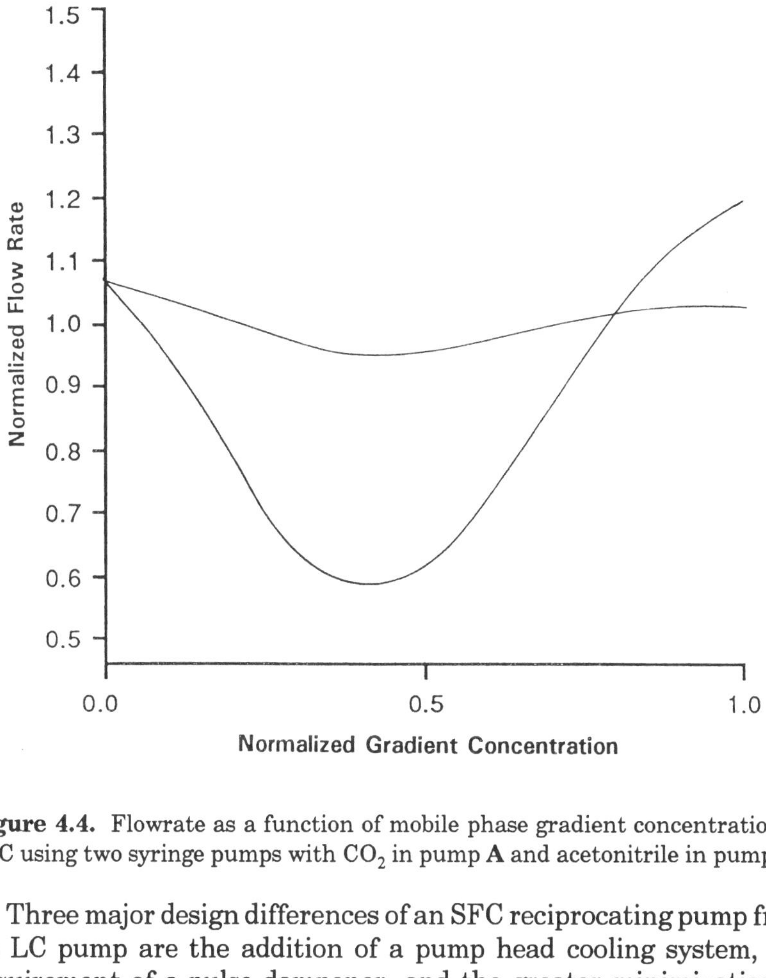

Figure 4.4. Flowrate as a function of mobile phase gradient concentration in SFC using two syringe pumps with CO_2 in pump **A** and acetonitrile in pump **B**.

Three major design differences of an SFC reciprocating pump from an LC pump are the addition of a pump head cooling system, the requirement of a pulse dampener, and the greater minimization of post-pump interface volume. As discussed earlier, the compressibility factor for CO_2 can be reduced by 60% by decreasing the CO_2 temperature from 25°C to 10°C. The addition of a coolant system at the pump head to reduce the temperature from 25°C to 10°C results in an improvement in pumping efficiency.

A reciprocating piston pump containing single, dual, or triple pump heads may be modified for SFC applications. A single pump head system with a pulse dampener is suitable for isocratic open tubular column and microbore packed column SFC applications where low flow rates and small pressure pulsations can easily be dampened with a small damper and a column.

For composition gradient elution, a dual or triple piston pump with a microvolume pulse dampener is the system of choice. In the design of such a pump for open tubular column or microbore packed column gradient elution, it is important to optimize the pump stroke volume and the post pump volume. With the use of a low pressure multi-port proportional valve, the pump stroke volume determines the minimum obtainable gradient step, and thus affects the gradient slope and accuracy. The minimum stroke volume should be large enough to compensate for the effect of fluid compressibility, and the leakage of check valves inside the pump head. With a 5°C pump head temperature, a pump stroke volume of 20-30 μL may be used.

For open tubular column and microbore packed column SFC where the mobile phase flow rate is on the order of 10 to 50 μL min^{-1}, it is important to reduce the interface volume between the pump head and the column inlet. The ratio of this post pump volume to the flow rate determines the gradient delay time to the column inlet, and thus the effectiveness and throughput of the composition gradient. A large post-pump volume can also affect the gradient linearity due to the effect of mixing of the proportioned solvents in the post-pump volume. A gradient profile of nonlinear curvature at both low and high solvent B concentration can be observed.

4.3 Oven Considerations

The chromatographic oven for SFC should meet the same requirements as conventional GC ovens. The oven should be designed so that thermal gradients between any two points in the area where the column is placed are less than ±0.1°C. This is especially crucial in open tubular column SFC, because open tubular columns have small diameters and thin walls which easily transfer any temperature variations inside the column, and any slight changes in temperature can have a pronounced effect on the density of the mobile phase, especially near the critical temperature. Localized cold or hot spots can lead to peak broadening, peak tailing, irreproducibility in retention, and peak splitting. The chromatographic oven should be capable of both positive and negative temperature programming for selected applications.

4.4 Fundamentals of Sample Introduction

For a chromatographic technique to be viable as an analytical method, the sample must be delivered to the column inlet reproducibly and with a minimum amount of discrimination. The sample solvent composition as well as the injection pressure and temperature can all affect sample introduction. Higher solute diffusion and lower viscosity, which favor supercritical fluid over liquid mobile phases, can cause problems in injection. Back diffusion can take place, causing broad solvent peaks and poor solute peak shape. There can be some very complex phase behavior as well as solubility phenomena taking place due to the fact that one may have combinations of a supercritical fluid (neat or mixed with sample solvent), a subcritical liquefied gas, sample solvent(s), and solutes present simultaneously in the injector and column head.

The physical position of the column relative to the valve, as well as the type of injection used, can have a positive or negative impact on injection. The amount of sample and solvent introduced into the column can also cause undesirable effects, such as overloading or peak splitting. Any dead volume in the connections can cause poor transfer of the solutes into the column. The connections are not as problematic for packed columns as for capillary columns, but the fittings and connections for packed columns still cannot be ignored.

Injection volume. Theoretical considerations show that there are stringent requirements for the allowable injection volume in open tubular column SFC. It was shown in Chapter 2 that volumes less than 96 nL should be injected in order to prevent more than a 1% loss in resolution for a 10-m x 50-μm i.d. column. These values were obtained by assuming no solute focusing or solvent elimination, and they were calculated for k values of 2. It is obvious from these numbers that special precautions must be taken to minimize band broadening when injecting into an open tubular column.

For packed columns, the theoretically predicted volumes are, of course, larger than for open tubular columns. For a 25-cm x 1-mm i.d. packed column, the volume must be 1 μL or less to have no more than 5% increase in band broadening.[18] Packed columns present fewer injection challenges than capillary columns because of the larger allowable injection volume, but attention must still be paid to dead volumes in connecting tubing and fittings.

Injection plug profile. Except for the case of supercritical fluid extraction, all sample introduction methods make use of a high pressure valve with an internal sample loop. In addition to the loop volume, its diameter and shape can have an effect on the injection peak profile. Any sharp diameter changes or corners can harbor solvent or solutes for extended time periods during injection. As already mentioned, back-diffusion upstream from the valve can occur if there is no restriction in the flow stream from the pump. Furthermore, as the liquid or supercritical mobile phase enters the sample loop, the center of the flow stream moves faster than the solvent near the walls, creating a nonideal plug profile. All of the above conditions can lead to tailing solvent and analyte peaks, irreversible adsorption, and incomplete transfer of the sample from the sample loop into the analytical column.

Solute focusing. Solute focusing can be implemented regardless of the injection technique or column configuration used. This is of value when one desires to inject larger sample sizes than the theoretically allowable volumes. Lee *et al.*[19] have discussed the role of solute focusing in SFC. In that work, an equation was given that relates the velocity of the solute band through the column as a function of the partition coefficient and the phase ratio

$$v = u\left[\frac{V_M c_M}{(V_S c_S + V_M c_M)}\right] = u\left[\frac{1}{(1 + K/\beta)}\right] \quad (4.3)$$

where v is the velocity of the solute band, u is the linear velocity of the mobile phase, c_S and c_M are the concentrations of the solute in the stationary and mobile phases, respectively, V_S and V_M are the volumes of the stationary and mobile phases, respectively, K is the partition coefficient (c_S/c_M), and β is the phase ratio (V_M/V_S).

This equation predicts that there are two ways to focus a solute in a column: phase ratio focusing and partition coefficient focusing. By changing the phase ratio or the partition coefficient, the velocity of the solute band relative to the linear velocity of the mobile phase (and the solvent) can be changed. In practice, phase ratio focusing occurs when an uncoated section of column is connected to the inlet portion of a coated column. The use of an uncoated section of tubing ("retention gap") between the injection valve and the column has been reported to assist in solute focusing.[20,21] Also, this precolumn may be coated with a thinner stationary phase film than the column,[22] which also will focus the analytes due to the change in phase ratio. Alternatively, focusing can occur by using a column which has a continually increasing stationary phase film thickness.

Partition coefficient focusing results when the partition coefficient is changed. This is accomplished in SFC by using an increasing temperature gradient (or reducing the density in some other manner) at the inlet of the column. As the mobile phase carries the solutes into a region of higher temperature, the density of the supercritical fluid decreases, and the solutes become less soluble. This decrease in solubility causes the solutes to focus in the hotter region. Various groups have reported the use of temperature to vary the mobile phase density for solute focusing.[19,23,24] Unfortunately, the higher mobile phase linear velocity in these hotter segments and the increase in vapor pressures of the more volative analytes can cause band spreading, hence offsetting the opposite focusing effect.

Another approach to solute focusing was recently reported by Hirata and coworkers.[25,26] In these reports, mixing chambers were interposed between the injector valve and the head of the column. The rationale behind the use of mixing chambers stems from the fact that when injecting large volumes of organic solvents into supercritical CO_2, there will be a momentary increase in the solvent strength of the fluid. This would be analogous to injecting a solvent of greater strength than the mobile phase in LC. This causes the solutes to travel down the column with the solvent until it is sufficiently diluted that the solutes precipitate or partition into the stationary phase. With the mixing chambers, the CO_2 mixes quickly with the solvent and dilutes its solvating power so that the solutes are focused at or near the column inlet.

Solvent compression. An approach which also allows the injection of sample sizes larger than the theoretically allowable volume is solvent compression. In this technique, injection is made at a low pressure followed by a rapid increase in pressure to compress the width of the solvent peak.[20,21,23,24] There are several potential problems with this technique, the main one being that high molecular mass or polar materials may precipitate in the injector at the low density conditions during injection and lead to poor quantitative recovery or tailing peaks.

Solvent elimination. Solvent compression can help in reducing the width of the solvent peak when injecting large sample sizes, but there is a limit to how useful it may be when injecting 200-nL or larger volumes. If the solvent peak is broad, it can obscure or interfere with the determination of early eluting peaks. The elimination of the solvent before the chromatographic column is desirable and often necessary when using larger volumes of solvents. This topic is discussed in detail in Section 4.6.

4.5 Direct and Split Injection Methods

As SFC was developing in the early years, small volume, high pressure injection valves were not commercially available. As a result, the injection systems were built in-house. These injectors were most often modified, high pressure GC type injectors with septa,[27,28] "home-built" devices made from available high pressure valves and fittings,[29,30] or high pressure chambers into which the sample was loaded, and from which the "headspace" was sampled to perform the chromatography.[1] These injection techniques were often adequate for the initial studies of SFC, but they had limitations, such as low pressure ranges, difficulty of use, poor reproducibility of injection volume, carry-over from sample to sample, and short life time. The first reports of open tubular column SFC were based on work done with commercially available injection valves, such as those used in microbore LC.[11,31] The techniques and procedures reviewed in the following sections describe the use of commercially available injection valves, and improvements in injection systems based on these valves.

Dynamic split. The most widely used injection method in SFC, especially with open tubular columns, is dynamic split using a microvolume valve and a splitter assembly. This is basically a refinement of the concentric tube design first described by Ettre and Averill[32] for GC. In this injection technique, the capillary column is held in place inside a stainless steel tube, which is connected directly to the injection valve. Downstream of the injection valve, the stainless steel tube is connected to a tee piece to which is attached some restriction device. As the sample moves into the flow stream when the injection valve is actuated, part of the sample is transferred to the column and part goes around the outside of the column and exits out the split restrictor. The split ratio of the fluid (and not necessarily the analytes) is determined by the ratio of the flow out of the column to the flow out of the split restrictor. In other words, if the flow out of the split is 100 times faster than the flow out of the column, the split ratio is 1:100.

Unfortunately, split injection is nonlinear in SFC as it is in GC. This means that the relationship between the split ratio and chromatographic peak area is nonlinear. The split ratio also changes with the viscosity of the sample solvent and the density of the mobile phase. This can be bothersome in SFC because samples can be injected at various pressures and in different solvents which will change the quantitative reliability of the data.

It has been reported by Köhler and Schomburg[33] that split injection works better at high split ratios. They also observed that the absolute area counts are strongly influenced by the column flow rate at constant split ratio. Both of these effects can be explained by the fact that nonhomogeneous distributions of the sample and solvent exist when operating under low flow rate conditions in the splitter (low split ratios or low column flows). When the flow rates are high, better homogeneity exists, and the performance improves. However, these high split ratio conditions are the cause of one of the major drawbacks of split injection in SFC. With high split ratios, the major part of the sample goes to waste, thus making detection limits high, and the technique not suitable for trace analysis.

Work has been reported concerning the reproducibility achievable with dynamic split injection in SFC.[21,33-35] It has been shown that several physical parameters can have pronounced effects on the reproducibility obtained. Factors such as the geometry of the splitter,[33] the valve design,[21] the actuator used,[21] and the position of the column in the splitter[21] all have effects on precision. If optimized conditions are used, one can obtain very acceptable reproducibility with dynamic split injection. The relative standard deviation (RSD) generally falls in the range of 1.5 to 4% for raw peak areas. Schomburg *et al.*[34] observed that there was no difference in precision and accuracy between the GC and the SFC measurements when the same samples of triglycerides were analyzed by GC and SFC. However, because of the small sample sizes required when using open tubular columns in SFC, samples must be 3-4 times more concentrated than is necessary in GC to achieve the same sensitivity.

Another disadvantage of dynamic split injection is that, in practice, it is easier to obtain higher split ratios than lower ones. This is because the restrictor most often used is a piece of fused silica capillary tubing with an i.d. in the range of 5 to 25 μm. Split ratios can be changed by changing the length or diameter of the restrictor, or by using back pressure control. Bohm *et al.*[35] recently reported the use of a variable restrictor for split injection which was based on a stepper motor driven metering valve. The split ratio can easily be changed with this setup, however, it has very limited applicability at this time because the valve cannot be used at pressures above 200 atm.

Delayed split. Delayed split[19] is a modification of the basic split injection technique which allows larger sample sizes to be injected into a column while still maintaining a narrow solvent peak. In fact, the solvent peak can be entirely eliminated. The geometry of the splitter and valve assembly is the same for dynamic split, except that an on/off valve is added to the split line. For sample injection, this valve is closed, and the injection valve is switched to the inject position with the system set at a preselected pressure. After a period of time (usually 1-15 s), the injection valve is returned to the load position and the split vent is opened. This allows a portion of the solvent to be purged out of the system. If the pressure of the system is quickly reduced after the split vent is opened, and if longer delay times are used (the time before the split valve is opened), all of the solvent is purged out of the vent. (This is called solvent backflush, and is described in more detail in Section 4.6.) By increasing the injection valve actuation time and the split valve delay time, more sample can be injected. Furthermore, since the solvent peak is reduced or eliminated, analysis times are shorter and any interference of the solvent peak with early eluting peaks is minimized.

Timed-split. The chromatographic injection technique called timed-split injection can also be called the "moving injection technique," and was first described for LC in 1964 by J.C. Moore.[36] This work was done with gel permeation chromatography using manually actuated valves. Moore found that the material introduced onto the column was more representative of the actual composition of the sample than could be obtained using the dynamic split technique. More recently, Harvey and Stearns have refined the technique for use in microbore LC, first with electric actuation[37] and then with pneumatic actuation.[38] The first report of the use of this injection technique in SFC was by Richter.[39]

In timed-split injection, the column (packed or capillary) is connected directly to the injection valve. Generally, no splitter assembly is used, although Schomburg and Roeder have reported the use of the timed-split mode of operation in conjunction with a splitter.[24] High speed pneumatics and electronics are used along with a standard injection valve and actuator which allow rapid actuation of the valve from the load to the inject position and back again in millisecond time frames. Because of the fast valve switching requirements, a low-viscosity gas such as helium is preferred to air for powering the pneumatic actuator. This rapid movement of the valve permits only a portion of the sample loop to be swept out and into the column by the mobile phase. This is considered a split type of injection because not all of the sample loop contents are transferred into the column. Timing, and not flow, is used to provide the splitting, leading to the term "timed-split."

In addition to less discrimination, timed-split injection has other advantages over dynamic split injection. It is easy to change the amount of material injected into the column simply by changing the dwell time or actuation time of the timing circuit. This can be either a manual adjustment or a software command, depending on the instrument configuration. The relationship between actuation time and peak area is linear within a certain range,[34] generally from 25 ms up to about 3 s. The upper time limit corresponds to the point where approximately 80% of the sample loop has been expelled into the column. Beyond this period, the mobile phase mixes with the sample loop contents to cause significant dilution of the sample. Therefore, it takes relatively longer to transfer the remaining 20% of the sample loop contents into the column. The actual time at which this occurs depends on the initial starting pressure, the viscosity of the sample solvent, the solutes, and the injector temperature. This can easily be determined experimentally.

Reproducibility studies have been conducted using timed-split injection in comparison to dynamic split injection.[21,33,34] Schomburg and coworkers saw no significant difference between the precision obtained with dynamic split or timed-split injection.[33,34] They reported 2-4% RSD for absolute peak areas, and less than 2% for relative peak areas. However, Richter *et al.*[21] showed timed-split to be superior to dynamic split injection, with values of <2% for absolute peak areas, <1% for relative peak areas, and <0.6% for an internal standard method. They also reported that the performance in timed-split was a function of valve design, as it is in dynamic split injection.

The use of timed-split injection has been examined with packed columns by Porter *et al.*[40] They found that the performance of timed-split injection with packed columns was strongly dependent on the injection valve used. One valve gave very good precision (2% RSD for absolute peak areas and 0.3% RSD for internal standard), while a second valve with different internal geometry gave poorer precision (7% RSD for absolute peak areas and 1% for internal standard). This is compared to 1% for absolute peak areas and 0.3%

for internal standard using the same packed column, but by injecting the entire sample loop contents (*i.e.*, direct injection).

Direct injection. Of course, the most desirable method of injection would be to inject the sample loop contents directly onto the column. Problems with discrimination and precision should be minimized because no splitting is taking place. Reports have shown that, for packed columns, excellent reproducibility can be achieved by direct injection.[40,41] Percent RSD values of <1% for absolute peak areas and <0.3% using an internal standard are easily achieved. In practice, this is the preferred injection method for packed columns, and because of the higher capacity of these columns, it is straightforward to make injections of 0.2-1 μL without any special precautions. Because of the small allowable injection volumes when using open tubular columns, it is much more challenging to inject 0.2-1 μL without obtaining poor chromatographic performance, such as column overloading or split peaks from excess solvent.

Richter and coworkers[21] used a standard valve configuration for timed-split injection in conjunction with a "retention gap," and successfully injected 0.5- and 1.0-μL volumes into 50-μm i.d. columns. The reproducibility was not as good as would be expected for those volumes, but the results showed that it is possible to inject volumes much larger than those predicted by theory. (The %RSD values for 0.5 μL were 6% for absolute areas and 3% for internal standard; for 1.0 μL, they were 4% for absolute areas and 0.9% for internal standard.) Since that initial report, subsequent work has shown that values of 1-3% RSD (raw peak areas) for 0.5-μL injections can be routinely achieved.[42]

As previously mentioned, Hirata *et al.*[25,26] used mixing chambers between the injection valve and column inlet to dilute the sample solvent and thus focus the solutes when injecting large sample volumes directly into open tubular columns. This work reported injections of up to 1 μL without split peaks. Relative standard deviations of less than 3.7% for absolute peak heights were shown. Minimum detectable quantities (MDQ) of less than 0.2 ng for triglycerides (S/N = 2) were reported. However, the amount that can be injected without peak splitting was a function of sample solvent. The more polar the solvent, the less that could be injected without peak splitting. This is due to the differences in polarity between the CO_2 and the sample solvents.

Solvent compression has been used in conjunction with direct injection.[19] For this work, a "retention gap" was cooled to 20°C to help focus the sample. Also, the sample was injected at high pressure, and upon injection, the pressure was quickly lowered to the desired initial operating pressure. The solvent peak was narrow, but the early eluting peaks indicated that they were not adequately focused. The chromatograms did indicate that compounds of higher molecular mass could be successfully loaded into the column and eluted by direct injection than could be achieved with split injection. In other words, less discrimination occurs in direct injection than with split injection.

Hawthorne and Miller[43] reported the development of an "almost" on-column injector for capillary SFC which was done by modifying a four-port switching valve to accommodate a syringe used for on-column injection in GC. The sample (0.5 μL) was loaded into a retention gap (glass-lined stainless steel tubing) with the column completely depressurized. The syringe needle was removed, the valve was switched, and the system was pressurized. After 15 s, the solvent was vented for 5 s, the valve was returned to the normal chromatographic position, and the separation was performed. This approach showed good promise, but early eluting peaks ($<C_{20}$) were broad, and response factors indicated the loss of these species during the solvent venting step. Reproducibility values were disappointing (7-11% RSD for absolute area counts), but this is somewhat understandable since it is a completely manual operation, and there is error in accurately measuring 0.5 μL in the GC syringe. Improved solute focusing and more automated operation would undoubtedly enhance the performance.

Another means of introducing a sample directly into the column is by dissolving the sample in the mobile phase itself. This is not as easy as in LC because of the temperatures and pressures needed. Early reports of supercritical fluid injection or on-line sampling of supercritical fluid extraction,[44-47] demonstrated the potential advantages of this procedure. In the cases cited, the sample was introduced into the chromatographic system basically by one of the following methods. In the first case, the sample was loaded into a chamber which was connected to an injection valve sample loop. This chamber was then pressurized and heated appropriately. The sample loop, containing material solubilized by the supercritical fluid, was then switched into the flow of the mobile phase. Solutes were carried to the column where the chromatographic separation then occurred. In the second case, the sample was loaded onto a wire or capillary tube which was then put into the extraction chamber.[47] The solutes were carried directly from this chamber onto the column as they became soluble during the pressure ramp.

4.6 Sample Introduction with Solvent Elimination

In trace analysis, there is a need to be able to inject relatively large volumes of dilute sample solutions. The introduction of large solvent volumes, however, produces several distinct disadvantages: the elution of the solvent peak is time consuming, the solvent may cause band broadening and peak splitting, and the properties of the stationary phase may undergo changes due to absorption/adsorption of solvents. With 10-15 m open tubular columns, the solvent often takes more than 10 min to elute, which contributes significantly to the analysis time.

Solvent effects on peak shape. Although there are reports of sample solvents acting as weaker solvents compared to the supercritical fluid,[48] the usual effect, particularly when using supercritical CO_2, is that the solvent appears to have a higher solvent strength than the fluid. The band broadening effect of sample solvents, which is well known in LC, has also been demonstrated to exist in SFC[49] (see Table 4.1). The band broadening is directly related to the relative solubilities of the solutes in the solvent, and the solvent in the fluid. In Table 4.1, the strongest

Table 4.1. Column Efficiency (Theoretical Plates) and Peak Shape (Asymmetry Factor) on a Packed Microbore C_{18} Column (1.3-mm i.d.) as a Function of Solvent and Injection Volume in SFC. Pentadecylbenzene (k = 3.8) was Injected in CO_2 at 40°C and 170 atm[a]

	0.1-μL Inj. Volume		1-μL Inj. Volume	
Solvent	$N(10^3)$	A_s	$N(10^3)$	A_s
Acetonitrile	5.9	1.3	6.3	1.2
Methanol	-	-	5.9	1.2
Methyl *t*-butyl ether	5.9	1.3	5.4	1.1
n-Octane	6.1	1.3	4.9	1.2
Dichloromethane	6.3	1.3	4.6	1.1
Tetrahydrofuran	6.0	1.3	3.7	1.1
Carbon disulfide	6.0	1.3	3.0	1.0

[a]Reprinted with permission from Ref. 49.

effect was noticed with carbon disulfide, which is a very good solvent for hydrocarbon solutes; methanol had little effect. With more polar solutes, such as poly(ethylene glycols), greater band broadening was found with methanol. Split peaks are another effect of the injection of excessive volumes.[21,22,25]

Although peak splitting and peak shoulders may be caused by various phenomena in the injector or at the column inlet, as is known from GC and LC, an important mechanism in SFC seems to be incomplete mixing of the injection solvent with the fluid. Part of the sample is eluted with the solvent plug until complete mixing is obtained on the column, while the rest is slightly retained by partitioning in the first part of the column, as suggested by Berg and Greibrokk,[22] and by Hirata *et al.*[25] This is a process that has been shown to take place on the column/precolumn, since the splitting disappeared by raising the column temperature, apparently due to the increased diffusion at higher temperature.[22] The peak shape is also a function of the starting pressure. At lower pressure, the increased diffusion rate and the increased diffusion time result in better mixing, and peak splitting disappears. By including mixing chambers between the injection valve and the column, Hirata and Inomata[26] showed that the peak splitting, which otherwise resulted from 0.2-1 μL injections on 100-μm i.d. open tubular columns, could be removed. Improved mixing is probably also the reason why an uncoated retention gap was reported to result in less discrimination.[19]

Solvent venting with a precolumn. In order to avoid the problems and limitations caused by the solvents, methods for solvent elimination were first proposed by Greibrokk *et al.*[23] and later demonstrated with a solvent venting technique.[50] With this technique, the solvent is separated from the solutes on the precolumn by a partitioning mechanism, and the solutes are transferred to the main column and focused at the column inlet. The major part of the solvent is vented through a restrictor at a rapid (10-20 cm s^{-1}) but controlled rate. With a 2-m precolumn, the venting time is only 15-20 s. Too lengthy venting times lead to discrimination of the early eluting peaks and must be avoided, and too high starting pressures degrade the peak shape. Figure 4.5 shows a schematic diagram of a solvent venting injection system. Peak focusing at the column inlet is based on phase ratios (by using a thicker film on the column than on the precolumn), on partition coefficients (by using a more retaining stationary phase on the column), or on both. By choosing conditions whereby equilibrium was attained on the precolumn, the

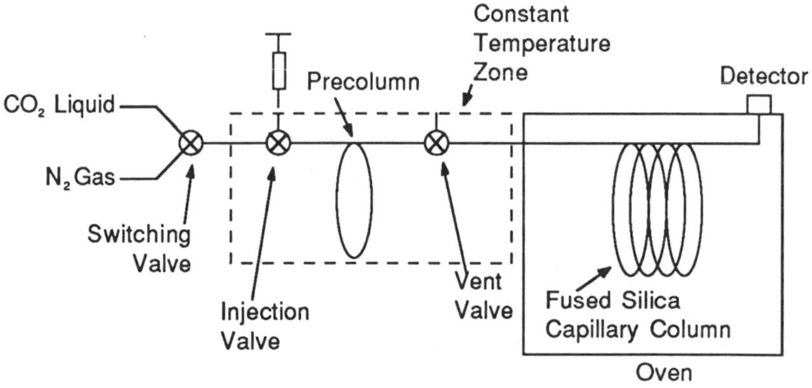

Figure 4.5. Schematic diagram of an SFC system with solvent vent injection.

sample recovery of palmitic acid, cholesterol, tripalmitine, and cholesteryl palmitate was 100%. With a 2-m x 50-μm i.d. precolumn, 1-μL injections resulted in 50% loss of the sample.[51] Thus, longer precolumns are required for injection of volumes of 1 μL or more. Based on absolute area measurements, the solvent venting technique resulted in good reproducibility (RSD = 2.5%). Addition of an internal standard reduced the relative standard deviation to less than 1%.[51]

This method of solvent elimination, which is based on a rough separation of solvent and solutes by partitioning on a precolumn, seems likely to be limited to sample volumes of 5 μL or less. In this range, sample concentrations as low as 0.1 ppm can be analyzed directly with flame ionization detection. The valve between the precolumn and the column causes some band broadening which must be corrected by peak focusing. The valve gives, however, additional opportunities, *i.e.* to use the system for coupled column separations. If only a part of the sample is of interest, the rest can be rapidly expelled through the precolumn.[51]

Solvent venting with gas purging. Another solvent elimination technique, based on gas purging, has been reported by Lee *et al.*[19] By purging the sample into a precolumn with an inert gas, *i.e.* as under GC conditions, the solvent is evaporated in the gas flow and carried out of the vent valve, which is located either between the precolumn and analytical column or between the analytical column and detector. After the solvent is evaporated and the solutes are precipitated on the walls of the precolumn, the purging valve is switched back to the CO_2 line, and liquid CO_2 is introduced. The solutes which dissolve in the flow are then focused by the temperature gradient at the column entrance. Reproducibility of 0.2-μL injections was found to be better than 0.9% (with internal standard). Alkanes could be analyzed in 0.5-ppm concentrations with 0.2-μL injections. With a 2.5-m precolumn, 0.5-μL volumes could be injected and purged several times before developing the chromatogram. Compounds that are soluble in the solvent, but less soluble in CO_2, may be lost, at least partially, by remaining adsorbed to the precolumn walls. Polar compounds may also lag or adsorb to the walls of the sample loop, a problem which can be solved by injecting the sample directly into the precolumn with a syringe.

Solvent backflush. Another method for complete elimination of the sample solvent is the backflush technique.[19] The instrumentation is based on the delayed split injection technique which includes an on/off valve in the vent line. The sample is injected with the split vent valve closed (off). After a time delay of approximately 1 min, the injection valve is closed and the split valve is opened simultaneously with a rapid negative pressure ramp. The rapid depressurization causes a density gradient to propagate along the column from the inlet, and a reversal in flow develops in the first part of the column. The solvent backflushes through the open vent line and the solutes precipitate on the column wall. By reversing again to a positive density program, the chromatogram is developed (Figure 4.6). Preliminary measurements from the injection of 0.2-μL volumes showed a recovery of 68% for naphthalene and 85% for coronene. The lower percentage for naphthalene is due to its higher vapor pressure and greater loss with the solvent during backflushing.

Another backflushing method, called post-injection solvent venting by Hawthorne and Miller,[43] making use of a short (7 cm), wide (0.3 mm) retention gap and syringe injection, allowed injection of 0.5-μL volumes on a depressurized column, with minimum detectable concentrations of 0.2 ppm of alkanes. Negative pressure ramping was not included, possibly due to the wide injection port and retention gap which facilitated rapid depressurization. Backflushing started 15 s after the pressure program was initiated, and lasted for 5 s. Peak area reproducibility was approximately 10% for raw areas and 0.9-2.5% with an internal standard. The appearance of several solvent peaks from the same solvent could be explained partly by the pressurization/depressurization processes. The solvent peaks or humps, which have been seen by many researchers when venting the solvent from large volumes of dilute solutions, and which affect the

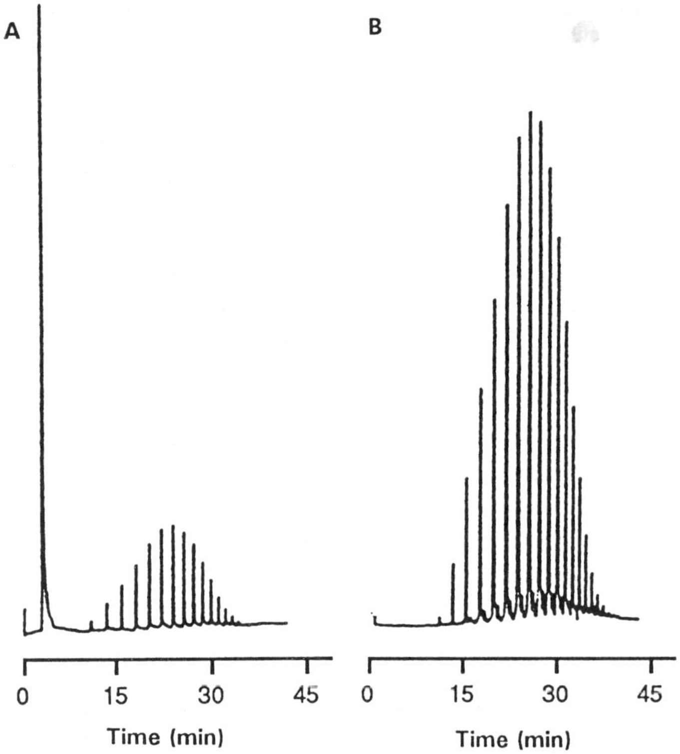

Figure 4.6. SFC chromatograms of Triton X-100 using (A) delayed split injection and (B) solvent backflush injection. Conditions: 5-m x 50-μm i.d. open tubular column, poly(30% biphenyl)methylsiloxane (d_f = 0.25 μm); CO_2; 120°C; 1-s delay time in (A) and 15-s delay time in (B); linear density program for (A) from 0.3 g mL^{-1} to 0.7 g mL^{-1} at 0.0075 g mL^{-1} min^{-1} after an initial 5-min isopycnic period, and for (B) from 0.3 g mL^{-1} to 0.1 g mL^{-1} at -0.096 g mL^{-1} min^{-1} after opening the split valve, then 0.1 g mL^{-1} to 0.3 g mL^{-1} at 0.0685 g mL^{-1} min^{-1}, and to 0.7 g mL^{-1} at 0.0075 g mL^{-1} min^{-1} (reprinted with permission from Ref. 19).

quality of the first part of the chromatograms, could be removed by increasing the injector temperature from 45 to 65°C.[43]

The major efforts so far in developing methods for solvent elimination have been in connection with the use of open tubular columns, but recent studies of similar techniques for packed microbore (1-mm i.d.) columns were initiated by Dean and Poole.[52] By purging the sample with gas into a steel retention gap, 1-10 μL injections of n-alkanes in 0.1-1 ppm concentrations resulted in large peaks with good peak shapes (Figure 4.7). The reproducibility of

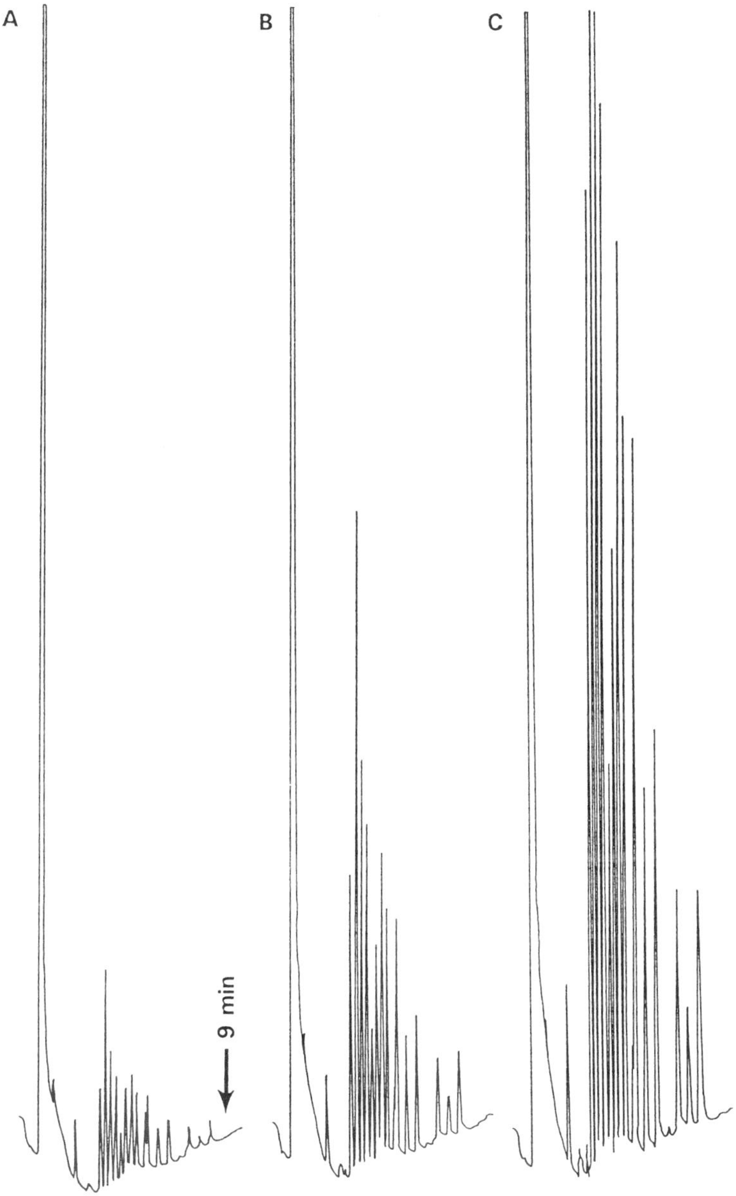

Figure 4.7. SFC chromatograms of *n*-alkanes (C_{19} - C_{40}) using gas purging-solvent venting of (A) 1-μL, (B) 6-μL, and (C) 10-μL samples of 1-10 ng μL^{-1} each in carbon disulfide. Conditions: 10-cm x 1-mm i.d. packed column, C_{18}-bonded silica particles; CO_2; 100°C; programmed from 100 atm to 200 atm at 50 atm min^{-1}, then to 400 atm at 20 atm min^{-1} (reprinted with permission from Ref. 52).

1-μL injections of 1-10 ppm concentrations was 0.7% by the internal standard method.

Both the solvent backflush and the gas purging/solvent venting methods are most effective if the solvent is much more volatile than the solutes of interest. Otherwise, loss of the more volatile components in the sample is possible. The solvent venting method, based on partitioning in a precolumn, can lead to loss of components with capacity factors close to the solvent. Since these components are almost always volatile, all of the solvent elimination methods suffer from essentially the same weakness. For the majority of applications in SFC, this weakness is not a serious problem, because the usual components of interest are compounds of relatively low volatility.

4.7 Fundamentals of Pressure Restriction and Column Flow

Mass flow in capillary columns is more complex than generally realized, and the means of controlling mass flow have been inadequate. These limitations are seldom recognized, but suggest that the technique can be improved. In this section, aspects of mass flow in capillaries are discussed. In Section 4.9, some recent efforts to improve instrumental control of mass flow are presented.

In open tubular column SFC, the optimum linear velocity (u_{opt}) is approximately 0.12 cm s^{-1} for a column i.d. of 50 μm, requiring mass flows as low as 2.4×10^{-7} g s^{-1}. There are no mechanical devices which allow adjustment of such low mass flow rates at such high pressures. Instead, fixed restrictors have been almost universally used to control mass flow at pressures dictated by the pump. In open tubular column SFC, the performance characteristics of these fixed restrictors dominate the system performance. Unfortunately, fixed restrictors produce undesirable variations in mass flow when pressure is changed.

Column linear velocity components dictated by the restrictor. The mass flow through all fixed restrictors increases rapidly with increasing head pressure. Mass flow usually increases more rapidly than density, resulting in increasing linear velocity (and decreasing efficiency) with increasing pressure in the column.[53-55]

Restrictor geometry changes the profile of column linear velocity *vs* pressure. Restrictors which create flows approaching turbulence produce less increase in linear velocity *vs* pressure than laminar flow restrictors[56], as shown in Figure 4.8. Characteristics which create turbulent flow in the restrictor include an aspect ratio (length/diameter) as small as possible and a restrictor temperature as low as feasible. At the very low mass flow rates required to produce near optimum column efficiencies, all restrictors tend to be laminar flow devices.

Changes in diffusion coefficient. Solute binary diffusion coefficients (D_m) are inversely proportional to both fluid density[58] and to the square root of molecular mass.[59] Thus, an increase in density by 4 times decreases the diffusion coefficient by 4 times; a change in molecular mass from 500 to 5,000, decreases D_m by an additional 3.2 times during a typical programmed run. Since efficiency is dictated by the ratio D_m/u, fixed restrictors produce substantial increases in u during density programming, when it would be much better to decrease u by more than an order of magnitude. From the beginning to the end of a programmed run, the efficiency might drop as much as 50 times, and the resolution by 7 times.

Peak compression. Peaks are narrower (in time) in pressure programmed runs, compared to peaks with the same retention times in isobaric runs. It has been suggested that this narrowing might be a manifestation of "peak focusing" caused by a density gradient across the peak. However, density gradients in open tubular columns are much too small to significantly focus peaks[57] in this manner. Instead, increasing mobile phase linear velocity causes column transit times, solute retention times, and peak widths to all decrease. It is important to realize that these narrower peaks do not represent increased efficiency, but quite the opposite.

Charging linear velocity. To increase the density of the fluid in the column, extra mass above the amount exiting the restrictor must enter the column inlet, producing a "charging linear velocity" (u_c)[53,57] that decreases along the length of the column. To raise the density in a 10-m x 50-μm i.d. open tubular column by 0.01 g mL^{-1} min^{-1} requires 3.3×10^{-6} g s^{-1} in addition to that exiting through the restrictor. For comparison, the steady state mass flow at a density of 0.2 g mL^{-1} and optimum linear velocity (0.12 cm s^{-1}) is 4.8×10^{-7} g s^{-1}. Thus, u_c at the head of the column is 6.9 times (0.83 cm s^{-1}) the steady state linear velocity caused by the flow through the restrictor, and the average charging linear velocity is more than

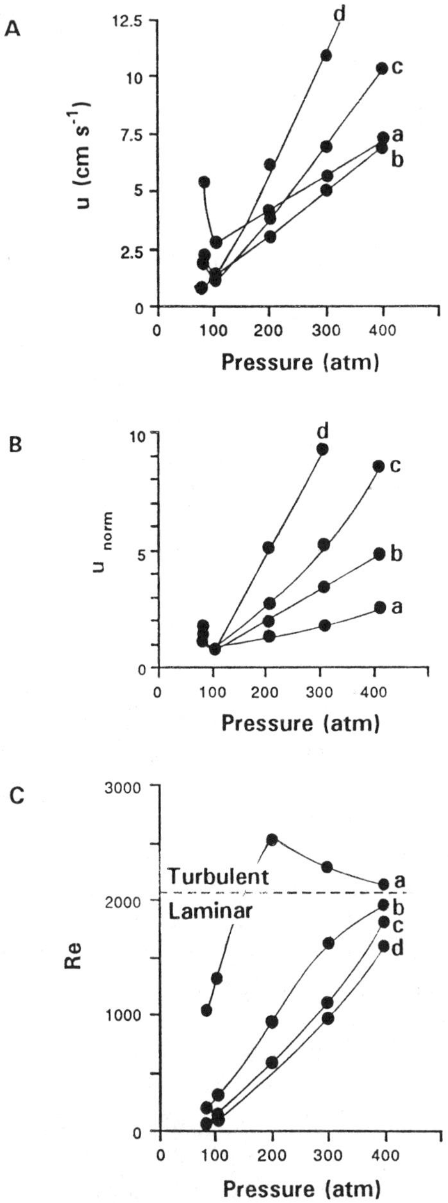

Figure 4.8. Effect of pressure and restrictor geometry on (A) linear velocity, (B) normalized linear velocity, and (C) Reynolds number. Conditions: 10-m x 50-μm i.d. open tubular column; CO_2; 40°C; 400°C restrictor temperature using (a) 37.5-μm x 1.5-μm i.d., (b) 250-μm x 2-μm i.d., (c) 5,000-μm x 4-μm i.d., (d) 50,000-μm x 8-μm i.d. restrictor sizes (reprinted with permission from Ref. 57).

3.4 times (0.42 cm s^{-1}) the steady state value. The efficiency expected from operation near the optimum linear velocity cannot be achieved unless the program rate is made much lower. At 0.0025 g mL^{-1} min^{-1}, the system efficiency is still less than half optimum (but still very high), while the time required to program from 0.2 to 0.8 g mL^{-1} increases from 60 min at 0.01 g mL^{-1} min^{-1} to 240 min at 0.0025 g mL^{-1} min^{-1}.

Very fast program rates have periodically been suggested.[60] It should be recognized that fast program rates make the restrictor flow irrelevant, and efficiency will always be very low. Charging linear velocity is directly proportional to the column volume. Thus, shorter, smaller diameter columns maintain a larger portion of their inherent efficiency than longer, larger diameter columns at a given program rate. Consequently, smaller, shorter columns can be used with higher program rates while retaining reasonable efficiency.

Other aspects of pressure reduction through restrictors. There are other desirable restrictor requirements in addition to controlling the system mass flow. One of the earliest restrictor types used was a long piece of very small, constant internal diameter tubing, usually referred to as a "linear" restrictor. The outlet of a linear restrictor is at atmospheric pressure. This means that somewhere inside the restrictor, the density drops to a very low value (*i.e.*, less than 0.005 g mL^{-1}). Solutes that are not very volatile and require significant solvation by the mobile phase could stick to the walls of such restrictors and not elute, or produce very broad peaks, independent of column-related band broadening. Linear restrictors represent the worst case of laminar flow restrictors. The opposite extreme to linear restrictors is an ideal pinhole of zero thickness through which fluids exhibit turbulent flow. The pressures inside turbulent flow restrictors never approach atmospheric pressure. Instead, a shock wave builds up downstream of the actual mechanical restriction, and much of the pressure drop (as much as half) occurs across this shock wave. Thus, the fluid throughout the restrictor can retain significant density, and solutes are much less likely to drop out of solution or stick to the walls. Real restrictors fall somewhere between the two extremes above.

Restrictor heating can also cause problems. The adiabatic cooling of the mobile phase, as it exits the restrictor, can cause crystals of CO_2 containing trapped solutes to form, which produces spikes in the output of downstream detectors. This problem can be avoided by heating the fluid to values above some characteristic temperature[53,56] related to its critical temperature. The manner in which this heating is accomplished is seldom considered. Restrictors are usually mounted, for example, in the base of flame ionization detectors originally designed for gas chromatography. In GC, substantial effort has been made over time to ensure that the transfer line inside the detector heated zone is at or near the set control temperature, but the length of this heated zone is otherwise unimportant. The FIDs designed for GC usually heat several inches of the restrictor, and would heat the fluid for a substantial distance in front of the location where the pressure drops. Such heating profiles can discriminate against the elution of very nonvolatile solutes.

4.8 Fixed Restrictors

At present, most SFC systems utilize some form of fixed capillary restrictor constructed of fused silica tubing, with the dimensions empirically selected to give the desired mobile phase linear velocity for the chromatographic temperature and pressure range. The ideal restrictor would: (a) provide for uniform pulse-free flow, (b) offer immunity from plugging, (c) be easily replaced, and (d) provide for complete transfer of labile or nonvolatile solutes to the detector without pyrolysis or formation of analyte particles.

Over the past several years, a number of restrictor designs have been investigated. Several designs are schematically illustrated in Figure 4.9.[61] The simplest restrictor, commonly called the "linear" restrictor (Figure 4.9A) is a 3-15 cm length of 5-10-μm i.d. straight-walled capillary tubing.[62] Due to its rather poor transmission of less volatile solutes, this type of restrictor has limited application. Tapered or drawn fused silica restrictors[20] (Figure 4.9B and C) provide improved performance, and are readily produced. They are made by a controlled drawing of a heated region of fused silica tubing from which the polyimide coating has been removed. A hairlike tip results with a long shallow taper of several cm, terminating with a 1-2-μm i.d. orifice.

The polished "integral" restrictor[63] (Figure 4.9D) is more robust than the tapered restrictor. It is fabricated by polishing the closed end of a piece of fused silica capillary (obtained by careful heating) until a small orifice, giving the desired flow rate is obtained. A similar variation, the converging-diverging restrictor (Figure 4.9E) is formed by compressing the column wall near the capillary exit at elevated temperatures, and retaining a short length at normal diameter to focus the expanding jet.[64] Another approach (Figure

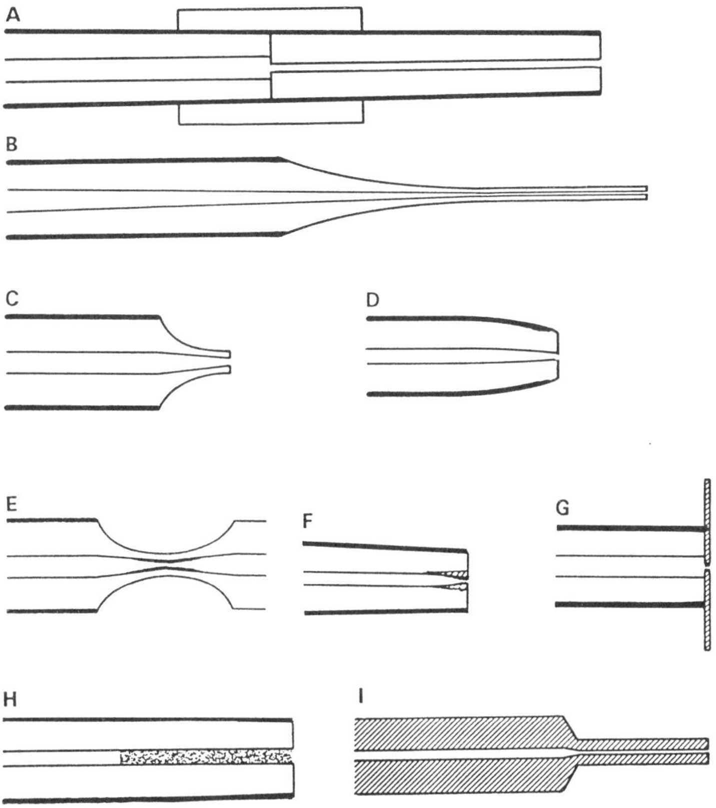

Figure 4.9. Schematic cross-section diagram of restrictor designs used in SFC. All restrictors except (G), which represents a pinhole laser-drilled orifice, and (I), a pinched restrictor, are fabricated from fused silica tubing coated with polyimide (dark lines) (reprinted with permission from Ref. 61).

4.9F) involves internal deposition of a material to partially close the end of a capillary.

The diaphragm, or laser drilled "pinhole" orifice (Figure 4.9G), was explored in earlier SFC-MS work and showed practical limitations for capillary interfaces due to the alignment and low-dead-volume connection requirements.[65] Properly used, this approach provides a nearly optimum expansion process, and is feasible with the higher flow rates of conventional-diameter packed columns. The thin diaphragm, however, must be carefully supported to withstand high SFC pressures.

The porous frit restrictor (Figure 4.9H) is significantly different from other restrictor designs.[66] The frit resembles a column packed with submicron particles, resulting in multiple fluid paths. The frit restrictor is relatively rugged and is not easily plugged by particles entrained in the fluid. The tortuous flow path and more intimate contact with the fritted material provides efficient heat transfer to the expanding gas. The disadvantage of the frit restrictor is the observed occasional poor transmission of high molecular mass, nonvolatile solutes or thermally labile materials.

The "pinched" restrictor (Figure 4.9I) is fabricated by carefully crimping approximately a 1-mm length of the termination of a segment of metal capillary tubing (50 to 100-μm i.d.) so as to obtain the desired flow rate.[65] This approach has the disadvantages of being irreproducible and giving unstable behavior, particularly at low flow rates. The advantage associated with these restrictors is their compatibility with higher temperatures and more polar fluids (*e.g.*, supercritical water or ammonia), as compared to fused silica restrictors.

The performance of the restrictor in a chromatographic system largely defines the ultimate capabilities and limitations of its operation. Although flow through even a simple capillary restrictor with subsequent expansion is an extremely complex process, significant insight into the performance of various restrictors can be obtained by considering the thermodynamic and kinetic processes occurring during expansion. These concepts are briefly reviewed here. More in-depth reviews are found elsewhere.[53-55,61]

The characteristics of an SFC restrictor and, to a significant degree, the efficiency for transfer of less volatile solutes to the gas phase during the expansion, depend on the state of the fluid before expansion, the dimensions of the restrictor, the method of restrictor heating (including the heat transfer properties of the restrictor), and the state of the fluid during and after the expansion. For this discussion, the simple case of a straight-walled capillary or linear restrictor is addressed. However, similar (but more complex) considerations apply to other restrictor types that are currently more frequently used. Figure 4.10 schematically illustrates an adiabatic process in which the enthalpy (ΔH) of the fluid is the same before and after the expansion process. Expansions relevant to SFC occur under conditions in which the Joule-Thomson coefficient, $(\partial T/\partial P)_H$, is generally positive, and net cooling results. If the restrictor is very short, the process may be considered adiabatic. In this case, the equilibrium state of the fluid after expansion (if it is also isolated from its surroundings) can be predicted directly from thermodynamic data. However, the state of the fluid during and shortly after expansion depends on the physical processes related to the expansion that are kinetically controlled and inherently not at equilibrium.

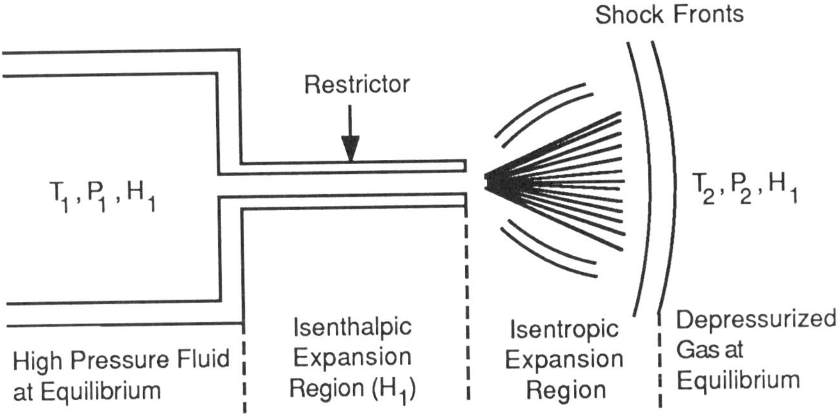

Figure 4.10. Schematic diagram of the supercritical fluid expansion in an adiabatic process where the initial and final states of the fluid are known (reprinted with permission from Ref. 53).

Flow through a short restrictor will be nearly isenthalpic until the end of the restrictor, where isentropic expansion to supersonic velocities occurs. Shock phenomena and interactions with the background gas in the expansion region largely convert the kinetic energy of the gas jet back to internal energy.[67] For CO_2, an initial enthalpy of less than 175 kcal g^{-1} results in the expanding fluid traversing the two-phase region (often manifested by a build-up of frozen CO_2 at the restrictor exit). In order to obtain good flow characteristics to avoid a two-phase region during CO_2 expansion, a fluid temperature of at least 80 to 100°C before entering the restrictor should be maintained. When an expansion occurs from pressures and temperatures that result in two phases, the assumption of adiabatic conditions allows the fraction of fluid in each phase to be calculated. Figure 4.11 gives the fraction of CO_2 in the condensed phase after expansion to low pressure as a function of reduced pressure and reduced temperature of the supercritical fluid. The solid line gives the minimum reduced pressure and temperature required to avoid passage through a two-phase region for any stage of an isenthalpic process. The dashed lines give the indicated fractions of fluid as a condensed phase after expansion to atmospheric pressure. The reduced conditions necessary to avoid a two-phase system given in this figure also provide guidance for other one-component fluid systems. For more complex fluids with greater numbers of internal degrees of freedom, the maximum temperature

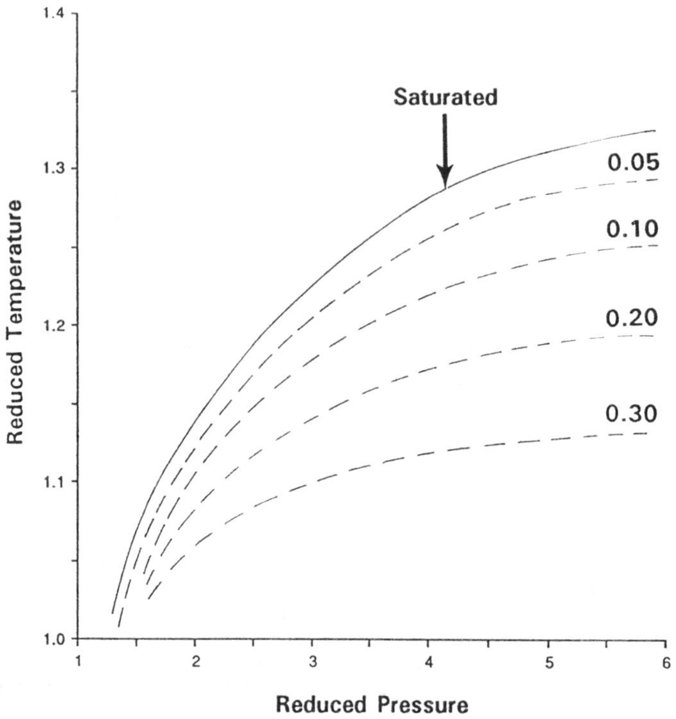

Figure 4.11. Fraction of CO_2 in the condensed phase after expansion to low pressure as a function of reduced temperature and pressure, showing constant vapor-liquid composition (---) after expansion to low pressures, and the minimum pressure and temperature conditions (—) necessary to avoid condensation during expansion (reprinted with permission from Ref. 53).

at which a two-phase separation occurs in an isenthalpic expansion is decreased.

In addition to providing analyte transfer to the detection region, a primary requirement for restrictors is to provide acceptable flow rates or linear velocities over the range of pressures relevant to a given separation. For short restrictors, the flow rate during an adiabatic expansion cannot exceed the speed of sound until within a few nozzle diameters of the end of the capillary.[68] For adiabatic expansions, the fluid temperature at the capillary exit can be calculated, which allows the fluid density at the capillary exit to be estimated.[68] The exit density can be surprisingly high. For example, a pentane mobile phase at 250°C and 240 atm [ρ = 0.47 g mL^{-1} at a flow rate of 150 μL min^{-1} (liquid)] that is expanded through a 6.4-mm x 9-μm i.d. restrictor has an exit density of approximatley 0.12 g mL^{-1}. Under these conditions, the average fluid density (0.29

g mL^{-1}) can be used to estimate the average velocity in the restrictor to be 8,400 cm s^{-1}, corresponding to a residence time of 7.5 x 10^{-5} s.

A one-dimensional compressible flow model[54] has been used to predict an incomplete pressure drop over tapered and straight-walled restrictors when operated under choked flow (sonic exit velocity) conditions. The remaining decompression takes place outside the restrictor, which maximizes the fluid density and minimizes analyte nucleation in the restrictor. The same model also predicts that the pressure drop will be lower (which gives a higher exit fluid density) for a tapered restrictor than for a straight-walled restrictor giving the same mass flow rate at constant temperature and pressure.

When long restrictors (straight-walled) are used to obtain the slower flow rates needed for more optimal capillary SFC, the assumption of an adiabatic process becomes invalid due to the longer restrictor residence times. However, the speed of sound limitation to maximum velocity still applies. The analyte will precipitate and ultimately plug a long restrictor if it is not heated to vaporize the analyte, or cooled to maintain sufficient density for solubility.

A vital property of SFC restrictors is their effectiveness in transporting "nonvolatile" compounds to various gas-phase detectors. Success depends on the restrictor design, fluid pressure and temperature, the particular demands imposed by the detection method (*e.g.*, flame ionization or mass spectrometry), and, under some conditions, the volatilities or melting points of the compounds. Heating of the fluid through relatively long restrictors by the FID block improves analyte detectability,[20] as does heating a region just ahead of the restrictor.[69] The performance of drawn restrictors is improved by heating of the fluid prior to the drawn restriction, and perhaps by better heat transfer through the thinner walls of the tapered portion of the restrictor.

For truly nonvolatile compounds, two distinct modes of operation appear feasible. In the first mode, the nonvolatile analyte tends to collect on the restrictor walls and, if the local temperature is above the analyte melting point, to flow toward the end of the restrictor. The liquid will collect at the end of the restrictor and be periodically entrained in the high shear gas flow and transported to the detector. This is a complex process with details that likely depend on the liquid viscosity and restrictor geometry. Solid particle agglomerates also appear to form by a related mechanism. These processes often lead to plugging of the restrictor. The second mode of operation, producing much smaller particles, occurs when analyte nucleation is delayed to near the end of the restrictor. These small particles evidently form rapidly ($<10^{-6}$ s) and apparently pose no difficulty for the FID. The major requirement for operation in the "nucleation mode" is to maintain solvating conditions to near the end of the restrictor, which is facilitated by the use of short restrictors and fluid conditions which enhance the solubility. It is important to recognize that this does not necessarily correspond to elevated restrictor temperatures.

Increased pressure always serves to enhance the transport of nonvolatile analytes, but in practice, pressure is limited by either the SFC conditions needed to yield sufficient retention, or the maximum pressure for the system. The effect of temperature on fluid phase solubility is more difficult to predict, but some guidance is provided by analyte vapor pressures. Heating of the fluid just prior to expansion, and heating of the restrictor itself, have been shown to improve and extend detection for compounds having some (low) volatility.[53] Volatility plays a major role in detection of such analytes, because fluid phase solubilities (at a given density) generally mirror analyte vapor pressures.[53,70] For detection of truly nonvolatile compounds, heating of the fluid before or during expansion can be counterproductive, because of lower fluid densities and resulting lower fluid phase solubilities.

Less volatile analytes will be more efficiently transported through a restrictor with a low length-to-diameter ratio, such as the tapered or integral restrictors. This allows an incomplete pressure drop to occur over the restrictor, which maximizes the fluid density at the restrictor exit and minimizes analyte nucleation in the restrictor. Operation at higher flow rates also improves detection of less volatile analytes. In a qualitative performance comparison of four restrictor types (linear, tapered, integral, and porous frit), using an oligomer mixture of relatively low volatility as test analytes, the tapered and integral restrictors more efficiently transported the higher molecular mass and less volatile oligomers.[61]

4.9 Variable Restrictors

Linear velocity changes during pressure (density) programming can significantly affect chromatographic efficiency. Better efficiencies are obtained at low pressures (densities) and high temperatures[71] (see Section 2.2). For example, solutes eluting at high pressures have smaller diffusion coefficients because of their larger sizes and because of the high mobile phase densities required for their elution. Therefore, the linear velocity should be greatly reduced at these pressures in order to maintain constant efficiency. Due to an increase in the pressure drop across the restrictor during pressure (density) programming, the mass flow rate and, consequently, the linear velocity increase with fixed restrictors. As a result, decreased resolution at high densities is observed. In packed column SFC, efficiency variations with linear velocity are more difficult to predict.[72,73] The pressure drop along the column increases as the linear velocity increases in packed column SFC, resulting in a changing density gradient in the column. This makes the elution process difficult to describe. However, density is the main factor affecting resolution,[72,74] and optimum resolution occurs at the lowest ratio of pressure drop to total change in flow rate along the column.[72] Klesper and Schmitz have reviewed the effects of linear velocity (flow) gradients in packed column SFC.[75]

Two methods of pressure programming have been used in SFC[76]: (a) programming the inlet pressure while the outlet pressure is controlled by a restrictor or needle valve, and (b) delivery of a constant flow of mobile phase while controlling the outlet pressure by using a variable valve. In open tubular column SFC, only the first approach has been applied. Fixed pressure restrictors were discussed previously (Section 4.8).

There are a number of easily built or commercially available devices to control pressure with the large flows (liters per minute of expanded flow) characteristic of conventional packed columns.[77-84] Most early efforts involved splitting the column effluent, with part going to the detector and part being pressure regulated. One design[85] had a split ratio of 1:50, with the bulk of the flow being controlled and regulated by a flowmeter with semiconducting elements and a motor-driven automatic expansion valve, heated to prevent condensation. Gere *et al.*[8] could independently vary the linear velocity while operating isobarically at pressures up to 425 bar by using a manually adjustable back-pressure regulator as the restrictor; a similarly designed back-pressure valve[81] gave a retention time reproducibility of 0.4 %RSD. Both inlet and outlet pressures could be programmed independently using a system with two pumps and two restrictors,[80] or a similar design with only one restrictor.[76] A newly developed regulating valve, based on high speed pulsing, is capable of controlling the column outlet pressure independent of the mass flow rate of the fluid.[82] With this valve, the retention time reproducibility during pressure programming on a 250-mm x 4.6-mm i.d. silica gel column was 1%. This new design is shown schematically in Figure 4.12, and is described in more detail in Section 4.23. A similar valve has been used with composition

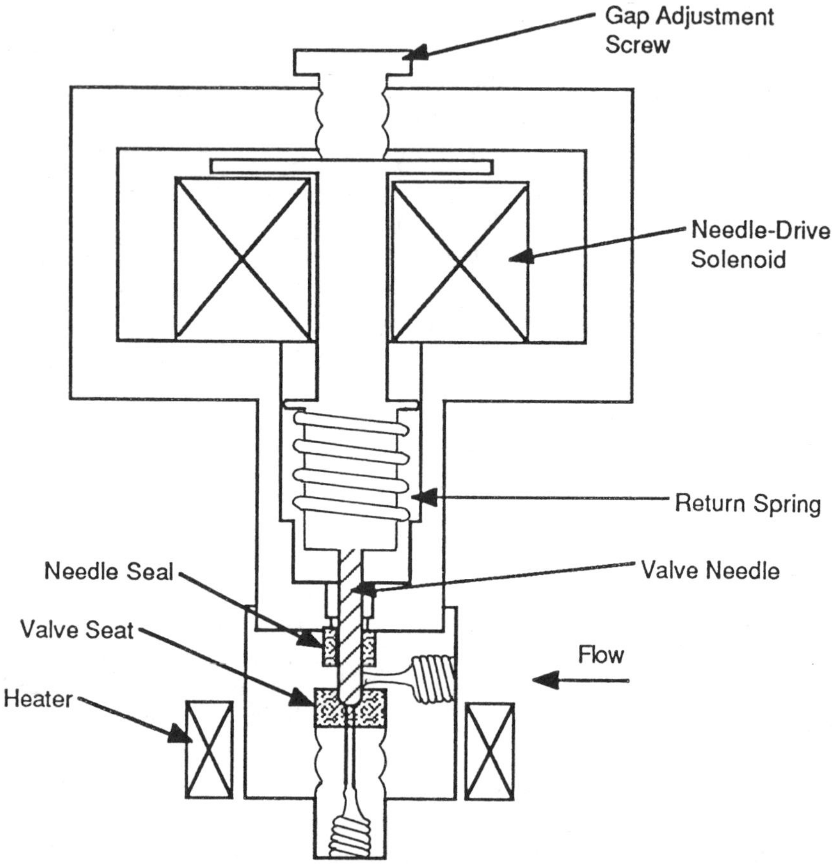

Figure 4.12. Schematic diagram of a pulsed back-pressure regulation valve (reprinted with permission from Ref. 82).

programming at constant pressure.[84] A crimped stainless steel capillary restrictor placed on a heated metal block can control restriction by thermal expansion of the restrictor.[23] Back-pressure regulation *via* valves or regulators appears to be the most prevalent and satisfactory method for controlled variation of flow rate in packed column SFC.

In the last few years, several techniques have emerged which may allow instrumental adjustment of the minute mass flows used in open tubular column SFC.[86-90] One early report of flow control in open tubular column SFC using back pressure control with a fluorescence detector has been published.[91] Raynie *et al.*[86] reported a flow control system that employed back-pressure regulation with a sheath-flow

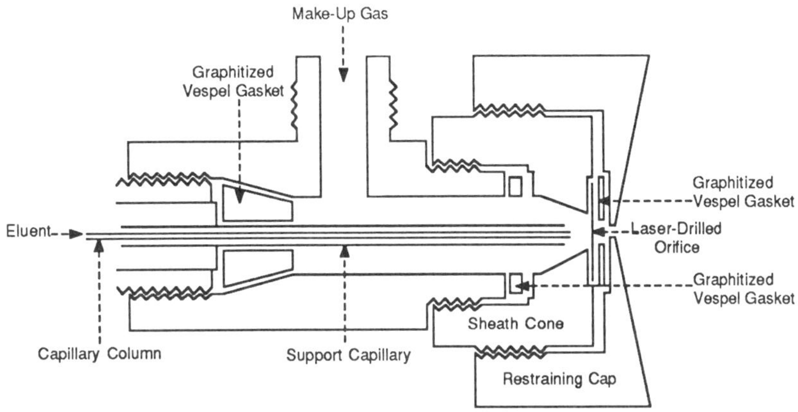

Figure 4.13. Schematic diagram of a sheath-flow nozzle used as a controllable back-pressure restrictor (reprinted with permission from Ref. 86).

nozzle depicted in Figure 4.13. The downstream restrictor consisted of a 1 to 50-μm i.d. pinhole, laser drilled into a 1/8-inch diameter metal disc. The disc was held in a metal screw cap and was replaceable. The column outlet (preferably containing a second restrictor) was centered behind the pinhole. The area around the column outlet and behind the pinhole was enclosed and swept with high pressure make-up gas. The mobile phase mass flow through the column was determined by the pressure drop across the column and column restrictor. A high pressure make-up gas was added to partially restrict and control the flow out of the column. The mass flow and linear velocity were easily adjusted by changing the pressure of the make-up gas relative to the column head pressure. The dynamic range of adjustment could be very large, but would require very accurate pressure transducers. Thus, the initial linear velocity (u_{init}) could be set, and u programmed to decrease with time. On the negative side, the make-up gas flow must be large compared to the column flow for accurate control and prevention of extracolumn band broadening. The total flow would also increase with column head pressure. Neither of these features is particularly compatible with downstream detection. Response factors and baseline slopes vary with head pressure.

Janssen *et al.*[88] reported another back-pressure regulated variable restrictor, based on splitting the column effluent in a low-dead volume T-piece at the column exit just prior to the detector. One stream from this T-piece went to the detector, while the other went to the back-pressure regulator. Control of the flow rate through this

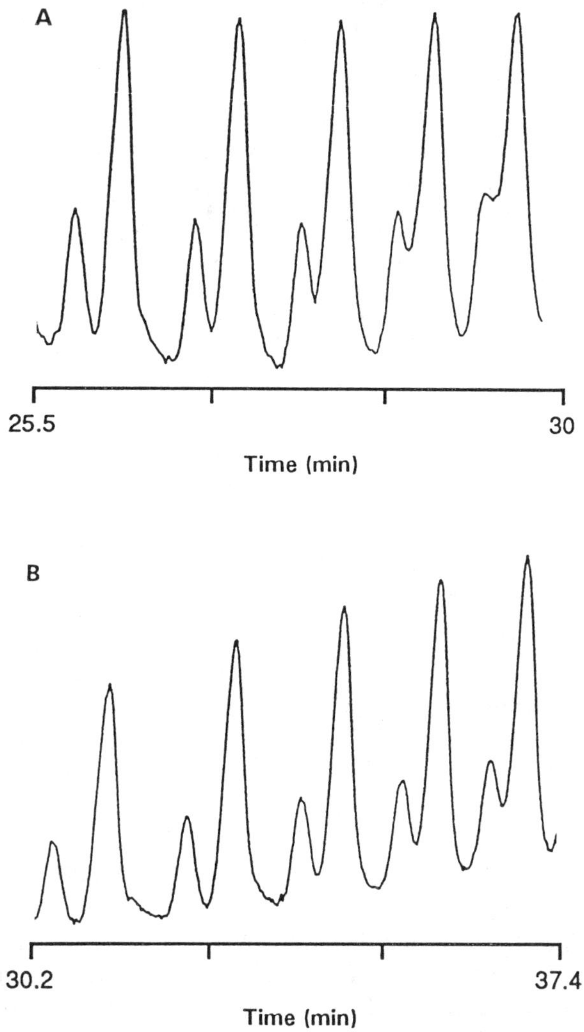

Figure 4.14. SFC chromatograms of the latter portion of an analysis of poly(hydro)methylsiloxane using (A) fixed and (B) variable flow restrictors. Conditions: 8.2-m x 107-μm i.d. open tubular column, polymethylsiloxane stationary phase; CO_2; 120°C; linear pressure program from 120 atm to 310 atm at 5 atm min^{-1} (reprinted with permission from Ref. 88).

second branch was accomplished by regulating the back-pressure from either a high pressure gas cylinder or a second syringe pump. An example of the resolution gain by linear velocity control is shown in Figure 4.14.

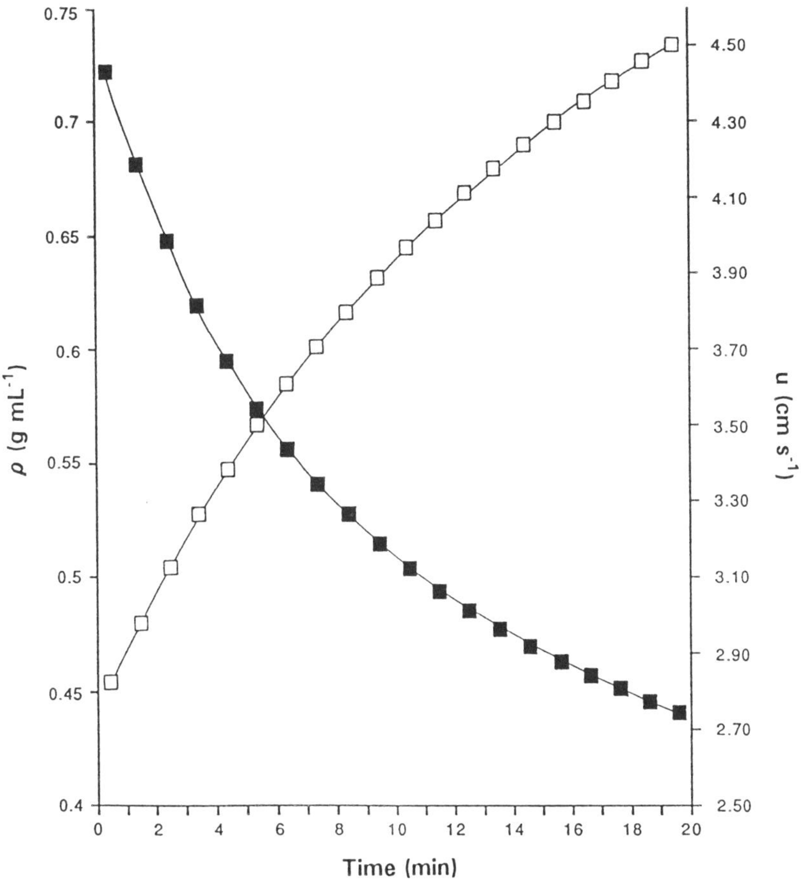

Figure 4.15. Linear velocity change (■) and density change (□) for constant mass flow conditions during linear pressure programming using restrictor temperature programming. Conditions: 50-μm i.d. open tubular column; CO_2; 100°C; linear pressure program from 190 atm to 380 atm at 9.5 atm min^{-1}; constant mass flow at 4.0 (±) 0.2 x 10^{-5} g s^{-1} (reprinted with permission from Ref. 92).

A more limited but simpler and less expensive approach to controlling mass flow was suggested by Smith *et al.*[53] and Berger and Toney.[87] The restrictor mass flow could be adjusted by changing its temperature, not for the purpose of changing the orifice size but to change the fluid characteristics. At any fixed pressure, the mass flow could be adjusted by a factor of about 5. Alternatively, within one pressure programmed run, the mass flow could be kept constant or even decreased to maintain constant efficiency or detector response. An example of maintaining constant mass flow by decreasing the linear velocity is shown in Figure 4.15. The available dynamic range does not allow control of both the initial linear velocity and the maintenance of constant efficiency during a run. To change u by more than a factor of 5, the restrictor must still be replaced.

Makishima and Griffiths[90] combined various aspects of these previously described approaches. The output of one pump was split into two parallel flow paths, one of which was the column and the other a bypass. The two paths were recombined after the column, and the total flow was directed into a fixed restrictor which was mounted inside an FID. A second restrictor was mounted in the bypass path, and the temperature of this second restrictor was programmed. The fixed restrictor in the detector determined the total mass flow, while the mass flow through the bypass was determined by the bypass restrictor temperature. The difference in flows between the two restrictors was the column mass flow.

The bypass restrictor can be made out of almost any material (*i.e.,* stainless steel tubing, frits, and pinholes) and can be easily changed to change the range of mass flow through the column. The dynamic range provided by each bypass restrictor is a limitation to this device, but the ease of replacement and the inexpensive nature of the hardware offset this limitation.

4.10 Fundamentals of Detection

The compatibility of SFC with a wide variety of detection methods offers analytical chemists an opportunity to utilize a variety of detection methods. However, this disparity of detectors presents a challenge for interdetector comparisons. As with most chromatographic approaches, detector nomenclature and concepts have been developed independently with their usage in gas or liquid chromatography. Even within the single discipline of either gas or liquid chromatography, tremendous differences in meaning can be found in the literature for terms such as sensitivity, noise, detection limit, minimum detectability, response factor, etc. Supercritical fluid chromatography offers the opportunity to unify concepts and nomenclature with respect to detector fundamentals. Descriptions and definitions of detector related concepts given below are a first attempt to unify detector concepts for SFC. More complete discussions of these definitions with respect to gas and liquid chromatography are provided elsewhere.[93-101] Most of the material presented here has been adapted to SFC from these references.

Mechanism of detection. The overall design of a chromatographic detector is determined by the detection mechanism, the detection cell type, and the pressure at which the detector is operated. There are two basic types of detectors for SFC, ionization detectors and optical detectors. Ionization detectors require an energy input sufficient to ionize the sample. The efficiency and selectivity of these ionization processes provide the sensitivity and selectivity of the detector for the compounds of interest. Optical detectors rely on either emission or absorption of light by the sample or its products to achieve response. For ionization detectors, ion collection and sometimes separation devices are necessary parts of the detector, while optical detection methods require photon detection with optical separation components.

Closed cell. In addition to the detection mechanism, detector design can depend on whether the solutes are to be detected in solution or not. If they are to be detected in solution, then the detector cell is said to be a "closed" cell, and generally it has a cross sectional area similar to that of the chromatographic column. For most closed cell detectors, the mobile phase can be maintained in the supercritical fluid state. In some cases, however, the pressure may be reduced to ambient, and by controlling the temperature, the mobile phase is converted to a liquid prior to passing through the detector.

Open cell. Open cell designs are employed when the mobile phase is decompressed prior to detection of the solute. Decompression of the mobile phase may be from supercritical to ambient pressure as in the flame ionization detector (FID) and the ion mobility detector (IMD), or it may be from supercritical to vacuum conditions as with mass spectrometry (MS) and supersonic jet spectroscopy (SJS). Whether the detector cell is of open or closed design, there are two paramount concerns: first, the solute to be detected must be introduced in the most efficient manner possible to permit maximum detector sensitivity; and second, the connecting tube, the decompression zone, and the cell volume must not significantly broaden the peak relative to the separation which was achieved during the chromatography.

Post-separation peak broadening. One primary design consideration for the detection system is to minimize peak broadening after the sample has emerged from the chromatographic column. Any broadening which occurs after the solute has left the column will result in a loss in chromatographic efficiency, and, therefore, a loss in resolution. Contributions from post-separation peak broadening can occur at several locations after separation. Assuming random peak broadening processes, the overall peak variance due to post-separation broadening can be described as the sum of the variances of the individual processes

$$\sigma_{P^2} = \sigma_{t^2} + \sigma_{v^2} + \sigma_{d^2} + \sigma_{e^2} \tag{4.4}$$

where σ_P^2 is the total post-separation peak variance, σ_t^2 is the peak variance due to the transfer lines, σ_b^2 is the peak variance from stagnant volume, σ_d^2 is the peak variance from the detector cell, and σ_e^2 is the variance from the detector electronics and recorder response time.

Transfer line broadening. For closed cell designs when detection is achieved under supercritical pressure conditions, peak broadening due to transfer line effects can be virtually eliminated by incorporating the sample cell within the column. With fused silica capillary columns, closed cell detection may be accomplished by removing the polyimide coating from a section near the end of the column, and using the bare fused silica as the boundaries of the detector cell. Under these conditions, the separation process continues throughout detection. For convenience or enhanced sensitivity, however, a closed cell is often placed in series after the column, and it is necessary to transfer the chromatographic eluent from the column to the detector cell *via* connecting tubes.

Several types of band broadening processes can occur during this transfer. In gas chromatography, the primary band broadening process is longitudinal diffusion. The faster the flow is swept through the transfer line, the less time diffusion contributes to the overall peak width. In this case, peak variance is simply

$$\sigma_{tg}^2 = \frac{2\,D_m\,L}{u} \tag{4.5}$$

where σ_{tg}^2 is the peak variance due to gaseous diffusion in the transfer line, D_m is the gas phase diffusivity of the solute, L is the length of the transfer line, and u is the velocity of the solute.

In LC, however, the primary dispersion process is radial diffusion, which develops from a parabolic velocity profile in which the velocity of the mobile phase varies over the cross-sectional area of the transfer tube. Solute molecules near the center of a transfer line move faster than those near the walls, leading to significant peak broadening. Mathematically, this effect can be expressed as follows

$$\sigma_{tl}^2 = \frac{\pi^2 r^6 L u}{24 D_m} \qquad (4.6)$$

where σ_{tl}^2 is the peak variance due to the liquid velocity profile in the transfer line, r is the tube radius, L is the length of the transfer line, u is the velocity of the solute, and D_m is the diffusivity of the solute in the mobile phase.

Broadening effects in transfer lines are more complicated for supercritical fluids, since both longitudinal and radial diffusion of the solute may be important, depending on conditions. Combining the gas and liquid band broadening descriptions above, an approximation of the band broadening caused by a supercritical fluid transfer line can be given by

$$\sigma_t^2 = \frac{2 D_m L}{u} + \frac{\pi^2 r^6 L u}{24 D_m} \qquad (4.7)$$

where σ_t^2 is the peak variance due to the transfer line under supercritical fluid conditions and D_m is the diffusivity of the solute in the supercritical fluid. When $u >> D_m$, radial diffusion controls the peak broadening process, and when $u << D_m$, longitudinal diffusion becomes the dominant process.

For ambient and low pressure detectors, a description of the solute transfer process becomes even more complicated. Solute diffusivity changes rapidly during the decompression stage of the transfer. Further, solutes lose solubility during the decompression stage, and adsorption and precipitation can spread the band before it reaches the detector.

Stagnant volume broadening. The term "dead volume" has come to mean that entire volume of space which is not involved in the separation process. However, in the most strict sense, that portion of space within a transfer line between the column and the detector is not dead since it performs a function and is continuously swept with mobile phase. "Stagnant volume" is defined as that volume of space which may exist in a transfer line or detector cell which is not swept by the mobile phase. Solutes that become trapped in a stagnant space must rely on diffusive processes to re-enter the main stream of the mobile phase, and this leads to significant peak broadening. For the most part, the existence of stagnant volumes results from plumbing and design failures. Careful attention to butt connections and detector interfaces can minimize problems associated with stagnant volumes.

Detector cell broadening. While the ideal transfer line has no volume, the detector cell must have a finite volume in order for detection to occur. Detector designs are inevitably a compromise between sensitivity and extra-column peak broadening. In many detectors, sensitivity can be maximized by increasing the cell volume, while peak broadening can be minimized by decreasing the cell volume.

Of course, both radial and longitudinal diffusion broadening occur in the detector in a manner similar to that in transfer lines. Also, peak broadening can occur if the response time of the detection mechanism is slow compared to the residence time of the solute in the detector. In general, however, response times for ionization and optical detection methods are extremely fast and do not limit separation efficiency in practice.

For an open cell detector such as the FID, it is not the entire internal volume of the detector which contributes to peak broadening, but rather only that portion where the detection process (flame ionization in this case) occurs. This is referred to as the effective detector volume.

Electronic broadening. Another nondispersive process which can contribute to detected peak width is the response time of the amplifier and recorder. Significant distortions of peak shape can occur if response times are too slow. The percent contribution to band broadening by the response time can be calculated from the equation

$$\%\sigma = 100\, \tau_d\, n^{1/2}\, t_R^{-1} \qquad (4.8)$$

where $\%\sigma$ is the percent contribution to band broadening by the response time of the electronics, τ_d is the response time of the electronics, n is the number of theoretical plates for the separation, and t_R is the retention time of the solute.

Five percent rule. In general, a maximum of a 10% decrease in separation efficiency, as judged by plate count, should be tolerated due to post-separation peak broadening processes. To achieve this goal, no more than 5% peak broadening can occur after separation is complete

$$\sigma^2 = \sigma_c^2 + \sigma_P^2 = (1.05\ \sigma_c)^2 \tag{4.9}$$

where σ^2 is the total variance of the peak, σ_c^2 is the peak variance from the separation process and σ_P^2 is the total post-separation variance. Thus, an apparent 5% increase in σ, due to σ_P, causes a 10% decrease in n

$$0.9\ n = \left(\frac{t_R}{1.05\ \sigma_c} \right)^2 \tag{4.10}$$

where n is the plate number due to the separation process only, t_R is the retention time of the solute, and σ_c is the standard deviation of the band in time units due to the column contribution only.

Assuming that a detector can be designed such that transfer line broadening, stagnant volume broadening, and electronic broadening can be eliminated or minimized, the 5% rule implies that a detector volume can be no larger than 1/20 of the volume of a separated solute zone. For microbore and open tubular columns, this is often a major challenge.

Figures of merit. Sensitivity, response factor, detection limit, detectability, and noise are a few, among many, figures of merit used to describe the response characteristics of various chromatographic detectors. Unfortunately, common usage of these terms does not always correspond to formal definitions. In some cases, even formal definitions of the terms vary, depending on the source. The following is an attempt to organize and define figures of merit for SFC detectors which will permit the unambiguous comparison of these detectors with each other and with nonchromatographic usage of these analytical devices. Adherence to IUPAC recommendations, and where possible, to common chromatographic usage has been practiced.

Mass sensitive detectors. Mass sensitive detectors are those detectors whose response is a function of the mass of solute which passes through the detector per unit time. The general response can be expressed as

$$R_d = S' F_m^i \tag{4.11}$$

where R_d is the instantaneous response of the detector, S' is a proportionality constant, F_m is the rate of mass flow through the detector, and i is the linearity index, discussed below. An example of a mass sensitive detector is the FID. With this detector, dilution of the solute with makeup gas flows does not affect the response as long as the mass flow rate of introduction is held constant.

Concentration sensitive detectors. Concentration sensitive detectors are those detectors whose response is a function of the concentration of the solute in the detector cell. The general response can be expressed as

$$R_d = S' c^i \tag{4.12}$$

where S' is a proportionality constant, c is the concentration of the solute in the detector cell, and i is the linearity index which is discussed below. An example of a concentration sensitive detector is the UV absorption detector. With this detector, variations in the rate of introduction of the solute into the detector will not change the response as long as the concentration in the detector cell remains constant.

Sensitivity. Perhaps the most misused term when characterizing a detector is the detector sensitivity (S). It is simply the slope of an analytical calibration curve of a solute. That is, the larger the change in response (R_d) of a detector as a result of a small change in concentration (c) or mass flow (F_m) of a solute, the larger the sensitivity of the detector

$$S = \frac{\Delta R_d}{\Delta c}$$

$$\text{or} \tag{4.13}$$

$$S = \frac{\Delta R_d}{\Delta F_m}$$

For example, the response of an FID is coulombs s^{-1}, and the solute introduction rate is mass s^{-1}. Thus the sensitivity of an FID has units of coulombs mass^{-1}. In some cases, sensitivity has been defined as the ratio of the integrated peak area to the quantity or concentration introduced into the detector. Because of the convenience of using peak area integration, this approach is desirable, but only approximate, and should be applied with caution because of the dynamic nature of chromatographic peaks.

For a concentration sensitive detector such as the UV detector, the dynamic response is absorbance units s^{-1} and the solute introduction rate into the detector is concentration s^{-1} (mmol mL^{-1} s^{-1}). Thus, the sensitivity of a concentration sensitive detector is AU mmol^{-1} mL^{-1}.

Noise. Noise (N) is defined as the standard deviation of the baseline when no sample is present in the detector, but all other conditions (*e.g.*, detector temperature, mobile phase flow rate, pressure, etc.) are identical to those used in an actual chromatographic run. In general, three types of noise can be defined: short term, long term, and drift. Short term noise is defined as baseline fluctuations which occur at a higher frequency than a solute peak. Short term noise usually arises from the electronics of the detector amplifier or recorder, and can often be eliminated or reduced by the use of electronic filters. Long term noise is defined as baseline fluctuations which occur in the same frequency range as the solute peak. These are often difficult to eliminate since they are often due to chemical responses in the chromatographic system. Finally, baseline perturbations which occur at a very long frequency are termed baseline drift. Drift can occur from relatively slow changes in the temperature, the power supplies, or the composition of the column eluent. Depending on the severity of the drift, this type of noise may, or may not, interfere with an analysis.

Limit of detection. The limit of detection, or detection limit (D_L) refers to the concentration or mass rate which provides the smallest detector response ($R_{d,lim}$) that can be measured with reasonable certainty. This smallest response is given by the equation

$$R_{d,lim} = R_{d,b} + a\,N \qquad (4.14)$$

where $R_{d,b}$ is the average baseline response, N is the standard deviation of the baseline, and **a** is a constant selected to establish a desired confidence level. A value of 3 for **a** is most often used and recommended, which provides that the minimum response be greater than 99% of all baseline perturbations, and includes allowances for nongaussian noise.

The limit of detection for a mass sensitive detector is defined as

$$F_{m,lim} = \frac{R_{d,lim} - R_{d,b}}{S} \quad (4.15)$$

and the limit of detection for a concentration sensitive detector is defined as

$$c_{lim} = \frac{R_{d,lim} - R_{d,b}}{S} \quad (4.16)$$

provided that S is constant for low values of c or F_m, where c is the concentration in the detector and F_m is the mass flow rate into the detector. Limits of detection are measurements from the detector itself, and are independent of the chromatographic conditions employed. For that reason, these values should be used to compare one detection method to another.

Minimum detectable quantity. From a practical analytical perspective, the smallest quantity of a solute which can be detected after a chromatographic separation is often desired. In this case, of course, the chromatographic system becomes an integral part of the overall ability of the detector to measure the presence of small quantities of solute. The minimum detectable quantity (MDQ) is the smallest amount of solute that can be injected into the chromatograph and be detected with a peak height of three times the baseline noise.

Selectivity. Selectivity (SEL) is the ratio of sensitivities of a detector for two different solutes, and is reported as relative molar response (RMR) or as relative weight response (RWR), depending on how the amount of solute passing through the detector was calculated.

$$SEL = RMR = \frac{S_1}{S_2} \quad (4.17)$$

Since SEL has no units, RMR or RWR must be specified. The difference between RWR and RMR is simply the ratio of the molecular weights of the two compounds compared.

Dynamic range. The concentration or mass flow rate range over which the detector's response is dependent on the concentration or mass flow rate of the solute in the detector is called the dynamic range of the detector. The minimum point of this range is the limit of detection, and the maximum point, called the detector saturation point, occurs when the detector can no longer respond to an increase in solute concentration or mass flow rate.

Index of linearity (i). The exponential term in the basic response equations 4.11 and 4.12 is called the index of linearity. When $i = 1.00$, the response is said to be linear and $S' = S$ (the detector sensitivity).

Linear dynamic range. The linear dynamic range (LDR) is that concentration or mass flow rate range over which the detector responds in a linear manner. No detection system is completely linear, and an acceptable tolerance for nonlinearity must be defined. One approach to determine the linear range of the detector is to plot detector sensitivity as a function of concentration or mass flow rate through the detector, depending on whether the detector is concentration or mass sensitive. When the sensitivity changes by more than 10%, then the detector is no longer operating in the linear dynamic range. An alternative method for determining the linear dynamic range is to plot the log of the response as a function of the log of the concentration or mass flow rate. From the response equations

$$\log R_d = \log S' + i \log c \tag{4.18}$$

$$\log R_d = \log S' + i \log F_m \tag{4.19}$$

the detector is out of the linear dynamic range when the slope of the plot is greater than 1.02, or less than 0.98.

Figure 4.16 shows an illustrative detector calibration graph with the relationships between noise, response, detection limit, concentration or mass flow rate, sensitivity, dynamic range, and linear dynamic range. It is important to notice that a detector calibration graph is not the same as an analytical calibration graph. The primary difference is that the x-axis represents the concentration or mass flow rate rather than the absolute amount of solute injected into the chromatograph. Reasonably good estimates

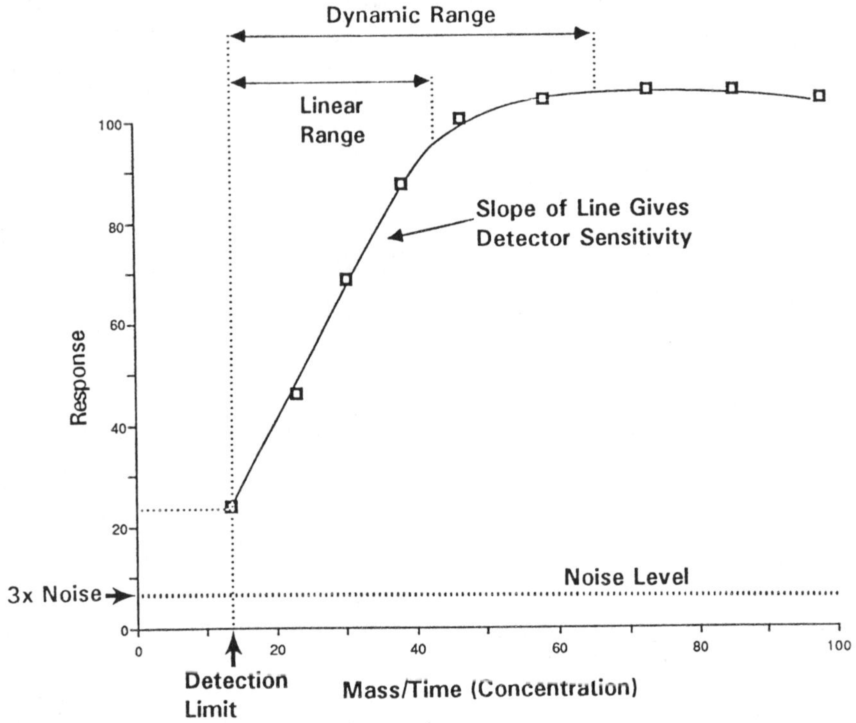

Figure 4.16. Detector calibration graph.

for figures of merit can be obtained from analytical calibrations, but the analyst should be aware of the differences.

4.11 Flame-based Detectors

Flame ionization detection. The flame ionization detector (FID) has become the most commonly used detector in SFC. Flame ionization detection was first used in SFC by Sie *et al.*[102] Fjeldsted *et al.*[62] applied the FID to open tubular column SFC, while Blu *et al.*[103] reported the possibility of using an FID under high pressure conditions. A comprehensive historical review of SFC/FID has been published.[104]

In an FID, organic compounds decompose and ionize in an H_2/air or H_2/O_2 flame. The key variables of the FID are the hydrogen flow rate, the air or oxygen flow rate, the carrier gas flow rate, and the collector voltage. When a small diameter flame jet, *ca.* 0.011-in. i.d., is used, typical conditions[105] for capillary GC are 20-50 mL min^{-1}

hydrogen and 200-500 mL min^{-1} air, with nitrogen carrier gas flow rates of 20-60 mL min^{-1}. In SFC, when 75 μL min^{-1} flow rates of supercritical CO_2 at 70°C and 102 atm are used, the optimum hydrogen and oxygen flow rates were 65 mL min^{-1} and 215 mL min^{-1}, respectively.[106]

The position and voltage of the collector may not be optimized for SFC in currently available flame ionization detectors. The position of the collector electrode relative to the flame, and the collector voltage, control the fraction of ions collected from the flame.[107] When supercritical CO_2 is used as the carrier gas, instead of nitrogen or helium, the flame structure is larger. When the flame size increases, gaseous blow-by of ions is increasingly more probable, causing < 100 % collection efficiency.

In addition to the variables described above, the kind of restrictor and its placement can affect the response of the FID. Furthermore, the low flow rates in capillary and microbore SFC allow complete expansion of the supercritical fluid into the flame; but in larger-scale packed column SFC, the effluent must be split to avoid extinguishing the flame.

Although the initial applications of SFC/FID showed poorer sensitivity when compared to GC methods, recent advancements have caused them to be comparable. Richter *et al.*[104] recently described a new FID detector that has detection limits of 5 pg carbon s^{-1} and a linear dynamic range of 10^6. Similar detection limits can also be achieved by adjusting the flame gas conditions of a standard GC/FID as demonstrated by Pekay and Olesik.[106]

Because the FID responds to organic carbon molecules, the use of modifiers with CO_2 is limited. Figure 4.17 shows the variation in the FID signal as a function of pressure when CO_2 modified with 5% methanol, methylene chloride, and hexane were used as mobile phases.[107] Figure 4.17 shows that quantification during pressure programming is also difficult when modifiers are present, due to the increasing baseline. To date, formic acid and water are the only two polar modifiers to CO_2 that allow ready quantification of compounds near the detection limit of the FID. In each case, the baseline varies minimally with pressure change.[23,108] Unfortunately, the solubility of these compounds in supercritical CO_2 is small (<1%).

Carrier gases other than CO_2 have been studied in SFC/FID. Lundanes *et al.*[109] compared the use of nitrous oxide and CO_2. Nitrous oxide causes significantly higher background signals than does CO_2. The combustion of nitrous oxide causes the formation of other oxides which ionize in the hydrogen-air flame.[110] The detection limit is increased by a factor of 1,000 when CO_2 is replaced

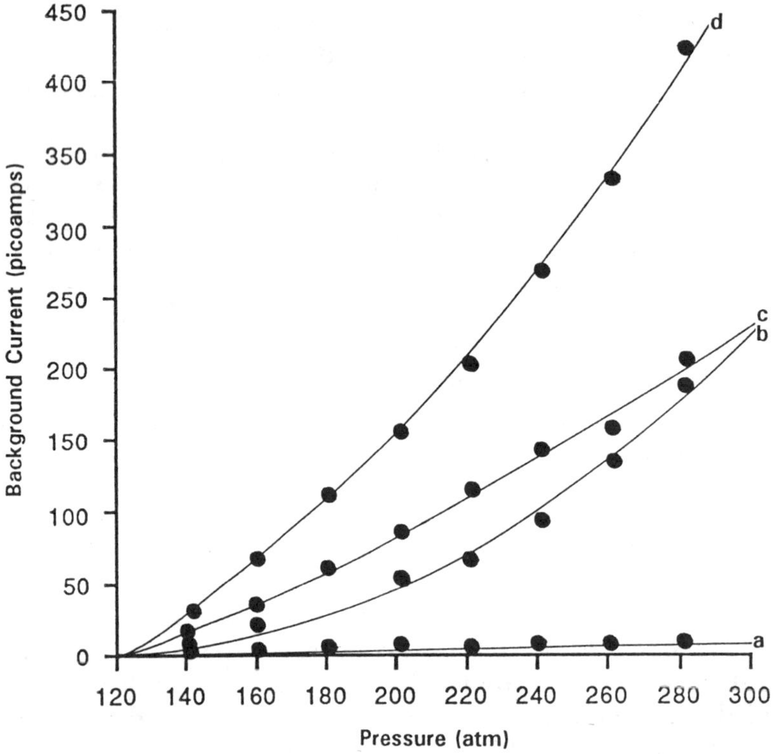

Figure 4.17. Background current of the FID as a function of mobile phase pressure for (a) neat CO_2, (b) 5% methanol in CO_2, (c) 5% hexane in CO_2, and (d) 5% methylene chloride in CO_2. All initial background readings were normalized to 0 pa (reprinted with permission from Ref. 107).

by nitrous oxide as the mobile phase. Apparently, the combustion of nitrous oxide causes clogging of metal restrictors, but not of fused silica restrictors. Sulfur hexafluoride[111-113] has also been used as a supercritical mobile phase in SFC/FID. Sulfur hexafluoride decomposes in a hydrogen-air flame, forming hydrogen fluoride. The HF corrodes the stainless steel collector in the FID. Accordingly, whenever SF_6 is used, the collector electrode should be plated with gold. Figure 4.18 shows a separation of primary alkyl amines using an SF_6 mobile phase. Minimal variation in background signal was observed with pressure variation in this chromatogram.

The high detection limits observed when using N_2O may be lowered considerably if the flame-gas chemistry is optimized for SFC/FID. Other analytical flames may also find use in addition to the standard hydrogen-air flame. Morrissey and Hill[114] recently

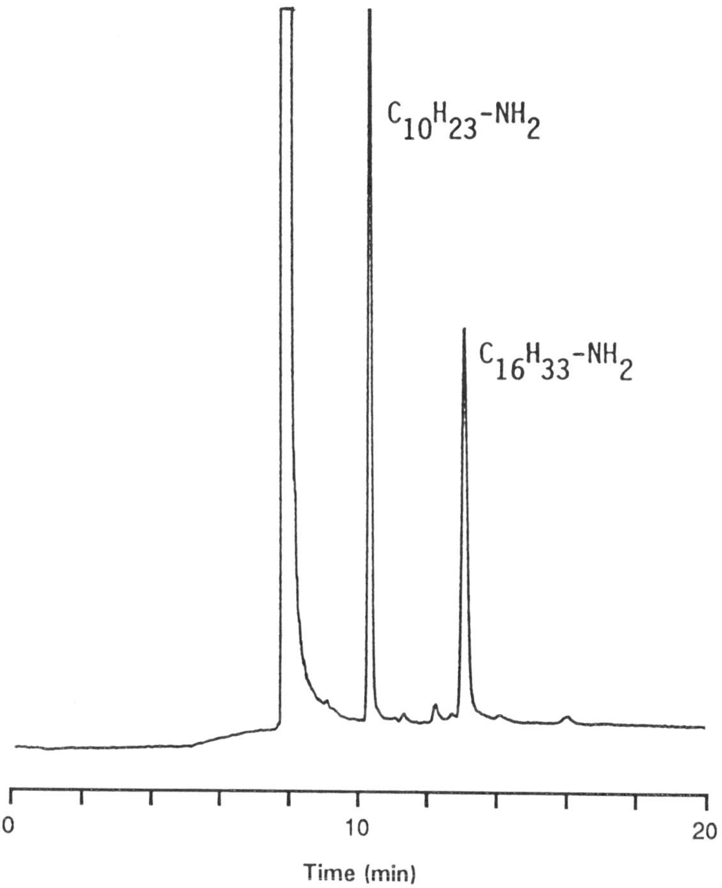

Figure 4.18. SFC chromatogram of primary alkyl amines. Conditions: 5-m x 100-μm i.d. open tubular column, polymethylsiloxane stationary phase; SF_6; 100°C; linear pressure program from 50 atm to 300 atm at 20 atm min^{-1} after a 5-min isobaric period (reprinted with permission from Ref. 113).

reported the use of the hydrogen-atmosphere FID for the analysis of organometallic compounds in supercritical solvents. Further applications of this detector can be expected.

Thermionic detection. Thermionic detectors (TIDs) selectively sense nitrogen- or phosphorus-containing organic compounds without interference from other coeluting hydrocarbons. In gas chromatographic analyses, TIDs are often used for trace analysis of environmental pollutants, such as organophosphorus pesticides or nitrogen substituted polycyclic aromatic hydrocarbons.

A conventional TID consists of a flame base and gas flow pneumatics similar to those used in a flame ionization detector. In the TID, a ceramic or glass bead is positioned above the flame jet. The bead has alkali metals such as rubidium or cesium coated on, or intermixed with, the ceramic. The bead is electrically heated to a temperature of 400-800°C. Hydrogen and oxygen gases flow around the bead. The proportion of hydrogen remains low so that a self-supporting flame cannot be maintained. Instead, a flame-like boundary layer forms around the heated bead. Analytes in the chromatographic effluent interact with the alkali metal, leading to the thermionic emission of ions when electronegative species such as nitrogen- and phosphorus-containing organics are present.[115] The ion current is the TID detection signal. Accordingly, the key variables of this detector are hydrogen and oxygen flow rates, bead temperature and composition, and the bias voltage on the bead. Under optimum conditions in GC analyses, the TID provides detection limits of 1×10^{-13} g N s^{-1} and 5×10^{-14} g P s^{-1}, specificity ratios of 4×10^4 g C/g N and 7×10^4 g C/g P, and a linear dynamic range of 10^4-10^5.[115]

Fjeldsted *et al.*[62] were the first to demonstrate the use of the TID in SFC analyses. Both supercritical CO_2 and nitrous oxide were used as mobile phases. The solvent peak was detected when nitrous oxide was used, which indicated the loss of some detector selectivity when using this supercritical solvent.

West and Lee[116] studied the use of the TID for the analysis of nitrated polycyclic aromatic compounds. Three different modes (TID-1-N_2, TID-2-H_2/air, and FTID-1) of operation of the TID were investigated. The minimum detectable quantity injected, the linear dynamic range, and the selectivity of each mode were compared. The standard compound used in the comparisons was 1-hydroxy-2-nitronaphthalene. The mobile phase was supercritical CO_2.

The TID-1-N_2 mode involved operation of the detector in an inert nitrogen environment. From gas chromatographic analyses, this mode is known to be useful for the analysis of nitro compounds because the sensitivity is excellent. The MDQ injected was 50 pg of compound. As in GC, the linear dynamic range in this mode was one order of magnitude. The specific concentration range that provided a linear response was also dependent on the bead current. High selectivity for nitro compounds was found. For 1-hydroxy-2-nitronaphthalene, a selectivity of 1×10^6 relative to naphthalene was reported.

In the TID-2-H_2/air mode, the bead is sheathed with hydrogen and oxygen under conditions of low flow. This is the conventional operation of a TID. The TID-2-H_2/air provided an MDQ injected of 2 ng of compound. The linear dynamic range was approximately one order of magnitude. The selectivity of this mode was 2×10^3 g C/g N for 1-hydroxy-2-nitronaphthalene. When the detector was operated in this mode, the baseline was very sensitive to variations in the density of the supercritical fluid. The FTID-1 mode involved placing the TID-1 detector downstream from a hydrogen/oxygen flame. This mode was unacceptable due to low sensitivity.

David and Novotny[117,118] recently published two papers showing some improvements in detector design, and various applications of the TID to the analysis of alpha-keto acids and steroidal thiophosphinic esters. Nitrous oxide was used as the mobile phase in each of these analyses. The alpha-keto acids were reacted with *o*-phenylenediamine to produce quinoxalinol derivatives. The steroid samples were reacted with dimethylthiophosphinic chloride to form the steroid thiophosphinic esters.

In both studies, the reported detection limits were high. The phosphorus detection limit was 120 fg P s^{-1} for the dithiophosphinic ester of pregnanediol, and the nitrogen detection limit was 2.0 pg N s^{-1} for *n*-hexylquinoxalinol. The linear dynamic range for both phosphorus and nitrogen was approximately 3-4 orders of magnitude. Figure 4.19 shows the selective detection of quinoxalinol derivatives of human urinary α-keto acids. Mathiasson *et al.*[119] recently optimized the key variables of the thermionic detector for the detection of amines by SFC. The flow rate of the nitrogen make-up gas and the bead current were held constant at values of 25 mL min^{-1} and 3.27 amps, respectively. The flow rates for hydrogen and oxygen and the bias voltage were optimized using a simplex optimization technique. The test mixture used in the optimization contained 2,6-toluenediethyldicarbamate and eicosane in concentrations of 100 ppm and 0.2%, respectively. The product of the sensitivity and the selectivity of the detector for the carbamate (compared to eicosane) was the parameter that was optimized. All measurements were done at a constant pressure of 120 atm and a nitrous oxide flow rate of approximately 6 μL min^{-1}. Under these conditions, the optimized key parameters were 1.2 mL min^{-1} hydrogen, 230 mL min^{-1} air, and a bias of 7.0 V. The measured selectivity for these optimized conditions was 200. Within the experimental error of the experiment, the sensitivity of the optimized detector was comparable in SFC to the GC values. Linear calibration curves were obtained over concentration ranges of 10-1,000 ppm for various aromatic amide and carbamate derivatives. No variation in the detector sensitivity was observed for amines and their derivatives

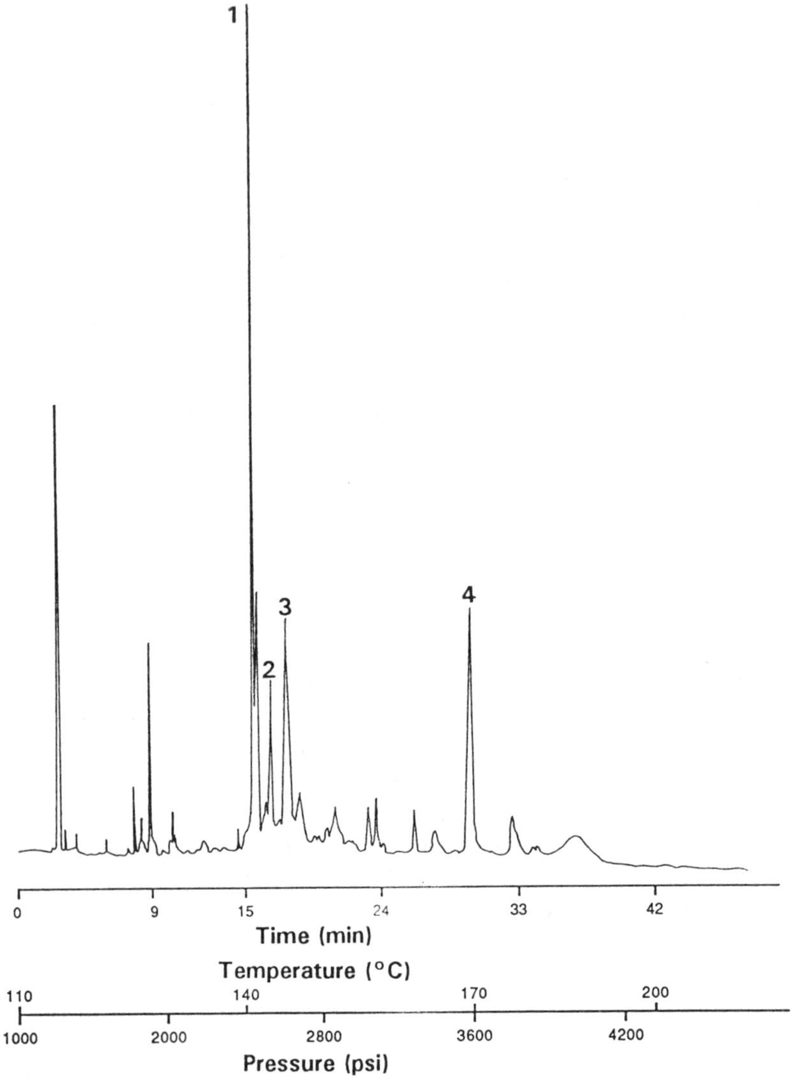

Figure 4.19. SFC chromatogram of quinoxalinol derivatives of human urinary α-keto acids. Conditions: 10-m x 50-μm i.d. open tubular column, polymethylsiloxane stationary phase; CO_2; linear pressure program from 68 atm to 268 atm at 5.5 atm min^{-1}; temperature program from 110°C to 200°C at 2°C min^{-1}; TID. Peak identifications: (1) methylquinoxalinol, (2) ethylquinoxalinol, (3) *p*-propylquinoxalinol, (4) 3-benzylquinoxalinol (reprinted with permission from Ref. 117).

when the system pressure was varied. The selectivity for nitrogen was low, 200 g C/g N. The loss in selectivity was attributed to increased nonspecific ionization caused by the use of nitrous oxide as the mobile phase.

The influence of modifiers on TID detector characteristics has only received moderate attention. Mathiasson *et al.*[119] reported no measurable solute signal when concentrations of methanol higher than 0.8% were added to the nitrous oxide mobile phase. However, Greibrokk *et al.*[23] recently demonstrated the successful detection of compounds with a TID using 7% methanol in CO_2.

Flame photometric detection. The flame photometric detector (FPD) is often used in GC analyses to selectively detect sulfur or phosphorus in organic compounds. This detector uses an H_2/O_2 or H_2/air diffusion flame. Organosulfur or organophosphorus compounds decompose in the flame. For sulfur detection, the chemiluminescence of the excited sulfur dimer, S_2, is monitored. For phosphorus detection, the chemiluminescence of HPO is monitored.

Markides *et al.*[120] used SFC/FPD for the separation and detection of polycyclic aromatic compounds with thiophenic rings, and pesticides containing sulfur or phosphorus. A dual-flame photometric detector was modified by shortening the flame base and adding a make-up gas line slightly above the air inlet. Using supercritical CO_2 as the mobile phase, MDQs of 25 ng and 0.5 ng injected were found for sulfur in benzo[*b*]thiophene, and phosphorus in parathion, respectively. The baseline of the FPD increased substantially when the CO_2 pressure was increased. Baseline correction was used in density programmed SFC.

Olesik *et al.*[121] studied and optimized the flame-gas chemistry in a single-flame FPD for detection of organosulfur compounds in supercritical CO_2. The flow rates of the flame gases and the detector temperature were optimized using a combination of simplex and factorial design optimizations. Response surfaces were determined for both the sulfur signal/noise ratio and the background as a function of the flame gas flow rates. The detector temperature did not affect the detected sulfur emission. The optimum conditions for sulfur detection under SFC conditions were 239 mL min^{-1} H_2, and 45 mL min^{-1} O_2 with a supercritical fluid CO_2 volumetric flow rate of 32 μL min^{-1}. The background signal was found to be low when the S/N ratio was high. In GC, the flame gas conditions were 70 mL min^{-1} H_2 and 24 mL min^{-1} O_2.

Olesik *et al.*[121] also studied the quenching effects of the coelution of a hydrocarbon with an organosulfur compound. In previous GC experiments, nanogram quantities of hydrocarbon were found to completely quench the sulfur emission from nanogram quantities of organosulfur compounds. The extent of quenching varied inversely with the concentration of the organosulfur compound in the earlier GC experiments. When the optimum flame gas conditions were used in the SFC/FPD studies, no quenching occurred until the concentration of hydrocarbon was much larger than that of the organosulfur compound. The extent of quenching did increase with concentration of the organosulfur compounds. This is opposite to the trends seen in GC/FPD studies. The quenching mechanisms are believed to be different in the GC and SFC studies.

A marked dependence of the sulfur emission on the position of the restrictor relative to the flame base has also been observed. Two optima were found: one approximately 1 cm beneath the flame base, and the other 0.5 cm into the flame. Similar to the results of Markides *et al.*,[120] the background signal increased with increasing CO_2 pressure. The detection limits for the optimized SFC/FPD were approximately equivalent to the lowest quoted GC/FPD values.

More recently, Pekay and Olesik[106] further characterized the SFC/FPD. White noise was found to be the predominant noise source. This characterization may allow further decreases in the detection limit through signal-to-noise enhancement techniques. The chromatogram shown in Figure 4.20 involved considerable variation in pressure. In contradiction to the previous study, the background signal did not vary with pressure. In this recent work, a sharply tapered restrictor was used instead of a linear restrictor. Sensitive detection of high molecular mass polymers containing sulfur was obtained. In addition, the use of SFC/FPD to monitor sulfur vulcanization of polymers was described.

4.12 UV Absorption Detection

Multi- or fixed-wavelength UV detectors. Conventional optical detection methods, such as UV absorbance, have been used in SFC since the first report of SFC appeared, and they are still among the most popular detection techniques. Unlike the FID, the UV detector only responds to solutes which have a UV absorbing functionality. At wavelengths lower than 210 nm, the number of organic molecules which absorb is quite large. Unmodified inorganic mobile phases, such as CO_2, exhibit little background absorbance in the 200-800 nm region and, therefore, produce low background for

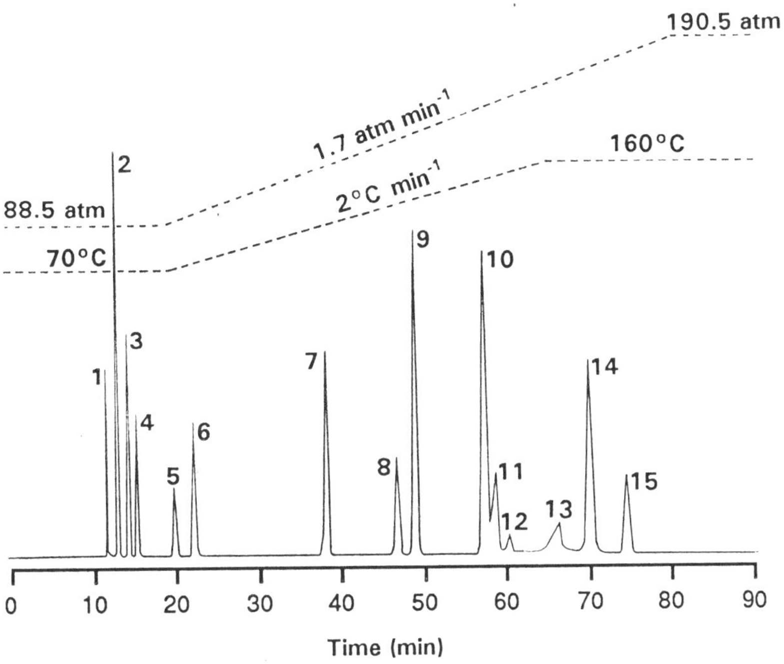

Figure 4.20. SFC chromatogram of organosulfur compounds. Conditions: 12-m x 220-μm i.d. open tubular column, poly(7% cyanopropyl, 7% phenyl)methylsiloxane stationary phase (0.25-μm film thickness); CO_2; linear pressure program from 88.5 atm to 190.5 atm at 1.7 atm min^{-1}; temperature program from 70°C to 160°C at 2°C min^{-1}; FPD. Peak identifications: (1) carbon disulfide, (2) tetrahydrothiophene, (3) cyclohexyl mercaptan, (4) thiophenol, (5) methyl *p*-tolyl sulfide, (6) benzo[*b*]thiophene, (7) phenyl sulfide, (8) dibenzothiophene, (9) phenyl disulfide, (10) quinomethionate, (11) ovex, (12) tetrathiafulvalene, (13) tolbutamide, (14) 1,4,8,11-tetrathiacyclotetradecane, (15) 1,5,9,13-tetrathiacyclohexadecane (reprinted with permission from Ref. 106).

highly sensitive absorbance detection. Unlike the FID, which is a very nonspecific and destructive detector, the UV detector can provide compound-specific information, and it is nondestructive. A UV detector can be tuned to the absorption wavelength of the analyte of interest in order to provide compound-specific information. Additionally, a scanning UV monitor may be employed to provide spectral information.

Packed columns. The UV detector has been used most frequently for detection in packed column SFC.[7,108,122,123] Interfacing the UV detector to an SFC instrument which uses

4.6-mm i.d. packed columns is straightforward. For these large volume columns, the major concern is not detector dead volume, but containment of the fluid at high pressure, because detection usually takes place in the condensed phase at up to 400 atm.

Optical design criteria also must be met in order to produce a high performance absorbance monitor. It is important that the excitation beam be focused into the flow cell so that interactions between the solute band and excitation light are maximized, while "wall effects" are minimized. These wall effects, in which the light reflects off of the inner walls of the flow cell, produce noise due to scattering and enhanced sensitivity to changes in the refractive index (RI) of the fluid in the cell. This RI perturbation is particularly bothersome in SFC if detection is carried out under supercritical fluid conditions, since RI changes result from density changes in the mobile phase. Fortunately, detection is usually accomplished under liquid CO_2 conditions.

Open tubular columns. The use of open tubular columns impose some of the same flow cell design requirements as for the use of packed columns, with the added dimension of stringent detector cell volume limitation. Peaden and Lee[124] and Fields *et al.*[125] have developed an expression to estimate the allowable detector volume in capillary SFC

$$V_d = 0.866\pi d_c^2 (L\,h)^{1/2} \left[\frac{1}{(1 - \Delta R_s)^2} - 1 \right]^{1/2} (1 + k) \frac{\rho_c}{\rho_d} \qquad (4.20)$$

where V_d is the volume of the detector, ΔR_s is the fractional resolution loss, L is the column length, h is the plate height, k is the capacity factor, d_c is the internal column diameter, and ρ_c and ρ_d are the fluid densities in the column and detector. This equation is identical to Equation 2.10, except for the addition of the contributions of density change from the column to the detector cell. As an example, the allowable detector volume which would produce no more than a 1% resolution loss for a 20-m x 100-μm i.d. column, with a plate height of 0.6 d_c, and for a k value of 1, is 270 nL. Employing the same parameters for a 50-μm i.d. column, the required cell volume is reduced to only 50 nL. These small volumes can be attained, but at the expense of detector path length. A more detailed discussion on sensitivity as it relates to path length is found later in this section.

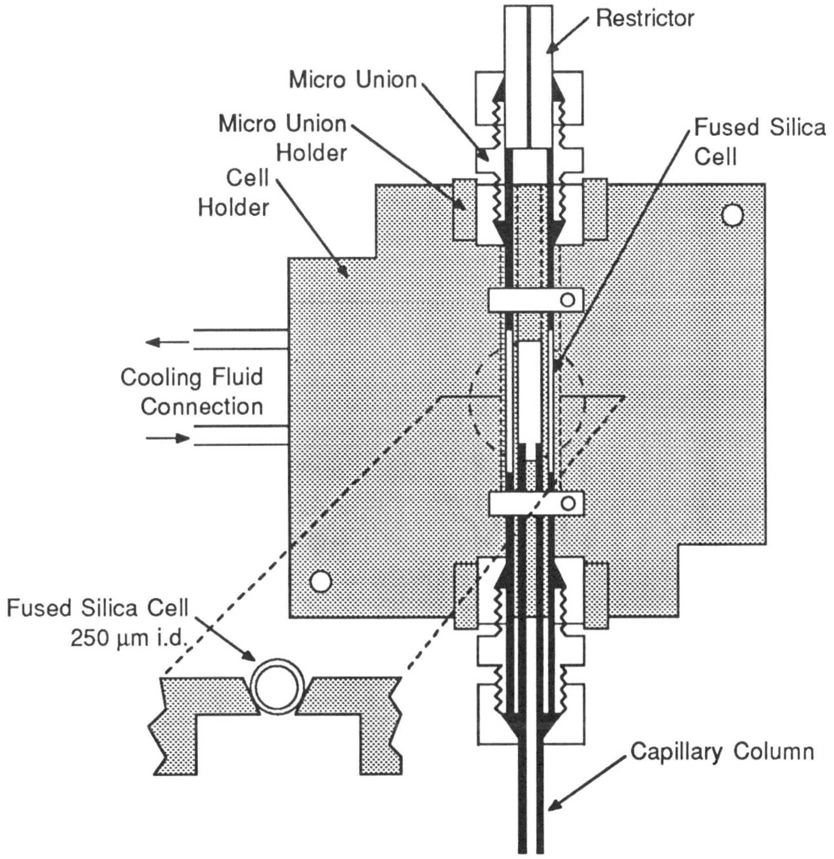

Figure 4.21. Schematic diagram of a "pseudo on-column" UV flow cell for open tubular column SFC (reprinted with permission from Ref. 126).

One approach taken for UV detection in open tubular column SFC[125,126] is similar to that employed for capillary LC.[127] In this technique,[126] a small section of the polymer coating is removed from a section of the analytical open tubular column, or from a short piece of capillary tubing, which serves as the detector cuvette. As shown in Figure 4.21, two butt connectors connect the detector capillary cuvette directly to the capillary column and to the restrictor. In this configuration, a 250-µm i.d. capillary is the detector cell. Certainly, a portion of the 50 or 100-µm i.d. separation column could be employed directly, but only at the expense of detector sensitivity. The use of a *ca.* 250-µm i.d. capillary gives a detector volume of about 60 nL, which is a good compromise between dead volume and acceptable sensitivity. Resolution losses due to detector band broadening when using the on-column detection approach have been calculated for microcolumn LC[128] and for open tubular column SFC.[125] In a study performed by Later *et al.*,[129] it was shown that the practical upper limit for the inner diameter of the capillary tubing used as the flow cell in conjunction with a 50-μm i.d. column is about 320 μm.

The measured noise for an open tubular column SFC/UV system using CO_2 at a density of 0.24 g mL^{-1} is *ca.* 2 x 10^{-5} AU; the static, short term noise is measured to be about 1.2 X 10^{-5} AU. The dependence of absorbance detector sensitivity on path length has been investigated by various scientists.[130,131] Poppe[130] derived an equation which suggests a somewhat surprising implication in that, at a specified cell volume, larger cell path lengths result in poorer detection limits. In his expression, movement to smaller path lengths, which would favor detectability, is limited by the instability of the source and the receptor. From the work of Baumann and Fresenius,[132] a volume of 10 μL and a path length of 1.0 cm was found to be the optimum for the existing technology. In particular, conventional deuterium and low pressure mercury lamps limit the number of photons per unit volume, and cells with a volume of less than 1.0 μL can only be used at the expense of detection sensitivity.

Ultimately, the analytical instrumentalist strives for the maximum signal-to-noise ratio (S/N) which can be obtained by the instrument. Because of various limitations, researchers have worked to develop absorbance detection technology which utilizes the high photon flux that a laser can produce in order to push the detection limits lower.[133-138] These nonconventional absorbance methods are not discussed here; yet it is important to note that thermal lens, one of these nonconventional absorbance methods, has been used both in LC[133-136] and in supercritical fluids.[137] Leach and Harris[137,138] found that near the critical point, a significant enhancement of the thermal lens signal is realized. Thus, it appears possible to design an absorbance detector with no path length dependence and superior sensitivity.

Despite the basic physical limitations of a conventional UV system, the on-column approach has proven quite successful. Because of the sensitivity enhancement as a result of the small column diameters and the low flow rates used in the capillary techniques, reasonable detection limits are obtained, even with the short path length provided by a capillary cuvette. Solutes such as anthracene and pyrene can be detected in the mid-picogram range[139,140] (*e.g.,* 5.8 pg of anthracene with a noise level of only 2 x 10^{-5} AU). The use of fiber optics has contributed to the ease of transfer of transmitted light from the small detector cell to the photomultiplier tube.[141]

Recently, Chervet *et al.*[142] reported a novel approach to microcolumn detection for LC and SFC. In their work, a 2-cm path length was obtained while keeping the detector volume to less than 90 nL by using a "Z" shaped capillary tube as the flow cell. A dynamic noise level of 1×10^{-4} AU with a 2-s detector response time and an MDQ of 7 pg of phenanthrene were reported. It is assumed that this high noise level was a result of light scattering at the exit and entrance points of the flow cell. Band broadening effects appear to be well controlled in this flow cell arrangement, and good chromatographic performance can be maintained. The maximum pressure rating for this cell was postulated to be about 500 atm.

Density programming. Density programming sometimes poses another problem for the UV detector, as manifested by the presence of a baseline shift. When CO_2 is the mobile phase,[125] this baseline offset is seen as a negative deflection in the chromatographic trace. Novotny[143] has reported a positive shift when using supercritical *n*-butane. It is postulated that density-related RI changes contribute largely to this effect.[125] The magnitude of this effect has been quantified for supercritical CO_2, giving typically encountered values of about 1.3×10^{-2} AU g^{-1} mL^{-1} when a 250-μm i.d. capillary was the detector cuvette. By regulating the temperature of the cell with cool H_2O, as much as a 55% improvement in RI sensitivity was realized. Another alternative to correct for the baseline shift is to program the detector to perform baseline compensation.

Baseline drift will decrease when modifiers are used in the CO_2, or for fluids having less RI change with density change. An example is given in Figure 4.22 with the analysis of a poly(50% phenyl)methylsiloxane sample[76] on a packed column using a 10% ethanol/*n*-hexane mobile phase. This analysis would have been quite difficult without the use of an organic modifier. Additionally, the analysis would have been impossible to perform had the FID been used as the detector.

Scanning UV detection. Multiwavelength (scanning) UV detection provides the advantages that tunable UV detection has in SFC, with the added benefit of providing additional spectral information about the analytes. As discussed previously, this additional gain in information often facilitates the identification of specific species through spectral matching. Collecting on-the-fly

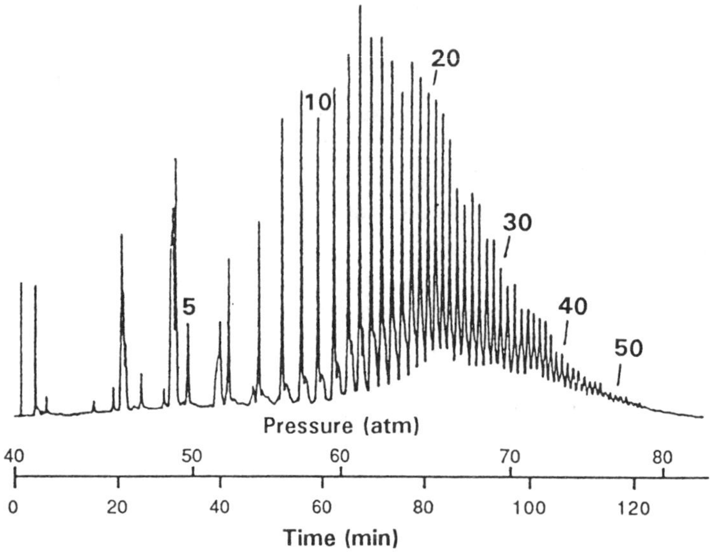

Figure 4.22. SFC chromatogram of poly(50% phenyl)methylsiloxane (OV-17). Conditions: 50-cm x 200-μm i.d. packed column, C_{18} bonded silica particles (10 μm, Finesil); 10% ethanol in n-hexane; 260°C; linear inlet pressure program from 40 atm to 80 atm at 0.32 atm min^{-1}; 15 atm ΔP; UV (210 nm) (reprinted with permission from Ref. 76).

spectra to assist in compound identification has been realized in SFC with the use of a photodiode-array UV (PDA/UV) detector and with a galvanometer driven monochromator (GDM) UV detector.

Open tubular columns. France and Voorhees[144] have used a scanning PDA/UV in open tubular column SFC for the analysis of some pesticides and herbicides. The reported noise level was found to be about 3 x 10^{-3} AU during broad band integration, resulting in a detection limit of only 3.8 ng of bendiocarb (S/N = 5). Although this approach was successful in obtaining spectral information, the modest sensitivity values, short linear dynamic range, and relatively high noise levels observed are due to excessive refractive index (RI) sensitivity which stems from positioning the flow cell at the beginning of the optical bench (reverse optics geometry). In the reverse optics design, alterations in the solvent composition and in the entry of solute bands into the flow cell change the refraction angle of transmitted light, altering the distribution of light on the dispersive element. The result is a change in light intensity at the level of the photodiode array. This intensity change manifests itself as excessive RI sensitivity to density programming and errors in measured absorbance, because the total measured absorbance by a PDA is equivalent to the sum of the light absorbed by the solute and the amount of light gained or lost due to the changing flow cell RI.[145]

An alternative approach to multiple wavelength detection on a capillary scale is obtained by the use of a forward bench in conjunction with a rapid-scanning, galvanometer-driven monochromator. Such a design minimizes propagational distances between the flow cell and phototransducer, thus limiting the overall variation in light distribution, secondary to flow cell RI change. Additionally, forward optical designs exhibit lower stray light, and they can utilize a more efficient double beam scheme, resulting in superior signal-to-noise performance when compared to reverse optical systems.[145] Past attempts to create a rapid scanning forward optical monitor have been hampered by poor wavelength reproducibility, resulting in typical multiple wavelength noise similar to that of a PDA. Only recently has a stable, reproducible movable grating system been developed.

Weinberger and Bornhop[140] used this recently developed monitor to successfully interface open tubular column SFC to a scanning UV detector. A pseudo on-column detection approach was used with a 250-μm i.d. capillary as the detector flow cell cuvette. In preliminary work, it was found that many of the theoretical advantages of the forward optical configuration could be realized. The noise level in the rapid scanning mode was found to be about 1×10^{-4} AU with CO_2 as the mobile phase in open tubular column SFC. An on-column MDQ of 32 pg for naphthalene while operating in the high speed scanning mode was reported, with a linear dynamic range which extended over more than 3 orders of magnitude in concentration.

Packed columns. Jinno *et al.*[146] utilized the PDA-based multichannel UV detector for the SFC analysis and identification of polycyclic aromatic hydrocarbons in diesel particulate matter. In their work, a conventional 4.6-mm i.d. packed column was used to provide the separation, thus eliminating the necessity to modify the flow cell volume.

In another report by Jinno *et al.*,[147] a multichannel UV detector was employed with computer enhancement in order to effect the separation of coeluting solutes. Packed columns with inner diameters of 4.6 and 1.0 mm were used to analyze toluene and perdeuterotoluene. Completely overlapping peaks were separated spectroscopically by the use of deconvolution techniques.

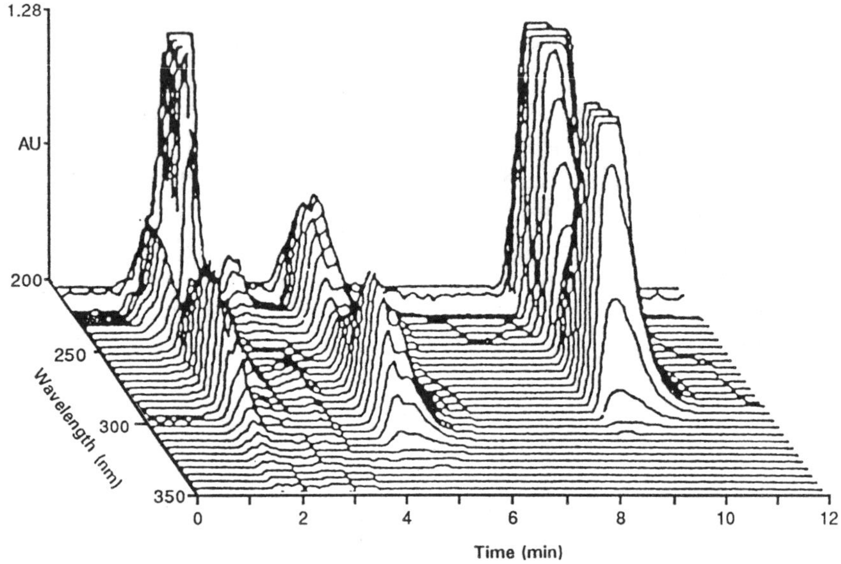

Figure 4.23. SFC chromatogram of a coffee extract using directly coupled SFE/SFC. SFE conditions: 20% (w/w) water added to the sample; CO_2; 40°C; 200 atm; 15 min. SFC conditions: 15-cm x 6-mm i.d. packed column, C_{18} bonded silica particles; methanol in CO_2 gradient; 40°C; 150 atm; 5 mL min^{-1} liquid flow rate; UV (200-350-nm) (reprinted with permission from Ref. 148).

The use of the PDA detector has not been limited to only single column chromatography, and in fact, multichannel detection has been used in various other supercritical fluid applications.[148,149] Sugiyama *et al.*[148] employed a PDA multiwavelength detector in a double-stage separation analysis method. They directly coupled a laboratory-scale SFE/SFC instrument to a commercial PDA instrument. Caffeine was extracted, chromatographed for separation from impurities, detected, and quantified. They found that the information from the PDA allowed for the positive identification of the caffeine solute. As shown in Figure 4.23, the three dimensional SFC chromatogram provides a wealth of solute information. Information can be extracted by plotting a specific wavelength trace and a "time slice" to obtain a spectrum of an eluting solute. Without this spectral information, efficient identification of caffeine would be very difficult.

4.13 Fluorescence Detection

Fluorescence detection for SFC has been developed as a selective and sensitive means of detecting samples capable of fluorescence.[150] Aside from the four major detectors used in SFC (*i.e.*, UV, FID, MS, and FTIR), fluorescence detection has been the next most popular detection method reported.

A wide variety of mobile phases have been shown to be applicable with fluorescence detection. These include CO_2, the light hydrocarbons (*i.e.*, propane, *n*-butane, and *n*-pentane), a halocarbon (dichlorofluoromethane), ammonia, *n*-pentane- and CO_2-based modifier mixtures, and reverse micelles (propane + AOT + water). An obvious requirement for the use of this detector is that the sample must contain a fluorophore. Sample types analyzed by this technique include polycylic aromatic compounds, vitamins, proteins, biphenylsiloxanes, and fluorescent whitening agents.

Flow cell designs for fluorescence detection in capillary SFC are refinements of previous developments for the use of spectroscopic detectors in capillary liquid chromatography.[127,151] Currently however, no commercial fluorescence detectors for SFC are available; consequently, spectrophotometers or detectors designed for liquid chromatography must be modified. A typical flow cell for fluorescence detection in capillary SFC is shown in Figure 4.24. This detector flow cell configuration consisted of a fused silica sleeve which also served as the flow cell (after the polyimide coating was removed from a section of tubing), an optical fiber (placed axially to the chromatographic column in the flow cell) for collection of emitted light, and the excitation light entering at a right angle to the flow cell.

With this fluorescence detection system,[150] a minimum detectable quantity of 2 pg for pyrene at a signal-to-noise ratio of 2 was reported, resulting in a detection limit of 500 fg s^{-1}. Jackson[152] noted that detection at ambient temperatures could conceivably result in the precipitation of marginally soluble solutes, but, with *n*-pentane as the mobile phase, he saw little change in detection limits for large polycyclic aromatic hydrocarbons, and nearly identical spectra for both sub- or supercritical detector temperatures. A small loss in sensitivity for pyrene at supercritical detection temperatures (>196°C) was attributed to the temperature dependence of the fluorescence quantum efficiency.

The feasibility of developing quantitative methods using capillary SFC-fluorescence was shown in the determination of coronene and

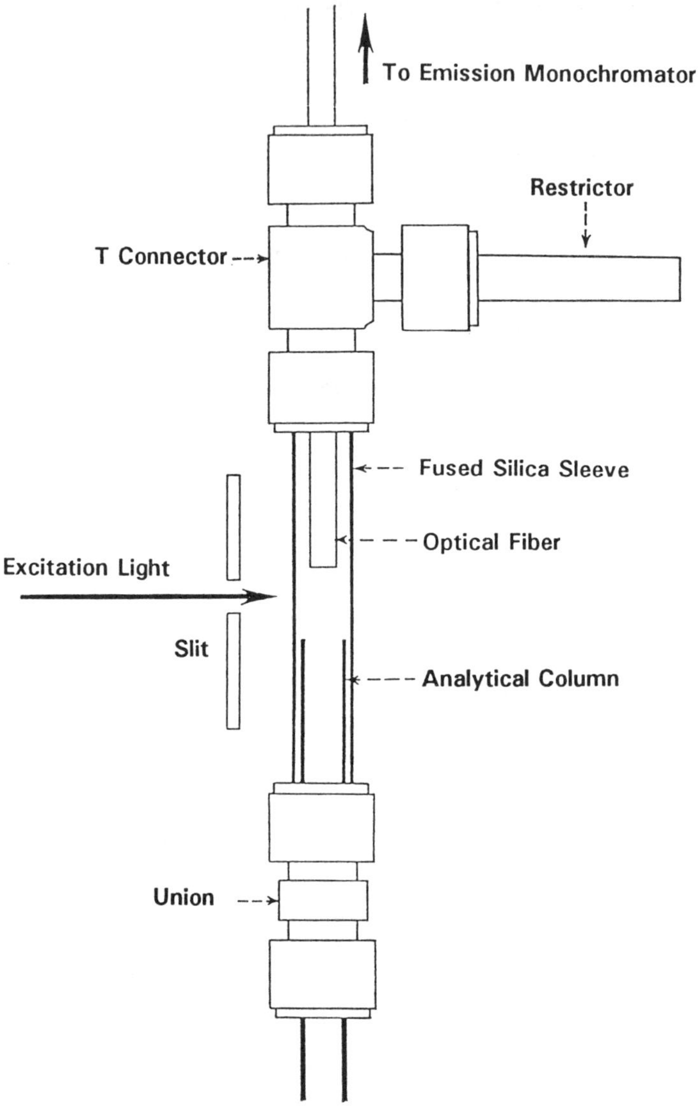

Figure 4.24. Schematic diagram of a flow cell for fluorescence detection in open tubular column SFC (reprinted with permission from Ref. 150).

ovalene in two hydrocracker recycle oils.[153] These results, listed in Table 4.2, are compared to liquid chromatography using UV detection. The two methods compared favorably, with the coefficient of variation for the SFC method ranging from 1.8 to 4.2%.

Table 4.2. Quantification of PAH in Hydrocracker Recycle Oils[a]

	Concentration (ppm)			
	Coronene		Ovalene	
Sample	LC/UV	SFC/Fluorescence	LC/UV	SFC/Fluorescence
A	220	225 (1.8%)[b]	264	264 (2.3%)
B	70	71 (4.2%)	68	65 (3.1%)

[a]Reprinted with permission from Ref. 153.
[b]Values listed in parentheses are the coefficient of variation for the SFC analysis performed in triplicate using the standard addition method.

Supersonic jet spectroscopy. Conventionally, gas phase molecular spectra are highly congested due to the thermal population of upper vibrational and rotational energy levels. Large compounds, which have many internal energy modes, give spectra consisting of broad contours of entirely unresolved bands. In the condensed phase, inhomogeneous broadening also contributes to the blurring of spectral features. There are several ways to increase the spectral resolution, but supersonic jet spectroscopy (SJS) is the only one which gives truly free-molecule spectra. It is also the most compatible technique for chromatographic detection. Analytical applications of supersonic jet spectroscopy have been well reviewed by Hayes,[154] and Goates and Sin.[155]

A supersonic jet is formed by expanding a gas through an orifice into a low pressure region. If the orifice size is much larger than the mean free path of the molecules, many two-body collisions occur during expansion, transferring both internal and external energies of the gas molecules into a directed forward motion. As a result, the velocity distribution of the molecules is narrowed, and a beam of "cold" molecules is produced. Spectra of molecules seeded into a jet expansion of a carrier gas will have highly resolved peaks. The degree of cooling is affected by the identity of the carrier gas. Monatomic gases give the most cooling because they have no internal degrees of freedom to contribute to the heat capacity. The spectral peak widths are typically less than one-tenth of a nanometer. Since the positions of the peaks are sensitive to molecular structure, this high resolution enables one to identify different compounds, even in the case of geometric isomers, which often cannot be easily resolved by chromatography or mass spectrometry.

Supersonic jet spectroscopy experiments are usually performed with lasers, so that enough sensitivity can be achieved. Laser-induced fluorescence (LIF) was used in the first GC/SJS experiment.[156] Detection of the total fluorescence allows for sensitive, selective detection of compounds in the sample. However, in this mode, selectivities are derived entirely from the ability to selectively excite the components of interest, and the identifying information of the fluorescence spectrum is forfeited. Dispersed fluorescence spectra can be obtained on the chromatographic time-scale with a photodiode array spectrometer, but there is a loss in sensitivity. As an alternative, resonant two-photon ionization (R2PI) can be used.[157] Molecules in the jet are ionized when the energy sum of two absorbed photons exceeds the ionization potential. The ion intensities are wavelength dependent, and resemble the absorption contour of the first photon. Mass spectrometric detection yields a dimension of data for compound identification. The R2PI approach is also useful for compounds with low fluorescence quantum yields. Picogram MDQs have been achieved in favorable cases for both LIF and R2PI detection schemes. A more detailed discussion of the challenges for detection is given by Goates and Sin.[155]

As a chromatographic detection method, SJS is more suitable for GC and SFC than for LC, because of the extensive formation of aerosols by the liquid solvents. The formation of aerosols, solvent clusters, and solute-solute clusters can occur in jet expansions of supercritical fluids and even gases. These problems can usually be eliminated by moderate heating of the nozzle. The processes involved in the expansion of supercritical fluids into a vacuum have been studied by several groups.[158-162] A fundamental conflict exists between the requirements for the column restrictor and the supersonic jet nozzle, requiring careful design of the nozzle interface. The remainder of this section will focus mainly on two types of interfaces recently developed for LIF supersonic jet detection in SFC.[163,164]

One approach is simply to use a fixed restrictor at the end of the SFC column as the supersonic jet nozzle. High restriction is needed to reduce the linear flow velocity in the column sufficiently for good separation of sample components. However, abrupt restriction is needed for supersonic expansion to occur, and to prevent the condensation of solute clusters. A direct expansion nozzle can be made by coupling a small, pinhole orifice to the end of a capillary transfer line.[163] In practice, the smallest orifice size one can use is about 5 µm; otherwise, frequent plugging of the nozzle occurs. For capillary SFC, the flow through a 5-µm orifice is too fast for an efficient separation, and it is also difficult to achieve a low enough dead volume at the junction of the orifice and the capillary. This nozzle works well for microbore column SFC, however, because total flow rates are higher. Good cooling of sample molecules is obtained with simple carriers such as supercritical CO_2 and N_2O, even though they are not as efficient supersonic jet coolants as are monatomic carrier gases. Spectral bandwidths are about 5 cm^{-1}. More complex supercritical fluids are not effective for direct expansion SJS. To avoid CO_2 clustering, it was necessary to heat the nozzle above 180°C. A 500-pg MDQ for perylene has been measured using this nozzle.

For capillary SFC/SJS, a sheath-flow nozzle has been found to be effective.[164] In a sheath-flow arrangement, the supercritical mobile phase is first introduced into a flowing sheath of Ar make-up gas as it exits the end of the column. Then the mixed flow is supersonically expanded into the vacuum through a 25-µm orifice. The restriction at the end of the column must still be abrupt to prevent condensation of the analyte, but because the expansion need not be supersonic at this point, a porous frit or a short tapered capillary restrictor can be used. The nozzle can withstand high pressures for back-pressure regulated restriction,[86] but for supersonic jet performance, the Ar pressure is usually set at 6 to 10 atm. At still lower make-up gas pressures, such as those used in the study of Anderson and Johnston,[165] large molecules are not transported through the nozzle into the jet. Clustering of mobile phase and sample molecules is inhibited by the introduction of Ar, enabling the nozzle to be operated at lower temperatures. Also, greater supersonic jet cooling occurs with the monatomic carrier. In comparison to results with perylene in a direct CO_2 expansion, spectral linewidths were reduced by about half, and vibrational hot bands were further suppressed. If He was used instead of Ar as the sheath-flow make-up gas, no signal was detected, probably because He does not have enough momentum to transport the eluent into the jet. An MDQ for perylene of 150 pg at a S/N ratio of three has been measured. The sample is swept cleanly and efficiently out of the nozzle, and the effective dead volume is negligible. With this nozzle, selective detection of perylene and benzo[*a*]pyrene in capillary SFC separations of complex samples was performed.

SJS provides highly selective detection for SFC, and shows great promise as a means for obtaining detailed information about complex samples. Better sensitivity is needed, especially for dispersed fluorescence detection, because it provides real time identification data for each chromatographic peak. Perhaps the greatest need for SJS to become a useful detection method is an extension of the SJS data base.

4.14 Chemiluminescence Detection

In a chemiluminescence reaction, a molecule in an excited energy state emits light while returning back to the ground state. Light can be emitted from vibrationally or electronically excited molecules. In a competitive reaction, where excited molecules collide with each other or with the reaction chamber walls, no light is emitted. In the chemiluminescence detector, the chromatographic effluent is mixed with the reagent gas, and the reaction is recorded using a photomultiplier tube. High efficiencies can be achieved, because individual photons are counted. The detector response is increased by operating at reduced pressures, because competitive reactions are minimized. In the sulfur chemiluminescence detector (SCD), low pressures are also needed for *in situ* reagent gas (O_3 or F_2) formation.

Detectors based on chemiluminescence reactions can be operated either as universal or as selective detectors. Selective detection has its potential in detecting single components in unresolved mixtures. Thus, the use of selective detection may reduce the need for extensive cleanup and prefractionation steps prior to analysis. Furthermore, since the solvent peak is reduced significantly using the SCD (Figure 4.25), the chromatographic conditions can be changed to provide faster elution than with the FID.[166]

The redox chemiluminescence detector (RCD) can be used to detect easily oxidizable analytes. The detection is based on a postcolumn reaction, in which the analytes reduce nitrous oxide (NO_2) to nitric oxide (NO) on a catalyst bed. Nitric oxide then reacts with ozone (O_3) to form an electrically excited nitrous oxide molecule.[167,168] The relaxation reaction is monitored at 600-800 nm.[169] Thus, the signal is proportional to the reducing activity of the analyte. The detection of nitrosamines with the thermal electron analyzer is based on the same reaction as used in the redox chemiluminescence detector, except that nitric oxide is generated by pyrolyzing the nitrosamines at 480-750°C, depending on the compound.[170]

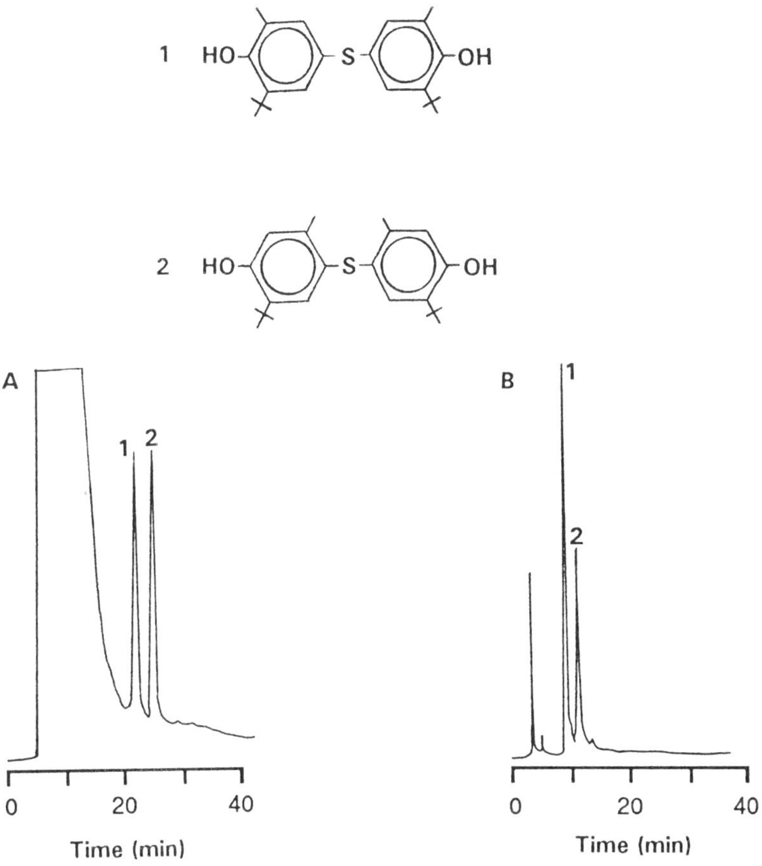

Figure 4.25. SFC chromatograms of phenolic anti-oxidant isomers detected by (A) FID and (B) F_2-SCD. Conditions: 3.5-m x 100-μm i.d. open tubular column, poly(5% phenyl)methylsiloxane stationary phase (d_f = 0.4 μm); CO_2; 80°C; linear pressure program from 150 atm to 250 atm at 2.5 atm min^{-1}; FID (100°C); SCD (80°C). Peak identifications: (1) 4,4′-thio-bis-(*t*-butyl-*o*-cresol), (2) 4,4′-thio-bis-(*t*-butyl-*m*-cresol) (reprinted with permission from Ref. 166).

Sulfur chemiluminescence detection is based on the reaction of sulfur monoxide (SO) with ozone, or on the reaction of a sulfur-containing compound with molecular fluorine (F_2):

$$\text{S-Compound} + H_2 + O_2 \xrightarrow{\Delta} SO + \text{other products}$$
$$SO + O_3 \longrightarrow SO_2^* + O_2$$
$$SO_2^* \longrightarrow SO_2 + h\nu \ (320\text{-}390 \text{ nm})$$

$$R_1\text{-S-}R_2 + F_2 \longrightarrow HF^* + \text{others}$$
$$HF^* \longrightarrow HF + h\nu \ (>450 \text{ nm})$$

In the latter case, either R_1 or R_2 must contain hydrogen to produce a vibrationally excited HF* molecule.[166,167] The highest response is achieved if there is a hydrogen in an α-position to the sulfur. Electronically excited HCF* and FCS* reaction products may also contribute to the signal.[166] The response in the F_2-SCD depends on the compound reactivity, while the response in the O_3-SCD is directly dependent on the amount of sulfur.

The configuration of the F_2-SCD reaction chamber is shown in Figure 4.26. The interface tube is heated to 100-200°C to prevent ice formation in the end of the restrictor,[166,171] and F_2 is generated *in situ* from SF_6. The same type of reaction chambers are used with other chemiluminescence detectors. In the RCD, the solutes must come in contact with a catalyst before entering the reaction chamber. The restrictor is placed a few centimeters upstream from the catalyst bed, and nitrous oxide is introduced along with the effluent. The catalyst can be glass beads that are coated with gold or palladium, gold being more selective in reducing compounds. If the catalyst is held below 300°C, the redox reaction is not complete, and the signal intensity decreases rapidly with decreasing temperatures.[168] At 440°C, the signal intensity is four times larger than at 350°C.[167] At temperatures above 400°C, the RCD is a universal detector.[168] The selectivity can be enhanced if filters are used in front of the photomultiplier tube.

Recently, a new SCD was developed, based on the chemiluminescence reaction of ozone with sulfur monoxide (SO), which is produced in an H_2/O_2 reducing flame that contains sulfur-containing analytes. In general, the SCD responds to all sulfur-containing compounds, which is not the case for the F_2-SCD. Equimolar response to sulfur compounds that have similar structures was obtained, and a selectivity (g S/g C) larger than 10^7 was observed in GC.[172,173] In addition, it was found that the SCD sensitivity was not affected significantly by CO_2 and water vapor.[173]

Without any modification, the SCD was used after capillary SFC, employing CO_2 as the mobile phase.[174] The SO was drawn from the FID to the SCD *via* a ceramic sampling probe and a transfer line

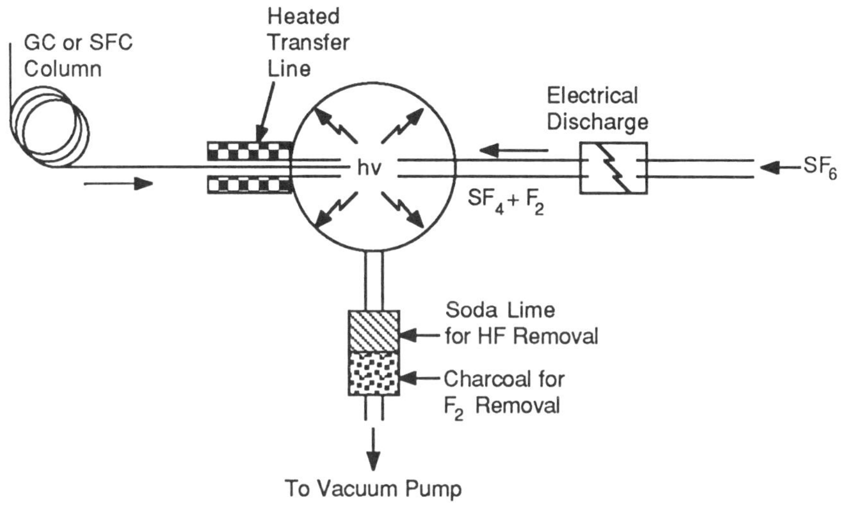

Figure 4.26. Schematic diagram of a sulfur chemiluminescence detector (reprinted with permission from Ref. 169).

using a vacuum pump (10-15 torr). The MDQ for dibenzothiophene was found to be 35 pg S (S/N = 3), and the detector linearity was at least three orders of magnitude.

There is a large variety of mobile phases that can be used with the chemiluminescence detector. The choice of mobile phase is not restricted, because most of the mobile phases used in SFC do not contain sulfur. In the redox chemiluminescence detector, mobile phases and modifiers which do not reduce NO_2 to NO can be used. These include CO_2, saturated hydrocarbons (*e.g.*, *n*-pentane), fully halogenated compounds (*e.g.*, CF_2Cl_2 and CCl_3F), diethyl ether, dioxane, tetrahydrofuran, SF_6, water, noble gases, and NO_2. Nitrous oxide can be used because it does not thermally decompose to NO at temperatures below 440°C.[167] It has been proposed that the redox chemiluminescence detector can be made selective for oxidizing compounds if ammonia is used as the reagent gas.[167]

The detection limits for SFC/chemiluminescence are higher than in GC/chemiluminescence, because CO_2 quenches the emission of light. With open tubular column flow rates, the signal is quenched approximately 15%.[167] Packed columns are rarely used with chemiluminescence detectors because the signals are reduced significantly by the high CO_2 flow rates.[167,168] In pyrolysis/chemiluminescence, signal quenching was prevented by trapping the other sample compounds in a cold trap (-150°C). If large amounts of organic compounds (*e.g.*, modifiers) are introduced, two cold traps should be used as in LC/chemiluminescence.[175] Grolimund *et al.*[170] were able to detect 0.1 ng of nitrosodimethylamine using their pyrolysis/chemiluminescence detector. Better detection limits were achieved with the F_2-SCD. Bornhop *et al.*[171] could detect 35 pg of isopropylthiol. The linear dynamic range was 10^3.

4.15 Photoionization Detection

The coupling of CO_2-based SFC systems with photoionization detection (PID) has now been demonstrated for both packed and open tubular columns.[176-178] The attractiveness of this detection approach for SFC arises in part from the same reasons which have made it valuable for gas chromatography. These include features such as reduced response to common sample solvents such as methanol, acetonitrile, and dichloromethane; wide linear dynamic range; concentration-sensitive response; and variable response selectivity. Since analyte ionization potential has been suggested as the single most important criterion for PID response,[179] it appears that SFC mobile phases which possess ionization potentials greater than the excitation source energy should not produce a substantial PID background current. Thus, the ability to couple CO_2 SFC systems with commercially-available single photon photoionization detectors is due to the relatively high ionization potential for CO_2 (13.79 eV) compared with the commonly used 10.2 eV excitation source. Thus it would appear that the PID offers an alternative to other detection strategies, provided that the ionization potentials of the mobile phase components are sufficiently high so as to preclude the production of a large background signal. It has been suggested that the PID be investigated specifically with respect to solvent-modified mobile phases[104] which are incompatible with flame ionization detection.

In packed column SFC using silica-based stationary phases, the addition of low concentrations (0.5-10 mole %) of low molecular weight organic solvents to the base supercritical fluid is routinely used in order to improve peak shape, reduce retention times, and alter chromatographic selectivity of polar analytes. However, the use of polar modifiers such as methanol immediately places constraints upon the type of detection option which is used. For example the most popular detection mode for SFC, the flame ionization detector, has been shown to exhibit substantial instability and background current in the presence of as little as 0.5% (w/w) methanol, thus making its use in solvent-modifed SFC exceedingly difficult. Not surprisingly, the UV absorbance detector has thus remained the detection method of choice when using methanol-modified eluents. Its use, however, depends on two criteria: the analyte must possess the necessary structural characteristics to facilitate UV absorption, and the carrier fluid components must not absorb strongly at the wavelength of interest. UV detection generally works best with compounds which possess the necessary π-structure to produce substantial absorbance in the 200 to 280 nm range.

On the other hand, compounds which tend to be relatively weak UV absorbers (*e.g.*, aldehydes, ketones, amines, and esters) are the same classes of compounds that generally require a modifier to give good peak shape on silica. These materials tend to display their maximum (albeit relatively weak) absorptivity in the 195-210 nm range, which unfortunately corresponds to the region of UV cut-off for methanol, the most popular polar modifier for SFC with CO_2. Photoionization detection does not rely on the presence of a chromophore to yield a response, in that an analyte need only have an ionization potential below the source lamp energy in order to produce a response. Most organic molecules have ionization potentials less than 10 eV which makes them amenable to detection using a PID with a 10.2 eV source lamp.

Experiments indicate that the presence of methanol at concentrations of 1 to 2 mole % in CO_2 results in substantial reductions in analytical sensitivity *vs* pure CO_2. This is due to a relative decrease in the number of photoions reaching the PID collector electrode, due to quenching mechanisms involving negatively-charged methanol. Since methanol has a large electron-capture cross-section, free electrons generated in the initial photoionization of an analyte AB may be "captured" by methanol and subsequently recombined with analyte photoions to yield neutral species:

$$AB + h\nu \longrightarrow AB^+ + e^-$$
$$CH_3OH + e^- \longrightarrow CH_3OH^-$$
$$CH_3OH^- + AB^+ \longrightarrow CH_3OH + AB$$

Reduction of the methanol concentration in the ion chamber by addition of make-up gases such as helium and argon has been shown to result in sensitivity improvements ranging between 40 and 200%.[180]

4.16 Electron Capture Detection

The electron capture detector (ECD) is probably the most sensitive detector in chromatography for compounds that have high electron affinities.[181] It has been widely used to analyze various environmentally interesting compounds, such as organic pesticides and chlorinated biphenyls that have been separated *via* GC. Because of its high sensitivity, the ECD is ideal for use in trace analysis. For this reason, extending its use to SFC is attractive.

Employing the ECD as a detector for SFC was attempted only recently. An MDQ of 35 pg for triazole was reported by Kennedy and Wall.[182] Although the ECD was not fully optimized in their study, the promise of using it in SFC was demonstrated. More recently, the use of a ^{63}Ni ECD after capillary SFC was reported by Chang and Taylor.[174] High femtogram MDQs were obtained for several halogenated compounds. In both reports, the commercially available ECD was employed without any modifications.

System optimization. Highly purified gases for both chromatographic separation and detector make-up are required when employing the ECD as a detector. SFC-grade CO_2 normally contains <5 ppm oxygen and <3 ppm water. Since water and oxygen have electron capturing properties, and their mass flows to the ECD will be increased by an increase in CO_2 density, it is necessary to further remove both contaminants in order to minimize background signal in the ECD. It was found that by employing high pressure oxygen traps and in-line moisture traps, their concentrations could be minimized.

CO_2 is a weak electron capturing gas, and it can act as an electron scavenger to produce a negatively charged CO_2 molecule by multibody collisions with other gases.[183,184] Consequently, CO_2 may consume a portion of the thermal electrons produced in the detector. In addition, the diffusion coefficients of thermal electrons in CO_2 have been found to be lower than in either methane or argon.[185] Furthermore, the velocities of thermal electrons would be expected to be lower at higher CO_2 densities where the relative CO_2 concentration is higher. This probably results in an incomplete collection of thermal electrons, since the applied pulse width of the detector is constant. The ECD response with respect to CO_2 density at several constant make-up gas (10% methane/argon) flow rates is demonstrated in Figure 4.27. This figure shows that the ECD background signal increased with increasing CO_2 density, and that

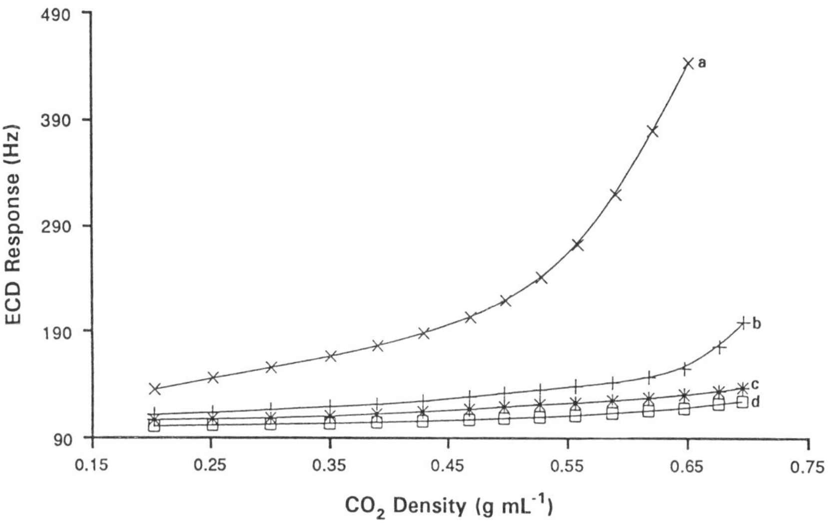

Figure 4.27. Response of CO_2 in an ECD as a function of density at different make-up gas (10% methane/argon) flow rates of (a) 10 mL min^{-1}, (b) 20 mL min^{-1}, (c) 32 mL min^{-1}, and (d) 49 mL min^{-1}. Conditions: 6-m x 50-μm i.d. open tubular column, poly(5% phenyl)methylsiloxane stationary phase (d_f = 0.25 μm); CO_2; 120°C; 0.2 g mL^{-1}; 1.5 cm s^{-1} linear velocity; ECD (350°C) (reprinted with permission from Ref. 174).

this increase was more significant at low make-up gas flow rates. Under the selected operating conditions, the CO_2 gas flow rate (measured at 25°C) increased from about 0.2 mL min^{-1} to 1.3 mL min^{-1} with density programming from 0.2 g mL^{-1} to 0.7 g mL^{-1}. This means that the ratio of CO_2 to make-up gas in the ECD cavity increased with increasing CO_2 density. This CO_2 flow increase gave rise to an increase in the ECD background signal.

At high ratios of CO_2 to make-up gas, the ECD apparently saturates or overloads, thus resulting in a nonlinear response to CO_2. This can be seen at CO_2 densities above 0.45 g mL^{-1} with 10 mL min^{-1} of make-up gas flow (Curve **a** in Figure 4.27). One consequence of this phenomenon is that if the amount of CO_2 entering the detector is too large, the linear dynamic range of the ECD is affected. Because the ECD is a concentration-sensitive detector, this effect can be compensated for by adjusting the ratio of CO_2 to make-up gas in the detector (*i.e.*, the concentration of CO_2 in the ECD cavity can be diluted by increasing the make-up gas flow rate). As shown in Figure 4.27 (Curves **c** and **d**), the ECD response was almost linear to increases in CO_2 density up to 0.7 g mL^{-1} with make-up gas flow rates above 30 mL min^{-1}. Therefore, adjusting the make-up gas flow rate in SFC/ECD is important not only to maintain chromatographic efficiency, but also to decrease the interference resulting from the CO_2.

The ECD sensitivity as a function of total gas flow rate (*i.e.*, make-up gas plus CO_2) has also been studied.[174] A make-up gas flow of 20-30 mL min^{-1} was recommended for most applications to achieve ample ECD sensitivity and to maintain detector linearity.

Minimum detectable quantities and linearity. The MDQs and linearities obtained for 2,3′,4′-trichlorobiphenyl, 2-bromobiphenyl, tri-allate, 4-nitrotoluene, and 3,4-dinitrotoluene are listed in Table 4.3. A make-up gas flow rate of 20 mL min^{-1} was used because all five compounds were eluted at CO_2 densities below 0.45 g mL^{-1} at an oven temperature of 100°C. Direct injection with a 10-s sample loading time was used as the sample introduction method. Under these conditions, it was assumed that more than 90% of the sample was injected into the column. The standard deviation using this injection method was found to be 5%. The detection limits of these studied compounds were comparable to those obtained from GC/ECD.[181] Linear response was obtained for at least four orders of magnitude.

Practical application. Separations of an Arochlor 1254 spiked fish oil using both the ECD and FID are given in Figure 4.28. As one can see, the polychlorinated biphenyls (PCBs) were easily detected by the ECD, but not by the FID. Naturally, the thermally labile triglycerides were detected using the FID. Therefore, direct

Table 4.3. Detection Limits and Detector Linearities of SFC/ECD for Selected Compounds

Compound	Detection limits[a](pg)	Linearity
Tri-allate	0.64	10^4
2,3',4'-Trichlorobiphenyl	0.27	10^4
2-Bromobiphenyl	0.30	10^4
4-Nitrotoluene	12.30	10^3-10^4
3,4-Dinitrotoluene	1.50	10^3-10^4

[a]Data were taken from Ref. 174; S/N was 3; Conditions: see Figure 4.27.

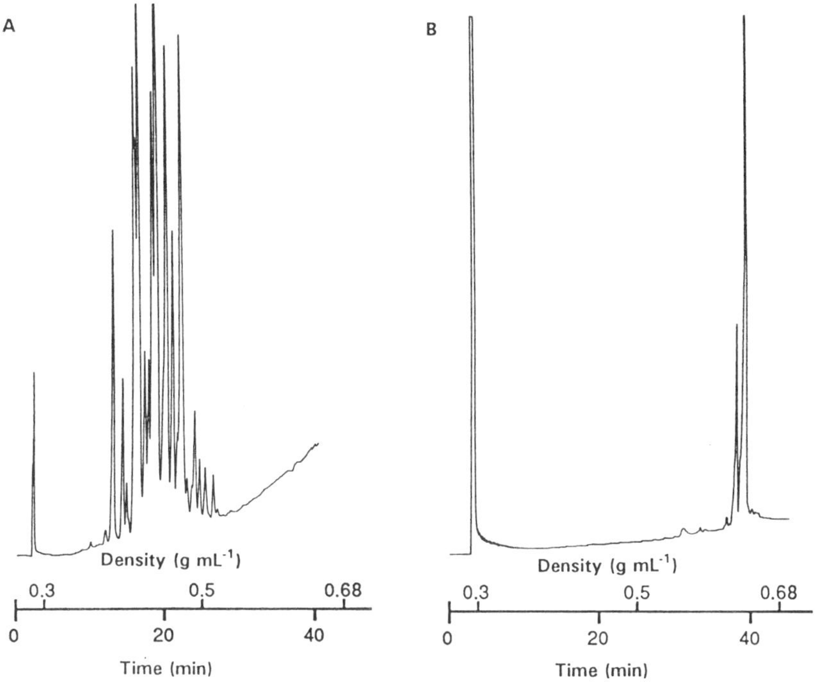

Figure 4.28. SFC chromatograms of fish oil spiked with Arochlor 1254 using (A) ECD and (B) FID. Conditions: 6-m x 50-μm i.d. open tubular column, poly(30% biphenyl)methylsiloxane stationary phase (d_f = 0.25 μm); CO_2; 120°C; density program from 0.3 g mL^{-1} to 0.68 g mL^{-1} at 0.01 g mL^{-1} min^{-1} after a 5-min isopycnic period (reprinted with permission from Ref. 174).

quantification of PCBs in fish oil can be achieved using SFC coupled with both the FID and ECD.

A major drawback of SFC/ECD is the insufficient vaporization of some high boiling compounds. Unlike GC/ECD, in which the compounds are vaporized in either the inlet region or in the chromatographic column prior to entering the detector, the eluted compounds in SFC/ECD are vaporized primarily in the detector region. Therefore, additional modifications may be necessary in the SFC/ECD system in order to extend the analysis to higher boiling compounds.

4.17 Element-selective Plasma Detection

Atomic emission spectroscopy (AES) is one of the oldest instrumental tools for elemental analysis. Over the years, emission sources, optics, and detectors have been refined so that AES is still one of the most widely used and versatile techniques for both bulk and trace elemental analysis. It is not surprising, then, that numerous spectroscopists and chromatographers have worked to combine the separation capabilities of chromatography and the sensitivity and element selectivity of AES. The efforts to combine GC and AES began almost 25 years ago.[186,187] The use of AES in conjunction with GC has been reviewed recently.[188]

An AES-based detector for chromatography offers the potential of several desirable performance characteristics, including versatility, selectivity, sensitivity, and a large linear dynamic range. The versatility arises from the fact that a different element can be selected by a simple change in detection wavelength. In principle, AES-based detectors could be used for almost any element, with the exclusion of several nonmetals. The selectivity has roots in the nature of atomic spectra, which consist of narrow lines isolated from their neighbors by large open spaces. The chances of line coincidences among the elements commonly encountered in chromatographic separations are small. AES is widely used for trace analysis; it is very sensitive. Many sources give linear responses over 5 to 6 decades of concentration.

In many respects, the potential of AES used for element-selective detection in GC has been realized. Such is not the case for SFC and other types of chromatography. The difference is in the effect that the mobile phase has on the emission sources. The addition of a few mL min^{-1} of He from a GC to the plasma sources most commonly used as detectors has no significant effect on the discharges. In contrast, the addition of small amounts of the molecular fluids most often used as mobile phases for SFC has profound negative effects. Sensitivity, linearity, and selectivity are all degraded. Not only does the mobile phase have a negative effect on the discharge, but this negative effect changes in magnitude as the mass flow rate of the eluent changes with a pressure or density program. The challenge facing scientists who wish to couple SFC and AES is to tailor new discharges that are less affected by the addition of SFC mobile phases than those used with GC, or to find new combinations of mobile phase composition and flow rates that will have fewer deleterious effects on existing sources. There has been some effort in both directions.

Inductively coupled plasmas,[189-191] microwave induced plasmas,[192-195] and radiofrequency plasmas[196] have been evaluated as detectors for SFC. All of the studies to date have been of a preliminary nature, and no single source has emerged as clearly superior to the others for a wide range of applications.

Inductively coupled plasmas. The principal advantage of the inductively coupled plasma (ICP) as an emission source for use with SFC is its high-power operation (1-2 kW) and attendant high tolerance for SFC mobile phases. ICPs were not extinguished by the addition of as much as 300 μL min^{-1} of supercritical CO_2, although the addition of CO_2 did result in significant decreases in atomic line intensities and an increase in background levels.[189,190] The argon ICP provides sensitive detection for most metallic elements. An MDQ of 60 pg of Fe was reported for samples of ferrocene dissolved in supercritical CO_2 and injected directly into an ICP.[189]

The ICP suffers some serious limitations as a detector. The initial purchase cost of the equipment is high and it is expensive to operate. A conventional argon ICP consumes between 8 and 20 L min^{-1} of high-purity argon. Commercially available torch housings are bulky, necessitating the use of heated transfer lines between the chromatographic oven and the torch. The excitation of nonmetals by the argon ICP is inefficient, resulting in poor sensitivity for these elements. Helium ICPs have shown better sensitivity for nonmetals,[197] and they may find some use as SFC detectors, but the purchase and operating costs remain high.

Microwave induced plasmas. Microwave induced plasmas (MIPs) are compact and, compared to the ICP, are relatively inexpensive. They can be sustained in resonant cavities[195] or in a surface wave device called a surfatron.[192-194] Plasmas sustained in He provide the best excitation characteristics for nonmetallic elements. As is the case with the ICP, the excitation efficiency decreases with increasing concentrations of mobile phase in the plasma. However, this degradation occurs with much lower levels of mobile phase. The flow rates reported as acceptable for the surfatron[194] are about one tenth of those tolerated by the ICP. The effect of CO_2 on sensitivity was shown dramatically by Rivière *et al.*[192] The MDQ for P in a reduced pressure He discharge increased from 170 pg to 9,000 pg as the CO_2 concentration was increased from 0.8 to 96%.

The interface between a supercritical fluid chromatograph and an MIP can be quite simple. The compactness of the source allows its placement very near to, or directly on, the chromatographic oven. In the interface described by Luffer *et al.*,[193] an integral restrictor was centered in the discharge tube with a quartz nozzle located just upstream from the plasma. A heated helium flow in the nozzle prevented spiking as the fluid expanded from the restrictor.

Radio frequency plasmas. The radiofrequency plasma is a discharge maintained in He between two electrodes at a frequency of approximately 330 KHz. Its spectroscopic characteristics are very similar to those of the MIP. The detection limits for sulfur in both detectors range between 25 and 180 pg s^{-1}, depending strongly on the flow rate of the CO_2 mobile phase. Perhaps the most important distinguishing characteristic of the radiofrequency plasma is its simplicity. It requires no resonant cavity and minimal tuning. Unlike the microwave plasma, it is self-igniting. In the work by Skelton *et al.*,[196] the detector was mounted on the top of the chromatographic oven, and the frit restrictor was placed directly in the discharge.

Figures 4.29 and 4.30 illustrate some of the positive and negative performance characteristics that are shared by the MIP and radiofrequency plasmas. Selectivities have not been quantitatively determined, but as shown by comparison of Figure 4.29A with Figures 4.29B and C, the selectivities for S and Cl are good. Figure 4.30 shows the selectivity used to good advantage, since the peak corresponding to DDT is cleanly isolated in a chromatogram of a DDT-spiked milk extract. A common negative feature of chromatograms from both the MIP and the radiofrequency plasma is the sloping baseline. This arises from changes in the spectral background as the flow rate of the mobile phase changes during a pressure or density program.

Optics. The plasma source is only part of a complete emission-based detector. Collection optics, a wavelength sorting device, a photodetector, and recording electronics are required to complete the system. Plasmas emit atomic lines at wavelengths ranging from the vacuum ultraviolet to the near infrared. No single optical system can efficiently cover that entire wavelength range, so the design of a practical system usually involves a series of compromises. Factors that must be considered are the wavelengths of the most intense emission lines, the structure and intensity of the spectral background, the efficiency of optical devices in a given spectral

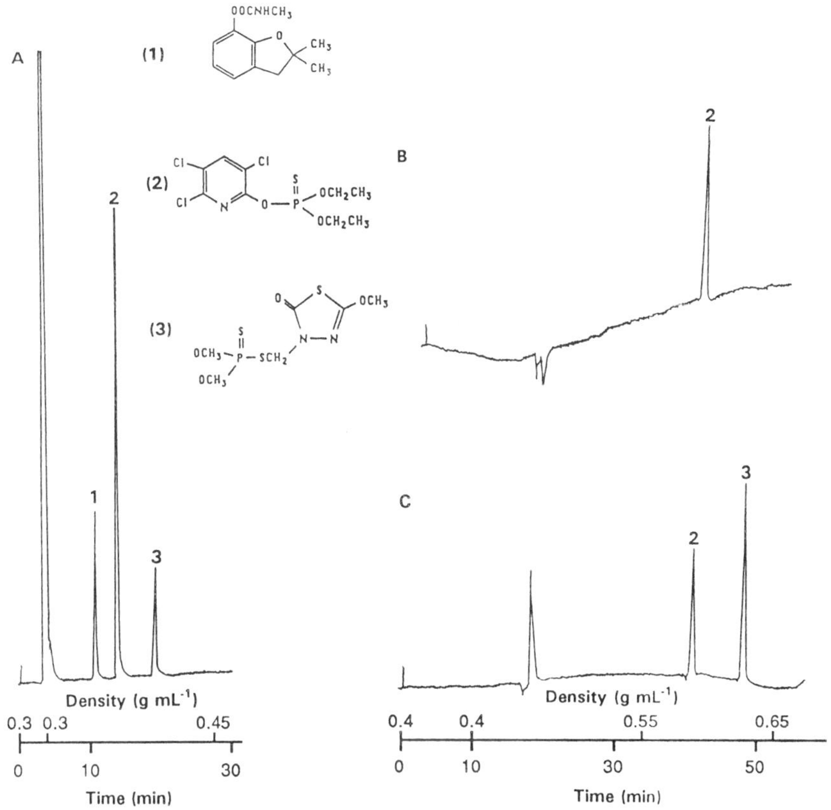

Figure 4.29. SFC chromatograms of a standard pesticide mixture detected by (A) FID, (B) RPD (837.6 nm for chlorine), and (C) RPD (921.3 nm for sulfur). Conditions: 2.5-m x 50-μm i.d. open tubular column, poly(30% biphenyl)methylsiloxane stationary phase (d_f = 0.5 μm); CO_2; 100°C; (A) linear density program from 0.3 g mL^{-1} to 0.45 g mL^{-1} at 0.006 g mL^{-1} min^{-1} after a 3-min isopycnic period, (B) and (C) linear density program from 0.4 g mL^{-1} to 0.65 g mL^{-1} at 0.006 g mL^{-1} min^{-1} after a 10-min isopycnic period (reprinted with permission from Ref. 196).

region, and the sensitivity of photodetectors to the chosen wavelength of radiation.

The resonance lines for an element are usually the most intense. For metallic elements, the resonance lines fall in the ultraviolet and visible regions of the spectrum, and those are unambiguously the regions of choice for detection of metals. The situation with nonmetals is more complicated. The resonance lines for these elements are in the vacuum ultraviolet. The design of optical

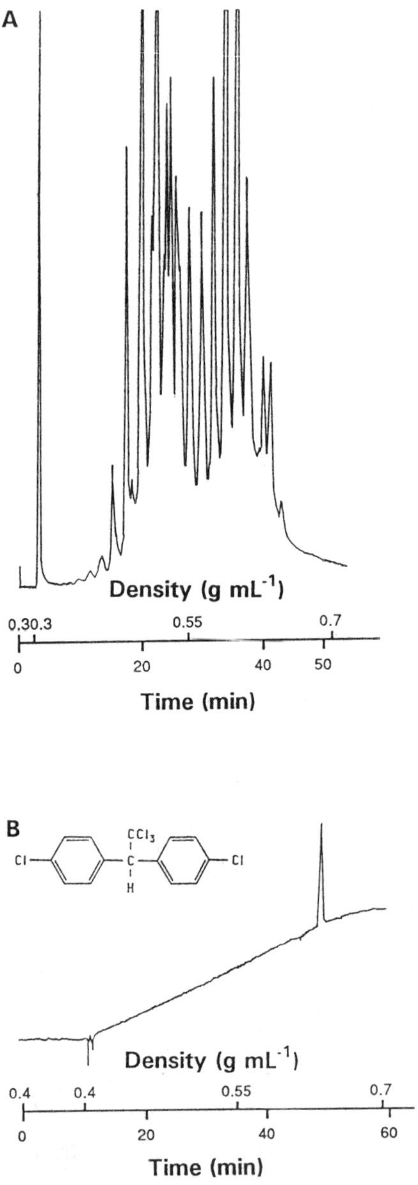

Figure 4.30. SFC chromatogram of a milk extract containing DDT detected by (A) FID and (B) RPD (837.6 nm for chlorine). Conditions: see Fig. 4.29 (reprinted with permission from Ref. 196).

systems for this wavelength region is complicated and expensive. At vacuum UV wavelengths, refractive materials transmit poorly, mirrors reflect poorly, and air is opaque. For these reasons, it is not practical to detect the resonance lines of nonmetals on a routine basis. Less intense emission lines from nonmetals are distributed throughout the UV-visible and near infrared regions of the spectrum. Light detection in the near infrared is significantly less efficient than it is at UV-visible wavelengths. This disadvantage is offset by the fact that the spectral background is generally less cluttered in the near infrared than it is at shorter wavelengths. Spectral simplicity translates directly to high throughput, compact optical systems, and gains in sensitivity. The choice of detection wavelengths for nonmetals with SFC is an active area of research, and it is likely to remain so since there are continuous advances in photodetector technology, particularly with devices that operate in the near infrared spectral region.

Background spectra. As indicated in the previous sections, the background spectrum influences the choice of detection wavelength and has a profound effect on overall detector performance. The atomic spectra produced by plasma-based detectors are relatively simple, and the probability of direct line overlap is small. However, there are broadband spectral features that can interfere with the detection of a selected atomic line, and these must be considered. The most important of these are emission bands from stable diatomic molecules, particularly CO and CN. CO bands are scattered through the spectrum over the wavelength range from 200 to 860 nm. They are constantly present when CO_2 is used as the mobile phase at levels that change with the CO_2 flow rate. They generate a nonspecific response to hydrocarbons in any discharge containing even a trace of oxygen. CN bands are spread from 350 to 2,000 nm and are particularly troublesome when N_2O is used as a mobile phase. An awareness of these features is essential for the design or use of an SFC detector based on atomic emission from an electrical discharge.

4.18 Light Scattering Detection

The majority of detectors in GC, LC, and SFC are based on principles of selective detection. Selective detection is definitely an advantage when only one component or one group of components is to be determined in a complicated sample. Nonselective detection, on the other hand, is required when all or most components of the sample need to be determined. The light scattering detectors (LSD) are based on a nonselective detection principle for nonvolatile compounds.

In 1966, Ford and Kennard[198] introduced a new detector for LC called the evaporative analyzer. The principle consisted of nebulizing the column effluent, evaporating the solvent from droplets generated in the nebulizer, illuminating the solute particles, and measuring the intensity of the scattered light. Charlesworth[199] presented the theoretical basis for a detector similar to the evaporative analyzer in 1978, and this detector, known as the mass detector, became commercially available a few years later. In 1983 Stolyhwo *et al.*[200] described a light scattering detector which replaced the continuous wavelength light source with a helium/neon laser, and this design formed the basis for another commercially available instrument. In addition to the light source, the main difference between the two instruments is the angle of detection relative to the incident light, 120° and 90°, respectively. Another instrument with a continuous wavelength light source was developed by Lafosse *et al.*[201] A schematic diagram of a light scattering detector with 90° configuration is shown in Figure 4.31.

Light scattering. The scattering of light can occur[199] by four different processes: Rayleigh scattering, Mie scattering, reflection, and refraction. For very small particles or droplets with particle diameter/wavelength ratios smaller than 0.1, Rayleigh scattering dominates. The particles behave as point sources of oscillating dipoles which are induced by the incident light, and which radiate light in all directions. Provided that the incident light is polarized perpendicular to the plane of observation, the intensity of the scattered light is independent of the detection angle.

For the larger particles, Mie scattering is the dominant process. Different points on the particles are exposed to light, and the particles behave as multipoint sources, irradiating light of different amplitudes and phases. The intensity of the light will be higher in the forward direction, *i.e.*, the intensity of scattered light from Mie scattering depends on the angle of detection. Both the refractive index of the solute and the polarization of light affect the intensity of the scattered light from Mie processes.

For even larger particles, where the particle diameter approaches the wavelength of the incident light, reflection and refraction prevail, giving rise to a more clearly defined forward scattering of the light

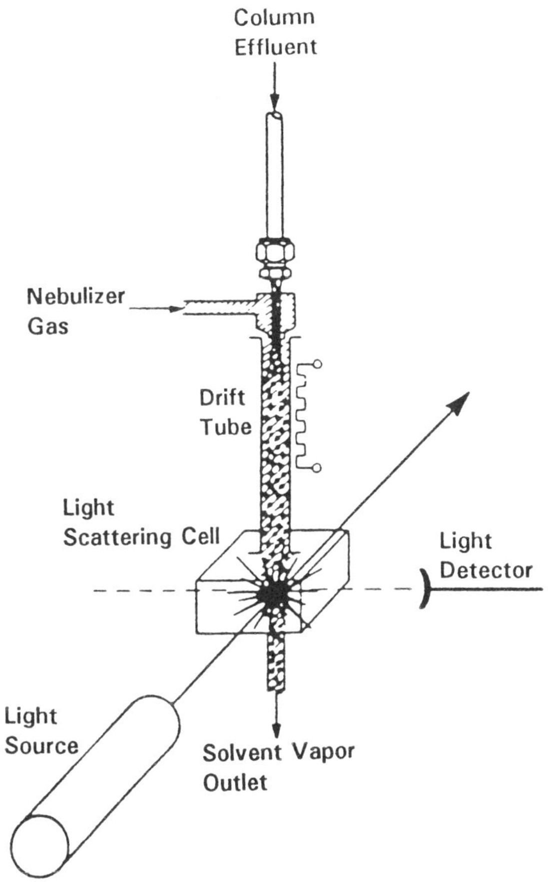

Figure 4.31. Schematic diagram of a light scattering detector with 90° detection angle.

beam. The intensity of the scattered light increases with increasing particle size in a nonlinear manner.

In the Rayleigh scattering region, the amount of scattered light is proportional to the sixth power of the particle diameter, and in the Mie region, to the fourth power. Since the particle volume (mass) is proportional to the third power of the diameter, the scattering intensity will be proportional to the second power of the mass in the Rayleigh region, and to the 4/3 power in the Mie region. For Rayleigh scattering, the intensity of scattered light is also inversely proportional to the fourth power of the wavelength, while for Mie scattering, the intensity is proportional to the second power of the wavelength.

Nebulization and mobile phase evaporation. The effluent is usually nebulized with air, nitrogen, or CO_2 in a concentric nebulizer. The droplets from the nebulizer are transferred to the drift tube where the major part of the mobile phase in each droplet is evaporated by preheated nebulizer gas. If the temperature of the desolvation zone becomes too high, loss of solutes, particularly the more volatile ones, may occur by evaporation from dry particles. Some detectors require considerable volumes of gases, while others are based on small bore columns with less consumption of nebulizer gas. With packed capillary columns and flow rates of only 3-5 μL min^{-1}, some instrument modifications were needed in order to achieve uniform nebulization.[202] The most important feature of the nebulizer is the size distribution of the generated droplets. The ratio between the total aerosol volume and the total aerosol surface area, the Sauter mean diameter, is a common way to describe an aerosol. The Sauter mean diameter (D_o) can be estimated[203] by an equation which includes the surface tension of the liquid, the density of the liquid, the viscosity of the liquid, and the liquid and gas flows. The equation has been shown to give a good approximation of D_o in the 15-90 μm range for liquids with densities of 0.7-1.2 g mL^{-1}, surface tensions of 19-73 dyne cm^{-1}, viscosities of 0.003-0.5 poise, and a nebulizer gas at sonic speed.

Mourey and Oppenheimer[203] showed that the size distribution based on volume, *i.e.*, the Sauter diameter, was different from the distribution based on numbers of droplets. From the droplet distribution, D_o was calculated to be 7.82 μm, and the number average droplet diameter was calculated to be 0.77 μm. With a monodisperse aerosol and a monochromatic light source, Oppenheimer and Mourey[204] calculated that an increase in the droplet size from 1 to 10 μm would increase the sensitivity by a factor of 10^6.

With a supercritical fluid as the mobile phase, there is, in principle, no need for the nebulizer gas.[201] However, with the large volumetric flow rates (5-8 mL min^{-1}) of fluids containing modifiers from conventional packed LC columns and linear restrictors, the use of a nebulizer gas was reintroduced in order to enhance the evaporation rate.[205] With the flow rates of narrow bore columns, integral type restrictors were used.[206] With 0.32-mm i.d. packed columns, the original nebulizer was removed, the column was placed in the drift tube, which only functioned as a column oven, and the restrictor was connected to the heated detector block with only a 15-mm desolvation zone. With 1-mm i.d. columns, flow-rates of 0.1-0.2 mL min^{-1} of methanol-modified CO_2 required the drift tube to be reinstalled as the desolvation zone, but without nebulizer gas.[207] A separate heater was then installed at the top of the drift tube. Good temperature control of the nebulization process was required to maintain a reproducible strong signal, without spiking.

Signal. With a mass sensitive detector, the mass flow (F_m) is given by

$$F_m = c\ F_v \tag{4.21}$$

where c is the concentration and F_v is the volumetric flow. When a nebulizer gas is added to the light scattering detector, the volumetric flow through the detector is mainly a function of the flow-rate of the nebulizer gas, which is large and constant; but the mass flow increases with the column flow. In SFC, with no nebulizer gas, the flow through the detector is only a function of the volumetric flow in the column, and the flow increases with increasing density. This may lead to an increasing signal, but only if the higher flow does not significantly reduce the size of the droplets.

Due to the different light scattering mechanisms, the peak area has been related to the sample mass by a nonlinear relationship[208]

$$A_p = R_f\ M^x \tag{4.22}$$

where R_f is the response factor, M is the sample mass, and x is the coefficient of regression (slope) of the response curve. Since the response is nonlinear, the peaks from the light scattering detector appear more narrow compared to the peaks from a linear detector. On a logarithmic scale, the equation above becomes

$$\log A_p = \log R_f + x \log M \tag{4.23}$$

where log R_f is the intercept on the y-axis. Unless the particle size distribution is narrow and monochromatic light is used, the coefficient of regression is not constant over a wide area. Linear log-log plots have, however, been demonstrated within limited ranges of injected sample amounts.

Composition gradient elution. A major benefit of the light scattering detector is the ability to use composition gradient elution without distorting the baseline, since solvent vapors do not scatter light. As long as sufficient heat is supplied and the solvent is free from solid impurities after evaporation, the baseline will remain straight regardless of the solvent composition. Furthermore, solvents (such as acetone and toluene) which are incompatible with UV detection, but are otherwise interesting ingredients in mobile phases, can be used without any negative effects on the background signal. This is also the case in SFC, where density and modifier gradients have no effect on the baseline.

The reported effects of density gradients in SFC on the signal are not completely consistent. Carraud *et al.*[209] found that the response factor variation precluded the use of CO_2 pressure programming, while Hoffman and Greibrokk[206] obtained good reproducibility for both peak areas and retention times. The difference probably can be attributed to the large difference in volumetric flow in the two systems. In order to achieve good quantitative data from the LSD, standard curves are, however, needed for at least one component in a compound group. Reproducible gradient formation and good temperature control is also required.

Minimum detection quantity. The MDQ of the "mass analyzer" has been reported to be approximately 1 μg.[210] Carraud *et al.*[209] found MDQs of 75 ng and 40 ng (S/N = 3) with their detector, but this was later improved to 12 ng.[205] The MDQ of the laser based detector has improved from 0.2 μg to 3 ng, calculated at S/N = 2,[211] and 6 ng, demonstrated at S/N = 20.[206] The latter, which equals a value of 2 ng, due to nonlinear response, was obtained with a packed capillary column and a modified nebulizer,[202] and corresponds to a mass flow of 0.5 ng s^{-1}. The high peaks at low mass flow, together with the calculations of Oppenheimer and Mourey,[204] indicate that the potential MDQ of the light scattering detector is better than shown by the results so far.

Whether SFC/LSD results in better MDQ than LC/LSD is still an open question. Carraud *et al.*[209] reported five times improvement in detection limits, but Nizery *et al.*[205] showed an increasing peak area when the temperature was decreased from supercritical to subcritical levels. In fact, several of the results above were obtained under subcritical conditions. Hoffmann and Greibrokk[206] obtained similar detection limits with SFC and LC, but also observed higher peaks with SFC than with LC at higher sample loads, probably due to more efficient mass transfer of large particles in the short desolvation zone.

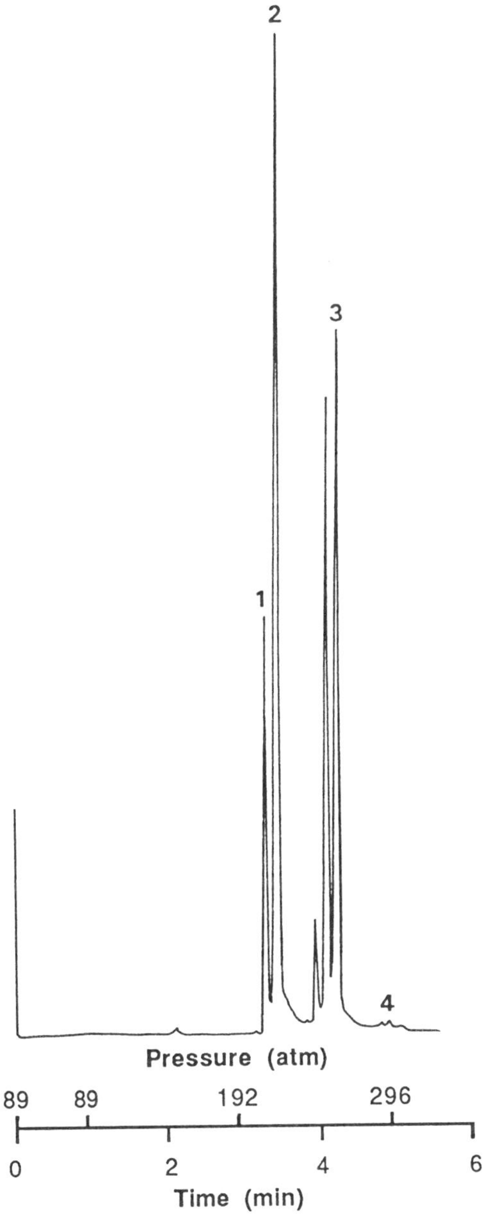

Figure 4.32. SFC chromatogram of technical glyceryl monostearate for purity test using an LSD. Conditions: 10-cm x 320-μm i.d. packed column, C_{18} bonded silica particles (4 μm, Nova-Pak); 2.0 mol% 1-propanol in CO_2; 85°C; pressure program from 89 atm to 296 atm at 52 atm min^{-1} after a 1-min isobaric period. Peak identifications: (1) monoglycerides; (2) glyceryl monostearate, (3) diglycerides, (4) triglycerides (reprinted with permission from Ref. 206).

Restrictors. The choice of restrictor is an important detail in SFC, particularly when the restrictor is located at the detector inlet, as is the case with the LSD. With very high flow-rates, linear 75-μm i.d. restrictors have been utilized.[205,209] The advantage of wide restrictors is the absence of plugging problems. With low flow rates and no nebulizer gas, integral restrictors were utilized.[206] In order to prevent occasional plugging, a 1-mm ceramic frit, which functioned as a filter, was placed at the restrictor inlet.

Practical application. Since the combination of SFC and LSD has been accomplished quite recently, only a few applications have been published. These include lipids,[206] polymer additives[206], and carbohydrates[211] (see Figure 4.32).

4.19 Ion Mobility Detection

Ion mobility spectrometry has been demonstrated for the qualitative and quantitative detection of organic compounds after separation by gas,[212] supercritical fluid,[213,214] and liquid chromatography.[215] The most simple way to view the design of an ion mobility detector is as an electron capture detector (ECD) to which an ion separation chamber has been added. An example of the design of an ion mobility detector used for SFC is shown in Figure 4.33. The electron capture-like portion of the detector is shown at the top of the figure marked "^{63}Ni Foil in Ceramic Insulator." In this region, the radioactive foil produces energetic beta particles which ionize the surrounding gas, producing electrons and primary positive ions. When electropositive and electronegative compounds, such as water and oxygen, are present in the detector gas, ion-molecule and electron capture reactions occur to form secondary ions. With the typical operating conditions discussed in this chapter, both water and oxygen are present in the detector gas, and the major secondary ions produced are $(H_2O)_nH^+$ and O_2. These ions are called the "reactant ions."

Positive and negative ions are extracted from the ionization region when a repeller voltage, labeled "High Voltage" in the diagram, is placed on the top plate of the detector. To detect negative ions, a negative potential of 3,000 volts is placed on the top plate, repelling the negative ions into the ion separation region of the detector. Similarly, to monitor positive ions, a positive potential is placed on the repeller plate. Under the influence of the electric field, these

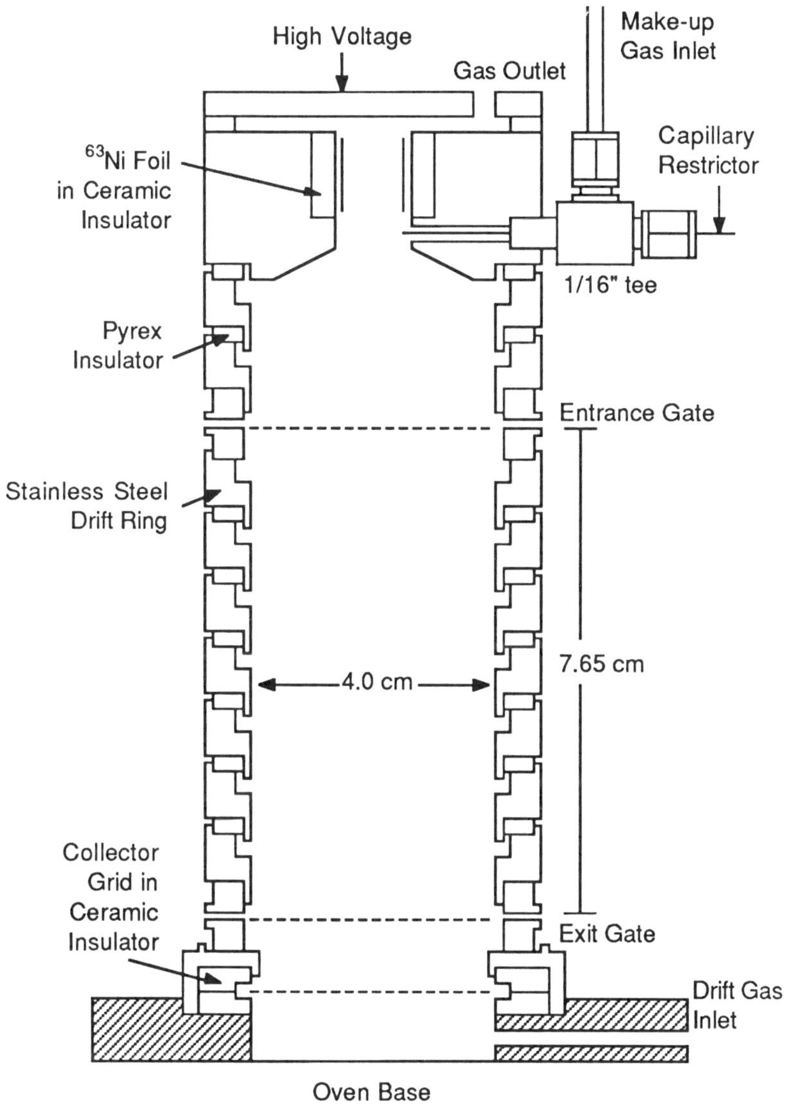

Figure 4.33. Schematic diagram of an ion mobility detector for SFC (reprinted with permission from Ref. 216).

ions migrate through the detector gas at a velocity proportional to the strength of the electric field. The ratio of the velocity of an ion (v_i) in cm s^{-1} to the strength of the electric field (E_t) in v cm^{-1} produces a constant called the mobility constant, with units of cm^2 v^{-1} s^{-1}. When this mobility constant is corrected for standard temperature and pressure, it becomes independent of the operating conditions and provides qualitative information about the ions. This reduced mobility constant (K_o) is calculated in the following manner

$$K_o = \left(\frac{v_i}{E_t}\right)\left(\frac{273}{T}\right)\left(\frac{P}{760}\right) \qquad (4.24)$$

where T is in °C and P is in torr.

The uniformity of the electric field in the ion separation portion of this detector is important, and it is controlled by stainless steel drift rings separated by Pyrex insulators and connected by matched megaohm resistors (not shown in the figure). The ion separation portion of the detector is defined by two ion gates, the entrance gate and the exit gate. Under microprocessor control, the entrance and exit of ions to and from the drift region are timed as they traverse the 7.65-cm distance between the ion gates. From these data, ion velocities and mobility constants are calculated. Ions which pass through the exit gate are collected and measured on the "Collector Grid" shown at the bottom of Figure 4.33.

The interface to the supercritical fluid chromatograph is provided through a capillary restrictor placed between the ionization region and the ion separation region of the detector. All types of capillary restrictors commonly used with SFC have been utilized successfully with the IMD. Of particular importance in the design of the IMD is the flow direction of the detector gas. For efficient chromatographic detection, a unidirectional flow of the detector gas, which is countercurrent to the ion migration is desired. In this design, the detector gas, called the drift gas in Figure 4.33, is introduced into the detector at the base of the detector below the collector grid. It passes in a countercurrent direction through the ion separation region, sweeps the SFC effluent from the capillary restrictor into and through the more narrow ionization region, and then exits the detector through an off-axis orifice in the repeller plate.

When the neutrals from the SFC are swept through the ionization region of the detector, ion-molecule reactions occur with the "reactant ions" to produce a second set of stable secondary ions which are called "product ions." Thus the general reaction for ion mobility detection is

Reactant Ions + Sample ——> Product Ions + Neutrals

The ion mobility detector has two basic methods of operation: a scanning method and a mobility monitoring method. In the scanning method, both the entrance and exit gates are opened and closed simultaneously while the frequency at which they are operated is varied from 20 to 5,020 Hz. As the frequency changes, ions traveling through the ion separation region come in and out of phase with the open gates, producing an ion mobility interferogram as shown in Figure 4.34A. Using a fast Fourier transform, this frequency domain spectrum can be converted to a time domain spectrum as shown in Figure 4.34B. A complete scan can be obtained in less than one second to capture "on the fly" qualitative information as a compound elutes from the chromatograph.

In the mobility monitoring method, the exit gate is pulsed open at some predetermined delay time after the entrance gate, permitting ions of only selected mobilities to pass through the ion separation region and be detected at the collector grid. In this mode, the gate open times are typically 0.2 ms, and the gates are pulsed open at a constant frequency of 50 Hz. When the phase delay between the two gates is equal to the drift time of an ion, then that ion will be detected continuously.

Chromatographic detection is generally performed using the ion mobility monitoring mode. It provides six different modes of detection: (a) negative reactant ion monitoring (-RIM), (b) negative product ion monitoring (-PIM), (c) selective negative product ion monitoring (S-PIM), (d) positive reactant ion monitoring (+RIM), (e) positive product ion monitoring (+PIM), and (f) selective positive product ion monitoring (S+PIM).

In the -RIM mode, the phase delay between the two gates is set to match the drift time of the negative reactant ion, O_2^-. When this ion undergoes an ion-molecule reaction with compounds from the SFC, the signal due to the reactant ion species is depleted and a response is observed as a decrease in detected ion current. In general, all electronegative compounds respond in this mode, providing chromatograms similar to those expected with an electron capture detector.

The -PIM mode is just the opposite of the -RIM mode with the exit gate open, except for the time which corresponds to the drift time of the reactant ion. Thus, all of the product ions from the

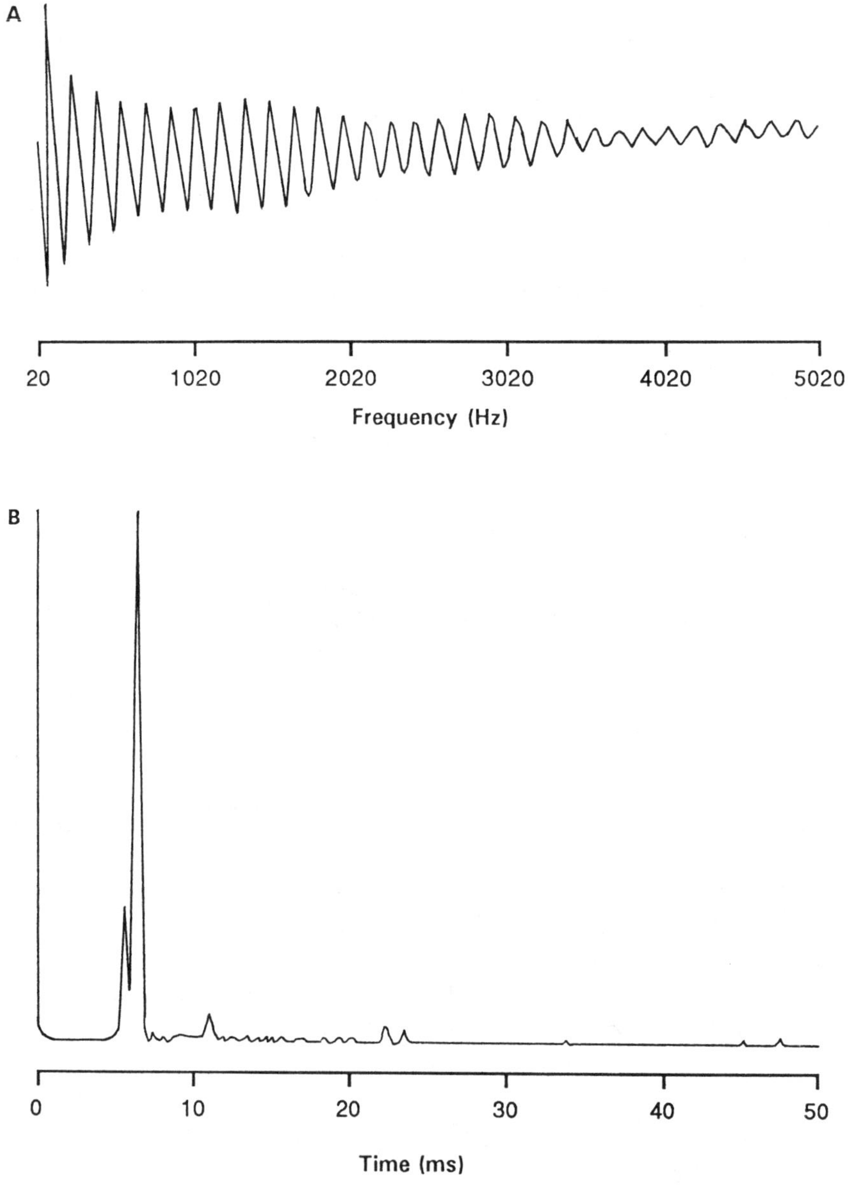

Figure 4.34. Typical ion mobility interferometry of positive reactant ions in a nitrogen drift gas showing (A) ion mobility interferogram and (B) time domain ion mobility spectrum (reprinted with permission from Ref. 213).

ion-molecule reactions of O_2 with compounds from the SFC are monitored. In general, the -PIM mode provides chromatograms of positive current that mirror those obtained in the -RIM mode. Noise, however, is generally less in the -PIM mode, since the noise of the reactant ions is gated out.

In the S-PIM mode, specific product ions are monitored, providing selective detector response. For example, when dissociative electron capture occurs, the Cl^- ion can be monitored to provide a selective response for chlorinated compounds, or the Br^- ion can be monitored to provide a selected response for brominated compounds, even in the presence of chlorinated ones. When associative capture occurs, molecular selective responses can be obtained from these electronegative compounds.

Analogous detection modes are possible when positive ions are monitored. +RIM provides reactant ion depletion, while +PIM monitors the total product ions produced from the ion-molecule reactions. Both these detection modes respond to compounds which are electropositive and which react with the positive ion cluster. S+PIM provides molecular selective responses for these electropositive compounds.

Effects of the mobile phase. Early studies in which CO_2 was investigated as the drift gas suggest that large clusters of CO_2 are formed around the ion and that the size of these clusters is dependent on temperature and CO_2 concentration.[217] Thus, when CO_2 is used as the mobile phase, care must be taken to ensure that the product or reactant ions do not form clusters with CO_2, changing the normal mobility of the ion. The use of a non-CO_2 containing drift gas, detector temperatures greater than 150°C, and a unidirectional gas flow pattern, has been found to eliminate problems associated with CO_2 cluster formation.[218]

Modified CO_2 mobile phases and non-CO_2 mobile phases may also be used with ion mobility detection. Figure 4.35 shows the SFC separation with +RIM mode detection of polydimethylsilicone. Chromatogram A was obtained with neat CO_2, while Chromatogram B was obtained using CO_2 modified with 5% (v/v) of methanol. More work is needed to fully characterize the IMD response with multiple mobile phases, but from initial results, it appears that the IMD is compatible with a variety of mobile phases.

Practical application. In a quantitative sense, the information content obtainable from an ion mobility detector compares favorably with that from a UV detector. In a qualitative sense, the detectors

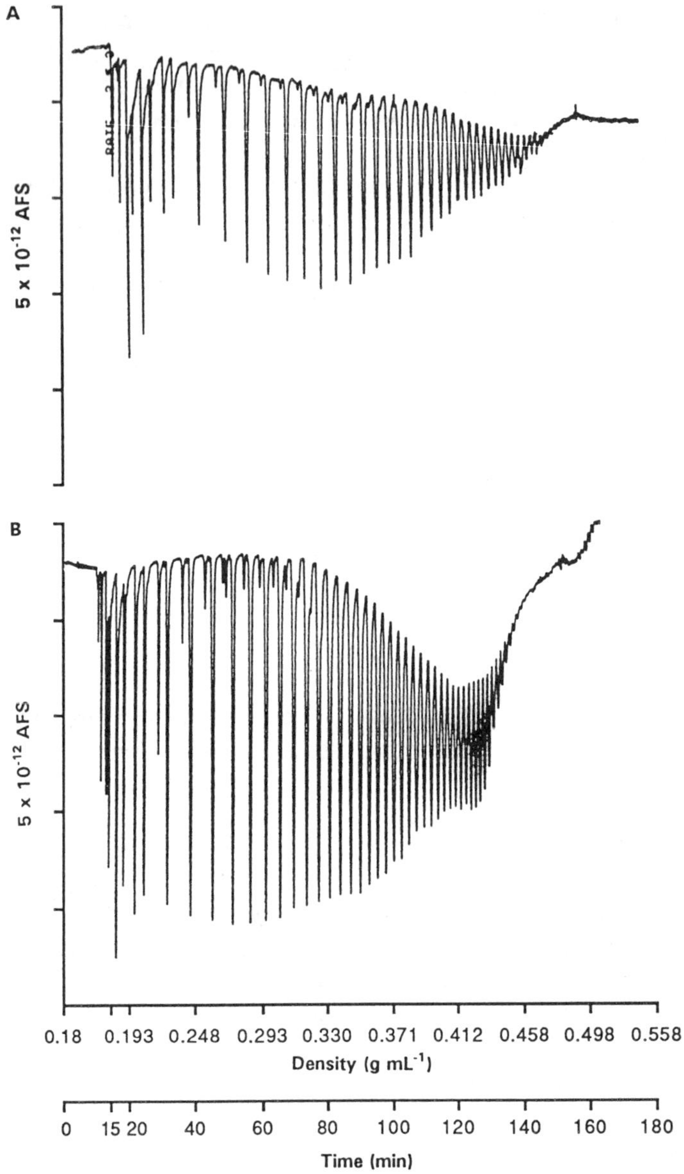

Figure 4.35. SFC chromatograms of polydimethylsiloxane using the +RIM mode of ion mobility detection with (A) neat CO_2 and (B) 5% (v/v) methanol in CO_2 as mobile phases. Conditions: 20-m x 100-μm i.d. open tubular column; polymethylsiloxane stationary phase; 250°C; linear density program from 0.18 g mL^{-1} to 0.50 g mL^{-1} at 0.003 g mL^{-1} min^{-1} after a 15-min isopycnic period; temperature program from 100 to 140°C at 0.25°C min^{-1} after an initial 20-min isothermal period. Density values shown in the figure are corrected for increasing temperature (reprinted with permission from Ref. 107).

are complementary, with the ion mobility detector providing responses for compounds which do not contain chromophores. Selective detection is demonstrated in Figure 4.36 in which specific oligomers of Triton X-100 were monitored.

Due to the lack of a commercial IMD for SFC, example applications of the detector are few. Nevertheless, this detector provides sensitive and selective response. The multimode nature of the detector, coupled with sub-picomole MDQ and wide ranging mobile phase compatibility makes the IMD an excellent detector for SFC separations.

4.20 Fourier Transform Infrared Spectroscopy

Molecular vibrational transitions give rise to the absorption of energy in the infrared region of the electromagnetic spectrum. These absorption frequencies can be correlated with the structure of the molecule and indicate the presence of specific functional groups. Infrared spectroscopy (IR) can, therefore, be used for identifying isomeric compounds, although it is poor for distinguishing between homologs, and in this respect is complementary to mass spectrometry (MS). In recent years, chemists have seen the need to develop combined systems where highly informative detectors (such as IR) are coupled with chromatographs to obtain both qualitative and quantitative information regarding the eluting components, a feat beyond the scope of either technique individually.[219]

Dispersive (grating or prism) infrared spectrophotometers are too slow and insensitive for on-line detection, since they have to scan through the IR region, which may take minutes. In Fourier transform infrared spectrometry (FTIR), infrared radiation which has been modulated is passed through the sampling area and is detected by a highly sensitive mercury cadmium telluride (MCT) detector. The interferograms, which are plots of infrared intensity *vs* time are signal averaged at intervals of less than one second and stored on the hard disc of the spectrometer's computer system. When required, the data system executes a Fourier transform of the interferograms, which may then be compared against a background to produce a transmittance (or absorbance) *vs* wavenumber IR spectrum. Because all the resolution elements of the IR spectrum are acquired concurrently in FTIR, rapid spectral acquisition and averaging are possible. Hence FTIR can be used as a highly informative, nondestructive, and universal chromatographic detector.[220]

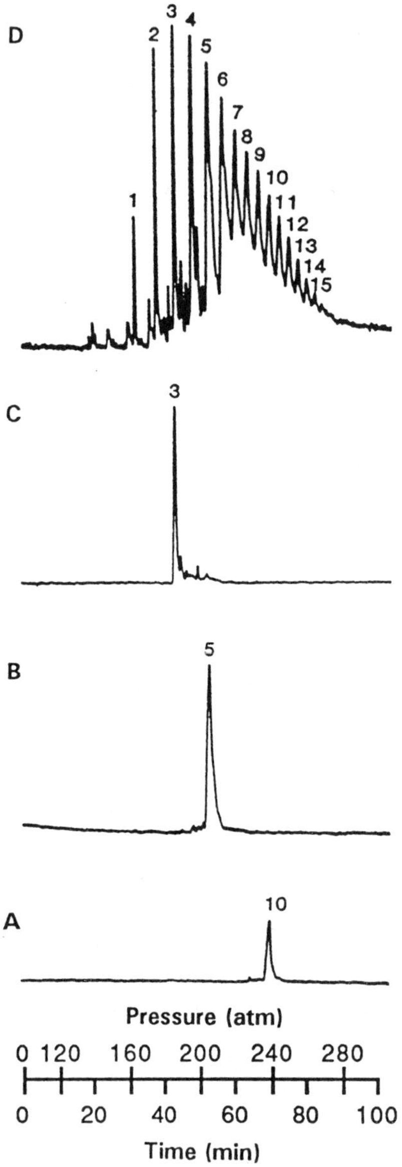

Figure 4.36. SFC chromatograms of Triton X-100 using +PIM and S+PIM modes of IMD, monitored for reduced mobilities of (A) 0.739, (B) 0.923, (C) 1.02, and (D) for all product ion mobilities. Conditions: 18-m x 100-μm i.d. open tubular column; polymethylsiloxane stationary phase; 110°C; neat CO_2; pressure program from 120 atm to 280 atm at 2 atm min^{-1} (reprinted with permission from Ref. 213).

Research into interfacing SFC with FTIR has been actively pursued, mainly because of the problems involved in analyzing thermally labile compounds, and those too involatile or polar to analyze by GC/FTIR.[221-223] Traditionally, these compounds were analyzed by LC, which virtually excluded the use of FTIR detection. Most LC solvents absorb strongly in the IR, so that flow cells with very short pathlengths (typically 100 μm) must be used to prevent the solvent bands from obscuring wide regions of the spectrum. As a result, detection limits are high (>20 times higher than in GC/FTIR). Furthermore, information about the solute cannot be obtained in regions where the solvent absorbs all of the energy, and solvent subtraction cannot be performed successfully with composition gradient eluents. Solvent elimination in which the solvent is evaporated and the solute concentrated onto an IR transparent support is also rendered difficult in LC/FTIR by the loss, spreading, or chemical change of the sample during evaporation.

One of the many advantages of supercritical mobile phases, such as CO_2, is that they have extensive regions of transparency in the IR region of the spectrum, making them compatible in flow cell SFC/FTIR.[224] CO_2 has strong absorption bands which obscure the IR spectrum from 3504-3822 cm^{-1}, 2137-2551 cm^{-1}, and below 800 cm^{-1} (Figure 4.37). As the density of the CO_2 is programmed upwards, Fermi-resonance bands between 1225 and 1475 cm^{-1} also increase in intensity, but these bands may be subtracted from the spectra of eluting components if short cell pathlengths are used. Supercritical xenon is a unique mobile phase for SFC/FTIR because it is totally transparent, allowing the transmission of energy at all frequencies through the flow cell, while having appreciable solvating power.[225] Other fluids, such as supercritical propane and dichlorodifluoromethane, have also been investigated, but are of little practical use since they become completely opaque at pressures commonly used in SFC.[226] Similar effects are seen when low levels of organic modifiers are added to CO_2 to increase the solvating power of the mobile phase or to deactivate column packings. In these cases, flow cell SFC/FTIR becomes as limiting as LC/FTIR.[227,228]

The basic principle behind flow cell SFC/FTIR is shown in Figure 4.38. A specially designed flow cell (which can withstand high pressures) with IR transparent windows is positioned in the focused IR beam. The FTIR is therefore able to monitor the column effluent from the chromatograph in real time, the signal being directly proportional to the amount of sample in the chromatographic peak flowing through the cell at any respective time. Since detection is

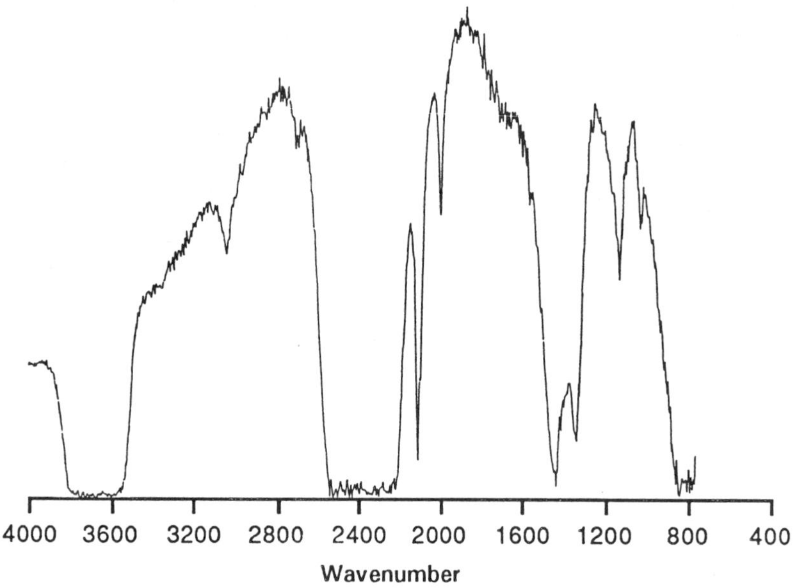

Figure 4.37. Infrared spectrum of CO_2 at 300 atm and 25°C using an optical pathlength of 10 mm.

nondestructive, additional detectors, such as an FID, may be connected in series after the flow cell. The advantage of on-line detection is that chromatograms based on total IR absorbance of organic molecules can be reconstructed using the Gram-Schmidt algorithm. The retention time data obtained from the Gram-Schmidt reconstructed (GSR) chromatogram is complemented by structural information provided by interpreting the IR spectra of the separated components, or by comparing them with a spectral library, often performed by computer searching facilities. The major disadvantage of this interface is the low detection sensitivity obtained; since it is a dynamic system, only a limited number of spectra can be accumulated as a peak flows through the cell.

There has, therefore, been much research into the development of mobile phase elimination interfaces for SFC/FTIR because of their greater potential detector sensitivity and compatibility with IR absorbing fluids.[219] The principle behind this interface is the use of a mobile phase which is easily eliminated once it emerges from the flow restrictor at the end of the chromatographic column. The boiling points of most SFC mobile phases are well below ambient, and they evaporate on decompression. A moving substrate, on which each separated component can be trapped, is positioned below the

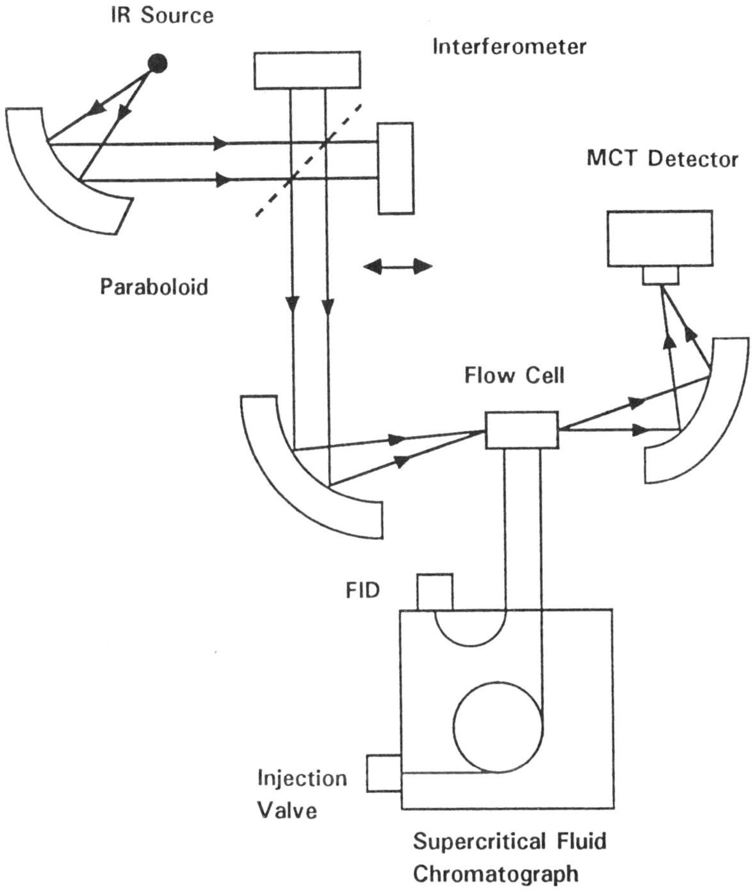

Figure 4.38. Schematic diagram of SFC/FTIR with a flow cell interface.

restrictor, and the sequence of trapped solutes is then passed through the beam of an FTIR spectrometer, the spectrum of each being measured in turn (Figure 4.39).[229-231]

Obviously, the substrate on which the compounds are deposited influences the type of IR spectrum which can be measured. If metallic substrates are used, reflection-absorption spectra are measured, while conventional transmission spectra are measured from KBr or ZnSe windows. If the analytes are deposited on the surface of KBr powder, then diffuse reflectance spectra can be measured. The major advantage of mobile phase elimination is that the sample is immobilized so that a large number of scans can be acquired. The signal-to-noise ratio is considerably improved as a result, and consequently, identifiable spectra can potentially be

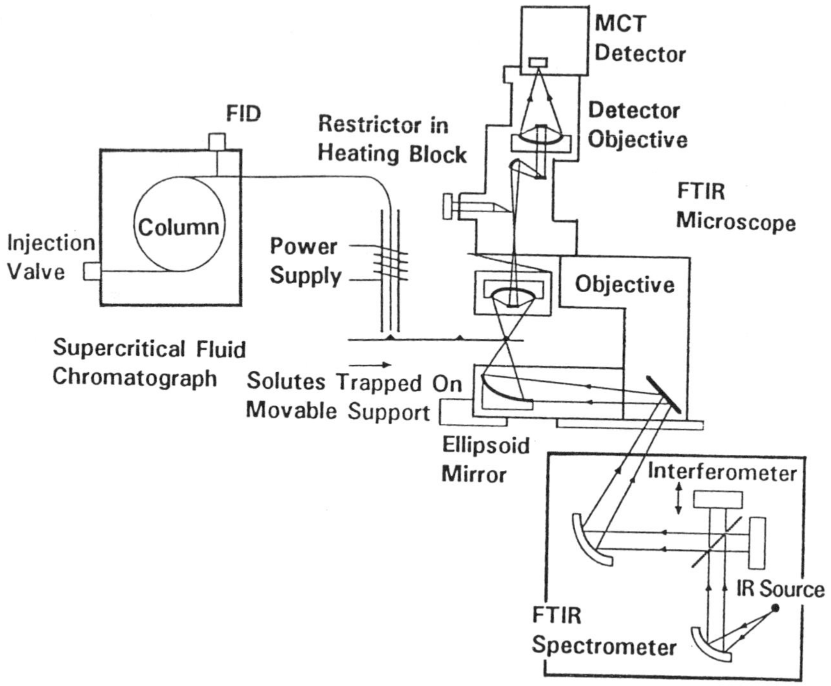

Figure 4.39. Schematic diagram of SFC/FTIR with mobile phase elimination.

measured from compounds deposited at the subnanogram level. However, the IR sensitivity which can be achieved depends greatly on the way in which the samples are deposited. Detection limits improve as the diameter of the trapped analyte is decreased. A concentrated spot (100-200 μm) produces the best results for transmission work, but this may not be possible if the compound is deposited as a liquid which may spread over a wide area. At high restrictor flow rates, sample may also be blown off the window. In order to match the IR beam size to the diameter of the spot, beam condensing optics of an FTIR microscope are required. The following sections deal with the flow cell and mobile phase elimination interfaces in more detail. Hardware and factors influencing the design of the interfaces are presented with operating conditions and applications unique to SFC/FTIR.

SFC/FTIR with a flow cell interface. A high pressure flow cell may be connected on-line to an SFC column as long as the mobile phase has sufficient transparency in the infrared. Real time monitoring of the column effluent is then possible. The cell design is most usually based on a cylinder, polished to maximize internal reflection, with IR transparent windows of calcium fluoride, zinc sulphide, or zinc selenide, and located so that the focussed beam of the FTIR instrument passes along its axis. High pressure is maintained in the cell by means of a flow restrictor. A compromise between the chromatographic and spectroscopic requirements is necessary for deciding the cell dimensions.[228,232,233] A number of variables must be considered: packed column *vs* open tubular column separation; flow cell design, dimensions, and construction; IR transparency of the mobile phase; and detection conditions.

The major chromatographic-related requirements of the cell are that it must withstand the high pressures (typically up to 400 atm) used in SFC and that it have a small volume in relation to the volume of a typical peak eluting from the column. The cell volume should not exceed 0.3 standard deviation or about 25% of the volume of a Gaussian peak, but the peak volume is dependent on the column type. For SFC/FTIR work with 4.6-mm i.d. packed columns, peak volumes are in excess of 40 μL, and cell volumes of less than 10 μL are optimal. Flow cells based on those developed for UV detectors in liquid chromatography have been generally employed (fitted with IR transparent windows), and Table 4.4 compares some of the important features of those reported in the literature.

The dimensions of these cells also take into account the requirement of maximum IR sensitivity. Absorbance detection methods such as FTIR are based on the Beer-Lambert law, where the amount of absorbance (A_b) is dependent on the sample molar extinction coefficient (ϵ), the sample concentration (c), and the pathlength (ℓ). Of these parameters, ℓ is the only flow cell variable affecting absorbance in packed column SFC/FTIR. Increasing the pathlength should improve the signal-to-noise (S/N) ratio of the spectra obtained. However, as ℓ increases, the light pipe diameter must be decreased in order to maintain a constant volume. This leads to a drop in optical efficiency and a decrease in the width of the IR window region. Jordan and Taylor compared 4-μL (ℓ = 5 mm) and 8-μL (ℓ = 10 mm) flow cells, and concluded that although the 10-mm cell afforded greater S/N, the Fermi resonance bands of the CO_2 from 1225 to 1475 cm^{-1} (see later discussion) became too intense to allow an adequately low noise level after spectral subtraction.[227] Consequently, 5 mm is the maximum usable pathlength if the mobile phase must be programmed to 400 atm. Figure 4.40 shows a good example of a packed column SFC/FTIR analysis of the thermally labile sesquiterpene hydrocarbons from ylang-ylang oil using CO_2 under isobaric conditions at 140 atm. An 8-μL flow cell (1-mm i.d. x

Table 4.4. Comparison of Selected Packed Column SFC/FTIR Studies

Flow cell parameters	Johnson *et al.*[232]	Hughes and Fasching[234]	Jordan and Taylor[227]	Morin *et. al.*[224]
Column	15-cm x 4.6-mm i.d.; aminopropyl bonded silica particles (5 μm)	25-cm x 4.6-mm i.d.; cyanopropyl bonded silica particles (5 μm)	25-cm x 4.6-mm i.d.; phenyl bonded silica particles (5 μm)	15-cm x 4.6-mm i.d.; silica particles (5 μm)
Flow rate/- linear velocity	0.4 cm s^{-1}	-	3 mL min^{-1}	-
Pathlength	5 mm	8 mm	10 mm	10 mm
Volume	4 μL	8 μL	8 μL	8 μL
Window	ZnSe	ZnSe	CaF_2	ZnSe
Pressure	272 atm	231 atm	200 atm	282 atm
Beam condenser	Yes, 3 x	No	Yes, 3 x	Yes, 4 x
Scans	2 s^{-1}	2.5 s^{-1}	2 s^{-1}	3 s^{-1}
Resolution	4 cm^{-1}	8 cm^{-1}	4 cm^{-1}	8 cm^{-1}

10-mm pathlength) was used, and at this pressure the CO_2 Fermi bands could be subtracted from the component spectra. The mild chromatographic conditions (40°C) preserved the structural information and enabled the differentiation of methyl and methylene groups of closely eluting components. Peaks 6 and 7 were therefore identified as α-farnesene and germacrene D, respectively.[235]

The compromise between chromatographic and spectroscopic requirements becomes especially critical when open tubular columns are used. A typical peak has a width at half-height of approximately 15 s during operation of a 50-μm i.d. column at 150 atm with a flow rate of 1.5 μL min^{-1}. This corresponds to a volume of 400 nL. The flow cell volume should therefore not exceed 100 nL. A similar result was obtained by substitution into the more rigorous expression (Equation 4.20).[228] If the density in the flow cell is greater than in the column (which is the case if the flow cell is at ambient temperature), the required detector volume for the same resolution is decreased. Equation 4.20 takes this into account by including a fractional density factor ($\Delta\rho$) for the column verses the detector. For a resolution loss of 1%, CO_2 as mobile phase, and a 20-m x 50-μm i.d. open tubular column which produces 10^5 theoretical plates at a k value of 1, the calculated cell volume is 68 nL for a column density

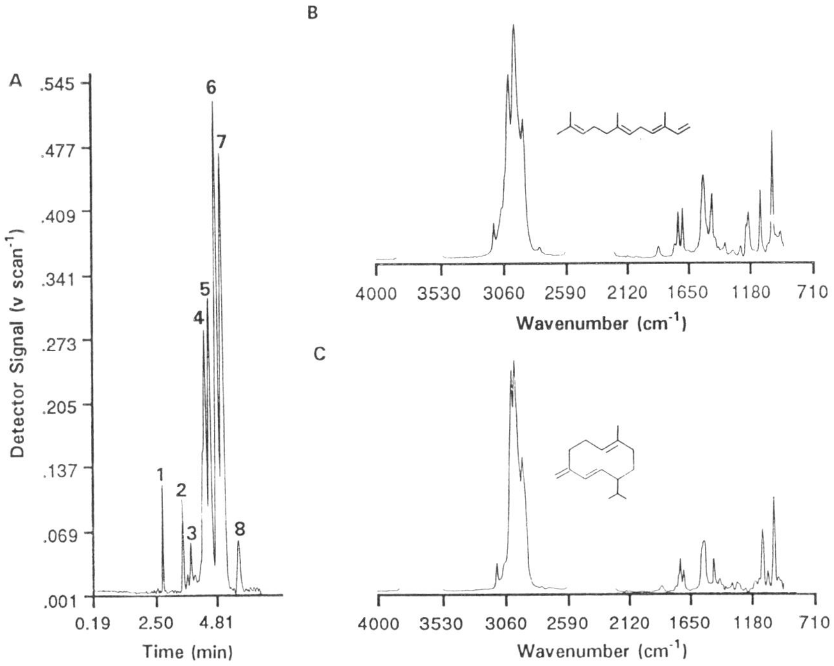

Figure 4.40. (A) SFC chromatogram of sesquiterpene hydrocarbons from ylang-ylang oil and FTIR spectra of (B) α-farnesene and (C) germacrene D. Conditions: two 25-cm x 4.6-mm i.d. packed columns in series, silica particles (Nucleosil 100 and Spherisorb; 5 μm); CO_2; 40°C; 4 mL min^{-1} flow rate; FTIR, 33°C cell temperature, 8 cm^{-1} optical resolution. Peak identifications: (1) hexane, (2) α-copaene, (3) and (4) unknown, (5) β-caryophyllene, (6) α-farnesene, (7) germacrene D, (8) unknown (reprinted with permission from Ref. 235).

at 90°C of 0.53 g mL^{-1} and a cell density at 20°C of 0.94 g mL^{-1}, and 95 nL for a column density at 90°C of 0.79 g mL^{-1} and a cell density at 20°C of 1.0 g mL^{-1}.

Unfortunately, such small volumes are in conflict with the requirements of the FTIR. Since the minimum attainable beam diameter is only slightly less than 1 mm using state-of-the-art optics, any value of the cell diameter less than this must reduce the available energy transmitted through the cell. Based on this fact, the minimum practical cell diameter is 0.5 mm. If it decreased any further, the reduced cell transmission would result in an unacceptable noise level in the measured spectra. Hence, since the cell diameter is set at 0.5 mm, ℓ must be reduced to achieve a small cell volume. However, the cylindrical cell design must allow for zero

Table 4.5. Comparison of Selected Capillary Flow Cell SFC-FTIR Studies

Flow cell parameter	Shafer and Griffiths[236]	Novotny *et al.*[225,226]	Wieboldt *et al.*[228]	Taylor *et al.*[232,238]	Raynor *et al.*[233,239]
Mobile phase	CO_2	CO_2, Xe, freons, propane	CO_2	CO_2	CO_2, Xe
Column	60-m x 0.33-mm i.d.	20-m x 150-μm i.d.	10-m x 100-μm i.d.	10-m x 100-μm i.d. and 25-cm x 1-mm i.d. packed column	10-m x 50-μm i.d.
Flow rate/-linear velocity	5 cm s^{-1}	-	-	-	3.5 mL min^{-1} gaseous (col. press., 150 atm)
Cell volume	8 μL	1 μL	1.4 μL	1.4 μL	0.8 μL
Pathlength	10 mm	1 mm	5 mm	5 mm	4 mm
Window	CaF_2	ZnS	-	-	ZnSe
Pressure	140 atm	200 atm	400 atm	400 atm	400 atm
Temperature	Ambient	Ambient to 50°C	Ambient to 50°C	33°C	Ambient
Beam condensing	No	Yes	Yes	Yes	Yes
Scan rate	1.35 s $scan^{-1}$	2.5 scans s^{-1}	4 scans s^{-1}	4 scans s^{-1}	3 scans s^{-1}

dead volume fittings (for a ferrule and bushing) for the inlet and outlet capillaries, and this limits ℓ to at least 4 mm.

Construction details of some cells reported in the literature are summarized in Table 4.5. For a 4-mm pathlength and 0.5-mm cell diameter, the cell volume becomes 800 nL, and the transmission of beam energy is 20%. With such a cell (Figure 4.41), spectra can be recorded with better than 10:1 signal-to-noise (S/N) ratio by recording approximately 60 scans during the elution of a typical peak from amounts of sample varying between 10 ng (strong IR absorbing compounds containing C=O groups, such as esters) and 100 ng (weak IR absorbers such as polycyclic aromatic hydrocarbons). An S/N value of 3:1 was observed in a similar flow cell with slightly larger dimensions for the C=O absorption from 2.5 ng of caffeine with a linear Beer's law plot.[237] An MDQ of 10 ng for methyl palmitate has been reported by Wieboldt *et al.*[228]

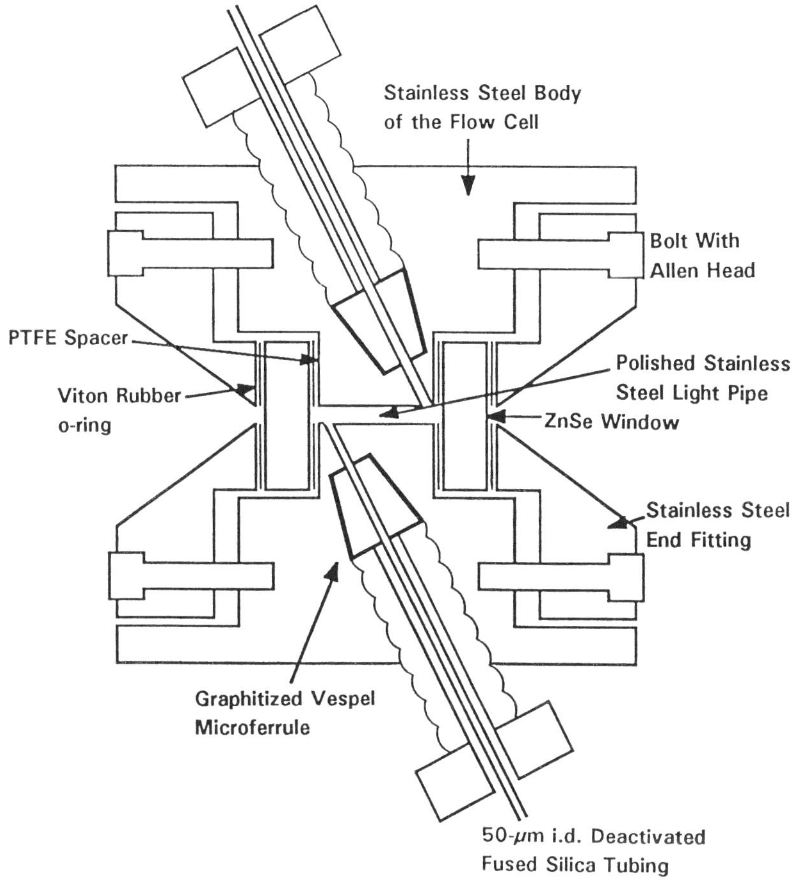

Figure 4.41. Schematic diagram of a high pressure flow cell for SFC/FTIR (reprinted with permission from Ref. 233).

Peak broadening, caused by the flow cell, originates in three ways.[233] Since the volume of the cell is not a negligible fraction of the peak volume, only a segment of each peak is seen by the spectrometer. Second, turbulance is introduced by various connections, introduction of make-up fluid, and any temperature changes. Finally, slow diffusion from corners introduces dead volumes. These effects can be reduced by the use of make-up fluid. A second length of capillary tubing with identical dimensions to those of the column can be installed in parallel. This doubles the rate of flow through the cell, which reduces tailing effects, and effectively halves the cell volume to 400 nL, while allowing the chromatography to be performed at close to optimum flow rates. Unfortunately, it lowers the IR sensitivity since the peaks are diluted by the make-up fluid. For an 800-nL flow cell operating with make-up fluid, quantitative considerations suggest a 30% loss of resolution for detection in liquid CO_2, with 50% peak broadening arising approximately equally from the cell volume and the combined effects of mixing and dead volume.

Stopped flow methods may be used to improve the sensitivity of the IR measurements obtained with a flow cell interface;[240] the lower diffusion coefficients of solutes in liquid CO_2 in the flow cell compared with those in supercritical fluids and gases facilitate this. In this procedure, the mobile phase flow in the column, make-up fluid line, and flow cell can be stopped by switching a valve, thereby isolating a component in the flow cell for FTIR analysis. Meanwhile, the mobile phase continues to flow through a by-pass capillary connected at one end to the mobile phase line prior to the on/off valve, and at the other end to the flow cell outlet using a butt connector and a graphite ferrule. The flow of CO_2 through the by-pass capillary effectively isolates conditions in the column and flow cell so that the peak apex is trapped in the light-pipe. At least ten times the number of spectra can be measured in comparison with a normal SFC/FTIR run, with the result that the FTIR sensitivity can be increased by three times. Once enough spectra have been accumulated, the on/off valve is switched on again so that the chromatographic run can be completed.

Application of the Gram-Schmidt orthogonalization procedure by Wieboldt and Hanna has been indispensible to the success of the flow cell interface and its use with CO_2 during density or pressure programming.[241] Normally, Gram-Schmidt software would produce a chromatogram with a steeply rising baseline (Figure 4.42A), because of the effects of density programming which reduce the energy throughput. However, when Gram-Schmidt orthogonalization is performed, vectors containing the information from the rising baseline are added to the basis set, and the chromatogram is then recalculated to remove the baseline drift and to enhance the signal intensity (Figure 4.42B). The IR spectra of the peaks contained in the associated data files can then be retrieved from computer memory, coadded and manipulated to remove the spectral features of CO_2. Figure 4.42C shows how the FTIR can be used as a chemical specific detector, by reconstructing data in a specific absorption region, in this case for an aromatic ketone from 1670 to 1690 cm^{-1}.

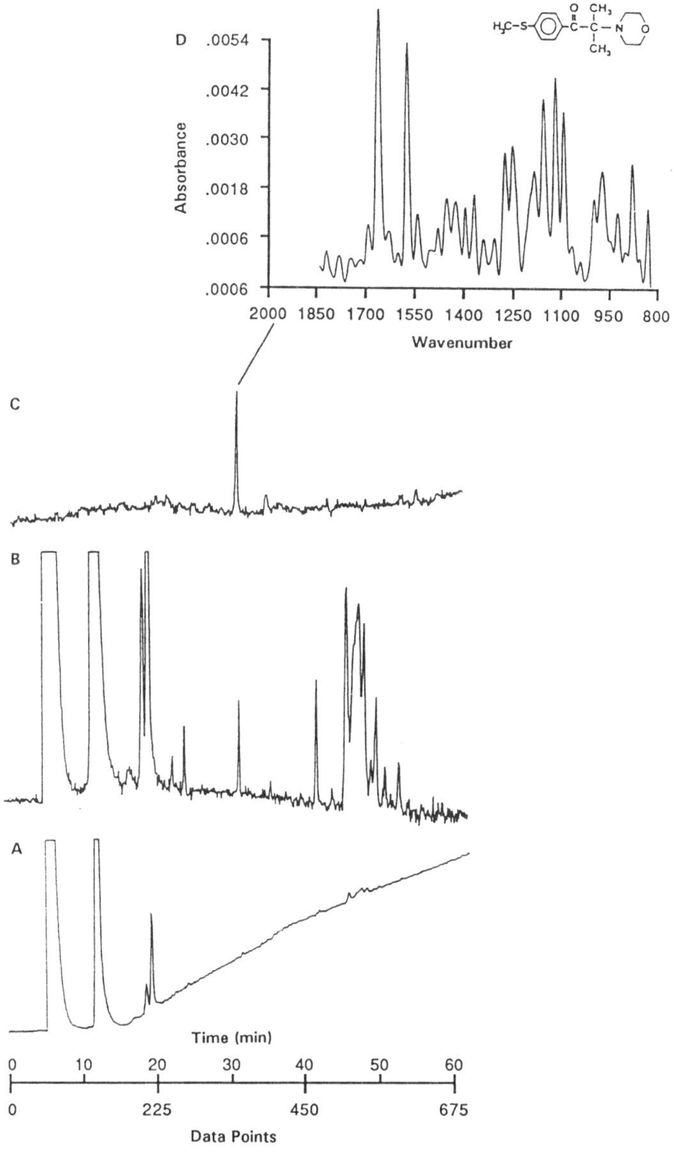

Figure 4.42. SFC chromatograms and FTIR spectra of a UV curing coating. (A) Gram-Schmidt reconstructed chromatogram of the total separation, (B) calculated with added basis vector 346 to remove the baseline drift from the increased CO_2 density, (C) 1670-1690 cm^{-1} selective detection of the carbonyl adsorption of an aromatic ketone, and (D) IR-spectrum of a photo-initiator detected in the 1670-1690 cm^{-1} reconstruction. Conditions: 10-m x 50-μm i.d. open tubular column; poly(50% biphenyl)methylsiloxane; 65°C; CO_2; pressure program from 100 atm to 250 atm at 2-5 atm min^{-1} following a 15-min isobaric hold (reprinted with permission from Ref. 233).

In this way, the photoinitiator in the UV curing coating was identified as Irgacure 907 from its infrared spectrum in Figure 4.42D.

A study of the effect of temperature and pressure on the solution IR spectra of compounds in liquid and supercritical CO_2 has shown variations in wavenumber with density of up to 10 cm^{-1}. Stretching vibrations are more sensitive than bending vibrations, and the shift is greater for polar groups such as carbonyl. Modification of intensity distribution and bandwidth variations may also occur. Changes in polarizability of the supercritical fluid lead to changes in dipole-induced dipole interactions with the solute.[242] In consequence, spectra recorded during SFC/FTIR with a flow cell interface may differ from those recorded in vapor or condensed phases. Mismatches with libraries of IR spectra can occur if comparisons with condensed phase spectra are made, and vapor phase libraries are preferred. However, SFC is generally applied to nonvolatile compounds, and a library of spectra recorded in supercritical or liquid CO_2 may be necessary.

SFC/FTIR has been applied to the analysis of a large number of thermally labile and nonvolatile samples. Normally, fatty acids are analyzed by GC as their methyl ester derivatives. Hellgeth and coworkers used packed column SFC/FTIR to separate and identify free fatty acids from butter, soap, soya bean oil, and coconut oil samples, without derivatization.[243] Morin *et al.* also used packed column SFC/FTIR to separate and identify polycyclic aromatic hydrocarbons[244] and sesquiterpene hydrocarbons[235] (Figure 4.40) in fractions from several essential oils.

Carbamate pesticides, stabilizers in polymers, and a pyrethrin extract have been analyzed by Wieboldt and coworkers with capillary SFC/FTIR using CO_2, a 100-μm i.d. open tubular column, and a 1.4-μL flow cell.[129,245] The method was particularly useful for analyzing the pyrethrin extract, since the separation was performed under much milder temperature conditions than in GC. IR spectra of cinerin II and pyrethrin II could be differentiated from each other without interference from degradation products by the out-of-plane C-H deformation at 912 cm^{-1}.

Raynor *et al.*[233] have also used an open tubular column SFC/FTIR system with an 800-nL flow cell to analyze UV curing coatings containing reactive diluents, thermally labile photoinitiators, and reactive oligomers (Figure 4.42). A 50-μm i.d. open tubular column was used with CO_2 as the mobile phase, and make-up fluid was added to the column effluent prior to the 800-nL flow cell to reduce any loss in chromatographic resolution. Taylor *et al.*[238,246] have used a 1.4-μL flow cell with both packed and open tubular columns for analyzing propellant stabilizer components and extracts of double base propellants. Since these compounds are thermally sensitive, SFC/FTIR was used to determine chemical changes in the stabilizer components. Nitrosoaniline derivatives were observed *via* FTIR to have been converted to nitro derivatives on storage. Mixtures of steroids have been analyzed in the same system with a 100-μm i.d. poly(cyanopropyl)methylsiloxane open tubular column. In this analysis, five steroids were separated, and the carbonyl absorption region was used for identification since the CO_2 masked the hydroxyl stretching absorptions of the steroids.[237] Similar problems were experienced in the analysis of peracetylated aldonitrile derivatives and byproducts from monosaccharides, because the presence of the nitrile functionality could not be confirmed in CO_2.[246] The aromatic C-H deformation absorptions of polycyclic aromatic hydrocarbons cannot be detected in CO_2 either, because of its 800 cm^{-1} cut off.

In these cases, xenon should be used. Although expensive, it has the advantage of IR transparency, so that the whole IR spectrum can be monitored, eliminating the need for background subtraction. Further, wavenumber shifts and changes in bandwidth are less problematic due to its monatomic nature. French and Novotny[225] showed that xenon has similar chromatographic properties as CO_2, and both calculations and measurements indicate similar or greater solute solubilities in xenon. Jenkins and coworkers have demonstrated that aromatic C-H deformation absorptions of PAH below 800 cm^{-1} can be detected in xenon.[247] Raynor analyzed an oligomeric sample containing aromatic isocyanates in xenon, since the N=C=O stretching absorption at 2270 cm^{-1} was masked by the CO_2 absorption from 2200-2500 cm^{-1} which prevented identification.[239]

SFC/FTIR using solvent elimination. SFC/FTIR interfaces in which the column effluent is passed continuously through a flow cell were discussed in the previous section. Although these interfaces are simple in concept and design, they have two primary disadvantages. First, the cells have a fixed volume, while the width of SFC peaks can vary considerably. The optimum sensitivity in any chromatography/FTIR measurement can only be achieved when the full width at half height (FWHH) of any peak is approximately equal to the cell volume.[248] Second, the pathlength of the cell is determined by the absorption spectrum of the mobile phase,[249] so that a single pathlength cannot be used for all measurements. A very long pathlength can obviously be employed if xenon (which is completely transparent in the infrared) is used as the mobile phase.

The typical pathlength for pure CO_2 is limited to 5 mm; even so, three spectral regions (3600-3400 cm^{-1}, 2500-2200 cm^{-1}, and 750-600 cm^{-1}) are completely lost, while the transmission in several other regions decreases as the density of the mobile phase is increased. Finally, if a polar modifier such as methanol is included in CO_2 or Xe, the cell pathlength must be less than 1 mm.

As mentioned earlier, the obvious way to eliminate these problems is to eliminate the mobile phase prior to measuring the spectrum of each solute. This approach was first used for LC/FTIR measurements in which absorption by the mobile phase is always very strong.[250] Whereas the construction of reliable LC/FTIR interfaces involving mobile phase elimination has proven difficult, the corresponding SFC/FTIR interface is relatively easy to design because the mobile phase becomes gaseous immediately when emerging from the restrictor. Nonvolatile solutes can be trapped on a suitable substrate which is mounted immediately below the restrictor. Several substrates and sampling techniques have been investigated, with the best results being by deposition on a ZnSe plate, followed by measurement of the transmission spectrum of each deposited component.[251]

From a spectroscopic standpoint, it is important to trap each eluate in as small an area as possible, since for a given sample quantity, the average thickness (and hence the absorbance of any band in the spectrum) varies inversely with the cross-sectional area. For about 10 μg of a typical organic molecule trapped as a 1-mm diameter cylinder, the pathlength is about 15 μm, which is sufficient to yield an infrared spectrum that would not require ordinate expansion (maximum absorbance of approximately 2). If the diameter of the cylinder is reduced to 100 μm, the amount of sample required to yield a spectrum of the same intensity is reduced to 100 ng. Reduction of the sample quantity to 1 ng without a further change in diameter would reduce the absorbance of the strongest bands in the spectrum to 0.02 (*i.e.*, the minimum transmittance is 95%). This absorbance is still sufficient for the spectrum to be measurable in one second with a signal-to-noise ratio of at least 20 with a state-of-the-art FTIR spectrometer equipped with microscope optics and a mercury cadmium telluride detector of approximately the same linear dimension (100 μm) as the sample.

In practice, if a substrate is held stationary about 50 μm below an integral restrictor, the thickness profile of the deposited spot is approximately Gaussian with a FWHH of less than 100 μm,[230] so that the spectral intensity is quite similar to that of the "ideal" 100-μm diameter cylinder. Many of the earlier SFC/FTIR measurements involving mobile-phase elimination were effected with a stationary substrate.[231,252] For measurements of this type, the chromatogram must be measured prior to the SFC/FTIR run, *e.g.*, with a flame ionization detector, in order to locate each spot readily. In general, however, it is preferable to make the deposition on a moving substrate, so that the train of spots reproduces the chromatogram, and a conventional detector is unnecessary.

For depositions on a moving window, the faster the substrate is moved, the better maintained is the integrity of the separation; however, the deposit is thinner at any point and, consequently, the absorbance of any band in the spectrum is lower. There is, therefore, a compromise between chromatographic resolution and spectrometric sensitivity that is similar to the equivalent compromise for flow-cell measurements made by varying the cell volume.[248] For a spot with a longitudinal FWHH of z μm, formed by depositing an SFC peak with an FWHH of t seconds, there is a loss of less than 20% in chromatographic resolution if the translational speed of the substrate is $2z/t$ μm s^{-1}.

SFC/FTIR measurements involving mobile-phase elimination can be performed off-line or on-line. Most measurements reported to date have been measured in the off-line mode, in which the eluates are first deposited on a moving window. The window is then removed from under the restrictor and mounted on the sampling stage of an FTIR microscope, and the spectrum of each spot is measured. Since each spot is located visually under the microscope, the substrate is rarely moved during the actual measurement, and each interferogram is usually signal-averaged for at least one minute. In on-line measurements, the restrictor is located a short distance before the infrared beam focus, as shown in Figure 4.43, so that the spectrum is measured only a few seconds after the sample is deposited and while the substrate continues to move. A GC/FTIR interface based on the same principle, in which the substrate is cooled to 77K,[253] has recently been introduced commercially. Corresponding SFC/FTIR data obtained using this device have not yet been reported, but they should be more easily obtained than the GC/FTIR measurements.

Chromatograms can be constructed directly from the measured series of interferograms using the Gram-Schmidt vector orthogonalization algorithm that gives a signal based on the total infrared absorption. Chromatograms can also be constructed after transforming the interferograms into spectra by integrating the absorbance in one or more defined spectral regions corresponding to the absorption by a given functional group. The coordinates of each point in the reconstructed chromatogram are recorded along with the

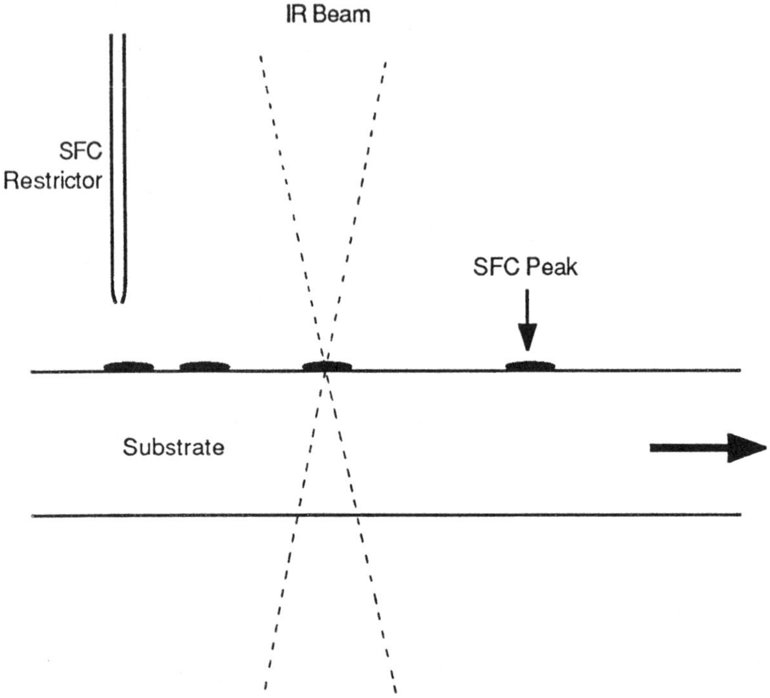

Figure 4.43. Schematic diagram of an SFC/FTIR interface with mobile phase elimination.

interferogram or spectrum. Since the coordinates of each maximum in the reconstructed chromatogram correspond to the location of each component, minor components can be returned to the infrared beam at the end of the chromatographic run, so that their spectra can be remeasured with more extensive signal averaging.[254]

SFC/FTIR interfaces involving mobile phase elimination have several advantages over the use of flow cells. First, the entire spectrum is measured, no matter what mobile phase or density program is employed. Second, the sensitivity of real-time measurements is greater (between 2 and 4 times) than the corresponding measurements made using flow cells in regions where the mobile phase is transparent (and by a much greater factor in those regions where the mobile phase absorbs strongly). The capability of returning any component into the beam for post run signal averaging to improve the signal-to-noise ratio is also important.

At the end of a density program, the width of the chromatographic peaks will normally increase. As the width increases, the concentration of a given quantity of a component decreases, so that the detection limits of flow cell measurements (for which the cell volume is constant) will increase. By reducing the speed at which the substrate moves in the mobile phase elimination process, the effect of the increasing peak width can be minimized.

Finally, and possibly of greatest importance, is the observation that in SFC/FTIR measurements in which the mobile phase is eliminated, the spectrum is directly comparable to the reference spectrum of the same material prepared in a KBr disk, so that over 100,000 relevant spectra are available in infrared data bases. Conversely, the spectra of materials dissolved in supercritical fluids are not comparable to reference spectra of the samples in other media, and they even show shifts with pressure and temperature.[242]

Although the solvent elimination SFC/FTIR technique is relatively new, it has been applied to the study of several types of samples, including barbiturates,[254,255] quinones,[256] anti-oxidants,[231,252] polyphenyls,[229] substituted indoles,[254] chlorinated pesticides,[257] polychlorinated biphenyls,[257] polycyclic aromatic hydrocarbons from lamp black,[257] siloxane oligomers,[221,258] and the residue from a still in which a herbicide (2,4,5-trichlorophenoxyacetic acid) was distilled.[219] The majority of these separations were performed using open tubular column SFC with neat CO_2 as the mobile phase, although one[256] involved a separation on a packed silica column with methanol-modified CO_2 as the mobile phase. The successful use of CF_2Cl_2 as the mobile phase for an SFC/FTIR measurement has also been reported.[254]

A few limitations to the technique should be mentioned. First, the components must be sufficiently nonvolatile to be deposited on the window and to remain there for several minutes. Most reported measurements have been made with the window at ambient temperature, but it can be cooled thermoelectrically to assist the retention of more volatile compounds.[229] Additionally, if the components are liquids, they may be blown away by the gas emerging from the restrictor; again, cooling of the window can alleviate this problem. Since most volatile or liquid components can be separated faster and at higher resolution by GC than by open tubular column SFC, these problems may be expected to be rare.

4.21 Mass Spectrometry

Considerable progress has been made in the design, application, and fundamental understanding of SFC/MS over the past eight years.[259-262] For many applications today, SFC/MS has been found to be superior to LC/MS due to the wider applicability and higher sensitivity of the former.[263] Also, new applications of SFC/MS have been found for which it is uniquely suited. For example, Figure 4.44 shows an electron impact (EI) mass spectrum of diphytanylglycerol diether obtained by Holzer *et al.*[264] using SFC/MS. This is the first EI mass spectrum ever obtained of diphytanylglycerol.

The requirements of an SFC/MS interface are as follows:

(a) The interface must be capable of handling the gas flow rates generated from the mobile phase. Most quadrupole instruments are capable of handling flow rates generated by open tubular column SFC. However, high capacity interfaces (moving belt, thermospray, etc.) must be employed for conventional packed column work or, alternatively, microcolumns must be used.

(b) The interface should maximize the transport of solute into the ion source without adsorption or thermal decomposition. The high capacity interfaces used for packed column SFC/MS easily fulfill these requirements, since they commonly operate at a high mobile phase velocity and at relatively low interface temperatures (~100-170°C). The heated restrictor approach to open tubular column SFC/MS interfacing is more critical. The solute/mobile phase velocities and related CO_2 gas volumes are lower. Therefore, effective transfer must be facilitated by the use of high temperatures (~150-550°C) precisely at the restrictor tip to increase the relative gas velocity, to prevent freezing at the tip, and to some degree, initiate vaporization.

(c) The interface must be constructed such that chromatographic integrity is maintained; this can be achieved by the use of minimal dead volume couplings.

Early studies. The earliest SFC/MS work was performed on sector-type instruments. Milne proposed the marriage[265] using a nozzle/skimmer arrangement with multi-stage pumping and electron impact ionization (EI) of the expanding molecular beam. He discussed the potential problems of solute-solvent and solute-solute associations. Giddings *et al.*[266] provided a theoretical treatment of the dissociation of solute-solvent clusters in a free jet expansion. They also provided experimental results. Beta-carotene was transported intact through an expansion from 1,000 atm to 1 atm at

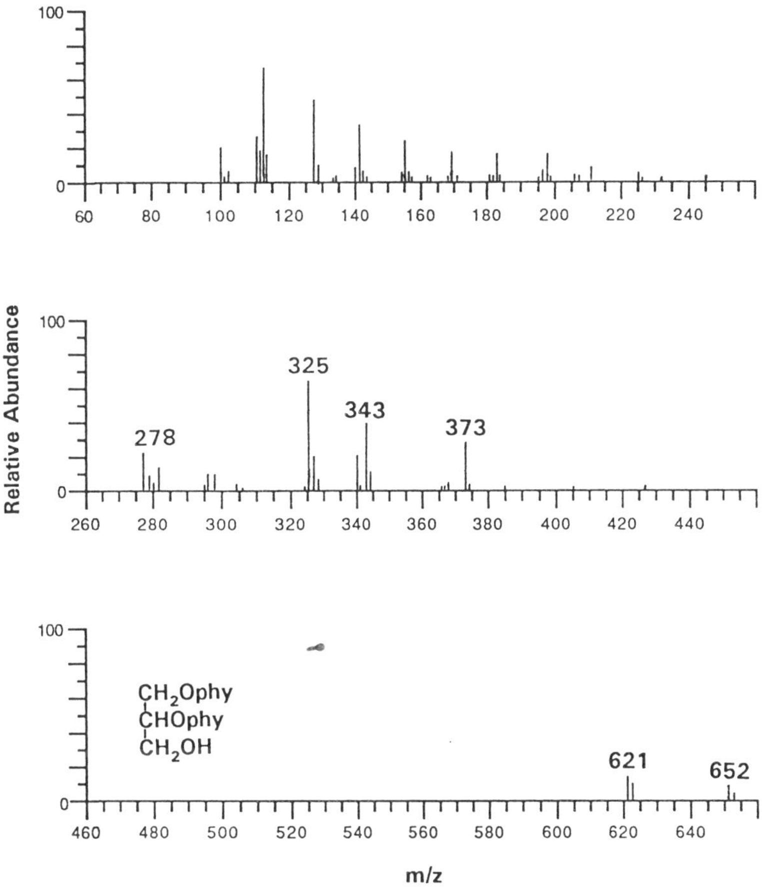

Figure 4.44. SFC/MS EI mass spectrum of diphytanylglycerol diether. Conditions: 10-m x 75-μm i.d. open tubular column, poly(50% phenyl)methylsiloxane stationary phase; CO_2; 120°C; 250 atm (reprinted with permission from Ref. 264).

40°C in CO_2. Squalane was injected into argon at 1,360 atm, transported through an expansion nozzle, and carried into the ion source through a capillary flow restrictor and an effusion separator held at 270°C. A significant proportion of the squalane condensed in the restrictor capillary.

The work of Giddings *et al.* provided the basis for the experiments of Randall and Wahrhaftig[160,267] using a nozzle/skimmer/collimator arrangement, multiple stages of pumping, and EI of the resulting molecular beam for interfacing SFC to MS. The nozzle aperture consisted of a 12.5-μm laser-drilled pinhole in nickel shim stock. They obtained EI spectra of small aromatic compounds introduced in ethylene, ethane, or CO_2. They observed that solvent cluster formation was a sensitive function of the temperature before the expansion, and the pressure before and in the expansion region. Solute-solvent clustering was not as prominent as expected. The complexity of the molecular beam system, its poor detectability (μg range), and problems such as recurrent plugging of the nozzle aperture discouraged further analytical applications, although separations were attempted using an uncoated, 6-m x 510-μm i.d. stainless steel open tubular column.

Gouw *et al.*[268] interfaced a sector instrument to a packed column SFC system using a crimped 0.01-inch metal capillary to restrict the flow into the ion source. The interface capillary was crimped at its inlet, rather than at its outlet. The column effluent was split between the mass spectrometer and a high pressure fraction collector. The authors were primarily interested in polynuclear aromatic hydrocarbons (PAH) isolated from automobile exhaust, although a coal oil was also analyzed. Tailing of the chromatographic peaks was minimized by heating the interface to 450°C.

SFC/MS interfaces. Since open tubular columns produce the lowest flow rates, SFC/MS interfaces with this column type are the most simple and are described here first. The most common open tubular column SFC/MS interface is the direct fluid interface (DFI). This interface allows the expansion of the supercritical fluid to occur at, or close to, the entrance to the ionization source. The interface body is typically heated to the same temperature as is used in the chromatographic separation. A second region of the interface near the point of expansion is heated to higher temperatures to assist cluster evaporation.

Open tubular column SFC/MS systems reported to date follow a common approach, with emphasis on the transfer of solute from the column restrictor to the ion source without solute precipitation or thermal decomposition. The restrictor design and interface heating methods are major considerations. There are numerous reviews that cover the subject.[260,262,269] Early studies on open tubular column SFC/MS utilized either a crimped platinum-iridium tube or a 0.5-2.0-μm laser drilled orifice as an interface restrictor.[65,260] Operation of these systems produced detection limits in the subnanogram range under CI conditions (full scan). Both restrictor designs performed equally; however, the latter was prone to plugging. Rapid analysis has been reported using this system with short open tubular columns and high density programming rates.[270] Attempts at generating EI spectra resulted in reduced sensitivity, and data containing both EI and CI contributions.[65] The incorporation of a heated expansion chamber in the ion source produced EI spectra and prevented CO_2 cluster formation.[271]

The ideal SFC/MS interface would control the flow in the chromatographic column while quantitatively transferring the nonvolatile analyte molecules into the ion source. Metal or fused silica capillary restrictors have been used in SFC/MS. In the early SFC/MS interfaces, long (10-15 cm) 10-15 μm i.d. capillaries were used to control the flow. Considerable plugging occurred when nonvolatile compounds were studied. These linear restrictors have been replaced by tapered or frit restrictors. The internal diameters of the tapered restrictors are typically 50-100 μm i.d. with the MS end drawn down to 1-4 μm i.d. Smith *et al.*[53] studied the flow dynamics of SFC restrictors, and Goates *et al.*[159] analyzed the formation of clusters in the expanding gases. Both studies showed that to eliminate the formation of large aerosol particles, additional heat must be applied to the end of the capillaries to compensate for the Joule-Thomson cooling. Reinhold recently modified a DFI probe which is capable of highly localized heating of the capillary restrictor tip[272] (Figure 4.45) which is described in greater detail later.

The restrictor tip temperatures used in SFC/MS are commonly 250-400°C. Owens *et al.*[273] optimized the tip temperature for the SFC/MS detection of Triton X-100. At temperatures below 300°C, the restrictor plugged before the separation was complete. The optimum temperature range was 325-350°C. For temperatures higher than this range, the column flow velocity decreased due to an increase in the expanding gas viscosity, the injection split ratio increased, and the signal diminished. The optimum restrictor tip temperature may be dependent on both the supercritical fluid and the analyte.

Zaugg *et al.*[274] and Pinkston *et al.*[275] recently described simple direct fluid interfaces that were effective for SFC/MS. In the DFI interface built by Zaugg *et al.*,[274] a quartz jet was attached to the end of a heated GC/MS interface. The jet fit easily into the GC port of the ionizer. Other groups[276,277,278] have also described similar interfaces that involved only minor changes in various GC/MS interfaces. Currently, expedient switching between GC/MS and SFC/MS is possible.

Special precautions must be taken for direct introduction from open tubular columns into sector mass spectrometers, since the ion sources of these instruments are generally held at high voltage (1 to 10 kV). These high voltages may lead to arcing of the source to ground if the source and interface are not designed to handle the

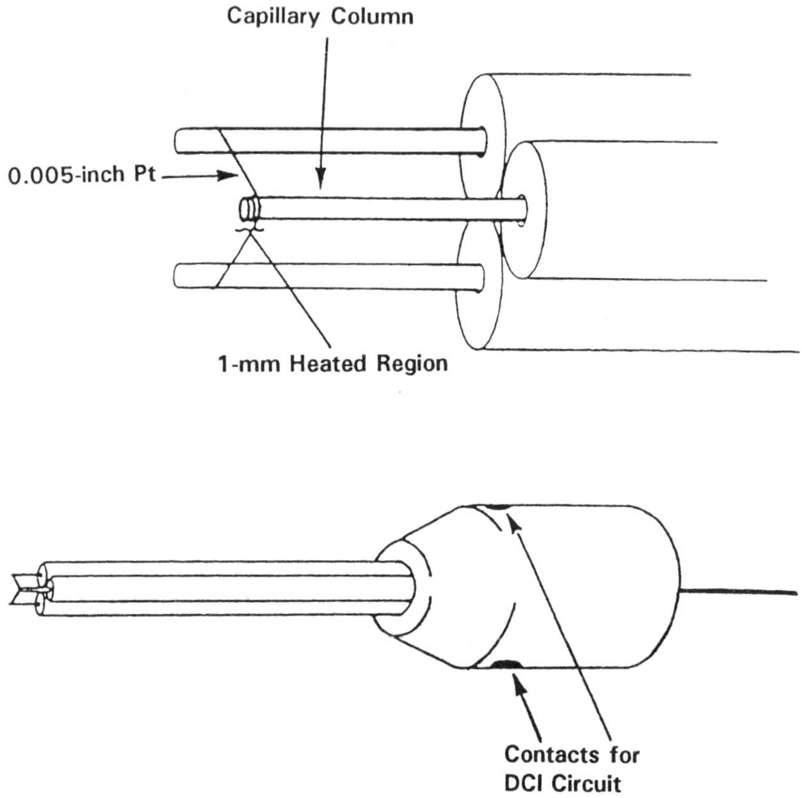

Figure 4.45. Schematic diagram of an SFC/MS probe constructed from a standard DFI probe tip with an additional ceramic feed-through for the SFC column. Contact points provide electrical coupling to taps on the side of the ion block for temperature control (reprinted with permission from Ref. 272).

pressures encountered during DFI SFC/MS, especially in the CI mode. Kalinoski *et al.*[279] described a DFI interface for high resolution (sector) instruments. The interface probe is shown in Figure 4.46. The Vespel barrel provides electrical isolation of the probe tip, which is operated at source potential (8 kV). The zero dead-volume seal on the restrictor eliminates one path to ground for the CI plasma, a possible source of arcing. The same research group[279] applied the high voltage DFI interface to the analysis of the trichothecene mycotoxin diacetoxyscirpenol (DAS, MW 366) and T-2 toxin (MW 466). These two toxins were spiked on wheat samples and extracted with a supercritical mixture of CO_2 and 2-propanol (5% v/v). The supercritical fluid extract was introduced directly into the

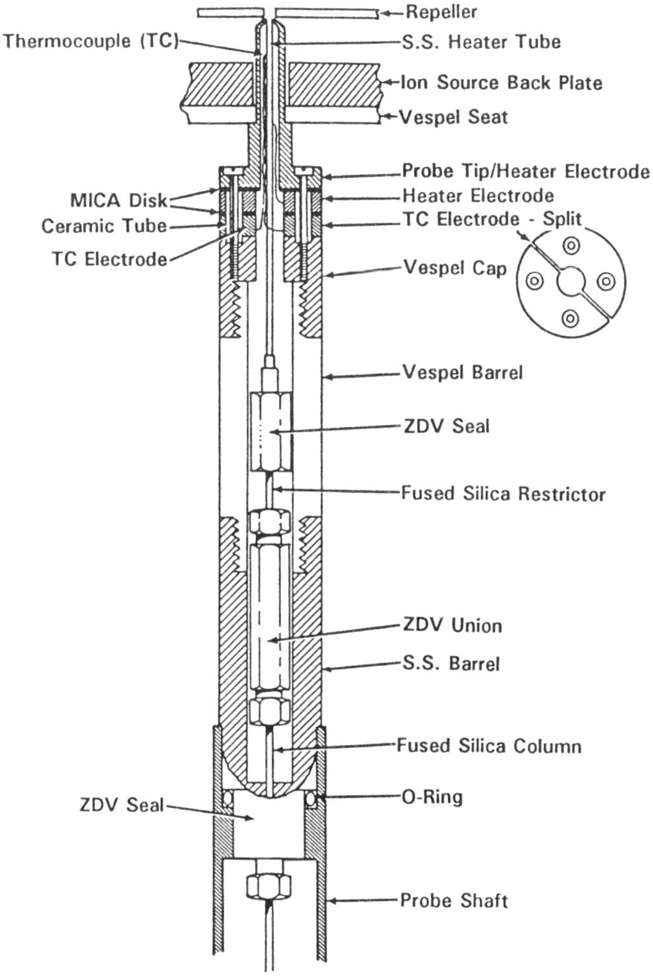

Figure 4.46. Schematic diagram of an SFC/MS interface for a high voltage CI source (reprinted with permission from Ref. 279).

CI source of a sector mass spectrometer operating with ammonia in the CI mode. Figure 4.47 shows the ammonium adduct ion region for DAS. It is clear that high resolution mass analysis is essential for determining DAS in the presence of coextracted matrix components of the same nominal mass. The high voltage interface has also been used for coupled SFC/MS.[280]

Huang *et al.*[281] initially described a simple DFI interface for a sector instrument in which the direct introduction probe (DIP) was

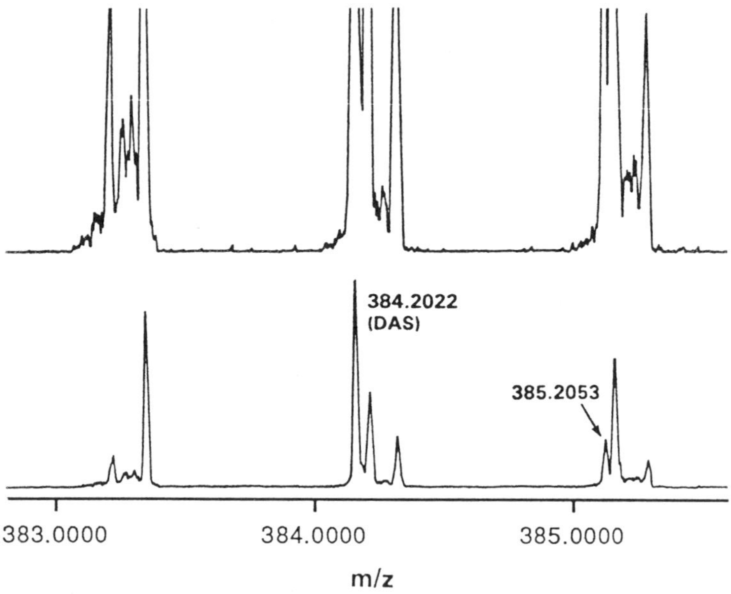

Figure 4.47. High resolution mass spectrum of an SFE extract of wheat containing diacetoxyscirpenol (DAS) and T-2 toxin. Ammonia CI was used at 7,400 resolution (5% valley) using the interface probe shown in Fig 4.46 (reprinted with permission from Ref. 279).

positioned opposite the SFC flow restrictor in the ion source. The DIP was heated to 400°C to provide additional heat to the ion source and the restrictor. Subsequently, a more sophisticated interface with direct heating was developed and applied to a variety of compounds using either EI or negative ion CI operation.[282,283] The frit restrictor was more dense than those typically used for flame ionization detection. This allowed flow rates and ion source pressures such that true EI spectra were obtained, with little or no CO_2 charge exchange.

A similar interface has recently been used in the work of Kallio *et al.*[284] Both flame ionization detection and EI mass spectral data were used to quantify the various triacylglycerols in Finnish butter. An integral-style flow restrictor was fashioned on the end of the open tubular column, which was then linked to the primary restrictor with a zero dead-volume union. The column restrictor was intended to protect the mass spectrometer in case of failure of the primary restrictor.

Reinhold *et al.*[272,285,286] have been interested in the structural elucidation of biopolymers, in particular oligosaccharides. They used a modified desorption/chemical ionization probe, shown in Figure 4.45, and an integral flow restrictor for the direct introduction of open tubular column effluents into the CI source of a sector instrument. The efficiency of ionization appears to have been maintained at least through the 15th oligomer of trimethylsilyl-derivatized poly(glucose). The ammonia CI spectrum of this derivatized oligosaccharide is dominated by the ammonium adduct ion at m/z 5850,[272] as is shown in Figure 4.48B. Figure 4.48A shows mass chromatograms for the ammonium adducts of the silylated oligosaccharides from degree of polymerization 2 through 15. Sheeley and Reinhold[285] have investigated a variety of derivatizing agents, and have found, for example, that addition of the pentafluorobenzyl ester of *p*-amino benzoic acid to the reducing end of the oligosaccharide significantly enhances its response in negative ion CI SFC/MS.

Many future applications will demand higher flow rates than those commonly used with conventional open tubular column SFC/MS. For example, microbore column SFC/MS may be desirable to increase the speed and the quantity of sample separated. The gas flux from an open tubular column can be easily handled by standard ion source pumping. However, the flow rates common in high speed open tubular column SFC, microbore SFC, or conventional packed column SFC are too large to allow the entire gas flux to enter the mass spectrometer.

In the first SFC/MS instrument, Randall and Wahrhaftig[267] interfaced analytical scale columns to a mass spectrometer by using three stages of differential pumping. Later, Crowther and Henion[287] interfaced conventional packed columns (*e.g.*, 2.1-mm i.d.) to the mass spectrometer by splitting the effluent of the column by approximately 1/12 at the mass spectrometer interface. The flow rate in the analytical column was typically 0.8 - 1.5 mL min^{-1}. This design used an LC direct liquid interface to which a 2.5-3.0-µm laser-drilled hole was added to act as the restricting orifice. The problems associated with this system included a short diaphragm lifetime, a high split ratio, and provision for only solvent mediated CI data. The MDQs using this interface were poor, typically in the range of ng of sample injected, mainly because of the loss of sample involved in flow splitting.

Matsumoto *et al.*[288,289] used a similar strategy to design an interface for flows as high as 40 µL min^{-1}. In their design, half of the effluent was vented. However, one unique characteristic was that a helium nebulizing gas sheathed the expanding supercritical fluid before entering the ion source. The sensitivity performance of this interface was not described. However, solvent mediated CI data were obtained for triglycerides[288] and water soluble vitamins[289] using

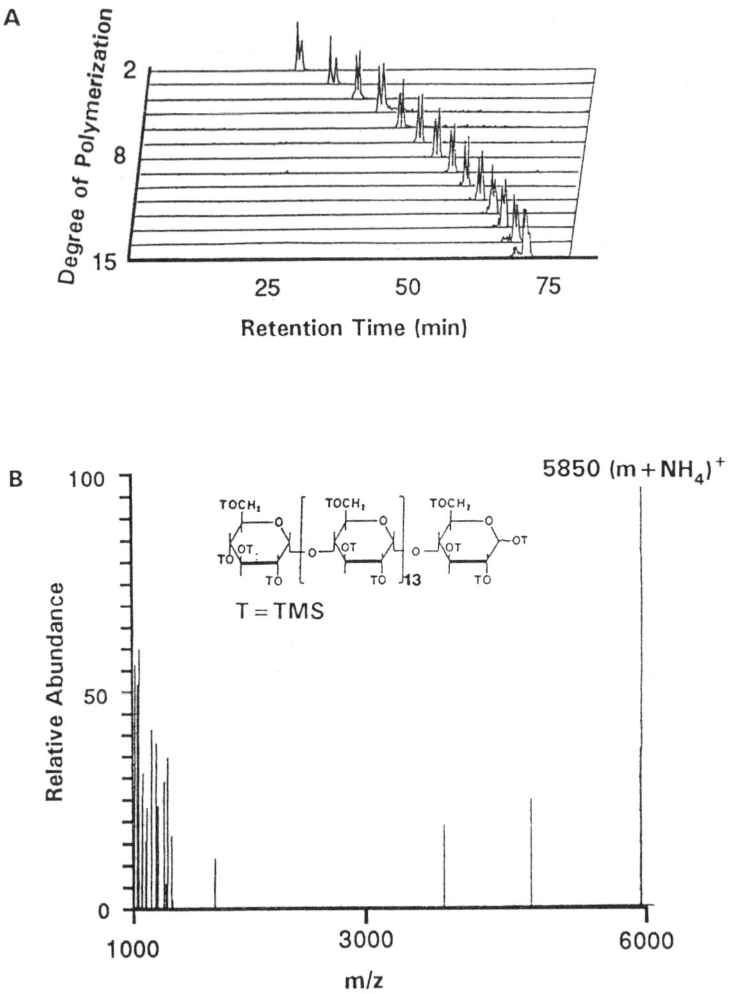

Figure 4.48. SFC/MS analysis of trimethylsilylated glucose oligomers from degree of polymerization 2 through 15 showing (A) selected ion chromatograms of ammonium adduct ions and (B) mass spectrum of the peak at degree of polymerization 15. Conditions: 10-m x 50-μm i.d. open tubular column, poly(5% phenyl)methylsiloxane stationary phase; CO_2; 90°C; pressure program from 115 atm to 400 atm at 3 atm min^{-1}; the interface probe shown in Fig 4.45 was used in the ammonia CI mode at 8 kV (reprinted with permission from Ref. 272).

packed fused silica columns (0.5-mm i.d.) with this vacuum nebulizing interface.

Smith *et al.*[290,291] designed an interface (Figure 4.49) that could handle flow rates of ca. 100-200 μL min^{-1} of a supercritical fluid. This interface was differentially pumped with a mechanical pump

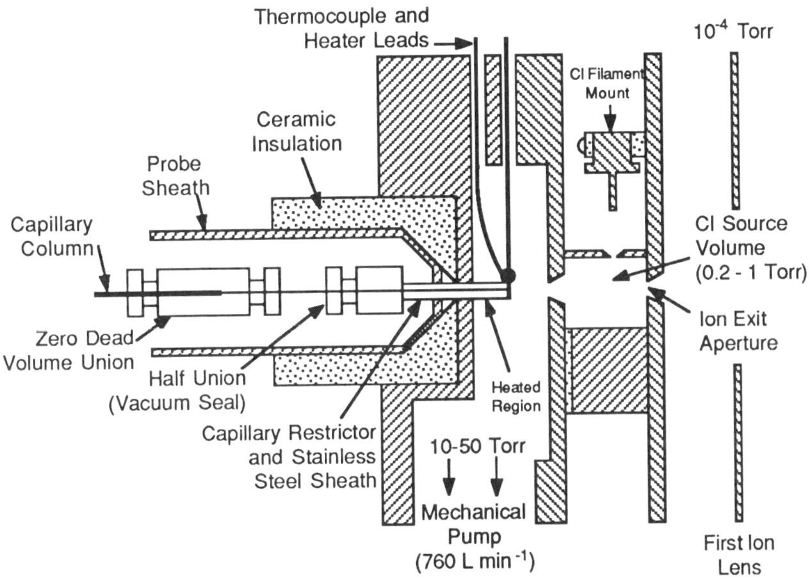

Figure 4.49. Schematic diagram of an SFC/MS interface that can handle flow rates of ≤200-μL min^{-1} of supercritical fluid (reprinted with permission from Ref. 291).

before the ionizer. Picogram detectabilities were achieved with this interface. Careful foreline trapping was recommended in the application of this interface to prevent contamination due to backstreaming of pump oil. This system utilized 1-mm i.d. columns and drawn fused silica restrictors in the interface. In studies of a range of organo-phosphorus insecticides using pressure-programmed, solvent-modified CO_2, solvent mediated CI data were obtained. The addition of ammonia reagent gas to the source demonstrated that conventional CI data could be generated.

Cousin and Arpino[292] tried a different strategy in interfacing high flow rate SFC to the mass spectrometer. Their design permitted flow rates of up to 50 μL min^{-1}. The fluid flow was not split or differentially pumped before the point of expansion, but was directly expanded into the CI ion source. Opposite to the point of expansion, a 12 m^3 h^{-1} mechanical pump was attached. The source chamber was evacuated with a 700 L s^{-1} diffusion pump. Poor detectability (ng injected) was achieved with this interface. This was thought to be caused by low ionization efficiency under the high pressure conditions in the source.

Another interface design for packed column SFC/MS was a modified LC thermospray device. Dätwyler *et al.*[293] were the first to demonstrate the use of a thermospray system for SFC/MS. In their study, the thermospray interface provided heat to desolvate the analytes, while a discharge electrode produced ions. Flow rates of 60-100 μL min^{-1} were readily employed with this interface. Solvent-mediated CI spectra were obtained. Chapman[294] reported a modification of this thermospray interface to link packed column SFC and sector MS to provide added sensitivity and flexibility. The thermospray interface was operated in the discharge ionization mode. The thermospray inlet probe was modified to allow the fused-silica flow restrictor to pass through the probe into the thermospray source, and also to direct the flow of various fluids into the thermospray source coaxially to the flow restrictor. The thermospray interface can easily handle the flow rates of packed column SFC, but, unlike the moving belt, the supercritical mobile phase dramatically affects the type of ionization observed. Chapman[294] saw indications of this effect when he observed that the coaxial addition of a solvent, such as methylene chloride, produced considerably enhanced response for carotenoid hydrocarbons, although it had little effect on the quality of the molecular ion-dominated spectra. The addition of diethylamine produced spectra dominated by the diethylammonium adduct ion for peracetylated disaccharides.

Games *et al.*[295-299] reported that similar results were obtained using a thermospray interface by incorporating a stainless steel hypodermic tube as a dual vaporizer and point restrictor in the ion source. The flow rate range that this interface could tolerate was approximately the same as that of Dätwyler's interface[293] (*ca.* 60-100 μL min^{-1} in the column). The system provided mediated CI spectra. However, additional fragment ions which might help with structural information could be induced by increasing either the vaporizer or source temperature, or by increasing the ion repeller voltage. Certainly, if neat CO_2 was used, EI type spectra could be obtained. Full scan spectra could be obtained at the ng to pg level, with low pg detection in the selected ion monitoring (SIM) mode.

The use of solvent modified supercritical fluids and subsequent cointroduction of solvent and analyte into the mass spectrometer *via* the interfaces described above primarily produces solvent mediated CI spectra. True EI data can be obtained if the concentration of analyte can be increased with respect to the mobile phase related gases before ionization. Two approaches using "enrichment" interfaces have been reported. The first utilizes ≤4.6-mm i.d. columns, from which the eluent passes from a deposition device onto a moving belt LC/MS transport interface.[299] A wide range of compounds were shown to be amenable to the technique, producing specific CI data or clean, full scan EI spectra in the mid to low ng range. The column effluent is spray-deposited on the belt, and the volatile mobile phase is removed as the analytes are carried to the ion source. In this way, the chromatograph and mass spectrometer are operated nearly independently. A packed SFC column with methanol- or methoxyethanol-modified CO_2 mobile phase was used while still obtaining EI and CI spectra. They reported separations of alkaloids, sulphonamides, carbamate and organophosphorus pesticides, steroids, and PAHs. The PAH total-ion-current trace showed good chromatographic fidelity when compared to simultaneous UV detection. Separations were faster and spectra were cleaner than when performing LC/MS on the same type of instrumentation and samples.

In a later publication, Balsevich *et al.*[300] used the moving belt interface for EI SFC/MS of indole alkaloids extracted from leaves of *Catharanthus roseus*. The utility of EI spectra is clearly demonstrated in this work, since high quality library matches were obtained for 14 alkaloids, representing over 95% (by weight) of the alkaloids of mature leaves. Ramsey *et al.*[301] have recently used this versatile interface to successfully identify three veterinary drugs in freeze-dried pig's kidney by SFE/SFC/MS. They demonstrated this instrument to be a powerful combination for the determination of relatively nonpolar substances in complex matrices. Niessen *et al.*[302] have used the moving belt interface for the SFC/MS determination of diuron, a phenylurea herbicide, in plasma.

The moving belt system can be used on sector instruments,[300] it is easily converted to LC/MS use, and the potential for using beams of atoms or ions, or lasers to effect desorption ionization of substances on the belt after SFC has yet to be fully realized. This interface provides flexibility in the type of ionization method used, and facile use of packed or open tubular columns. The disadvantages of the moving belt interface are similar to those found when it is used as an LC interface. Berry *et al.*[295] reported that problems are sometimes encountered with thermally labile compounds when using the belt interface. Mechanical complexity is another disadvantage of the belt interface when compared to the direct introduction approach. The interface is cumbersome, and in addition, poor spectra are often obtained for polar compounds, due to thermal decomposition. Quantitative transfer of volatile samples, such as benzene and naphthalene, is not possible.

The most recently studied high capacity SFC/MS interface is the momentum separator.[303] This device has proven effective in enhancing the sensitivity in LC/MS.[304] The design has some similarities to that of Matsumoto and coworkers[288,289] as described above. In the momentum separator, helium nebulization gas sheaths the expanding supercritical fluid. Most of the helium and CO_2 is pumped away as it passes through the momentum separator. A beam of solute particles passes through the momentum separator and into the ion source where it collides with the heated wall of the ionizer. This interface could tolerate flow rates up to 2.0 mL min^{-1}. Unfortunately, the gain in flow rate tolerance is at the expense of sensitivity. The MDQ with this interface is in the range of micrograms injected. Various drugs, metabolites, and pesticides have been investigated using the system, providing EI spectra under solvent modified CO_2 conditions.

Ionization methods. Chemical ionization is the most frequently used ionization technique in SFC/MS. MDQs of *ca.* 1 pg and 100 pg were reported for pyrene in the first report of capillary SFC/MS[65] under selected ion monitoring and full scan conditions, respectively. MDQs in the range of tens of picograms have been obtained in various laboratories.[275,277,290] These detectabilities are comparable to those found in GC/MS analyses.

Even when large partial pressures of CO_2 are present in the ion source, the CI sensitivity is not strongly changed, due to the low proton affinity of CO_2.[290] Accordingly, standard CI reagent gases, such as ammonia, isobutane, methane, and methanol, can be used effectively in CI SFC/MS.

When mild reagent gases, such as ammonia or isobutane, are used in CI GC/MS, the amount of internal energy transferred to the product ion is low. The internal energy of the product ion can be increased through heat conduction from the ion source walls. Hence, by increasing the ion source temperature, the extent of fragmentation can be increased.[305] When stronger reagent gases such as methane are used, the effect of the source temperature on fragmentation is minimal. Games *et al.*[296] showed that in SFC/MS with thermospray injection, the extent of fragmentation was increased by raising the temperature of the vaporizer. However, the internal energy of the molecular ions produced in most cases must be low, because fragmentation is often minimal.[276]

The ion source heater has an additional role in SFC/MS. Heat is needed to evaporate the CO_2 solvent sphere that encompasses the analyte molecule. Typical ion source temperatures used in CI

SFC/MS are 220-300°C.[272,281,292] Source heaters that can provide more power may be needed in SFC/MS as compared to GC/MS. Cousin and Arpino[292] found that a 10-Watt source heater was able to maintain the source temperature at 260°C for CO_2 flow rates of 50 μL min^{-1}.

Owens *et al.*[273] studied the effect of source chamber pressure on the resulting mass spectra in SFC/MS. The CI mass spectra of silylated ethylene glycol oligomers (PEG 600) were obtained as a function of pressure. In the first experiments, the source pressure conditions rose to over 3 x 10^{-4} torr for a final column pressure of 365 atm. Next, cryopumping was added to the source chamber which caused the source chamber to reach a pressure of *ca.* 8 x 10^{-5} torr for a pressure of 365 atm in the column. The mass spectral S/N improved by a factor of two overall, but for the high molecular mass oligomers that eluted toward the end of the pressure program, the S/N increase was a factor of four.

Cousin and Arpino *et al.*[292] showed that the loss in sensitivity associated with increasing source pressure was greatest for high molecular mass compounds. When CO_2 flow rates were increased from 50 μL min^{-1} to 100 μL min^{-1}, the ion transmission for ions with m/z $\geq$300 was practically zero. Because the departing ions had approximately the same kinetic energy, the higher molecular mass ions remained in the drift region between the ion source and the mass filter longer than the lighter molecules. Accordingly, the heavy molecules collisionally scattered to a greater extent than the lighter molecules. The detection efficiency for heavier ions was markedly diminished with increasing source pressure. Hence, before a CI SFC/MS technique is used for quantitative analysis, the constancy of the ion transmission with source pressure variation should be ensured.

Charge exchange (CE) ionization is actually a type of chemical ionization. Odd electron molecular ions are formed by an electron-transfer reaction between the reagent ion and the sample molecule. CO_2 and nitrous oxide have ionization potentials of 13.6 and 12.9 eV, respectively. Organic molecules have ionization potentials that typically range from 8-10 eV. Therefore, CO_2 and nitrous oxide can both be used as reagent molecules for charge exchange ionization of organic molecules. Lee *et al.*[306] showed that searchable EI-like spectra could be obtained when CO_2 mobile phase was combined with helium make-up gas in a low conductance CI source. In charge exchange ionization, the extent of fragmentation depends on the amount of excess energy deposited in the molecular ion in the charge transfer reaction. Consequently, this difference is dependent on the ionization potentials of both the reagent and the sample. An exact match of a charge-exchange mass spectrum with an electron impact mass spectrum is, therefore, not assured.

Cousin and Arpino *et al.*[292] showed the existence of competition between charge-exchange ionization and ammonia chemical ionization SFC/MS. The proportion of the CE mechanism increased with increasing flow of CO_2 during a pressure-programmed SFC/MS run. The later eluting peaks had fragmentation patterns that were characteristic of CE, while the earlier peaks had both CI and CE fragmentation patterns. For complex matrices, a mixed ionization mechanism could be used to provide further structural information.

Smith *et al.*[271] first reported EI spectra from a DFI SFC/MS. Vorhees and coworkers[264,274] further characterized DFI SFC/MS with electron impact ionization and found reduced sensitivity compared to CI SFC/MS. Others have now reported MDQs for electron impact SFC/MS that are at least an order of magnitude higher than those found for CI SFC/MS.[306]

The high MDQ in EI SFC/MS may be caused by incomplete desolvation of the analyte molecules before the point of impact with the electrons in the ion source. Holzer *et al.*[264] studied the variation of CO_2 cluster ion intensities as a function of temperature and pressure. Their data indicated that to keep the quantity of CO_2 clusters low, the temperature of the ion source should be *ca.* 200°C, and the pressure in the ion source should be maintained at pressures less that 4×10^{-5} torr. Another strategy used to break-up the clusters before electron impact ionization was to expand the effluent from the restrictor in a heated region just prior to the ion source. Unfortunately, neither effort substantially improved the detection limits in EI SFC/MS.

One other possible cause of sensitivity loss in EI SFC/MS is that the high pressure in the ion source scatters the ions away from the entrance to the quadrupole mass analyzer. The ion source pressures in all of the EI SFC/MS to date were approximately an order of magnitude higher than those used in EI GC/MS. If higher capacity pumps are added to the ion source region, precise focusing of the sample beam into the ion source will be required to gain an advantage in S/N.

With more fundamental studies of the interactions of the ion beam with CO_2 in a high pressure source, improvements in EI SFC/MS may be possible in the future. New ionization sources that make maximum use of the properties of the expanding gas clusters should also be expected.[307]

Atmospheric pressure ionization (API) is one of the more recent mass spectrometric techniques investigated for coupling to chromatography. Horning *et al.*[308] were among the first to report on API with mass spectrometry, and to suggest its potential for GC/MS and LC/MS. They reported impressive LC/MS detection limits as early as 1974,[309] while a wide variety of applications and developments have appeared more recently.[310,311] The commercial availability of instrumentation, as well as recent developments, suggest that the API technique will be widely applicable in the future. In fact, there are various ways to form ions at atmospheric pressure, and more than one may be suitable for SFC/MS applications.

Historically, ions were formed in API by emission of beta particles from a ^{63}Ni foil, but this was later simplified by replacement with a corona discharge. The relatively low energy ions formed under these conditions are sampled through a tiny orifice (10-100 microns) by means of a potential difference from atmospheric pressure conditions into the high vacuum region of the mass analyzer. The nonequilibrium thermodynamic conditions present in the free jet expansion on the high vacuum side of this orifice usually produces numerous solvent or gaseous "adduct" ions which can complicate the interpretation of results. Modern API systems, however, can effectively preclude the formation of these unwanted adduct ions, and thus more efficiently produce protonated or deprotonated molecular ions with positive or negative charges, respectively. Analyte ions from these processes may be detected from extremely small quantities of material, making API well suited for trace analysis.

Alternatively, liquids or fluids passing through a capillary held in the neighborhood of 3 kV produce charged droplets which emit ions into the gas phase by "ion evaporation." This electrospray approach has been described by Bruins *et al.*[312] and others[313,314] for the characterization of a wide variety of applications, ranging from polar drugs to peptides, proteins, and polymers. These approaches to form ions at atmospheric pressure provide unusual flexibility and opportunity for combining inlet separation methods with mass spectrometry.

A heated pneumatic nebulizer-type interface can be used[315] for coupling both packed and capillary SFC with an API source. This system combines a post-column restrictor necessary for maintaining supercritical conditions *via* a small pinhole diaphragm housed in a direct liquid introduction LC/MS probe described previously.[316] The expanding gases exiting the restrictor are directed into a heated quartz liner at atmospheric pressure along with a nitrogen make-up gas. The volatilized analytes eluting from the packed or open tubular column maintained under supercritical CO_2 conditions are then ionized *via* a corona discharge needle maintained at 6 kV.

The API approach does not directly provide structural information for ionized compounds, because little or no fragmentation occurs under the very mild ionization conditions. Although molecular mass information is usually provided, identification of unknown compounds eluting from the SFC system requires plausible fragmentation processes to produce a more useful mass spectrum.

The technique of tandem mass spectrometry (MS/MS) can provide this necessary fragmentation. The process involves selecting parent ions, such as the protonated molecular ions generated under APCI, by a first stage of mass separation.[317] These ions are then subjected to ion-molecule interactions which impart excess energy to the parent ions. This energy rapidly distributes itself throughout the excited ions and causes breakage of weak bonds, resulting in a family of product, or "daughter," ions. A second stage of mass analysis produces a full-scan daughter ion mass spectrum which can be very useful for structural characterization.

The on-line packed-column API SFC/MS analysis of a synthetic mixture containing 250 ng of 19-nortestosterone (MW 274) and 500 ng of methandrostenolone (Dianabol, MW 300) is shown in Figure 4.50. The total ion current (TIC) from API SFC/MS analysis of this mixture is shown in Figure 4.50A, while the corresponding APCI mass spectra for 19-nortestosterone and Dianabol are shown in Figures 4.50B and 4.50C, respectively. These spectra are dominated by protonated molecular ions which easily provide the molecular weights for these compounds, plus adduct ions at $(M + 33)^+$, which presumably are proton-bound adducts between the analyte and methanol used as modifier for the CO_2 mobile phase. No fragment ions are observed in these mass spectra due to the extremely mild ionization conditions.

As an example of open tubular column SFC/MS using the API technique, the total ion current for a synthetic mixture of anabolic and estrogenic compounds is shown in Figure 4.51. Approximately 20 ng of each component were injected into the fused silica open tubular column, whereupon a linear pressure program from 100-400 atm in 45 min produced the separation shown in Figure 4.51. Although these preliminary results can be improved, the TIC chromatogram shown here was obtained easily and reveals a smooth, stable ion current throughout the pressure program. The mass spectrum for each component observed in Figure 4.51 revealed an abundant protonated molecular ion with little or no fragmentation.

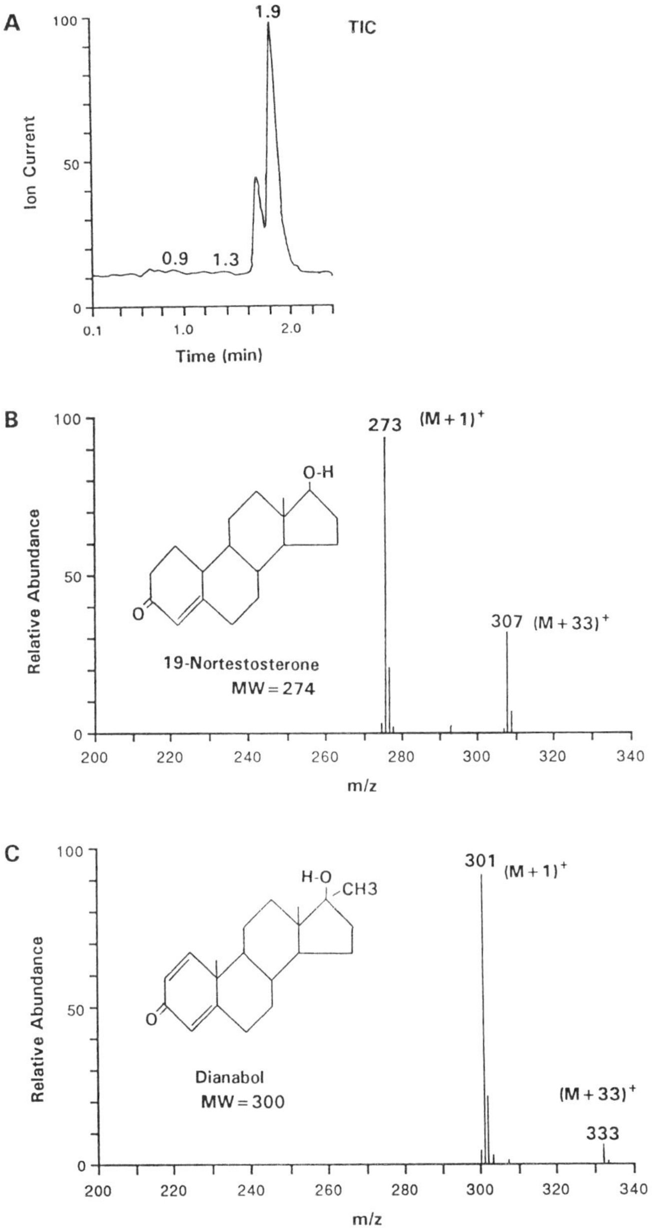

Figure 4.50. SFC/MS analysis of a synthetic mixture containing 19-nortestosterone (250 ng) and Dianabol (500 ng). Conditions: 10-cm x 4.6-mm i.d. packed column, cyanopropyl bonded silica particles (3 μm); 18% methanol in CO_2; 80°C; 1.5 mL min^{-1} flow rate; 231 atm inlet pressure; 210 atm outlet pressure; 20-μm i.d. pinhole restrictor; API.

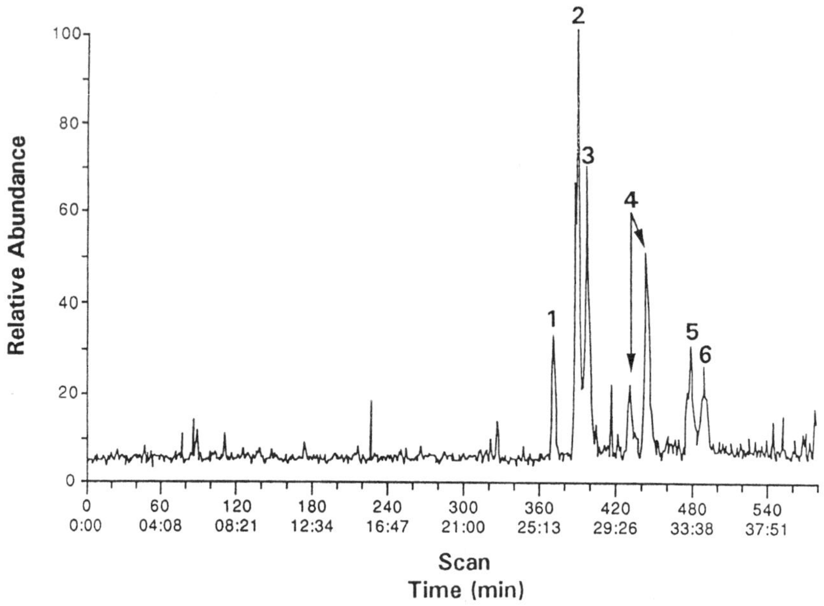

Figure 4.51. SFC/MS analysis of a synthetic mixture containing anabolic and estrogenic compounds (~20 ng each). Conditions: 5-m x 50-μm i.d. open tubular column, poly(50% cyanopropyl)methylsiloxane stationary phase; CO_2; linear pressure program from 100 to 400 atm at 6.7 atm min^{-1}. Peak identifications: (1) melengestrol acetate, (2) medroxyprogesterone, (3) trenbolone, (4) diethylstilbestrol, (5) zeranol, (6) triamcinolone acetonide.

For more helpful structural information, the benefits of tandem mass spectrometry (MS/MS) may be utilized.[318]

Mass analyzer considerations. Quadrupole mass spectrometers operate at ground or low source potential, depending on the model and type of instrument. By virtue of this, the design considerations of an SFC/MS interface are less restrictive than those experienced with sector instrument sources operating at high potential.

Most SFC/MS has been performed using quadrupole mass spectrometers. They are able to operate at relatively high pressures, are easily controlled by computers, and are relatively inexpensive. However, the mass range of quadrupole instruments, in which "unit" mass resolution can be achieved with good sensitivity, is limited by both electronics (RF power supplies) and machining (quadrupole rods). Also, quadrupoles are limited in their ability to perform high resolution exact mass measurements at part-per-million accuracies, an important tool in structure elucidation and targeted compound analysis. These are both areas where sector instruments perform better than quadrupoles.

Significant advances have also been made in the development of interfaces for bench top quadrupole instruments. The first system described was based on a small EI system which was modified with a CI source.[276] The system provided data on a range of compounds, including fatty acid esters and xanthenes. Some spiking and peak shape anomalies were evident, due to the lack of specific restrictor heating.

Sector mass spectrometers rely on magnetic and electric fields to disperse the ion beam according to momentum and energy, respectively, in order to achieve high mass resolving power. Since ion beams require high vacuum, sector instruments are not able to perform as well as quadrupoles at the high analyzer pressures ($>10^{-5}$ torr) which often result in the union of a high pressure technique, such as SFC, with mass spectrometry.

On the other hand, sector SFC/MS offers many potential advantages over quadrupole SFC/MS. Due to the increased mass resolving power of the sector instruments, accurate mass and relative-intensity measurements can be obtained in the presence of interfering isobaric ions. The mass range of the analysis can also be extended. State-of-the-art sector instruments are capable of detecting ions with m/z values up to 10,000. Since one object of SFC/MS is to provide mass spectral characterization of high molecular mass, less volatile substances, the use of sector instruments for SFC/MS has increased in recent years. Various laboratories and vendors are pursuing a variety of approaches in interfacing SFC to sector MS. The additional concern in the development of interfaces for sector instruments is that the high voltage of the ion source must be isolated from the interface probe.

The ion trap detector (ITD) is a compact, robust detector that is fully automated. This detector was originally developed for routine capillary GC/MS analysis. Ions can either be generated in the ion trap by chemical ionization or electron ionization. Alternatively, the ions can be generated externally and then injected into the ion trap for mass filtering. Todd and coworkers showed that high pressures of helium (1×10^{-3} torr) collisionally forced the ions to slow down and move to the center of the trap.[319] The effect of this ion focusing was to increase the sensitivity and resolution of the analysis. A mass spectrometer, such as the ITD, that functions well under high pressure conditions would seem ideally suited for interfacing to SFC.

The first SFC-ITD system was recently described by Todd *et al.*[320] The column and restrictor were connected inside the ITD transfer line, and the effluent from the frit was expanded into the ITD. The mass spectra in this study were of microgram quantities of butylbenzene. Although the specific detection limits were not discussed, the S/N of the total ion chromatograms indicated that the detection limits were substantially higher than for SFC/MS using a quadrupole mass filter. Todd *et al.*[320] also described the ion trap MS/MS analysis of *n*-butylbenzene.

The demand for SFC/ITD is expected to increase. Hemberger *et al.*[321] recently substantially extended the mass range of the ITD. By reducing the electrode size, reducing the RF frequency, and changing the resonance ejection conditions, the mass range of the ITD was extended to 45,000 m/z. As stated earlier, the detection limits of the SFC/ITD must also reach low values in the future. One possible means of lowering the detection limit would be to form the ions in a separate chamber, and then inject the ion packets into the ITD. Pedder *et al.*[322] recently described the successful interfacing of a high pressure CI source to an ITD by an ion injection technique.

Fourier transform mass spectrometry (FTMS) is capable of ultra high mass resolving power at low pressures, and has the potential for high mass range.[323-325] It has been successfully linked to gas chromatography,[325] and two reports exploring SFC/MS using FTMS have appeared.[326,327] High pressure ($>10^{-7}$ torr) limits the mass resolving power available in FTMS. Unlike quadrupole SFC/MS, the gas load from a capillary SFC is a considerable burden to an FTMS, which functions best under vacuum conditions of $\leq 1 \times 10^{-8}$ torr. For this reason, both studies used a dual cell FTMS to provide differential pumping. The approximate MDQ was low nanogram, and searchable EI-like spectra were obtained.

Lee *et al.*[326] were able to measure the mass of the molecular ion of caffeine using SFC/FTMS to within 0.036 ppm of the calculated value. A mass resolving power of 8,300 was obtained during "self-CI" of caffeine. In these preliminary experiments, SFC/MS of higher molecular mass compounds was not possible since the 1-m oven-to-cell transfer line was not heated uniformly, and the restrictor was not heated above 100°C.

Laude *et al.*[327] attempted to overcome the dynamic range limitations of FTMS by using 13 eV electrons for ionization (to limit ionization of CO_2), and by continuously ejecting the CO_2 ions that were produced. They ran standard mixtures of PAHs, barbiturates, and pesticides, and estimated that detection limits in the 1-ng range should be routine. However, sensitivity decreased 10-fold, and spectral resolution and line shape degraded at SFC pressures above 200 atm due to space-charge effects. One improvement suggested by the authors would be to use a 50-μm i.d., rather than 100-μm i.d., column to reduce the flow into the mass spectrometer.

It is clear from these reports that the FTMS performance of even the differentially-pumped, dual-cell instrument was compromised during direct introduction of the capillary SFC effluents. An alternative which might be more successful is ionization of the SFC effluent external to the electromagnet and injection of the ions into the high vacuum FTMS cell.[323]

The FTMS has many desirable features, such as high resolution mass analysis and facile MS/MS analysis. For SFC/FTMS to be competitive with quadrupole SFC/MS, the detection limits must be lowered by at least two orders of magnitude. Also, long transfer lines are unacceptable. Considerable loss in chromatographic efficiency occurs due to the large dead volume of the transfer line.

Time-of-flight (TOF) mass spectrometers are simple, inexpensive, and can operate at relatively high pressures.[328] They also have a potentially unlimited mass range and fast scan rates. Their most serious disadvantage is low mass resolving power. Lubman *et al.*[329-331] have explored supercritical fluid injection/MS of substances of low volatility using TOFMS. They have taken advantage of the selectivity of resonant two-photon ionization to overcome the mass resolution limitations of their TOF instrument. Rotational/vibrational cooling in the supercritical CO_2 expansion provides sharp spectral features, although cooling is not as complete as in Ar expansions. These studies began with relatively nonpolar materials, such as PAHs[331] and derivatized aromatic acids,[329] and were extended to underivatized peptides, nucleosides, and steroids by the use of supercritical ammonia.[332] Despite these successes, the future use of TOFMS for more general SFC/MS applications will hinge on improvements in the mass resolving power of the mass spectrometer.

4.22 Multidimensional Chromatography

Concepts of multidimensional chromatography. Many advantages are gained by on-line coupling of two chromatographic columns of different selectivities in series, an approach commonly known as multidimensional chromatography. The secondary column is fed a fraction from the eluting stream of the primary column of duration Δt. The peak capacity of a coupled-column system depends on the efficiency of each individual separation and on the width of the primary column fraction. Peak capacity is increased because the number of components transferred at one time from the primary to the secondary column is controlled, thereby statistically limiting the chances of peak overlap. In addition, the different selectivities of the two stationary phases are chosen to optimize component resolution.

Two criteria have been defined for multidimensional separations.[333,334] First, components should be subjected to two or more independent and selective displacements; second, the separated components must remain resolved throughout the process. A tandem column arrangement, in which all of the components present in the mixture pass directly through both columns (*i.e.*, when Δt is large), may result in the second criterion not being upheld; tandem column chromatography, therefore, has only a limited value compared with true multidimensional chromatography. In two-dimensional planar separations, a sample is subjected to two displacement processes oriented at right angles to each other. If the peak capacity along the primary (z) axis is defined as N_z', and the peak capacity along the secondary (y) axis is defined as N_y', then the total theoretical peak capacity of the system is given by $N_z'N_y'$, because both displacements occur simultaneously (whereas coupled-column chromatography relies on sequential separations). The peak capacity of a multidimensional coupled-column arrangement, therefore, ranges in value according to its design. If the peak capacity of a single column is defined as N', then coupling two similar columns of the same selectivity will yield a peak capacity of only 1.4 times the value of N'. Multidimensional coupled column systems range in their peak capacities from $2N'$ for two similar columns of correlated selectivities up to a theoretical maximum value of N'^2 for two similar columns of independent selectivites.

Multidimensional coupled-column systems may consist of packed/packed, packed/open tubular, or open tubular/open tubular column combinations, depending on the sample type, the analysis required, and the mobile phase used. A highly efficient open tubular column is generally desirable as the secondary column to maximize resolution, and the use of such columns in multidimensional chromatography involving GC and SFC as the second stage is usual. The primary column may be either an open tubular column or a packed column; if the latter is used, sample loading is increased and analysis times are reduced.

SFC/SFC. Initial attempts at multidimensional SFC were based on column-switching reversed-phase/normal-phase LC methods, using supercritical CO_2 as a common mobile phase. Multidimensional chromatography using supercritical CO_2 for the separation of polycyclic aromatic compounds (PAC) in a coal tar was reported in 1985 by Christensen.[335] Two 4.6-mm i.d. packed columns (aminosilane-bonded silica and octadecylsilane-bonded silica) were arranged in series with two 6-port switching valves so that they could be switched in or out of the mobile phase flow. This system was adapted from the on-line normal-phase LC/reversed-phase LC method described by Sonnefeld *et al.*[336] Carrying out both separations using the same mobile phase avoided the difficulties previously experienced with the intermediate solvent exchange procedure. When the secondary column was switched out of the flow, the aminosilane primary column eluted coal tar PAC at 250 atm and 50°C, in the order of increasing aromatic ring number. Shortly before the elution of the "chrysene" fraction of interest, the secondary octadecylsilane column was switched into the flow path and the primary column was switched out. PAC within this fraction were eluted at 100 atm using a negative temperature ramp (50°C to 35°C at 4°C min^{-1}). The "chrysene" fraction was resolved into triphenylene, benz[*a*]anthracene, and chrysene on the secondary column. Although the use of conventional packed columns in this system resulted in low efficiencies, the high mobile phase flow rates provided fast analysis times.

Lundanes and Greibrokk[337] used two 6-port valves and three microbore columns packed with cyano-silica, silica, and silver-loaded silica, with supercritical fluid CO_2 mobile phase and both UV and FID in series. The combination of silica and silver-loaded silica packed columns separated the saturates from the aromatics, and the trapped polar compounds were backflushed from the cyano column. Quantitative results were obtained directly from the chromatograms. A later publication[338] reported an investigation of the use of nitrous oxide (N_2O) as mobile phase for multidimensional SFC. However N_2O gave a higher background signal in the FID than did CO_2.

In 1988, Campbell *et al.*[112] described the use of SFC/SFC for the quantitative determination of saturates, olefins, and aromatics in gasolines and middle distillate fuels in a single chromatographic run (Figure 4.52) In order to adequately separate the saturate group from the olefinic group, two 1-mm i.d. microbore columns were used in a column-switching configuration with a mixed mobile phase of 10% CO_2 in sulfur hexafluoride (SF_6). The aromatic fraction was isolated on a silica gel column at 50°C, while a silver-loaded cation

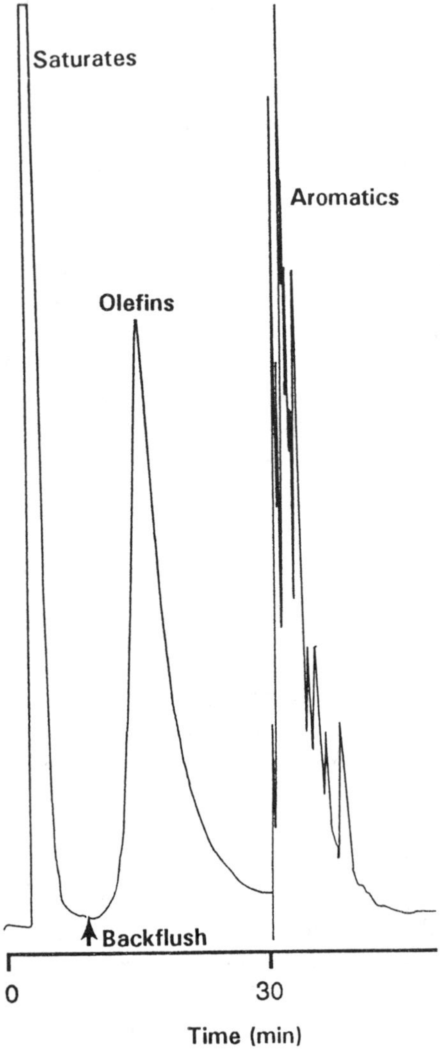

Figure 4.52. SFC/SFC chromatogram of a gasoline sample showing hydrocarbon group-type separation. Conditions: 25-cm x 1-mm i.d. packed silica and silver-loaded sulphonic acid silica microbore columns; 10 mol% CO_2 in SF_6; FID (reprinted with permission from Ref. 112).

exchange column at 40°C was used to separate the saturates from the olefins. Group quantification was achieved with an FID, which provided nearly uniform linear response for the hydrocarbons.

Levy *et al.*[339,340] recently described a multicolumn SFC system with conventional 4.6-mm i.d. packed columns in series (silver-, cyano-, and amino-silica) for the group-type separation of petroleum fuels. Packed column SFC was also interfaced to open tubular column SFC through the standard rotary injector valve of the SFC instrument. In this way, crude SFC separations of aliphatics/-aromatics were carried out isobarically on the packed column, and a 4-ring aromatic fraction was then transferred into the SFC which contained a 100-μm i.d. liquid crystalline coated open tubular column.

Microcolumn multidimensional SFC/SFC was recently reported[341-343] using either two 50-μm i.d. open tubular columns or two 250-μm i.d. packed capillary columns in series, CO_2 as mobile phase, and a flow-switching interface. This technique was used to resolve isomeric PAC in fossil fuel derived materials, and to isolate a labile pesticide (carbofuran) from a bird extract (Figure 4.53). A solvent-vent injection method was incorporated, which allowed sample injections of microliter volumes. By means of a bypass line and either a T-piece interface or an offset-cross interface, solutes could be directed from the primary column into the secondary column. The solutes in the primary column, when the mobile phase flow was diverted to the secondary column, remained "parked" there until flow was once again resumed through both columns.

The major advantages of the open tubular column SFC/SFC system was the excellent resolution of PAC structural isomers by the liquid crystalline stationary phase of the secondary column; whereas the packed capillary column system provided a faster PAC ring-number primary separation. Unfortunately, the flow-switching interface was unsuitable for coupling two columns of different dimensions; differences in back pressures and volumetric flow rates interfered with the balance of mobile phase splitting within the interface.

The concept of using a rotary switching valve interface for microcolumn SFC/SFC was discussed by David *et al.*[344] Systems were proposed for either 100-μm i.d. open tubular columns or 250-μm i.d. packed capillary columns. One configuration proposed for the direct transfer of selected solutes from the primary to the secondary column used a 4-port valve interface. Pressure restrictors leading to FIDs would allow components eluting from each column to be detected. The authors suggested refocusing the transferred solutes by operating the secondary column at a lower density than the primary column. The major problem with this design was thought to be the difficulty in balancing the pressure at the interface during switching.

David *et al.*[344] also proposed valve-switching in SFC/SFC, incorporating a cold trap at the interface. Solutes from the first column could be transferred into the trap by switching a 3-port valve, while the end of the second column is open to the atmosphere

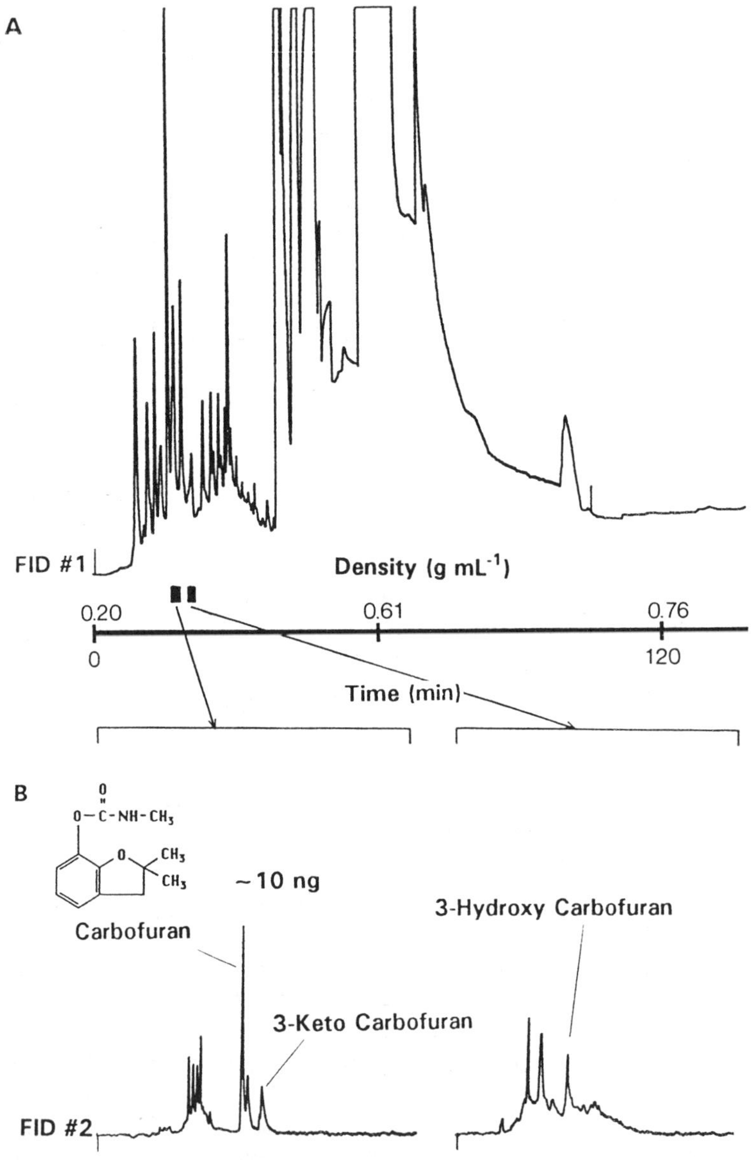

Figure 4.53. SFC/SFC chromatograms of carbofuran insecticide and its 3-keto and 3-hydroxy metabolites in a bird extract (ethyl acetate). Conditions: 10-m x 50-μm i.d. primary open tubular column, poly(30% biphenyl)methylsiloxane stationary phase; 4-m x 50-μm i.d. secondary open tubular column, poly(50% oligoethylene oxide)methylsiloxane stationary phase; 1-μL sample injected (reprinted with permission from Ref. 341).

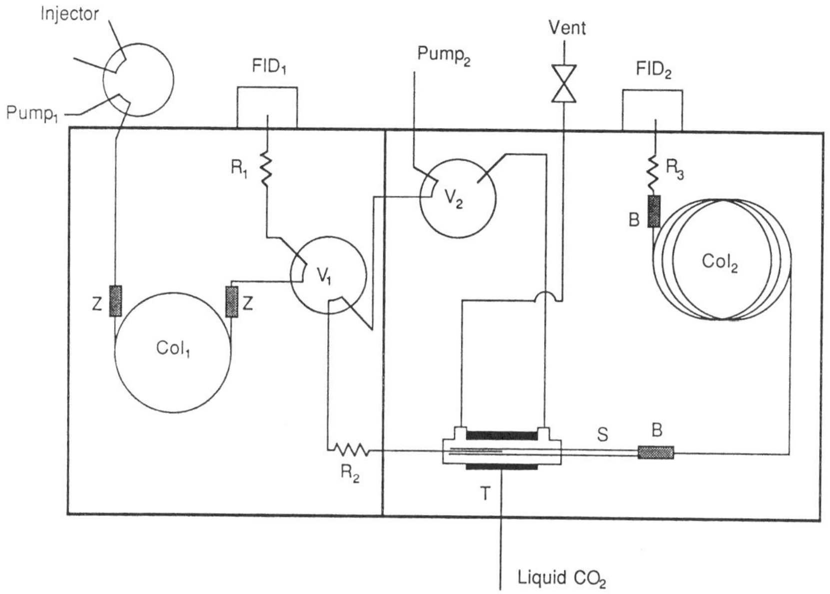

Figure 4.54. Schematic diagram of a multidimensional packed capillary to open tubular column SFC/SFC system. Abbreviations: Col_1 = primary packed capillary column; Col_2 = secondary open tubular column; V_1 and V_2 = rotary valves; T = cold trap; S = capillary solute concentrator; R_1 and R_3 = frit restrictors; R_2 = linear restrictor; Z = zero dead volume union; B = butt connector (reprinted with permission from Ref. 345).

through another 3-port valve. After transfer, the two 3-port valves would be turned simultaneously back to their respective FID restrictor positions. No results were shown to support these SFC/SFC proposals.

An SFC/SFC system (Figure 4.54) utilizing a rotary valve interface to provide independent flow control of the two dimensions was reported by Juvancz *et al.*.[345] A cold trap was employed to refocus the solutes from single or multiple fractional cuts, after being transferred to the second dimension. Improved performance, including time savings, was achieved with a packed capillary to open tubular column arrangement, compared to earlier open tubular column SFC/SFC and packed capillary column SFC/SFC systems. The packed capillary in the first dimension provided chemical class separation, while the open tubular column in the second dimension provided high resolution of closely related isomers. Sequential heart-cuts could be trapped for later resolution in the second dimension (Figure 4.55).

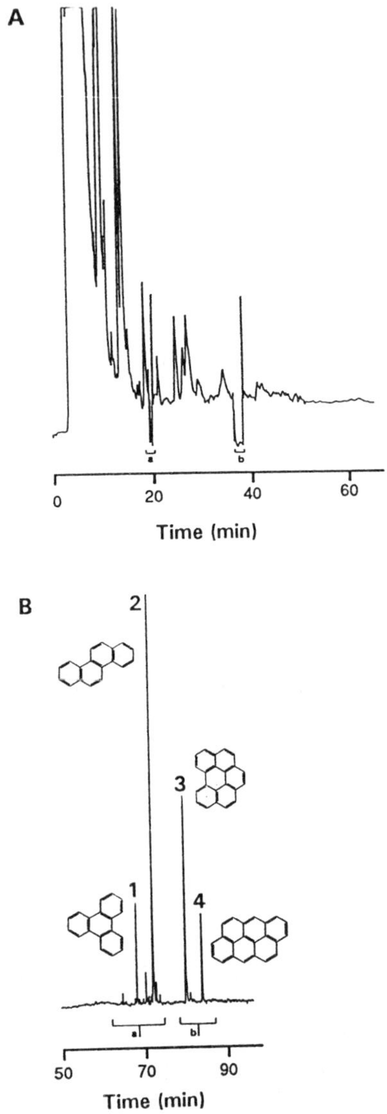

Figure 4.55. Two-dimensional SFC chromatograms of a standard coal tar (SRM 1597). Two fractions were collected from the first dimension, and then analyzed simultaneously in the second dimension. Cut **a** was collected between 20.2 and 21.2 min, and cut **b** between 38.7 and 40.2 min. Conditions: (A) CO_2; 75°C; pressure program from 200 atm to 414 atm at 5 atm min^{-1}; (B) CO_2; 110°C; pressure program from 70 atm to 120 atm at 20 atm min^{-1}, then 120 atm to 414 atm at 8 atm min^{-1}, after a 2-min isobaric period. Peak identifications: (1) triphenylene, (2) chrysene, (3) benzo[*ghi*]perylene, (4) anthanthrene (reprinted with permission from Ref. 345).

An on-line SFE/SFC/SFC/GC system[343,346] was also proposed which would utilize a packed capillary primary column and a secondary open tubular column. The system would use 12-port rotary valve interfaces, a backflush valve, and cold traps between dimensions.

SFC/GC. Levy *et al.*[339,340,347,348] were the first to report coupled SFC/GC using conventional packed column SFC on-line with open tubular column GC. A heated capillary restrictor was passed directly from the end of the SFC column into a GC split vaporization injector. Examples of SFC/GC applications reported by this group included the use of combined packed cyano-silica and silica SFC columns for the separation of an aviation fuel into saturates and 1-ring and 2-ring aromatics, and a C_{18} silica packed SFC column for the isolation of PAC from complex hydrocarbon mixtures, and group-type separation of high-olefin gasoline fuels (saturates, olefins, and aromatics).[347,348] The SFC separations were performed with CO_2 using several 4.6-mm i.d. packed columns in series. Heartcut fractions were transferred into a 0.25-mm i.d. open tubular GC column coated with a methyl polysiloxane stationary phase. Cryofocusing was used during transfer into the GC.

LC/SFC. Size exclusion chromatography (SEC) was coupled on-line to capillary SFC by Lurie[349] for the determination of high molecular mass trace impurities in cocaine. Since both SEC and SFC are used to analyze high molecular mass species, their combination may be advantageous, particularly since SFC can provide the high resolution analytical step after SEC fractionation. The two techniques were coupled *via* a 10-port switching valve, which directed the SEC eluent through a 200-nL injection valve on the SFC instrument. Conventional SEC columns were used with dichloromethane mobile phase, and the SFC separation was performed on a 50-μm i.d. biphenyl open tubular column. Unfortunately, due to the nature of the interface and the use of conventional-size SEC columns, only small portions of the SEC fractions could be analyzed by SFC; packed capillary SEC/open tubular column SFC would be an obvious way to improve the compatibility between dimensions.

4.23 Fraction Collection

The problems in trapping compounds emerging from a hot gas chromatographic column outlet are well known. Quantitative recoveries are very difficult to achieve because of a tendency for these materials to form a fine fog that cannot be easily condensed or trapped. The collection of analytes can be accomplished in much better yields using SFC than by using GC.

The problem of fog formation is obviously absent in LC. The trapping of peaks in modern LC is quite convenient; most modern LC instruments provide a collection outlet or port. However, in this technique, higher molecular mass solvents are generally used, and complete solvent removal can constitute a somewhat onerous assignment, especially if the analyte of interest is volatile or thermally labile.[350,351]

The practical advantage of fraction collection from SFC with CO_2 is the ease of solvent removal. This is more difficult, of course, if a mobile phase with a higher critical temperature is used, such as pentane or methanol. However, removal of these solvents from high molecular mass solutes is still not difficult. From a practical point of view, one would use low boiling fluids as the mobile phase if the solutes were volatile or thermally unstable.

There are two general approaches to fraction collection in SFC: dynamic and static. In dynamic collection, depressurization occurs as the sample is expanded from a flow restrictor, and the solutes are collected by various means, including on the internal wall of a small i.d. capillary tube, on an adsorbent, or in a solvent through which the sample is bubbled. This general method of collection has the advantage that continuous collection for long periods of time can be performed without accumulation of large volumes of solvent which eventually must be removed. The disadvantage is that some solutes of interest may be too volatile to ensure near quantitative collection.

With static collection, the total volume of the fraction (sample plus mobile phase) must be collected prior to depressurization. The sample size is limited by the volume of the collection vessel. This method generally provides better quantitative recoveries, but it is less sensitive and is more restrictive in its application to different problems.

Jentoft and Gouw[352] used a static high-pressure fraction collector system to collect the eluant from a home-built SFC system (Figure 4.56). In this sample collection system, which was essentially a fraction collector mounted in a high-pressure container, the

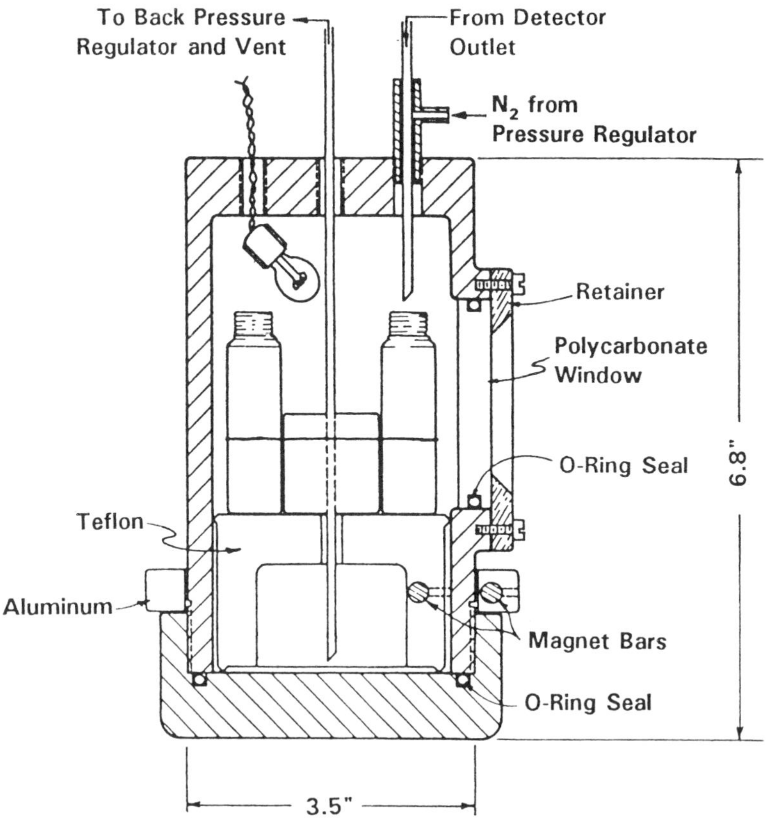

Figure 4.56. Schematic diagram of a high-pressure fraction collection system to collect the eluent from an SFC (reprinted with permission from Ref. 352).

environment was maintained at sufficiently high pressure to allow the chromatographic effluent to remain liquid. Otherwise, the mobile phase would be decompressed to a gas, and the solutes would be dispersed inside the unit. If the bleed through the regulator was properly adjusted, the mobile phase could be evaporated without "bumping" from the test vials. The solutes were left behind as a residue in the vials. This high-pressure fraction collector could be used in those cases in which a low-boiling fluid such as CO_2 or nitrous oxide was used as the mobile phase.

Both dynamic and static collection of the effluent from a packed column SFC system were reported for a variety of coal- and petroleum-derived samples.[353] Figure 4.57 shows a schematic diagram of the SFE/SFC/fraction collection instrumentation that was used for this study. The collection vessels (125-mL stainless steel)

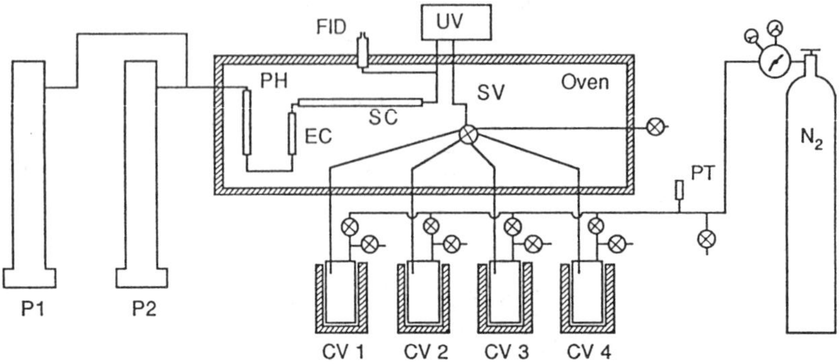

Figure 4.57. Schematic diagram of a supercritical fluid extraction/fractionation/collection system. P1 and P2, syringe pumps 1 and 2; PH, solvent preheater/equilibrator; EC, extraction column; SC, separation column; SV, six-port switching valve; FID, flame ionization detector; PT, pressure tranducer; N_2, high-pressure cylinder of nitrogen gas; CV 1-4, pressurized fraction collection vessels equipped with cooling jackets (reprinted with permission from Ref. 353).

were fitted with cooling jackets which were cooled to ~3°C *via* a circulating cooling bath during collection. At this temperature, and with pressurization of the vessels with nitrogen gas, the fractions were collected in liquid CO_2. A six-port switching valve was used to effect collection of successive fractions in different collection vessels. After collection, the vessels could be reduced to atmospheric pressure through a micrometering valve.

For the collection of volatile components, one of the collection vessels was replaced with a solvent bubbler or a 60-cm length of 1/16 in. i.d. Teflon tubing which was immersed in a dry ice/isopropyl alcohol bath (-78°C). To recover the collected materials, the Teflon tubing was removed from the bath and rinsed with a small amount of an appropriate solvent.

The collection system just described was used for the collection of polycyclic aromatic compounds in coal-derived materials according to the number of aromatic rings. Furthermore, it was successfully applied to boiling point fractionation of a crude petroleum oil. Figure 4.58 shows gas chromatograms of the resolved fractions; amazingly little overlap of isolated fractions was observed.

For dynamic collection, a simple restrictor has generally been used for depressurization of the sample before collection. In order to provide the capability of controlling the pressure independent of the mass flow rate, Saito *et al.*[82] developed a pulsed nozzle system that has proven to be very effective. The regulating valve is an ordinary solenoid valve consisting of a needle which is driven by a solenoid

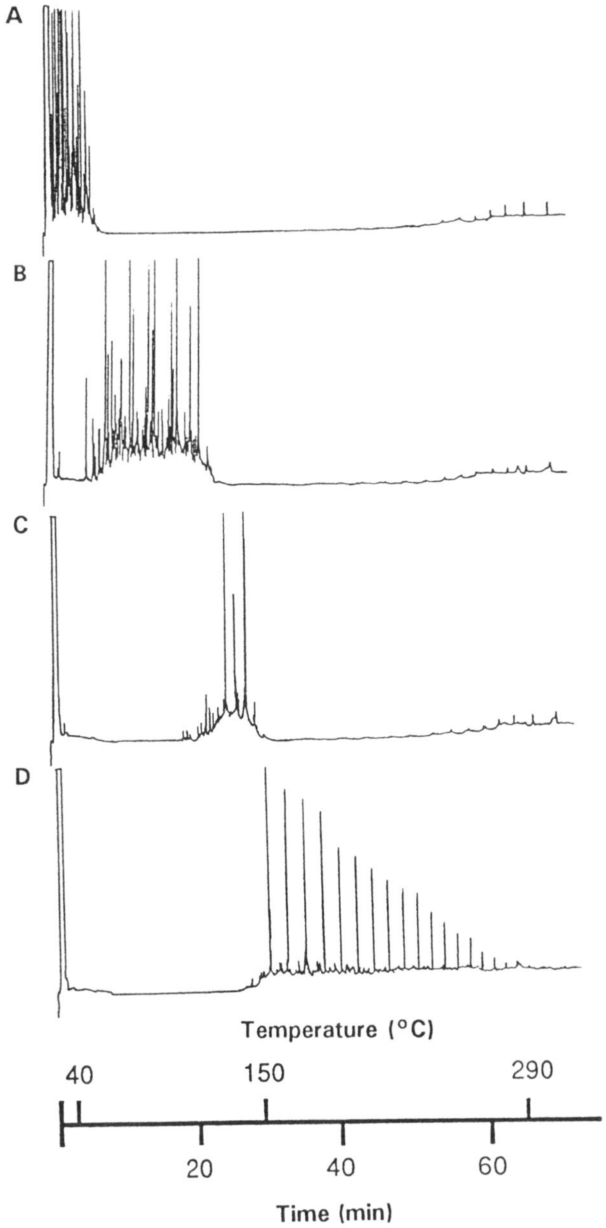

Figure 4.58. GC chromatograms of fractions obtained during the supercritical fluid fractionation of a Wyoming Recluse crude oil (reprinted with permission from Ref. 353).

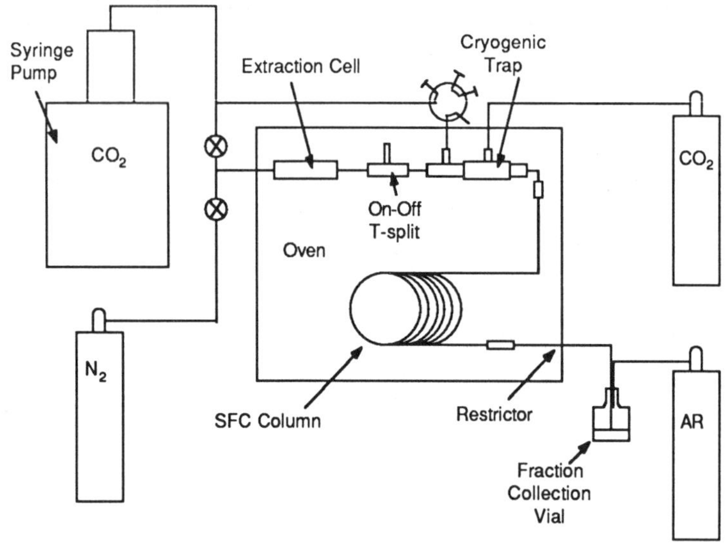

Figure 4.59. Schematic diagram of a microvolume SFE/SFC/FC system (reprinted with permission from Ref. 355).

and a valve seat (see Figure 4.12). The restriction is based on high speed pulsing of the needle in the valve seat to open and close the effluent flow path. For collection, any suitable container can be connected to the outlet port of the regulating valve merely by spring force. As the sample emerges from the nozzle in short bursts, the CO_2 dissipates and the sample falls to the bottom of the collection vessel as a fine powder.

In an interesting application of dynamic fraction collection, Norris and Rawdon[354] used a restrictor constructed from a stainless steel capillary tube that was crimped just enough to achieve the desired pressure. They noted that the expanding gas formed a plume of small solid CO_2 particles which precipitated out at the apex of the expanding CO_2 cone. By holding a mass spectrometer direct-inlet probe filament in this cone as the peak was displayed on a printer-plotter from UV detection, they collected enough sample for off-line MS analysis. Alternatively, fractions were collected in small tapered, screw-capped vials, preferably fitted with Teflon liners to reduce contamination. These vials were carefully cleaned before use, because "as-received" vials often contain impurities that significantly contaminate peaks of interest.

Xie *et al.*[355] constructed a supercritical fluid extraction-supercritical fluid chromatography-fraction collection (SFE/SFC/FC) system (Figure 4.59) to provide extraction, separation, and fraction collection of biologically active compounds on an analytical scale. The supercritical CO_2 extract was decompressed through a small i.d. linear restrictor and deposited in a deactivated capillary solute concentrator within a cryogenic trap. The capillary solute concentrator was directly connected to an SFC column. During SFC, fractions of the effluent were collected from a frit restrictor at the column outlet in vials that contained a preselected solvent. The collected fractions were then sent out immediately for biological activity assay.

The examples reported here are only representative of the possibilities of fraction collection from SFC. As the technique develops further and is applied to different problems, new and innovative methods are certain to develop.

References

1. M.N. Myers and J.C. Giddings, *Sep. Sci.* **1**, 761 (1966).
2. L. McLaren, M.N. Myers, and J.C. Giddings, *Science* **159**, 197 (1968).
3. J.C. Giddings, M.N. Myers, and J.W. King, *J. Chromatogr. Sci.* **7**, 276 (1969).
4. Z. Balenovic, M.N. Myers, and J.C. Giddings, *J. Chem. Phys.* **52**, 915 (1969).
5. F.P. Schmitz and E. Klesper, *Polym. Bull.* **5**, 603 (1981).
6. F.P. Schmitz, H. Hilgers, and E. Klesper, *J. Chromatogr.* **267**, 267 (1983).
7. D.R. Gere, *Science* **222**, 253 (1983).
8. D.R. Gere, R. Board, and D. McManigill, *Anal. Chem.* **54**, 736 (1982).
9. T. Greibrokk, A.L. Blilie, E.J. Johansen, and E. Lundanes, *Anal. Chem.* **56**, 2681 (1984).
10. H. Lauer, D. McManigill, and R. Board, *Anal. Chem.* **55**, 1370 (1983).
11. M. Novotny, S.R. Springston, P.A. Peaden, J.C. Fieldsted, and M.L. Lee, *Anal. Chem.* **53**, 407A (1981).
12. S.R. Abbott, J.R. Berg, P. Achener, and R.L. Stevenson, *J. Chromatogr.* **126**, 421 (1976).
13. P. Achener, S.R. Abbott, and R.L. Stevenson, *J. Chromatogr.* **130**, 29 (1977).
14. M. Martin, G. Blu, C. Eon, and G. Guiochon, *J. Chromatogr.* **112**, 399 (1975).
15. M. Martin and G. Guiochon, *J. Chromatogr.* **151**, 267 (1978).
16. H. Schwartz and R.G. Brownlee, *Am. Lab.* **16**(10), 43 (1984).
17. R.L. Stevenson, R. Henry, H. Magnussen, and P. Mansfield, *Am. Lab.* **10**(1), 41 (1978).
18. P. Kucera, in *Microcolumn High Performance Liquid Chromatography*, P. Kucera, Ed. (Elsevier, Amsterdam, 1984), p. 39.

19. M.L. Lee, B. Xu, E.C. Huang, N.M. Djordjevic, H-C.K. Chang, and K.E. Markides, *J. Microcol. Sep.* **1**, 7 (1989).
20. T.L. Chester, D.P. Innis, and G.D. Owens, *Anal. Chem.* **57**, 2243 (1985).
21. B.E. Richter, D.E. Knowles, M.R. Andersen, N.L. Porter, E.R. Campbell, and D.W. Later, *J. High Resolut. Chromatogr. Chromatogr. Commun.* **11**, 29 (1988).
22. B.E. Berg and T. Greibrokk, *J. High Resolut. Chromatogr.* **12**, 322 (1989).
23. T. Greibrokk, B.E. Berg, A.L. Blilie, J. Doehl, A. Farbrot, and E. Lundanes, *J. Chromatogr.* **394**, 429 (1987).
24. G. Schomburg and W. Roeder, *J. High Resolut. Chromatogr.* **12**, 218 (1989).
25. Y. Hirata, M. Tanaka, and K. Inomata, *J. Chromatogr. Sci.* **27**, 395 (1989).
26. Y. Hirata and K. Inomata, *J. Microcol. Sep.* **1**, 242 (1989).
27. T.H. Gouw and R.E. Jentoft, *Adv. Chromatogr.* **13**, 1 (1975).
28. B. Pierce and W.Ll. Thomas. *Anal. Chem.* **44**, 1107 (1972).
29. T. Beugeling and H.D. ter Stal, *Proc. Fifth Int. Symp. Sep. Meth.: Col. Chromatogr. (Supplement to Chimia)*, Oct. 7-10, 1969, p. 201.
30. T.L. Kwa, *J. Chromatogr.* **270**, 105 (1983).
31. P.A. Peaden, J.C. Fjeldsted, M.L. Lee, S.R. Springston, and M. Novotny, *Anal. Chem.* **54**, 1090 (1982).
32. L.S. Ettre and W. Averill, *Anal. Chem.* **33**, 33 (1961).
33. J. Köhler and G. Schomburg, *J. High Resolut. Chromatogr. Chromatogr. Commun.* **11**, 191 (1988).
34. G. Schomburg, H. Behlau, U. Häusig, B. Hoening, and W. Roeder, *J. High Resolut. Chromatogr.* **12**, 142 (1989).
35. M. Bohm, F. Umland, and B. Wenclawiak, *J. Chromatogr.* **465**, 17 (1989).
36. J.C. Moore, *J. Polym. Sci.* **A2**, 835 (1964).
37. M.D. Harvey and S.D. Stearns, *J. Chromatogr. Sci.* **21**, 473 (1983).
38. M.D. Harvey and S.D. Stearns, *Anal. Chem.* **56**, 837 (1984).
39. B.E. Richter, *1986 Pittsburgh Conference and Exposition on Analytical Chemistry and Applied Spectroscopy*, Atlantic City, NJ, March 10-14, 1986, Paper 514.
40. N.L. Porter, B.E. Richter, D.E. Knowles, M.R. Andersen, and D.W. Later, *1988 Pittsburgh Conference and Exposition on Analytical Chemistry and Applied Spectroscopy*, New Orleans, LA, February 22-26, 1988, Paper 426.
41. *Chromatographic Newsletter*, Brownlee Labs, Spring (1987).
42. B.J. Murphy, Lee Scientific, unpublished results.
43. S.B. Hawthorne and D.J. Miller, *J. Chromatogr. Sci.* **27**, 197 (1989).
44. B.E. Richter, D.E. Knowles, M.R. Andersen, and D.W. Later, *1986 Eastern Analytical Symposium*, New York City, NY, October 20-24, 1986, Paper 102.
45. W.P. Jackson, K.E. Markides, and M.L. Lee, *J. High Resolut. Chromatogr. Chromatogr. Commun.* **9**, 213 (1986).
46. R.J. Skelton, Jr., C.C. Johnson, and L.T. Taylor, *Chromatographia* **21**, 3 (1986).

47. Y. Hirata, F. Nakata, and M. Horihata, *J. High Resolut. Chromatogr. Chromatogr. Commun.* **11**, 81 (1988).
48. Y. Hirata and F. Nakata, *J. Chromatogr.* **295**, 315 (1984).
49. A.L. Blilie and T. Greibrokk, *Anal. Chem.* **57**, 2239 (1985).
50. A. Farbrot Buskhe, B.E. Berg, O. Gyllenhaal, and T. Greibrokk, *J. High Resolut. Chromatogr. Chromatogr. Commun.* **11**, 16 (1988).
51. B.E. Berg, J. Paus, and T. Greibrokk, University of Oslo, unpublished results.
52. T.A. Dean and C.F. Poole, *J. High Resolut. Chromatogr.* **12**, 773 (1989).
53. R.D. Smith, J.L. Fulton, R.C. Petersen, A.J. Kopriva, and B.W. Wright, *Anal. Chem.* **58**, 2057 (1986).
54. R.W. Bally and C.A. Cramers, *J. High Resolut. Chromatogr. Chromatogr. Commun.* **9**, 626 (1986).
55. H.E. Schwartz, P.J. Barthel, S.E. Moring, and H.H. Lauer, *LC-GC* **5**, 490 (1987).
56. T.A. Berger, *Anal. Chem.* **61**, 356 (1989).
57. T.A. Berger, *J. High Resolut. Chromatogr.* **12**, 96 (1989).
58. G.M. Schneider, *Ber. Bunsenges. Phys. Chem.* **88**, 841 (1984).
59. P.E. Liley and W.R. Gambill, in *Chemical Engineers Handbook*, 5th Ed., R.H. Perry and C.H. Chilton, Eds. (McGraw-Hill, New York, 1973), Section 3.
60. R.D. Smith, E.G. Chapman, and B.W. Wright, *Anal. Chem.* **57**, 2829 (1985).
61. B.W. Wright and R.D. Smith, in *Modern Supercritical Fluid Chromatography*, C.M. White, Ed. (Huethig, Heidelberg, 1988), p. 189.
62. J.C. Fjeldsted, R.C. Kong, and M.L. Lee, *J. Chromatogr.* **279**, 449 (1983).
63. E.J. Guthrie and H.E. Schwartz, *J. Chromatogr. Sci.* **24**, 236 (1986).
64. C.M. White, D.R. Gere, D. Boyer, F. Pacholec, and L.K. Wong, *J. High Resolut. Chromatogr. Chromatogr. Commun.* **11**, 94 (1988).
65. R.D. Smith, J.C. Fjeldsted, and M.L. Lee, *J. Chromatogr.* **247**, 241 (1982).
66. K.E. Markides, S.M. Fields, and M.L. Lee, *J. Chromatogr. Sci.* **24**, 254 (1986).
67. R.D. Smith and H.R. Udseth, *Anal. Chem.* **55**, 2266 (1983).
68. C.E. Lapple, *Trans. Am. Inst. Chem. Eng.* **39**, 385 (1943).
69. B.E. Richter, *J. High Resolut. Chromatogr. Chromatogr. Commun.* **8**, 297 (1985).
70. R.D. Smith, H.R. Udseth, and B.W. Wright, in *Supercritical Fluid Technology*, J.M.L. Penninger, M. Radosz, M.A. McHugh, and V.J. Krukonis, Eds. (Elsevier, Amsterdam, 1985), p. 191.
71. S.M. Fields and M.L. Lee, *J. Chromatogr.* **349**, 305 (1985).
72. J.A. Graham and L.B. Rogers, *J. Chromatogr. Sci.* **18**, 75 (1980).
73. P.A. Mourier, M.H. Caude, and R.H. Rosset, *Chromatographia* **23**, 21 (1987).
74. E. Klesper and W. Hartmann, *Eur. Polym. J.* **14**, 77 (1978).
75. E. Klesper and F.P. Schmitz, *J. Supercrit. Fluids* **1**, 45 (1988).

76. Y. Hirata, F. Nakata, and M. Kawasaki, *J. High Resolut. Chromatogr. Chromatogr. Commun.* **9**, 633 (1986).
77. R.E. Jentoft and T.H. Gouw, *J. Chromatogr. Sci.* **8**, 138 (1970).
78. J.A. Nieman and L.B. Rogers, *Sep. Sci.* **10**, 517 (1975).
79. E. Klesper, *Angew. Chem. Int. Ed. Engl.* **17**, 738 (1978).
80. Y. Hirata and F. Nakata, *Chromatographia* **21**, 627 (1986).
81. K.R. Jahn and B.W. Wenclawiak, *Anal. Chem.* **59**, 382 (1987).
82. M. Saito, Y. Yamauchi, H. Kashiwazaki, and M. Sugawara, *Chromatographia* **25**, 801 (1988).
83. K. Anton, N. Pericles, S.M. Fields, and H.M. Widmer, *Chromatographia* **26**, 224 (1989).
84. S. Küppers, B. Lorenschat, F.P. Schmitz, and E. Klesper, *J. Chromatogr.* **475**, 89 (1989).
85. D. Bartmann, *Ber. Bunsenges. Phys. Chem.* **76**, 336 (1972).
86. D.E. Raynie, K.E. Markides, M.L. Lee, and S.R. Goates, *Anal. Chem.* **61**, 1178 (1989).
87. T.A. Berger and C. Toney, *J. Chromatogr.* **465**, 157 (1989).
88. H-G. Janssen, J.A. Rijks, and C.A. Cramers, *J. Microcol. Sep.* **2**, 26 (1990).
89. K. Anton, N. Pericles, and H.M. Widmer, *Proc. Tenth Int. Symp. Cap. Chromatogr.*, P. Sandra, G. Redant, and F. David, Eds. (Huethig, Heidelberg, 1989), p. 1324.
90. P.T. Griffiths, *1989 Symposium/Workshop on Supercritical Fluid Chromatography*, Snowbird, UT, June 12-15, 1989. See also H. Makishima, MS Thesis, University of California, Riverside, 1989.
91. J.C. Fjeldsted, W.P. Jackson, P.A. Peaden, and M.L. Lee, *J. Chromatogr. Sci.* **21**, 222 (1983).
92. T.A. Berger and C. Nathan, Hewlett Packard, unpublished work.
93. R.P.W. Scott, *Liquid Chromatography Detectors, Journal of Chromatography Library*, Vol. 11 (Elsevier Scientific Publishing Company, Amsterdam, 1977).
94. R.C. Denney, *A Dictionary of Chromatography* (John Wiley & Sons, New York, 1982).
95. M.L. Lee, F.J. Yang, and K.D. Bartle, *Open Tubular Column Gas Chromatography* (John Wiley & Sons, New York, 1984).
96. E.D. Katz, in *Small Bore Liquid Chromatography Columns*, Vol. 72, R.P.W. Scott, Ed. (John Wiley & Sons, New York, 1984), p. 23.
97. ASTM E260-85, *Standard Practice for Packed Column Gas Chromatography*.
98. M. Dressler, *Selective Gas Chromatographic Detectors, Journal of Chromatography Library*, Vol. 36 (Elsevier, Amsterdam, 1986).
99. H. Freiser and G.H. Nancollas, *IUPAC Compendium of Analytical Nomenclature, Definitive Rules 1987* (Blackwell Scientific Publications, Oxford, 1987).
100. H.A. Strobel and W.R. Heineman, *Chemical Instrumentation: A Systematic Approach* (John Wiley & Sons, New York, 1988).

101. Th. Noij, J.A. Rijks, A.J. Van Es, and C.A. Cramers, *J. High Resolut. Chromatogr. Chromatogr. Commun.* **11**, 862 (1988).
102. S.T. Sie, W. van Beersum, and G.W.A. Rijnders, *Sep. Sci.* **1**, 459 (1966).
103. G. Blu, F. Lazarre, and G. Guiochon, *Anal. Chem.* **45**, 1375 (1973).
104. B.E. Richter, D.J. Bornhop, J.T. Swanson, J.G. Wangsgaard, and M.R. Andersen, *J. Chromatogr. Sci.* **27**, 303 (1989).
105. R. Buffington and M.K. Wilson, in *Detectors for Gas Chromatography: A Practical Primer* (Hewlett-Packard, Avondale, PA, 1987), p. 27.
106. L.A. Pekay and S.V. Olesik, *Anal. Chem.* **61**, 2616 (1989).
107. H.H. Hill, Jr. and M. Morrissey, in *Modern Supercritical Fluid Chromatography*, C.M. White, Ed. (Huethig, Heidelberg, 1988), p. 95.
108. J.M. Levy and W.M. Ritchey, *J. Chromatogr. Sci.* **24**, 242 (1986).
109. E. Lundanes, B. Iversen, and T. Greibrokk, *J. Chromatogr.* **366**, 391 (1986).
110. L.L. Reigle, W.J. McCarthy, and A.C. Ling, *J. Chem. Eng. Data* **18**, 79 (1973).
111. H.E. Schwartz and R.G. Brownlee, *J. Chromatogr.* **353**, 77 (1986).
112. R.M. Campbell, N.M. Djordjevic, K.E. Markides, and M.L. Lee, *Anal. Chem.* **60**, 356 (1988).
113. S. Fields and K. Grolimund, *J. Chromatogr.* **472**, 197 (1989).
114. M.A. Morrissey and H.H. Hill, Jr., *J. High Resolut. Chromatogr. Chromatogr. Commun.* **11**, 375 (1988).
115. S.O. Farwell, D.R. Gage, and R.A. Kagel, *J. Chromatogr. Sci.* **19**, 358 (1981).
116. W.R. West and M.L. Lee, *J. High Resolut. Chromatogr. Chromatogr. Commun.* **9**, 161 (1986).
117. P.A. David and M. Novotny, *J. Chromatogr.* **452**, 623 (1988).
118. P.A. David and M. Novotny, *J. Chromatogr.* **461**, 111 (1989).
119. L. Mathiasson, J.A. Jonsson, and L. Karlsson, *J. Chromatogr.* **467**, 61 (1989).
120. K.E. Markides, E.D. Lee, R. Bolick, and M.L. Lee, *Anal. Chem.* **58**, 740 (1986).
121. S.V. Olesik, L.A. Pekay, and E.A. Paliwoda, *Anal. Chem.* **61**, 58 (1988).
122. S.K. Yarbro and D.R. Gere, *Chromatography Forum* **2**(3), 49 (1987).
123. F.P. Schmitz, H. Hilgers, B. Lorenschat, and E. Klesper, *J. Chromatogr.* **346**, 69 (1985).
124. P.A. Peaden and M.L. Lee, *J. Chromatogr.* **259**, 1 (1983).
125. S.M. Fields, K.E. Markides, and M.L. Lee, *Anal. Chem.* **60**, 802 (1988).
126. J.C. Kuei, K.E. Markides, and M.L. Lee, *J. High Resolut. Chromatogr. Chromatogr. Commun.* **10**, 257 (1987).
127. F.J. Yang, *J. High Resolut. Chromatogr. Chromatogr. Commun.* **4**, 83 (1981).
128. F.J. Yang, *J. Chromatogr.* **236**, 265 (1982).
129. D.W. Later, D.J. Bornhop, E.D. Lee, J.D. Henion, and R.C. Wieboldt, *LC/GC* **5**, 804 (1987).
130. H. Poppe, *Anal. Chim. Acta.* **145**, 17 (1983).

131. M.V. Novotny and P. Kucera, Ed., *Microcolumn High-Performance Liquid Chromatography* (Elsevier, Amsterdam, 1984), p. 194.
132. W. Baumann and Z. Fresenius, *Anal. Chem.* **31**, 284 (1977).
133. J.M. Harris and N.J. Dovichi, *Anal. Chem.* **52**, 695A (1980).
134. M.D. Morris and K. Peck, *Anal. Chem.* **58**, 811A (1986).
135. T.G. Nolan, B.K. Hart, and N.J. Dovichi, *Anal. Chem.* **57**, 2703 (1985).
136. D.J. Bornhop and N.J. Dovichi, *Anal. Chem.* **58**, 1632 (1987).
137. R.A. Leach and J.M. Harris, *J. Chromatogr.* **218**, 15 (1981).
138. R.A. Leach and J.M. Harris, *Anal. Chem.* **56**, 1481 (1984).
139. D.J. Bornhop, S. Schmidt, and N.L. Porter, *J. Chromatogr.* **459**, 193 (1988).
140. S.R. Weinberger and D.J. Bornhop, *J. Microcol. Sep.* **1**, 90 (1989).
141. A.E. Bruno, E. Gassmann, N. Pericles, and K. Anton, *Anal. Chem.* **61**, 876 (1989).
142. J.P. Chervet, M. Ursem, J.P. Salzmann, and R.W. Vannoort, *J. High Resolut. Chromatogr.* **12**, 278 (1989).
143. M. Novotny, *J. High Resolut. Chromatogr. Chromatogr. Commun.* **9**, 137 (1986).
144. J.E. France and K.J. Voorhees, *J. High Resolut. Chromatogr. Chromatogr. Commun.* **11**, 692 (1988).
145. S.R. Weinberger and L. Hlousek, *Int. Labmate* **9**, 9 (1988).
146. K. Jinno, T. Hoshino, T. Hondo, M. Saito, and M. Senda, *Anal. Chem.* **58**, 2696 (1986).
147. K. Jinno, T. Hoshino, T. Hondo, M. Saito, and M. Senda, *Anal. Lett.* **19**, 1001 (1986).
148. K. Sugiyama, M. Saito, T. Hondo, and M. Senda, *J. Chromatogr.* **332**, 107 (1985).
149. M. Saito, T. Hondo, and T. Yamauchi, in *Supercritical Fluid Chromatography*, R.M. Smith, Ed. (The Royal Society of Chemistry, London, 1988), Chap. 8.
150. J.C. Fjeldsted, B.E. Richter, W.P. Jackson, and M.L. Lee, *J. Chromatogr.* **279**, 423 (1983).
151. M.J. Sepaniak and E.S. Yeung, *J. Chromatogr.* **190**, 377 (1980).
152. W.P. Jackson, Ph.D. Dissertation, Brigham Young University, 1985.
153. D.E. Raynie, Ph.D. Dissertation, Brigham Young University, in preparation.
154. J.M. Hayes, *Chem. Rev.* **87**, 745 (1987).
155. S.R. Goates and C.H. Sin, *Appl. Spectrosc. Rev.* **20**, 81 (1989).
156. J.M. Hayes and G.J. Small, *Anal. Chem.* **54**, 1202 (1982).
157. D.M. Lubman, *Anal. Chem.* **59**, 31A (1987).
158. J.C. Giddings, M.N. Meyers, L. McLaren, and R.A. Keller, *Science* **162**, 67 (1968).
159. S.R. Goates, N.A. Zabriskie, J.K. Simons, and B. Khoobehi, *Anal. Chem.* **59**, 2927 (1987).
160. L.G. Randall and A.L. Wahrhaftig, *Anal. Chem.* **50**, 1705 (1978).
161. R.D. Smith, H.T. Kalinoski, and H.R. Udseth, *Anal. Chem.* **56**, 5971 (1984).

162. E.C. Huang, B.J. Jackson, K.E. Markides, and M.L. Lee, *Anal. Chem.* **60**, 2715 (1988).
163. J.K. Simons, C.H. Sin, N.A. Zabriskie, S.M. Fields, M.L. Lee, and S.R. Goates, *J. Microcol. Sep.* **1**, 200 (1989).
164. S.R. Goates, C.H. Sin, J.K. Simons, K.E. Markides, and M.L. Lee, *J. Microcol. Sep.* **1**, 207 (1989).
165. B.D. Anderson and M.V. Johnston, *Appl. Spectrosc.* **41**, 1358 (1987).
166. W.T. Foreman, C.L. Shellum, J.W. Birks, and R.E. Sievers, *J. Chromatogr.* **465**, 23 (1989).
167. W.T. Foreman, R.E. Sievers, and B.W. Wenclawiak, *Fres. Z. Anal. Chem.* **330**, 231 (1988).
168. B.W. Wenclawiak, R.E. Sievers, and W.T. Foreman, *Proc. Int. Symp. Supercr. Fluids*, M. Perrut, Ed. (Institut National Polytechnique de Lorraine, Vandoeuvre Cedex, 1988), p. 433.
169. A.A. Turnipseed and J.W. Birks, in *Chemiluminescence and Photochemical Reaction Detection in Chromatography*, J.W. Birks, Ed. (VCH Publishers Inc., New York, 1989), Chap. 2.
170. K. Grolimund, W.P. Jackson, M. Joppich, W. Nussbaum, K. Anton, and H.M. Widmer, *Proc. Seventh Int. Symp. Cap. Chromatogr.*, P. Sandra, Ed. (Huethig, Heidelberg, 1986), p. 625.
171. D.J. Bornhop, B.J. Murphy, and L. Krieger-Jones, *Anal. Chem.* **61**, 797 (1989).
172. R.L. Benner and D.H. Stedman, *Anal. Chem.* **61**, 1268 (1989).
173. R.L. Shearer, D.L. O'Neal, R. Rios, and M.D. Baker, *J. Chromatogr. Sci.* **28**, 24 (1990).
174. H-C.K. Chang and L.T. Taylor, *J. Chromatogr. Sci.* **28**, 29 (1990).
175. R.A. Scanlan, in *Analysis of Food Contaminants*, J. Gilbert, Ed. (Elsevier, London, 1984), p. 321.
176. W. Gmur, J.O. Bosset, and E. Plattner, *Chromatographia* **23**, 199 (1987).
177. P.G. Sim, C.M. Elson, and M.A. Quilliam, *J. Chromatogr.* **445**, 239 (1988).
178. J. Hayhurst and B. Magill, *Int. Labmate* **14**, 1 (1989).
179. A.N. Freedman, *J. Chromatogr.* **190**, 263 (1980).
180. M.S. Davis, *1990 Pittsburgh Conference and Exposition on Analytical Chemistry and Applied Spectroscopy*, New York, NY, March 6-10, 1990, Paper 672.
181. A. Zlatkis and C.F. Poole, *Electron Capture: Theory and Practice in Chromatography* (Elsevier, New York, 1981).
182. S. Kennedy and R.J. Wall, *LC/GC* **6**, 930 (1988).
183. N.H. Sagert, *Can. J. Chem.* **46**, 336 (1968).
184. J.N. Bardsley, *J. Chem. Phys.* **51**, 3384 (1969).
185. D.R. Nelson and F.J. Davis, *J. Chem. Phys.* **51**, 2323 (1969).
186. A.J. McCormack, S.C. Tong, and W.D. Cooke, *Anal. Chem.* **37**, 1470 (1965).
187. C.A. Bache and O.J. Lisk, *Anal. Chem.* **37**, 1477 (1965).
188. P.C. Uden, *Tr. Anal. Chem.* **6**, 238 (1987).
189. J.W. Olesik and S.V. Olesik, *Anal. Chem.* **59**, 799 (1987).
190. C. Fujimoto, H. Yoshida, and K. Jinno, *J. Chromatogr.* **411**, 213 (1987).

191. C. Fujimoto, H. Yoshida, and K. Jinno, *J. Microcol. Sep.* **1**, 19 (1989).
192. B. Rivière, J.M. Mermet, and D. Derauz, *J. Anal. At. Spectrom.* **3**, 551 (1988).
193. D.R. Luffer, L.J. Galante, P.A. David, M. Novotny, and G.M. Hieftje, *Anal. Chem.* **60**, 1365 (1988).
194. L.J. Galante, M. Selby, D.R. Luffer, G.M. Hieftje, and M. Novotny, *Anal. Chem.* **60**, 1370 (1988).
195. C.B. Motley, M. Ashraf-Khorassani, and G.L. Long, *Appl. Spectrosc.* **43**, 737 (1989).
196. R.J. Skelton, Jr., P.B. Farnsworth, K.E. Markides, and M.L. Lee, *Anal. Chem.* **61**, 1815 (1989).
197. S.K. Chan and A. Montaser, *Spectrochim. Acta* **42B**, 591 (1987).
198. D.L. Ford and W. Kennard, *J. Oil Colour Chem. Assoc.* **49**, 299 (1966).
199. J.M. Charlesworth, *Anal. Chem.* **50**, 1414 (1978).
200. A. Stolyhwo, H. Colin, and G. Guiochon, *J. Chromatogr.* **265**, 1 (1983).
201. M. Lafosse, M. Dreux, and L. Morin-Allory, *J. Chromatogr.* **404**, 95 (1987).
202. S. Hoffmann, H.R. Norli, and T. Greibrokk, *J. High Resolut. Chromatogr.* **12**, 260 (1989).
203. T.H. Mourey and L.E. Oppenheimer, *Anal. Chem.* **56**, 2427 (1984).
204. L.E. Oppenheimer and T.H. Mourey, *J. Chromatogr.* **323**, 297 (1985).
205. D. Nizery, D. Thiebaut, M. Caude, R. Rosset, M. Lafosse, and M. Dreux, *J. Chromatogr.* **467**, 49 (1989).
206. S. Hoffmann and T. Greibrokk, *J. Microcol. Sep.* **1**, 35 (1989).
207. H.M. Hagen, S. Hoffmann, and T. Greibrokk, University of Oslo, unpublished results.
208. A. Stolyhwo, H. Colin, M. Martin, and G. Guiochon, *J. Chromatogr.* **288**, 253 (1984).
209. P. Carraud, D. Thiebaut, M. Caude, R. Rosset, M. Lafosse, and M. Dreux, *J. Chromatogr. Sci.* **25**, 395 (1987).
210. J.L. Robinson, M. Tsimidou, and R. Macrae, *J. Chromatogr.* **324**, 35 (1985).
211. B. Herbreteau, M. Lafosse, L. Morin-Allory, and M. Dreux, Université d'Orléans, unpublished results.
212. M.A. Baim and H.H. Hill, Jr., *Anal. Chem.* **54**, 38 (1982).
213. R.L. Eatherton, M.A. Morrissey, W.F. Siems, and H.H. Hill, Jr., *J. High Resolut. Chromatogr. Chromatogr. Commun.* **9**, 154 (1986).
214. S. Rokushika, H. Hatano, and H.H. Hill, Jr., *Anal. Chem.* **59**, 8 (1987).
215. C.B. Shumate and H.H. Hill, Jr., *Anal. Chem.* **61**, 601 (1989).
216. M.A. Morrissey, W.F. Siems, and H.H. Hill, Jr., *J. Chromatogr.* **505**, 215 (1990).
217. S. Rokushika, H. Hatano, and H.H. Hill, Jr., *Anal. Chem.* **58**, 361 (1986).
218. R.L. Eatherton, M.A. Morrissey, and H.H. Hill, Jr., *Anal. Chem.* **60**, 2240 (1988).
219. P.R. Griffiths, S.L. Pentoney, Jr., A. Giorgetti, and K.H. Shafer, *Anal. Chem.* **58**, 1349A (1986).

220. P.R. Griffiths and J.A. de Haseth, *Fourier Transform Infrared Spectrometry* (Wiley & Sons, New York, 1986).
221. K. Jinno, *Chromatographia* **23**, 55 (1987).
222. L.T. Taylor and E.M. Calvey, *Chem. Rev.* **89**, 321 (1989).
223. K.D. Bartle, M.W. Raynor, A.A. Clifford, I.L. Davies, J.P. Kithinji, G.F. Shilstone, J.M. Chalmers, and B.W. Cook, *J. Chromatogr. Sci.* **27**, 283 (1989).
224. P. Morin, M. Caude, H. Richard, and R. Rosset, *Chromatographia* **21**, 523 (1986).
225. S.B. French and M. Novotny, *Anal. Chem.* **58**, 164 (1986).
226. S.V. Olesik, S.B. French, and M. Novotny, *Chromatographia* **18**, 489 (1984).
227. J.W. Jordan and L.T. Taylor, *J. Chromatogr. Sci.* **24**, 82 (1986).
228. R.C. Wieboldt, G.E. Adams, and D.W. Later, *Anal. Chem.* **60**, 2422 (1988).
229. S.L. Pentoney, Jr., K.H. Shafer, and P.R. Griffiths, *J. Chromatogr. Sci.* **24**, 230 (1986).
230. S.L. Pentoney, Jr., K.H. Shafer, P.R. Griffiths, and R. Fuoco, *J. High Resolut. Chromatogr. Chromatogr. Commun.* **9**, 168 (1986).
231. M.W. Raynor, K.D. Bartle, I.L. Davies, A. Williams, A.A. Clifford, J.M. Chalmers, and B.W. Cook, *Anal. Chem.* **60**, 427 (1988).
232. C.C. Johnson, J.W. Jordan, L.T. Taylor, and D.W. Vidrine, *Chromatographia* **20**, 717 (1985).
233. M.W. Raynor, A.A. Clifford, K.D. Bartle, C. Reyner, A. Williams, and B.W. Cook, *J. Microcol. Sep.* **1**, 101 (1989).
234. M.E. Hughes and J.L. Fasching, *J. Chromatogr. Sci.* **24**, 537 (1985).
235. P. Morin, H. Pichard, H. Richard, M. Caude, and R. Rosset, *J. Chromatogr.* **464**, 125 (1989).
236. K.H. Shafer and P.R. Griffiths, *Anal. Chem.* **55**, 1939 (1983).
237. S. Shah, M. Ashraf-Khorassani, and L.T. Taylor, *Chromatographia* **25**, 631 (1988).
238. M. Ashraf-Khorassani and L.T. Taylor, *Anal. Chem.* **61**, 145 (1989).
239. M.W. Raynor, G.F. Shilstone, K.D. Bartle, A.A. Clifford, M. Cleary, and B.W. Cook, *J. High Resolut. Chromatogr.* **12**, 300 (1989).
240. M.W. Raynor, K.D. Bartle, A.A. Clifford, M. Cleary, and B.W. Cook, University of Leeds, UK, unpublished results.
241. R.C. Wieboldt and D.A. Hanna, *Anal. Chem.* **59**, 1255 (1987).
242. P. Morin, B. Beccard, M. Caude, and R. Rosset, *J. High Resolut. Chromatogr. Chromatogr. Commun.* **11**, 697 (1988).
243. J.W. Hellgeth, J.W. Jordan, L.T. Taylor, and M. Ashraf-Khorassani, *J. Chromatogr. Sci.* **24**, 183 (1986).
244. P. Morin, M. Caude, H. Richard, and R. Rosset, *Analusis* **15**, 117 (1987).
245. R.C. Wieboldt, K.D. Kempfert, D.W. Later, and E.R. Campbell, *J. High Resolut. Chromatogr.* **12**, 106 (1989).
246. E.M. Calvey, L.T. Taylor, and J.K. Palmer, *J. High Resolut. Chromatogr. Chromatogr. Commun.* **11**, 739 (1988).

247. T.A. Jenkins and M.A. Healy, in *SFC Applications, 1989 Workshop/Symposium on Supercritical Fluid Chromatography*, K.E. Markides and M.L. Lee, Comps. (Brigham Young University Press, Provo, 1989).
248. P.R. Griffiths, *Appl. Spectrosc.* **31**, 284 (1977).
249. D.W. Vidrine, in *Fourier Transform Infrared Spectroscopy: Applications to Chemical Systems*, Vol. 2, J.R. Ferraro and L.J. Basile, Eds. (Academic Press, New York, 1979), p. 82.
250. D. Kuehl and P.R. Griffiths, *Anal. Chem.* **52**, 1394 (1980).
251. R. Fuoco, S.L. Pentoney, and P.R. Griffiths, *Anal. Chem.* **61**, 2212 (1989).
252. M.W. Raynor, I.L. Davies, K.D. Bartle, A. Williams, J.M. Chalmers, and B.W. Cook, *Eur. Chromatogr. News* **1**, 18 (1987).
253. A.M. Haefner, K.L. Norton, P.R. Griffiths, S. Bourne, and R. Curbelo, *Anal. Chem.* **60**, 2441 (1988).
254. P.R. Griffiths, S.L. Pentoney, G.L. Pariente, and K.L. Norton, *Mikochim. Acta (Wien)* **III**, 47 (1988).
255. A.M. Haefner, K.L. Norton, H. Makishima, and P.R. Griffiths, *1989 Pittsburgh Conference and Exposition on Analytical Chemistry and Applied Spectroscopy*, Atlanta, GA, March 6-10, 1989, Paper 459.
256. K.H. Shafer, S.L. Pentoney, and P.R. Griffiths, *Anal. Chem.* **58**, 58 (1986).
257. H. Makishima, M.S. Thesis, University of California, Riverside, 1989.
258. C. Fujimoto, Y. Hirata, and K. Jinno, *J. Chromatogr.* **332**, 47 (1985).
259. L.G. Randall, *Sep. Sci. Tech.* **17**, 1 (1982).
260. B.W. Wright, H.T. Kalinoski, H.R. Udseth, and R.D. Smith, *J. High Resolut. Chromatogr. Chromatogr. Commun.* **9**, 145 (1986).
261. J. Cousin and P.J. Arpino, *Analusis* **14**, 215 (1986).
262. R.D. Smith, H.T. Kalinoski, and H.R. Udseth, *Mass Spectrom. Rev.* **6**, 445 (1987).
263. H.M. Widmer, in *Modern Supercritical Fluid Chromatography*, C.M. White, Ed. (Huethig, Heidelberg, 1988), p. 211.
264. G. Holzer, S. Deluca, and K.J. Voorhees, *J. High Resolut. Chromatogr. Chromatogr. Commun.* **8**, 528 (1985).
265. T.A. Milne, *Int. J. Mass Spectrom. Ion Phys.* **3**, 153 (1969).
266. J.C. Giddings, M.N. Myers, and A.L. Wahrhaftig, *Int. J. Mass Spectrom. Ion Phys.* **4**, 9 (1970).
267. L.G. Randall and A.L. Wahrhaftig, *Rev. Sci. Instrum.* **52**, 1283 (1981).
268. T.H. Gouw, R.E. Jentoft, and E.J. Gallegos, *J. High Pressure Sci. Technol.*, 6th AIRAPT Conf., 583 (1979).
269. R.D. Smith, W.D. Felix, J.C. Fjeldsted, and M.L. Lee, *Anal. Chem.* **54**, 1883 (1982).
270. R.D. Smith, H.T. Kalinoski, H.R. Udseth, and B.W. Wright, *Anal. Chem.* **56**, 2476 (1984).
271. R.D. Smith, H.R. Udseth, and H.T. Kalinoski, *Anal. Chem.* **56**, 2971 (1984).
272. V.N. Reinhold, D.M. Sheeley, J. Kuei, and G. Her, *Anal. Chem.* **60**, 2719 (1988).

273. G.D. Owens, L.J. Burkes, J.D. Pinkston, T. Keough, J.R. Simms, and M.P. Lacey, in *Supercritical Fluid Chromatography and Extraction: Technology and Applications, ACS Symposium Series 366*, B.A. Charpentier, M.R. Sevenants, Eds. (American Chemical Society, Washington DC, 1988), p. 191.
274. S.D. Zaugg, S.J. Deluca, G.U. Holzer, and K.J. Voorhees, *J. High Resolut. Chromatogr. Chromatogr. Commun.* **10**, 100 (1987).
275. J.D. Pinkston, G.D. Owens, L.J. Burkes, T.E. Delaney, D.S. Millington, and D.A. Maltby, *Anal. Chem.* **60**, 962 (1988).
276. E.D. Lee and J.D. Henion, *J. High Resolut. Chromatogr. Chromatogr. Commun.* **9**, 172 (1986).
277. S.B. Hawthorne and D.J. Miller, *J. Chromatogr.* **468**, 115 (1989).
278. W. Blum, K. Grolimund, P.E. Jordi, and P. Ramstein, *J. High Resolut. Chromatogr. Chromatogr. Commun.* **11**, 441 (1988).
279. H.T. Kalinoski, H.R. Udseth, E.K. Chess, and R.D. Smith, *J. Chromatogr.* **394**, 3 (1987).
280. H.T. Kalinoski, H.R. Udseth, B.W. Wright, and R.D. Smith, *J. Chromatogr.* **400**, 307 (1987).
281. E.C. Huang, B.J. Jackson, K.E. Markides, and M.L. Lee, *Chromatographia* **25**, 51 (1988).
282. E.C. Huang, B.J. Jackson, K.E. Markides, and M.L. Lee, *Anal. Chem.* **60**, 2715 (1988).
283. E.C. Huang, B.J. Jackson, K.E. Markides, and M.L. Lee, *J. Microcol. Sep.* **2**, 88 (1990).
284. H. Kallio, P. Laakso, R. Huopalahti, R.R. Linko, and P. Oksman, *Anal. Chem.* **61**, 698 (1989).
285. D.M. Sheeley and V.N. Reinhold, *Proc. Thirty-seventh ASMS Conf. Mass Spectrom. Allied Topics*, Miami, FL, May 21-26, 1989.
286. V.N. Reinhold, D.M. Sheeley, J.C.H. Kuei, and G.R. Her, *Proc. Eleventh Int. Mass Spectrom. Conf.*, Bordeaux, France, August 29-September 2, 1988.
287. J.B. Crowther and J.D. Henion, *Anal. Chem.* **57**, 2711 (1985).
288. K. Matsumoto, S. Tsuge, and Y. Hirata, *Anal. Sci.* **2**, 3 (1986).
289. K. Matsumoto, S. Tsuge, and Y. Hirata, *Chromatographia* **21**, 617 (1986).
290. R.D. Smith and H.R. Udseth, *Anal. Chem.* **59**, 13 (1987).
291. H.T. Kalinoski and R.D. Smith, *Anal. Chem.* **60**, 529 (1988).
292. J. Cousin and P.J. Arpino, *J. Chromatogr.* **398**, 125 (1987).
293. P. Dätwyler, H.J. Walther, and P. Hirter, *Proc. Fourth Symp. Liq. Chromatogr. Mass Spectr. LC/MS MS/MS*, Montreux, October, 1986.
294. J.R. Chapman, *Rapid Commun. Mass Spectrom.* **2**, 6 (1988).
295. A.J. Berry, D.E. Games, I.C. Mylchreest, J.R. Perkins, and S. Pleasance, *J. High Resolut. Chromatogr. Chromatogr. Commun.* **11**, 61 (1988).
296. D.E. Games, A.J. Berry, I.C. Mylchreest, J.R. Perkins, and S. Pleasance, *Eur. Chromatogr. News* **1**, 10 (1987).
297. A.J. Berry, D.E. Games, I.C. Mylchreest, J.R. Perkins, and S. Pleasance, *Biomed. Environ. Mass Spec.* **15**, 105 (1988).

298. D.E. Games, I.C. Mylchreest, M.W. Raynor, J.P. Kithinji, K.D. Bartle, E.D. Morgan, and I.D. Wilson, *J. Chromatogr.* **467**, 292 (1989).
299. A.J. Berry, D.E. Games, and J.R. Perkins, *J. Chromatogr.* **363**, 147 (1986).
300. J. Balsevich, L.R. Hogge, A.J. Berry, D.E. Games, and I.C. Mylchreest, *J. Nat. Prod.* **51**, 1173 (1988).
301. E.D. Ramsey, J.R. Perkins, D.E. Games, and J.R. Startin, *J. Chromatogr.* **464**, 353 (1989).
302. W.M.A. Niessen, M.A.G. de Kraa, E.R. Verheij, P.J.M. Bergers, G.F. La Vos, U.R. Tjaden, and J. van der Greef, *Rapid Commun. Mass Spectrom.* **3**, 1 (1989).
303. P.O. Edlund and J.D. Henion, *J. Chromatogr. Sci.* **27**, 274 (1989).
304. P.C. Winker, D.D. Perkin, W.K. Williams, and R.F. Browner, *Anal. Chem.* **60**, 489 (1988).
305. J.R. Chapman, *Practical Organic Mass Spectrometry* (John Wiley, New York, 1985), p. 56.
306. E.D. Lee, S. Hsu, and J.D. Henion, *Anal. Chem.* **60**, 1990 (1988).
307. S.V. Olesik and W. Larkins, Jr., *Proc. Thirty-seventh ASMS Conf. Mass Spectr. Allied Topics,* Miami Beach, FL, May 21-26, 1989, pp. 1091, 1516.
308. D.I. Carroll, I. Dzidic, R.N. Stillwell, M.G. Horning, and E.C. Horning, *Anal. Chem.* **46**, 706 (1974).
309. E.C. Horning, D.I. Carroll, I. Dzidic, K.D. Haegel, M.G. Horning, and R.N. Stillwell, *J. Chromatogr. Sci.* **12**, 725 (1974).
310. B.A. Thomson, J.V. Iribarne, and P.J. Dziedic, *Anal. Chem.* **54**, 2219 (1982).
311. T.R. Covey, E.D. Lee, A.P. Bruins, and J.D. Hension, *Anal. Chem.* **58**, 1451A (1986).
312. A.P. Bruins, T.R. Covey, and J.D. Henion, *Anal. Chem.* **59**, 2642 (1987).
313. C.M. Whitehouse, R.N. Dreyer, M. Yamashita, and J.B. Fenn, *Anal. Chem.* **57**, 675 (1985).
314. J.A. Olivares, N.T. Nguyen, C.R. Yonker, and R.D. Smith, *Anal. Chem.* 59, 1232 (1987).
315. E. Huang and J. Henion, Cornell University, unpublished results.
316. J.D. Henion and T. Wachs, *Anal. Chem.* **53**, 1963 (1981).
317. K. Levsen, in *Tandem Mass Spectrometry*, F.W. McLafferty, Ed. (John Wiley, New York, 1983), p. 1.
318. R. Huopalahti, E. Huang, and J. Henion, Cornell University, in preparation.
319. G.C. Stafford, Jr., P.E. Kelley, J.E.P. Syka, W.E. Reynolds, and J.F.J. Todd, *Int. J. Mass Spectrom. Ion Proc.* **60**, 85 (1984).
320. J.F.J. Todd, I.C. Mylchreest, A.J. Berry, D.E. Games, and R.M. Smith, *Rapid Commun. Mass Spectrom.* **3**, 55 (1988).
321. P.H. Hemberger, J.D. Moss, R.E. Kaiser, J.N. Louris, J.W. Amy, R.G. Cooks, J.E.P. Syka, and G.C. Stafford, *Proc. Thirty-seventh ASMS Conf. Mass Spectrom. Allied Topics*, Miami Beach, FL, May 21-26, 1989, p. 60.

322. R.E. Pedder, R.E. Yost, and M. Weber-Grabau, *Proc. Thirty-seventh ASMS Conf. Mass Spectrom. Allied Topics*, Miami Beach, FL, May 21-26, 1989, p. 468.
323. C.L. Wilkins, A. Chowdhury, L.M. Nuwaysir, and M.L. Coates, *Mass Spectrom. Rev.* **8**, 67 (1989).
324. B.S. Freiser, *Tech. Chem. (N.Y.)* **20**, 61 (1988).
325. M.P. Chiarelli and M.L. Gross, *Analytical Applications of Spectroscopy*, C.S. Creaser and A.M.C. Davies, Eds. (Royal Society of Chemistry, London, 1988), p. 263.
326. E.D. Lee, J.D. Henion, R.B. Cody, and J.A. Kinsinger, *Anal. Chem.* **59**, 1309 (1987).
327. D.A. Laude, Jr., S.L. Pentoney, P.R. Griffiths, and C.L. Wilkins, *Anal. Chem.* **59**, 2283 (1987).
328. J.F. Holland, C.G. Enke, J. Allison, J.T. Stults, J.D. Pinkston, B. Newcome, and J.T. Watson, *Anal. Chem.* **55**, 997A (1983).
329. H. Pang, C. Sin, and D.M. Lubman, *Spectrochim. Acta* **43B**, 671 (1987).
330. H. Pang, C. Sin, D.M. Lubman, and J. Zorn, *Anal. Chem.* **58**, 1581 (1986).
331. C. Sin, H. Pang, D.M. Lubman, and J. Zorn, *Anal. Chem.* **58**, 487 (1986).
332. H.M. Pang and D.M. Lubman, *Proc. Thirty-seventh ASMS Conf. Mass Spectrom. Allied Topics*, Miami, FL, May 21-26, 1989.
333. J.C. Giddings, *Anal. Chem.* **56**, 1258A (1984).
334. J.C. Giddings, *J. High Resolut. Chromatogr. Chromatogr. Commun.* **10**, 319 (1987).
335. R.G. Christensen, *J. High Resolut. Chromatogr. Chromatogr. Commun.* **8**, 824 (1985).
336. W.J. Sonnefeld, W.H. Zoller, W.E. May, and S.A. Wise, *Anal. Chem.* **54**, 723 (1982).
337. E. Lundanes and T. Greibrokk, *J. Chromatogr.* **349**, 439 (1985).
338. E. Lundanes, B. Iversen, and T. Greibrokk, *J. Chromatogr.* **366**, 391 (1986).
339. J.M. Levy, R.A. Cavalier, T.N. Bosch, A.F. Rynaski, and W.E. Huhak, *J. Chromatogr. Sci.* **27**, 341 (1989).
340. J.M. Levy, J.P. Guzowski, and W.E. Huhak, *J. High Resolut. Chromatogr. Chromatogr. Commun.* **10**, 337 (1987).
341. I.L. Davies, B. Xu, K.E. Markides, K.D. Bartle, and M.L. Lee, *J. Microcol. Sep.* **1**, 71 (1989).
342. K.M. Payne, I.L. Davies, K.D. Bartle, K.E. Markides, and M.L. Lee, *J. Chromatogr.* **477**, 161 (1989).
343. I.L. Davies, Ph.D. Dissertation, University of Leeds, UK, 1989.
344. F. David, P. Sandra, and C. van Tilburg, *Proc. Ninth Int. Conf. Cap. Chromatogr.*, P. Sandra. Ed. (Huethig, Heidelberg, 1988), p. 288.
345. Z. Juvancz, K.M. Payne, K.E. Markides, and M.L. Lee, *Anal. Chem.* **62**, 1384 (1990).
346. I.L. Davies, Z. Juvancz, K.M. Payne, K.E. Markides, M.L. Lee, and K.D. Bartle, *1989 Symposium/Workshop on Supercritical Fluid Chromatography*, Snowbird, UT, June 13-15, 1989.

347. J.M. Levy and J.P. Guzowski, *Fres. Z. Anal. Chem.* **330**, 207 (1988).
348. H.E. Schwartz, J.M. Levy, and J.P. Guzowski, in *Modern Supercritical Fluid Chromatography*, C.M. White, Ed. (Huethig, Heidelberg, 1988), p. 135.
349. I.S. Lurie, *LC/GC* **6**, 1066 (1988).
350. L.R. Snyder and J.J. Kirkland, *Introduction to Modern Liquid Chromatography*, 2nd Ed. (John Wiley & Sons, Inc., New York, 1979), p. 593.
351. T.H. Gouw and R.E. Jentoft, *Adv. Chromatogr.* **13**, 1 (1975).
352. R.E. Jentoft and T.H. Gouw, *Anal. Chem.* **44**, 681 (1972).
353. R.M. Campbell and M.L. Lee, *Anal. Chem.* **58**, 2247 (1986).
354. T. Norris and M. Rawdon, Texaco Research Center, unpublished results.
355. Q.L. Xie, K.E. Markides, and M.L. Lee, *J. Chromatogr. Sci.* **27**, 365 (1989).

CHAPTER 5

ANALYTICAL SUPERCRITICAL FLUID EXTRACTION

5.1 Introduction

Supercritical fluid extraction (SFE) is a relatively new technique in the field of analytical chemistry, having evolved in the last decade as an alternative method of preparing samples prior to analysis. SFE offers many advantages to the analyst that are not inherent in other sample preparation techniques, such as distillation, extraction with liquid solvents, or low resolution liquid chromatography. The most unique property of supercritical fluids for extraction purposes is the ability to adjust their "solubilizing power" primarily *via* mechanical compression (and additionally *via* temperature), thereby providing the possibility of using one supercritical fluid to extract a host of analytes of varying polarity and molecular size. In addition, solute-fluid binary diffusion coefficients are much greater in supercritical fluid media than in liquid-liquid systems, thereby facilitating fast extraction from a variety of sample matrices.

The proper choice of supercritical fluid can also provide specific advantages when applied in sample workup prior to analysis. For example, the low critical temperature of supercritical CO_2 makes it an excellent candidate for extracting thermally labile compounds under conditions slightly above room temperature. In addition, CO_2 provides an extraction environment free from molecular oxygen, thereby limiting potential oxidation of the extracted solutes. Supercritical CO_2, unlike many liquid extraction solvents, is a nontoxic extraction medium; hence, its use in a laboratory environment can eliminate the cost and problems associated with solvent disposal as well as long term exposure of laboratory personnel to potential toxic vapors.

In practice, SFE can provide an appreciable savings in time and cost associated with sample preparation. As will be shown later in this chapter, on-line coupling of SFE to microcolumn chromatographic instrumentation permits the extraction and characterization of very small samples. In general, large polar compounds exhibit almost no solubility in supercritical CO_2,[1] making it an excellent extraction medium for the separation of nonpolar to moderately polar solutes from such matrices as inorganic solids. However, the solubility of polar analytes can be enhanced in many supercritical fluids by the addition of co-solvents, or modifiers,[2] at low levels to the dense gaseous phase.

By far, the most widely used extraction fluid has been supercritical CO_2; however, the extractability of polar solutes can be improved by using a more polar supercritical fluid. Table 3.1 is a tabulation of various supercritical fluid media and their associated critical properties which have been used in performing supercritical fluid extractions as well as SFC. Many of the listed fluids would not be suitable for practical extractions due to their unfavorable physical properties, costs, or reactivities. For example, ethylene which exhibits a subambient critical temperature has been widely investigated in the laboratory as an extractant. However, its flammability limits its application in many analytical problems. Conversely, most polar fluids have high critical temperatures which can prove destructive to both the analyte and extraction system. The isoelectronic analogue of CO_2, N_2O, has been shown to be a useful extracting fluid;[3] however, it exhibits a high reactivity toward many compounds and can cause dangerous physiological effects. Other fluids, like fluoroform (HCF_3), are unique in their ability to solubilize basic solutes through intermolecular hydrogen bonding in the supercritical fluid state,[4] but the exorbitant cost of the fluid limits its use for SFE.

It is useful to compare the physical properties exhibited by CO_2 under SFE conditions to those associated with liquid solvents under ambient conditions in order to gain a better understanding of the advantages which are attendant to conducting extractions in the supercritical fluid state. Table 5.1 compares the physical properties of CO_2 under typical SFE conditions (200 atm and 55°C) with parameters calculated for three liquid solvents: *n*-hexane, methylene chloride, and methanol at ambient conditions. The density of CO_2 at the above conditions is greater than the corresponding value for *n*-hexane, but lower than the densities exhibited by methanol or methylene chloride. Although density is only an approximate measure of intermolecular attraction, the value for CO_2 suggests that

Table 5.1. Comparison of Physical Properties of Supercritical CO_2 with Liquid Solvents at 25°C

	CO_2[a]	*n*-Hexane	Methylene chloride	Methanol
Density ($g\ mL^{-1}$)	0.746	0.660	1.326	0.791
Kinematic viscosity ($m^2\ s^{-1} \times 10^7$)	1.00	4.45	3.09	6.91
Diffusivity of benzoic acid ($m^2\ s^{-1} \times 10^9$)	6.0	4.0	2.9	1.8
$\left(\frac{P_{v,sat}\ \text{solvent}}{P_{v,sat}\ \text{solute}}\right)^b$	1.4×10^5	4.2×10^2	1.2×10^3	3.6×10^2

[a]At 200 atm and 55°C
[b]Solute is phenol at 25°C

near liquid-like densities can be achieved for this gas in its supercritical fluid state.

Likewise, kinetic-based properties such as viscosity and solute diffusivity, for CO_2, have values that are more typical of gases than those of the liquid state. These gas-like transport parameters contribute to improved rates of mass transfer for solutes in supercritical fluid media, resulting in faster extraction. The ratio of the saturated vapor pressures of the extraction solvents to that exhibited by a typical solute, phenol, at 25°C are also tabulated in Table 5.1. The vapor pressure for the CO_2/phenol case is 2-3 orders of magnitude larger than the corresponding ratios for the liquid solvent/phenol pairs. This accounts for the ease by which the dissolved solute (phenol) can be separated from CO_2 upon decompression, a phenomenon which is in stark contrast to the miscible liquid solute-solvent systems.

The intermediate properties exhibited by supercritical fluids permit the interfacing of SFE with GC, SFC, LC, and MS. Details of tandem arrangements of SFE with GC and SFC are presented in Sections 5.4 and 5.5 of this chapter; however, in such systems in general, the supercritical fluid is delivered by a pump into an extraction cell where the extraction takes place under controlled pressure and temperature conditions. The extract is then swept into an injection system where it can be transferred to the eluent stream of the chromatographic system. Separation of the extract components takes place on the chromatographic column with subsequent detection by a general or selective method. The coupling of SFE with a suitable chromatographic method can also impart an extra degree of selectivity into the overall separation problem. Hence, by varying the extraction pressure and temperature in the cell, the analyst can change the composition of the extract and achieve additional fractionation of the sample. In addition, SFE may provide a solvent-free injection method for open tubular column chromatography where injection volume otherwise limits the minimum detectable level of trace solutes.

5.2 Fundamental Thermodynamic and Kinetic Parameters

The effective utilization of SFE in analytical chemistry requires an appreciation of the fundamental thermodynamic and kinetic parameters which impact on the distribution of the analyte between the dense fluid phase and the substrate that is being extracted. Reviews of the phase behavior[5] and thermodynamics[6] pertaining to supercritical fluid systems exist; however, similar treatises on the kinetics and mass transfer of solutes in supercritical fluids are limited. A thorough discussion of all theories and experimental studies describing supercritical fluid phase equilibrium is beyond the scope of this analytical text; consequently, this section will describe the key physicochemical factors which the analyst should know for utilizing SFE.

Four properties are seminal to planning and executing successful analytical SFE. These are the detectable "threshold pressure," the appropriate conditions for fractionating solutes, the occurrence of solubility maxima in supercritical fluid systems, and when possible, knowledge of the physical properties of the extracted solutes. The term "threshold pressure" was first mentioned by Giddings[7] to define the pressure at which the solute partitioned into the supercritical fluid. Unfortunately, such a parameter is dependent on the measurement technique employed;[8] however, even within this context it is a useful concept. For example, if one is trying to extract a specific analyte from a matrix containing coextractable material, selective isolation of the target analyte may be possible if its threshold pressure is sufficiently removed from those of the coextractants.

The threshold pressure should not be confused with the pressure region where a significant increase occurs in the solubility of the solute in the supercritical fluid. This difference can be seen in Figure 5.1 where the solubility of naphthalene in CO_2 is plotted as a function of pressure. Here, the threshold pressure would appear to occur at values under 75 atm, depending on the experimental method used for its determination. The onset of the maximum differential solubility change appears to occur above 90 atm and parallels the increase in fluid density with pressure, a reflection of the increased molecular interaction between the solute and the solvent. The prediction of the maximum solubility change with pressure has been theoretically treated by Gitterman and Procaccia.[9]

Fractionation ranges for SFE can be theoretically estimated[10] or determined empirically by experimental methods. In general, the fractionation range potentially exists between the threshold pressure region of the solutes and the occurrence of solubility maxima for the dissolved solutes in the supercritical fluid. The fractionation of complex mixtures by SFE is frequently difficult unless appreciable differences exist in the molecular sizes, polarities, or volatilities of the mixture components. Enrichment of certain solute fractions can be affected by gradually increasing the pressure of the extracting fluid; however, this result is countered by a time-based fractionation effect. Enhancement of the SFE fractionation can be achieved by employing a thermal gradient or a chromatographic column downstream from the extraction module.

Solubility maxima occur for many solute types in supercritical fluid solvents.[11] Normally, the pressures associated with this phenomenon are quite high and thus would appear to be of little consequence to the analytical chemist. However, it has been recently demonstrated that the bulk removal of lipid phases by supercritical CO_2 is best effected at high pressures where lipid solubility is maximized.[12] Further increases in extraction fluid pressure can actually result in a "salting out" of the solute from the dense fluid phase as repulsive solute-solvent forces begin to predominate in the single phase system.

Although, not always available prior to SFE, a knowledge of the physical properties of the solute(s) can be of considerable aid in establishing optimal conditions for conducting the extraction. When extracting solid substrates, a knowledge of the solute or mixture

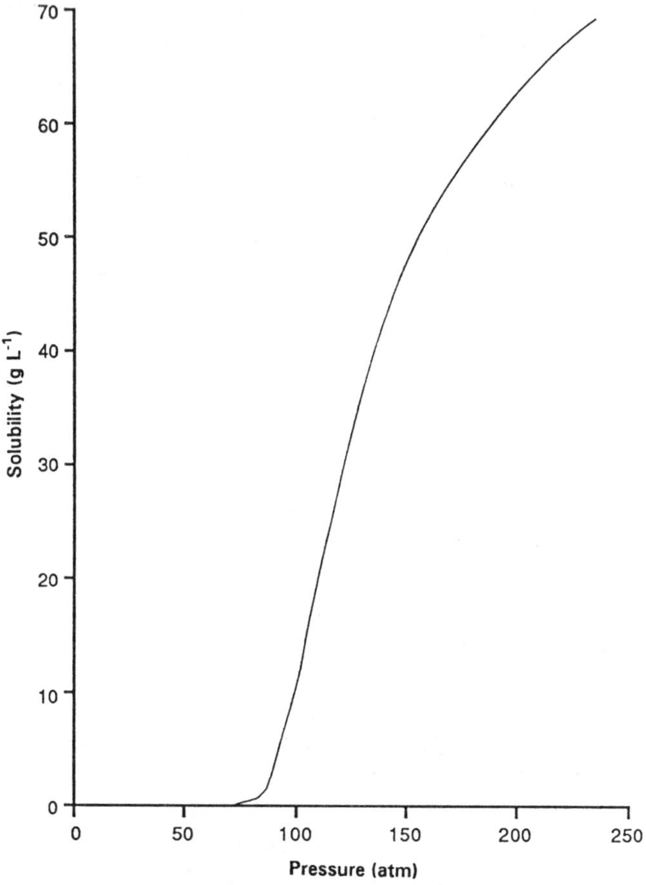

Figure 5.1. Solubility as a function of pressure for naphthalene in supercritical CO_2 at 45°C.

melting point is critical. In general, supercritical fluids are more effective extracting agents when the extraction is performed at a temperature above the melting point of the substrate. In this case, both mass transfer of the solute into the supercritical fluid is improved as well as solute solubility due to the weakening of the cohesive forces of the solid. Likewise, knowledge of the vapor pressure of the solute as a function of temperature can have a

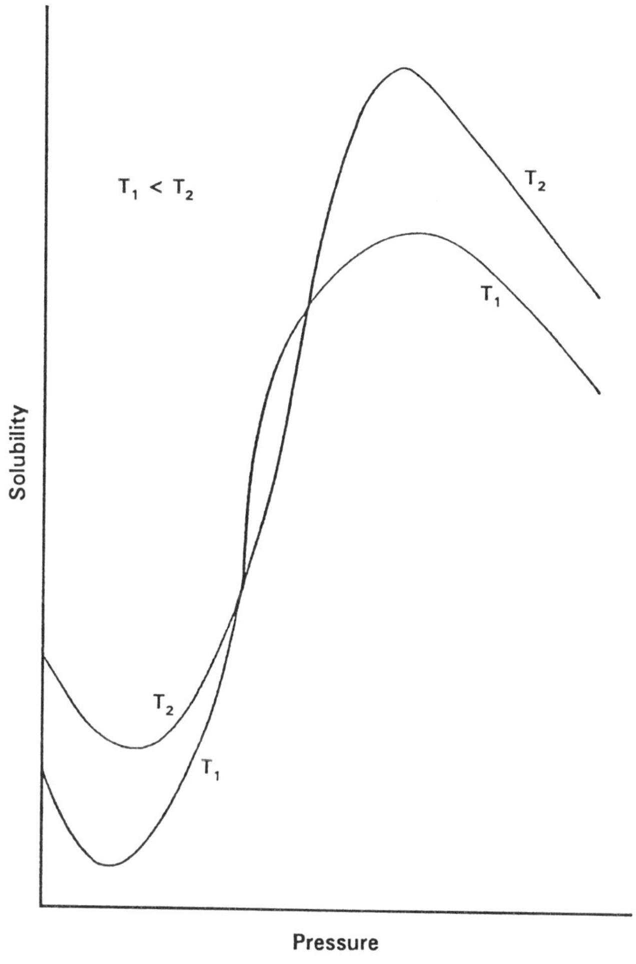

Figure 5.2. Generalized solubility isotherms as a function of pressure.

profound effect on both the recorded solubility and the separation factors that are obtained in multi-component solute separation schemes.[13]

The effect of varying both the pressure and temperature on the solubility of a solute during SFE is depicted in Figure 5.2. The recorded initial solubility in the noncompressed gas phase is a function of the vapor pressure of the solute; however, upon additional compression of the supercritical fluid phase, a solubility minimum is observed. After the occurrence of the solubility minimum, there is an exponential rise in the solubility of the solute with increasing gas pressure, and a solubility maximum is eventually attained at a pressure which is determined by the extraction temperature. Note that the effect of increasing temperature in this case results in an increase in solubility at both low and high pressures; however, at intermediate pressures, the reverse trend may be observed. The latter region has been termed the "cross-over region,"[14] and its occurrence permits selective fractionation of solutes into the supercritical fluid medium.

Solubility trends in supercritical fluids. Today, there exists a substantial data base from which general rules regarding solute solubilities in supercritical fluids can be formulated. For the analyst employing SFE, the dependence of solubility on solute structure is, perhaps, the most important factor for predicting the relative effectiveness of SFE as a pre-analysis sample preparation method. In addition, an understanding of the importance of the absolute solubility of the solute is critical, since this will affect the time required for executing a particular extraction.

By far, the most extensive solubility data collected for solutes in a particular supercritical fluid are for binary CO_2/solute systems. Solubility trends in these systems suggest that nonpolar, lipophilic solutes exhibit the largest solubilities in supercritical CO_2, and the introduction of polar substituents into the molecular structure adversely affects the solubility of a compound in CO_2. Stahl,[15] on the basis of his extensive SFE/TLC studies, formulated generalized extraction rules which proved applicable to cases involving relatively low molecular mass solutes. These experimental observations are as follows:

(a) Hydrocarbons and other typically lipophilic organic compounds of relatively low polarity (*e.g.*, esters, ethers, lactones, and epoxides) can be extracted in the lower pressure range (*i.e.*, 70-100 atm).

(b) The introduction of strongly polar functional groups (*e.g.*, -OH and -COOH) makes the extraction more difficult. For the benzene derivatives, substances with three phenolic hydroxyls are still capable of extraction, as are compounds with one carboxyl and two hydroxyl groups. Phenols that cannot be extracted are those with one carboxyl and three or more hydroxyl groups.

(c) More strongly polar substances (*e.g.*, sugars and amino acids) cannot be extracted in the range up to 400 atm.

(d) Fractionation occurs in the pressure gradient when there are sufficient differences in the commencement of boiling or sublimation, *i.e.*, in the volatility and/or polarity, of the substances. The fractionation effects are most marked in the range where there is a sharp rise in the density and dielectric constant of CO_2.

Qualitative trends in solute solubilities have also been summarized by Hyatt[16] and Dandge,[17] and they have been quantitatively correlated with the aid of the reduced solubility parameter concept by King.[18] These correlations suggest that an increase in the molecular mass of a homologous or oligomeric series of solutes leads to a decrease in solubility in the supercritical fluid phase. An exception to this trend has been documented by Schultz and Randall[19] for the case of homologues partitioning between water and liquefied CO_2. Here, the addition of methylene groups accentuates the partitioning of the solute from the aqueous medium into the nonpolar phase, despite an increase in solute molecular mass. This trend has also been verified for solute partition into supercritical CO_2 from polar solvents.

It should be appreciated that high solute solubility in the extraction fluid is not always a prerequisite for applying SFE for analytical purposes. To illustrate this point, the solubility of the pesticide, alachlor, has been plotted (see Figure 5.3) over the pressure range of 50-300 atm. Note that a solubility in excess of 10 weight percent can be achieved in CO_2 at a compression level of approximately 270 atm. Such a solubility level is of critical importance in engineering applications of SFE where extractions are conducted for profit. However, for the analyst faced with extracting a 1-ppm level of alachlor from an environmental sample, the pressure requirements are not so stringent, and a rapid extraction can be achieved at much lower pressures despite the substantial reduction in alachlor solubility.

The use of co-solvents can also have a profound effect on increasing the solubility levels of polar solutes in supercritical fluids. For example, Wong and Johnston[20] have shown that the addition of 3.5 mol% of methanol to CO_2 at a pressure of 150 atm will increase the solubility of cholesterol seven-fold over that achieved using pure CO_2. In general, the addition of an entrainer to a supercritical fluid will enhance the solubility of a solute in the fluid phase as well as alter the separation factor between coextracted solutes.[21] Kurnik and Reid[22] have also shown that the presence of a coextracted solute can dramatically change the solubility level of a compound over that recorded for the simple binary system consisting only of the extracting fluid and the dissolved compound.

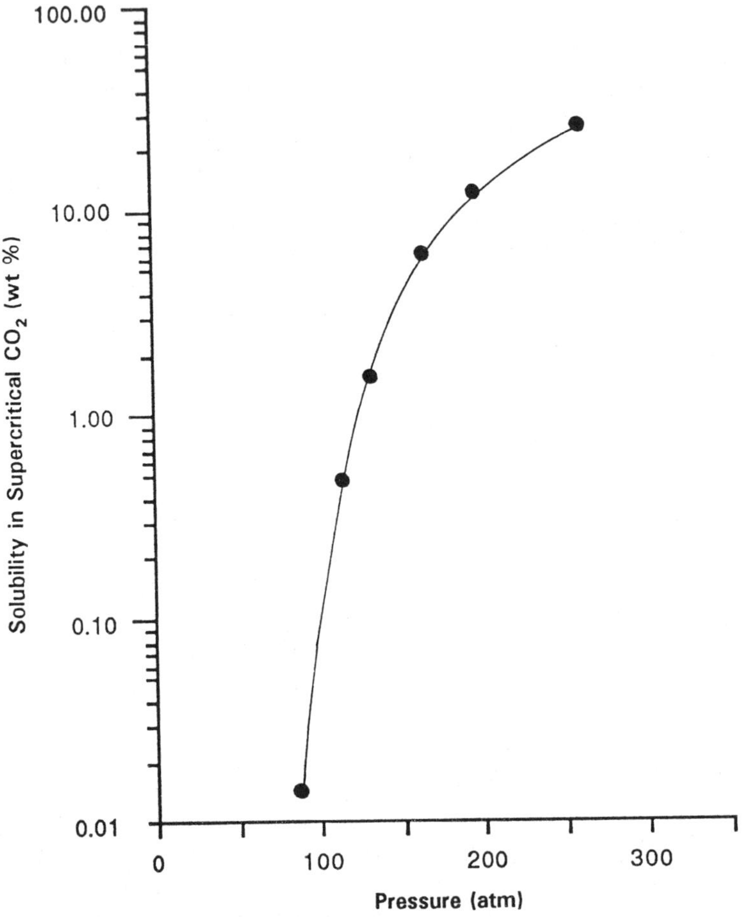

Figure 5.3. Solubility as a function of pressure for alachlor in supercritical CO_2 at 55°C.

Solubility theory. There have been few attempts to date to apply theory for the optimization of extraction conditions for analytical SFE or for the prediction of analyte solubilities in supercritical fluids. This is not surprising considering the relative immaturity of analytical SFE. Hence, the analyst usually selects extraction conditions on an empirical basis, particularly when little is known about the sample matrix prior to extraction. In addition, the rapid demand for results in many applied analytical chemistry situations leaves limited time to apply complex theoretical calculations to the problem at hand.

A plethora of phase equilibrium studies and attempts at modeling supercritical fluid mixtures exists in the literature.[23] These methods, unfortunately, require a considerable number of physicochemical parameters for their application to even the simplest binary solute-solvent systems. Such theories can be difficult to apply to many situations in analytical SFE due to the lack of physicochemical data on the structurally complex solutes that are frequently encountered in applied analysis problems. A review of the complex phase behavior of supercritical mixtures is given by Schneider,[24] while Johnston *et al.*[25] have produced an excellent review of theoretical attempts to model the behavior of supercritical fluid mixtures.

Despite these reservations, several attempts have been made to provide theoretical guidelines for performing analytical SFE. For example, Bartle *et al.*[26] have utilized the Peng-Robinson equation of state for predicting the solubilities of model coal tar components in CO_2. Solubilities were computed by

$$\ln s = \ln\left(\frac{P_v}{P}\right) - \ln \phi - \ln \overline{V}_l + \frac{P\overline{V}_s}{RT} \qquad (5.1)$$

where s is the solute solubility in supercritial CO_2, P_v is the vapor pressure of the solute, P is the system pressure, ϕ is the fugacity of the solute in the supercritical fluid phase, T is the system temperature, $\overline{V}_s$ is the molar volume of the solute, and $\overline{V}_l$ is the molar volume of the supercritical fluid. The Peng-Robinson equation of state was then used to calculate the fugacity of the solute in the supercritical fluid phase as

$$\ln \phi = \left(\frac{b_1}{b_2}\right)(Z - 1) - \ln\left(\frac{Z - Pb_2}{RT}\right) - \left(\frac{a_{22}}{2\sqrt{2b_2}\ RT}\right)\left(\frac{2a_{12}}{a_{22} - \frac{b_1}{b_2}}\right) \ln\left[\frac{Z + \frac{(1 + \sqrt{2}b_2 P}{RT}}{Z - \frac{(1 - \sqrt{2})b_2 P}{RT}}\right] \qquad (5.2)$$

where Z is the fluid compressibility and a and b are constants dependent on the system pressure, the critical temperature, the vapor pressure of the solute or the fluid, and an adjustable interaction parameter. The solubilities obtained from this modestly complex equation of state were predicted to be reliable between 10 to 20% over the chosen experimental range of temperature and pressure.

Another approach that has seen wide use in predicting conditions for solubilizing solutes in supercritical fluids is the solubility parameter theory.[27] Historically, Giddings[3] has applied this concept to quantitatively describe the solvent power of dense gases according to

$$\delta_1 = 1.25\, P_c^{1/2} \left(\frac{\rho_{r,g}}{\rho_{r,l}} \right) \tag{5.3}$$

where δ_1 is the solubility parameter of the compressed gas, P_c is the fluid critical pressure, $\rho_{r,g}$ is the reduced density of the fluid, and $\rho_{r,l}$ is the reduced density of the extracting fluid in the quasi-liquid state. This concept has been further expanded by King[10] to calculate the conditions under which maximum solute solubility is realized in the extracting fluid phase. In this case, the solute-solvent interaction parameter, ψ, can be approximated by

$$\psi \cong \psi_H = \frac{\overline{V}_l (\delta_1 - \delta_2)^2}{RT} \tag{5.4}$$

where ψ_H is the enthalpic interaction parameter, δ_2 is the solubility parameter of the solute, and $\overline{V}_l$ is the molar volume of the supercritical fluid. The respective solubility parameters and $\overline{V}_l$ are functions of temperature and pressure, thereby, making ψ a function of these thermodynamic variables. Plots of ψ *vs* pressure show a minimum at a pressure corresponding to the maximum solubility of the solute in the supercritical fluid. The above approach, which combines the regular solution concept with Flory-Huggins theory, can also be used to predict the pressures required for solute miscibility with the supercritical fluid phase.[10]

Mass transfer in supercritical fluids. As noted previously, recorded solute diffusion coefficients in supercritical fluids have values between those attained in gaseous or liquid solvents. Although such data as a function of pressure are relatively scarce, solute diffusion coefficients tend to exhibit similar trends as recorded self-diffusivities of the extracting fluid. The variance of the self-

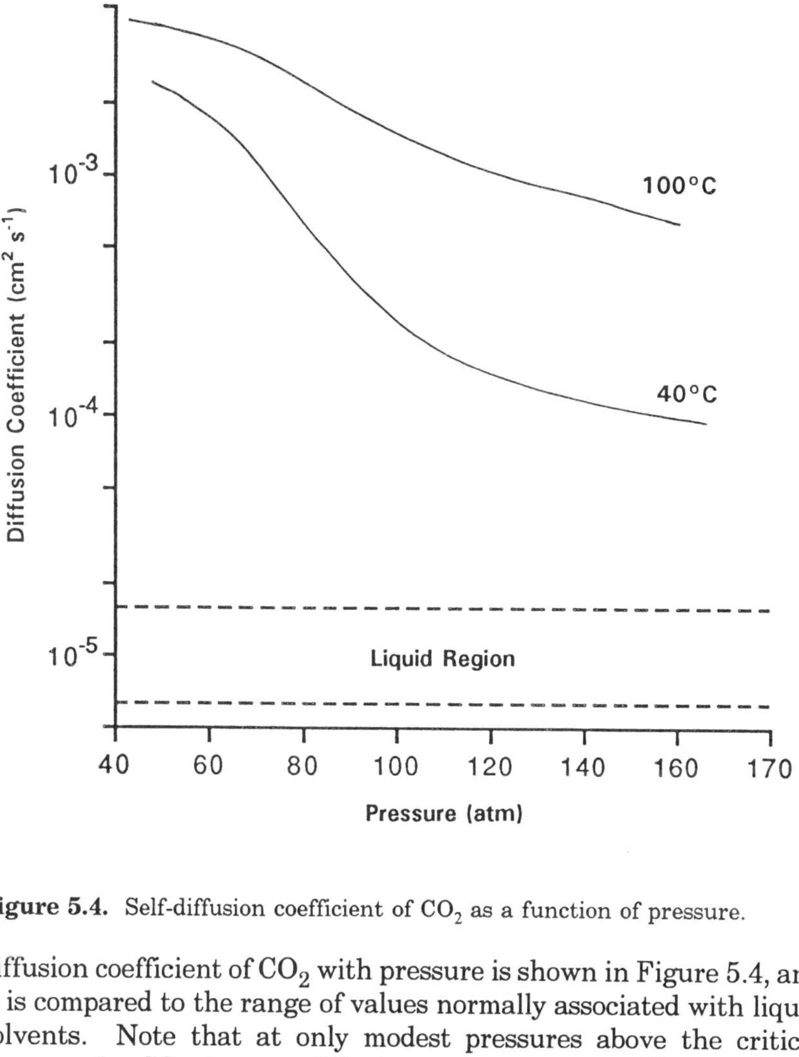

Figure 5.4. Self-diffusion coefficient of CO_2 as a function of pressure.

diffusion coefficient of CO_2 with pressure is shown in Figure 5.4, and it is compared to the range of values normally associated with liquid solvents. Note that at only modest pressures above the critical pressure for CO_2, the self-diffusivity for supercritical CO_2 is almost two orders of magnitude greater than the diffusivities recorded for the liquid state.

As the temperature of the supercritical CO_2 phase is lowered, the diffusivities begin to approach those associated with the liquid state; however, even at 40°C, the self-diffusivity of CO_2 is one order of magnitude greater than values for liquid CO_2. This trend suggests that solute interphase mass transfer rates in supercritical CO_2 will be significantly higher than those recorded in a liquefied gas. This observation has been verified by Krichevskii *et al.*[28] for the rate of naphthalene dissolution into both liquid and supercritical CO_2, and

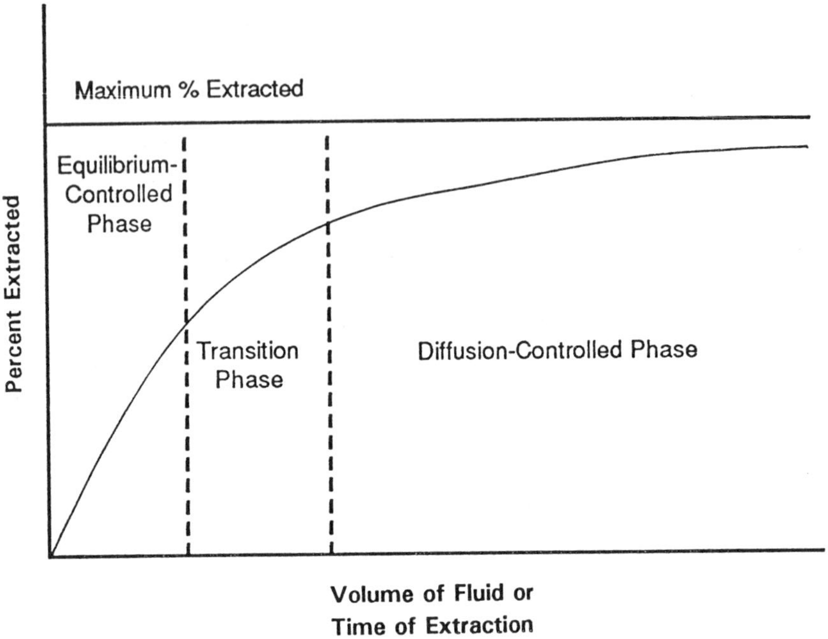

Figure 5.5. Generalized extraction curve of percent solute extracted as a function of volume of extraction fluid or time of extraction.

it accounts for the long extraction times frequently associated with Soxhlet extractions using liquid CO_2.[29]

It is interesting to examine several specific cases of SFE that exhibit rate limiting phenomena. For example, the extraction of solutes from a solid matrix enclosed in a tubular vessel using a supercritical fluid closely parallels the kinetics observed in liquid extractions. As shown in Figure 5.5, the initial portion of the extraction curve is linear, indicating that quasi-equilibrium conditions are governing the partition of the solute into the mobile dense fluid phase. After a finite time, the yield curve starts to become convex with respect to the time axis as the extraction experiences a transition from equilibrium to diffusion controlled kinetics. In the final stage of the extraction, the kinetics are dominated by diffusive mechanisms which may be quite complex, depending on the morphology of the substrate being extracted. For the case of the solid substrate, factors such as the degree of comminution or swelling of the substrate can have a profound effect on the yield curve.

Mathematical models have been formulated to account for the observed kinetics associated with specific supercritical fluid extractions of various substrates. For example, Bulley *et al.*[30] have shown that the SFE extraction of triglycerides from seeds can be described by differential equations giving the material balances in both the fluid and solid phases. The modeling of mass transfer effects for the SFE of organic solutes from water and the extraction of caffeine from coffee have been treated by Brunner.[31] Theoretical modeling for the SFE of a diverse number of substrates has been reported, including mushroom plugs,[32] mackerel powder,[33] and porous rods.[34]

Swanson *et al.*[35] have applied mass transfer theory for the case of extraction under turbulent flow conditions in a packed extractor cell. In this case, the concentration gradient is not known, so a modified mass transfer coefficient (K'_A) was utilized, defined as

$$K'_A = K_c (c_{A,i} - c_{A,f}) \tag{5.5}$$

where K_c is the mass transfer coefficient, $c_{A,i}$ is the concentration of solute A at the interface, and $c_{A,f}$ is the concentration in the adjacent bulk fluid. The factors which impact on the mass transfer coefficient, such as the fluid density or viscosity, or the geometric characteristics of the extraction cell, can be expressed as dimensionless numbers defined as

$$Sh = \frac{K_c\, d}{D_{AB}} \tag{5.6}$$

$$Re = \frac{\rho F_m}{\eta\, d} \tag{5.7}$$

$$Sc = \frac{\eta}{D_{AB}\, \rho} \tag{5.8}$$

where D_{AB} is the diffusion coefficient, ρ is the fluid density, F_m is the fluid velocity, η is the fluid viscosity, and d is the diameter of the extraction vessel. The Sherwood number can be shown to be a function of both the Reynolds and Schmidt numbers, *i.e.*, Sh = f(Re, Sc). Unfortunately, the empirical correlations that have been formulated between Sh, Re, and Sc for packed bed behavior in the

Table 5.2. Comparison of Rates of Extraction from Cells Having Two Different Diameters[a]

Diameter (mm)	Extraction (peak area counts x 10^{-3})	
	C_{18}	C_{20}
6	478	140
4	610	202

[a]CO_2, 80 atm, 70°C, 1-min extraction, equal volumes of alumina, and identical mass flow rates.

presence of dilute gases or liquids are not applicable to supercritical fluids because of large bouyant effects.[36]

The above definitions for dimensionless numbers can be used to optimize extraction conditions. For example, the Reynolds number indicates turbulence, and as the extraction cell diameter is decreased, the mass transfer within the cell should increase. The effect of the extraction cell diameter on the rate of extraction of *n*-alkanes, spiked on an alumina sorbent, is shown in Table 5.2. These samples were extracted at 80 atm and 70°C for 1 min, and subsequently cryotrapped at -70°C and analyzed by gas chromatography. The resultant peak area counts recorded upon elution of *n*-octadecane and *n*-eicosane from the extraction cell, for equal alumina volumes and CO_2 mass flow rates, clearly show the advantages of the smaller cell diameter for enhancing extraction.

Extraction rates can also be increased when turbulence in the fluid flow pattern is increased. For example, if diffusers are placed in the ends of a 1.0-cm diameter extraction cell, the recorded extraction rate (total area counts) for the above two hydrocarbons exceeds that recorded for the 0.6-cm diameter cell in Table 5.2 under equivalent experimental conditions. Alternative methods for increasing turbulence in packed beds have been reported, including pressure pulsations[37] and ultrasonic bombardment.[38]

5.3 Off-line SFE

Off-line *vs* On-line SFE. The analyst who hopes to exploit the attractive characteristics of supercritical fluids for rapid and quantitative extraction and recovery of target analytes from bulk matrices has essentially two different approaches to choose from. SFE can be performed purely for sample preparation by collecting the extracted analytes for subsequent analysis by a variety of techniques including (but not limited to) chromatographic, spectroscopic, and gravimetric methods. This "off-line" approach is contrasted to "on-line" (or "coupled") SFE techniques in which the SFE step replaces the normal sample injection process into the chromatograph; *i.e.*, with on-line SFE, extracted analytes are transferred to and collected in a chromatographic injection loop, a thermal or sorbent trap prior to the chromatographic column, or in the stationary phase at the head of the chromatographic column itself.

Off-line SFE is inherently simpler than on-line SFE since the analyst needs only to consider the extraction and analyte collection steps. In contrast, on-line SFE requires the SFE parameters, the analyte trapping conditions, and the chromatographic separation all to be understood before the analysis can be successfully completed. A sample extracted off-line can be analyzed by any appropriate technique, and is available for multiple analyses. A sample extracted on-line, however, is dedicated to the coupled chromatographic system. Once the on-line SFE analysis is completed, the extract is no longer available for evaluation using different techniques or chromatographic parameters.

Since off-line SFE is generally simpler to perform, does not require previous understanding of necessary chromatographic conditions, and allows the extract to be analyzed by any appropriate technique, off-line SFE should be the first choice of any analyst who desires to develop SFE as a sample preparation method. The principal advantages of on-line SFE, *i.e.*, the ability to quantitatively transfer all of the extracted analytes to the chromatographic system (which results in maximum sensitivity), and the elimination of any sample handling between extraction and chromatographic separation, can be more easily exploited after the analyst has used off-line SFE to become familiar with the extraction conditions and techniques required for a particular analyte/matrix combination.

While both off-line and on-line SFE are likely to be widely used in the future, practical operational aspects should also be considered when choosing between these two general approaches to SFE.

On-line SFE requires that during the extraction (typically 10 min to 1 h), the chromatograph is not being used for what it does best, *i.e.*, performing separations. Similarly, the SFE system lies idle (except in the special case of on-line SFE/SFC, where the pumping system can be utilized for both functions) during the chromatographic separation. Automation of on-line SFE, while much discussed, is likely to be in the developmental stages for several years. Hence, when high sample throughput is desired, the choice of off-line SFE can greatly increase productivity.

The size of samples that can be extracted using SFE varies greatly depending on the goals and limitations of the extraction and analyses to be performed; however, analytical-scale SFE has generally been applied to samples ranging from 1 mg to hundreds of grams.[12,39] The size of sample used depends on a variety of factors including the size of sample needed to ensure homogeneity (*e.g.*, mg for air particulates to hundreds of grams for meat samples), the size of sample needed to achieve the desired sensitivity (which in turn depends on the percentage of the extracted analytes that can be transferred to the chromatographic system), and the difficulty in collecting sufficient quantities of the samples.

Off-line SFE accommodates large sample sizes better than on-line SFE, simply because the methods used to trap the analytes using on-line techniques require lower flows of the supercritical fluid. In contrast, on-line SFE can yield similar sensitivities as off-line SFE with much smaller samples since the potential exists with on-line techniques to transfer all of the extracted analytes quantitatively to the chromatographic column.[39] For example, on-line SFE/GC using an on-column injector for a 1-mg sample yields the same sensitivity (in terms of analyte concentration in the bulk sample) as off-line SFE of a 1-g sample when the analytes are collected in 1 mL of solvent followed by analysis using 1-μL on-column injection into the GC.

Techniques for off-line SFE. Analytical-scale SFE is typically performed using syringe pumps which are similar (or identical) to those used for SFC, although less expensive alternatives are available since SFE is normally performed at constant pressure and sophisticated pressure/density ramp controllers are not generally required. Gas compressors are also useful, particularly for larger-scale extractions where the desired flow and volume of the supercritical fluid are higher than can be conveniently provided by a syringe pump.[12] The temperature of the extraction is normally controlled by placing the extraction cell in a chromatographic oven or in a simple thermostatted tube heater.

Off-line SFE has been performed using two different modes: dynamic (in which the supercritical fluid is continuously flowing through the cell) and static (in which the cell is pressurized with supercritical fluid, and the extraction is allowed to proceed without any outflow of the supercritical fluid until extraction is finished). Both static and dynamic SFE have been used to achieve quantitative results, but dynamic extraction might be expected to yield more rapid recoveries by continuously providing pure extraction fluid to the sample. Dynamic extractions can also be performed without any valves between the extraction cell and the collection medium, since the extraction cell outlet restrictor can be placed directly in (for example) the solvent used to collect the extracted analytes. The elimination of the valve between the extraction and collection media is attractive, particularly for the extraction of trace compounds, since the chances for analyte loss or contamination are reduced. Static extractions, however, have the advantage that known modifier concentrations can be prepared by simply adding a measured volume of the liquid modifier to the cell prior to extraction.[40] Static extraction is also advantageous when large samples (*e.g.*, larger than 10 g) must be extracted, since supercritical fluid consumption and required pumping capacity can be reduced. However, the limited volume of extracting fluid may lead to incomplete extraction.

Although not as important as for SFC, the method used to depressurize the supercritical fluid as it exits the extraction cell must be carefully selected to allow the extracted analytes to be quantitatively transferred into the collection medium. Extracts from static and dynamic extractions can be depressurized through a micrometering valve,[12] or through flow restrictors made from fused-silica or metal tubing, such as restrictors used in SFC.[41] More sophisticated variable valve flow controllers have also been proposed.[42]

The method used to collect the extracted analytes is, to a large extent, dependent on the extraction fluid flow and the depressurization method. Proposed methods include depressurization of the SFE extract into a small volume of liquid solvent,[41] thermal trapping in cooled vessels,[43] and collection of the analytes onto an accumulator resin, followed by secondary extraction with liquid solvents or supercritical fluids.[42,44] The success of the trapping method depends on the total flow and volume of the supercritical fluid and its expanded gas. For example, a dynamic CO_2 extraction using a 10-cm x 30-μm i.d. restrictor will result in a supercritical fluid flow rate of *ca.* 1.0 mL min^{-1} (measured at the pump) which, when depressurized, results in a gas flow rate of *ca.* 500 mL min^{-1}.

These high gas flow rates have been found to be responsible for aerosol formation, which can result in more than 90% loss of analytes when thermal trapping is used.[43,45] However, analyte loss does not appear to be a problem when small (a few mL) volumes of liquid solvents are used to collect the extracted analytes.[41,46,47] The evaporation of the collection solvent is also much less than expected because the depressurization of the supercritical fluid cools the solvent during the extraction. Collection of the extract in a few mL of an appropriate liquid solvent is attractive also because the sample is immediately ready for analysis by a variety of chromatographic and spectroscopic techniques. Depressurization of the supercritical fluid with a micro-metering valve into a receiving vessel has been successful, particularly when large quantities of bulk matrix are extracted, as in the case of SFE of fats from meats.[12]

Applications of off-line SFE. Analytical scale off-line SFE has been applied to a wide variety of matrix/analyte combinations ranging from the extraction of bulk fats from meat products to the extraction of trace dioxins from environmental solids. While off-line SFE has been applied to both qualitative and quantitative extractions, the attainment of quantitative methodology is obviously the more important and difficult goal. Table 5.3 lists several analyte/matrix combinations for which quantitative (*i.e.*, greater than 90%) extraction and recovery of the target analytes has been achieved using off-line SFE. (The list of analyte/matrix combinations in Table 5.3 is meant to be instructive, rather than exhaustive.) Note that off-line SFE extraction times are generally less than 1 hour, while the liquid solvent extraction methods typically utilized for the same analyte/matrix combinations generally require several hours or even days to perform. In general, SFE required the use of either no liquid solvent, or only a few milliliters for sample collection (often representing a considerable advantage in solvent purchase and disposal costs). For samples with trace analytes, the concentration steps which are normally required following liquid solvent extraction were eliminated or reduced to only a few minutes. Note also that, in spite of the fact that CO_2 has been the most popular supercritical fluid for SFE, many of the studies listed in Table 5.3 have demonstrated that either N_2O or modified CO_2 (*e.g.*, with 2-10% methanol) yields greater recovery of moderately to highly polar analytes than pure CO_2. N_2O appears to have particular advantages over CO_2 when analytes such as PAHs and dioxins are extracted from sorptive matrices including air particulates and fly ash. Although not shown in Table 5.3, the ability to alter the solvent

Table 5.3. Representative Applications of Off-line SFE for Quantitative Extraction

Extracted analytes	Sample matrix	Supercritical fluid	Extraction time (min)	Refs.
Biological samples				
Fat, pesticides	Meat, sausage	CO_2	30-60	10,12
Vitamin K_1	Powdered infant formulas	CO_2	15	48
Menadione (vitamin K_3)	Animal feed	CO_2	20	49
Terpenes, aldehydes, esters, alcohols	Lemon peel	CO_2	20-30	50
Environmental and other solids				
Diuron, linuron pesticides	Soil	CO_2/MeOH, CO_2/EtOH	35-50	40
Triazine and other pesticides	Soil, humic acid, vegetables	MeOH	120	51
PAHs, PCBs	Soil, fly ash, sediment, air & diesel particulates	CO_2, N_2O, CO_2/MeOH, N_2O/MeOH, ethane	1-60	41,43,46 47,52,53
Chlorinated dibenzodioxins	Sediment	CO_2/MeOH		54
Anthraquinone	Paper, plywood sawdust	CO_2	20	55
Oil hydrocarbons	Sedimentary rocks	CO_2	15-30	56
Sorbent resins and polymers				
PAHs, O-, S-, N-PACs, alkanes, PCBs	Polyurethane foam	CO_2, CO_2/MeOH	10-30	43,57
PAHs, O-, N-PACs	XAD-2	Isobutane, CO_2/MeOH	30-45	43
Pesticides, PAHs	Tenax	CO_2	15	47,58
Polymer additives	Polyethylene	CO_2	120	59
Ionic surfactants	Reactor sludges, soil	CO_2/MeOH	30	60

Table 5.4. Comparison of Off-line and Conventional Techniques for the Quantitation of PAHs in Urban Dust (SRM 1649)

	Conventional[a]	Off-line SFE
Sample size	1 g	20 mg
Liquid solvent required	450 mL	3 mL
Bench space (for sample preparation)	5 m	1 m
Supply cost per extraction	\$10	\$0.5
Extraction time	48 h	1 h
Extract concentration time	3 h	0-10 min
Shortest possible total analysis time (one sample)	2 days	2 h

[a]Estimates are from S.A. Wise, NIST, and are adapted from from Ref. 62.

strength of a supercritical fluid by simply changing the extraction pressure (and/or temperature) has also been exploited to achieve class-selective extractions from complex matrices ranging from the sequential extraction of alkanes and PAHs from diesel exhaust particulates, to the selective extraction of pesticides from meat fats.[10,47,61]

The practical advantages of SFE are further demonstrated in Tables 5.4 and 5.5 by a comparison of conventional methods with off-line SFE for the extraction and recovery of PAHs from urban air particulates (National Institute of Standards and Technology, NIST, Standard Reference Material 1649). Note that SFE yielded excellent agreement with the PAH concentrations certified by NIST, and that SFE reduced the extraction time and the total time required for the analysis by two orders of magnitude.

Off-line SFE has been demonstrated by several investigators to yield quantitative extraction and recovery of a broad variety of analytes from an equally broad range of sample matrices. Studies of the quantitative results reported in the literature indicate that, as a rule-of-thumb, trace organics that are amenable to analysis by gas

Table 5.5. Quantitation of PAHs from Urban Air Particulates (NIST SRM 1649)

PAH	PAH concentration ($\mu g\ g^{-1}$)			
	Certified value[a]	Off-line SFE[b]	Split SFE/GC[c]	On-column SFE/GC[d]
Fluoranthene	7.1±0.5	8.0±0.6	7.2±0.9	7.3±1.0
Benz[*a*]anthracene	2.6±0.3	2.9±0.5	2.6±0.8	2.6±0.8
Benzo[*a*]pyrene	2.9±0.5	3.2±0.3	2.7±0.4	2.8±0.5
Benzo[*ghi*]perylene	4.5±1.1	4.4±0.3	3.9±1.0	3.6±0.9
Indeno[1,2,3-*cd*]pyrene	3.3±0.5	3.1±0.2	3.4±0.6	3.0±0.5

[a]Value certified by the NIST based on 48-h Soxhlet extraction of a 1-g sample.
[b]Based on triplicate 60-min extractions of 20-mg samples at 350 atm with N_2O/5% MeOH. Values were adapted from Ref. 41.
[c]Based on 30-min extractions of 50-mg samples at 375 atm and 50°C with CO_2. Values were adapted from Ref. 63.
[d]Based on four replicate analyses of 2-mg samples using SFE/GC/MS with supercritical N_2O. Each extraction was performed at 350 atm and 45°C. Results are adapted from Ref. 39.

chromatography can be quantitatively extracted with neat supercritical CO_2 or N_2O, although N_2O appears to have advantages with sorptive matrices. As the polarity and molecular mass of the analytes increase, more polar supercritical fluids are required to obtain quantitative extraction, and the use of polarity modifiers becomes attractive. The simplicity and low cost of performing off-line SFE, as well as the ability to analyze resultant extracts by any appropriate method indicate that off-line SFE should be the choice of the analyst for initial methods development. After extraction conditions are developed and understood using off-line techniques, the analyst can better consider the relative advantages of off-line and on-line SFE for routine analyses.

5.4 On-line (Coupled) SFE/GC

Introduction. On-line SFE/GC has several potential advantages over conventional extraction/GC analysis methods for the qualitative and quantitative determination of minor and trace analytes in complex matrices. Quantitative extractions can often be achieved in a few minutes, and the ability to directly transfer the extracted analytes to the capillary GC reduces the potential for analyte loss and degradation, as well as eliminates sample handling steps between the extraction and chromatographic analysis. SFE/GC is particularly attractive when only limited quantities of samples are available and maximum sensitivity is desired, since, with appropriate coupling techniques, every extracted analyte molecule can be transferred into the GC column for separation and detection. With coupled SFE/GC, an entire analysis, including extraction, concentration, and gas chromatographic separation, can be completed in less than one hour. Since most SFE/GC developments have focused on open tubular column GC, the following discussions all refer to SFE coupled with open tubular GC columns.

Techniques for SFE/GC. The essential steps for performing on-line SFE/GC include supercritical fluid extraction, depressurization and venting of the extraction fluid with the associated collection and focusing of the extracted analytes, transferring the analytes to the GC column, and GC separation. Analogous to off-line SFE, the depressurization step is normally accomplished using a length of 5 to 30 μm i.d. fused silica tubing, or in some cases, crimped stainless steel tubing. Not surprisingly, the supercritical fluids used for SFE/GC are gases at ambient conditions, with CO_2 and, to a lesser extent, N_2O being the most popular. Since the supercritical fluids are converted to the gas phase for collection of the extracted analytes before GC separations can be performed, the relationship between the volume of the supercritical fluid needed for quantitative extraction, and the volume of the gas flow that results upon depressurization must be carefully considered for proper design of the SFE/GC system. For example, an extraction that uses 1 mL min^{-1} of supercritical CO_2 requires the analytes to be collected from a flowing gas stream of approximately 500 mL min^{-1}. Optimizing the relationship between the flow rate of the supercritical fluid needed to obtain quantitative extraction in a reasonable time, and the ability to quantitatively collect and focus the extracted analytes for subsequent GC analysis, is the fundamental problem to be solved by any proposed SFE/GC technique.

The SFE/GC techniques that have been reported vary primarily in the method used to collect and focus the analytes upon depressurization, and the associated procedure used to transfer the analytes to the GC column.[39,44,45,62-71] These approaches can be roughly divided into two categories: methods in which the analytes are collected in an accumulating device external to the GC, and methods utilizing the GC for collection. External collections of the extracted analytes have been accomplished by depressurizing the SFE effluent into a cold trap placed prior to the GC.[44,45,65] After extraction, the analytes are transferred to the GC column by heating the trap and sweeping with a carrier gas. SFE/GC methods that have utilized the GC as the trapping system include depressurizing the SFE effluent directly into a retention gap at the head of the GC column,[67,68] depositing the extracted analytes directly inside the GC column[39,47,62] with the use of a conventional on-column injector, and depressurizing the SFE effluent directly inside a split/splitless injection port.[63,69-71] The addition of a switching valve between the extraction cell and the GC has also been used to allow specific fractions of the extract to be selectively transferred to the GC.[63,67] The results from each of these approaches are discussed below.

Comparison of SFE/GC Methods. The use of cold traps placed external to the GC for the collection of the extracted analytes is attractive since the supercritical fluid does not have to be vented through (or into) the gas chromatographic system during the extraction, and any potential effect of the supercritical fluid (and modifiers) on the chromatographic column and detector can be ignored.[44,45,65] Unfortunately, this approach has been limited by the efficiencies of the cold trapping systems that have been employed, and quantitative collection of the extracted analytes has not been demonstrated. Since GC analysis is aimed at relatively volatile components, the use of cold traps may have limited utility compared to their potential applications for SFE/SFC and SFE/HPLC. However, cold traps containing a suitable sorbent (*e.g.*, Tenax) that can be thermally desorbed to recover the analytes may be suitable for increasing the range of analytes which can be quantitatively collected from the depressurized SFE effluent.

Methods utilizing the chromatographic column for collecting and focusing the extracted analytes have been more successful in achieving reproducible and quantitative SFE/GC results. These techniques are divided into "on-column" SFE/GC (where the SFE effluent is depressurized directly inside the open tubular GC column and all of the extracted analytes are deposited directly in the column stationary phase[39,47,62]) and "split" SFE/GC (where the depressurization occurs inside a conventional heated split GC injector[63,69-71]). Both techniques utilize the stationary phase and cooling to collect and focus the extracted analytes during the SFE step. These approaches allow the quantitative trapping of analytes

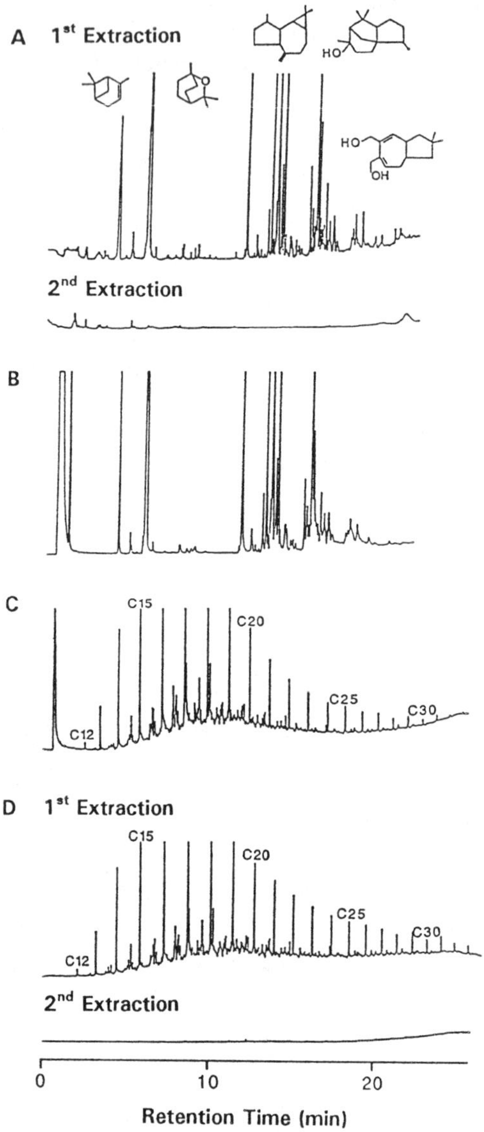

Figure 5.6. Comparisons of gas chromatograms generated using coupled SFE/GC/FID and conventional liquid solvent injections: Eucalyptus leaves analyzed by (A) on-column SFE/GC (10-min extraction with 380 atm CO_2 at 45°C) and (B) 1-μL on-column injection of a methylene chloride extract (reprinted with permission from Ref. 62); fuel-contaminated sediment analyzed by (C) split SFE/GC with a 10-min CO_2 extraction at 400 atm and 50°C, and (D) 1-μL split injection of a methylene chloride extract (reprinted with permission from Refs. 62 and 71).

less volatile than *n*-heptane and the monoterpenes,[62] and *n*-octane and benzene,[63] for on-column and split SFE/GC, respectively. As demonstrated in Figure 5.6, both techniques also yield chromatographic peak shapes that compare favorably with those obtained using conventional on-column and split injections of liquid solvent extracts.

A comparison of on-column SFE/GC with split SFE/GC is similar in many respects to a comparison between conventional GC on-column and split injections. With on-column SFE/GC, all of the extracted analytes are transferred into the GC column for analysis. This results in maximum sensitivity (*i.e.*, ppb detection limits with mg samples using conventional GC detectors); however, it also results in all of the extractable species being transferred into the column stationary phase. As with on-column injection of liquid solvents, this characteristic makes on-column SFE/GC unsuitable for samples that have high concentrations of extractable (but nonchromatographable) matrix components. In contrast, in split SFE/GC, only a fraction of the extracted analytes are transferred to the GC column (depending on the split ratio), but many of the nonvolatile matrix components are trapped on the injection port liner rather than being introduced directly into the stationary phase. With on-column SFE/GC, the analytes are never exposed to high injection port temperatures as they are in split SFE/GC. However, some matrices which contain a few percent or higher concentrations of water cannot be analyzed using on-column SFE/GC, because the water occasionally freezes in the restrictor outlet (or in the GC column) and prevents further transfer of the analytes from the extraction cell into the GC column.[62] Split SFE/GC appears to work well with wet samples since the additional heat supplied to the extraction cell restrictor by the injection port prevents freezing and subsequent plugging. Figure 5.7 demonstrates the split SFE/GC analysis of a wet polyurethane foam (PUF) sorbent plug which had been used for the solid phase extraction of organic pollutants from a coal gasification wastewater.[71] Split SFE/GC also has potential for extractions using polar modifiers as demonstrated in Figure 5.8 by the SFE/GC analysis of a solid hydrocarbon waste using formic acid-modified CO_2.[63]

As discussed above, the coupling technique used for SFE/GC determines the maximum flow rate of the supercritical fluid through the extraction cell which, in turn, affects the size of sample that can be extracted in a reasonable time. With on-column SFE/GC, the maximum flow rate that yields efficient focusing of the analytes is *ca.* 350 μL min^{-1} (for a 320-μm i.d. GC column). Assuming that a typical sample has a void fraction of 1/3 (*i.e.*, a minimum of three cell

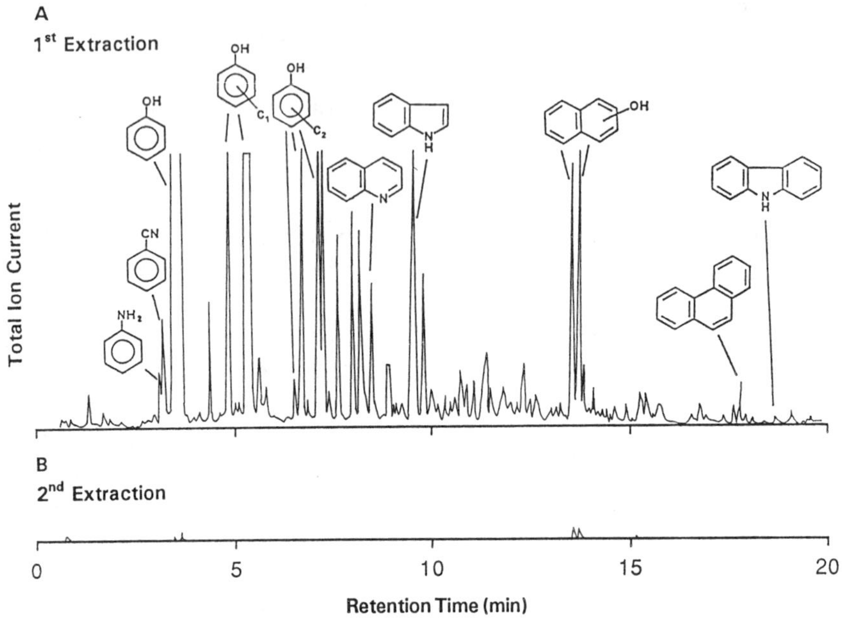

Figure 5.7. SFE/GC/MS analysis of a wet polyurethane foam (PUF) sorbent plug. Conditions: the plug had been soaked in 15 mL of a coal gasification wastewater, gently squeezed (but not dried), and extracted with CO_2 at 400 atm and 45°C. The GC oven was held at 5°C during the 10-min extraction, then the oven was rapidly heated to 70°C, followed by an 8°C min^{-1} temperature ramp to 320°C. Separations were performed with a 20-m x 250-μm i.d. open tubular column with poly(5% phenyl)methylsiloxane stationary phase (0.25-μm film thickness). The absence of significant peaks from the second SFE/GC analysis (B) indicates that the first 10-min extraction was sufficient to quantitatively recover the analytes from the PUF plug (reprinted with permission from Ref. 71).

volumes of supercritical fluid are needed for quantitative extraction) and that the extraction should be completed in 30 min, then on-column SFE/GC should be capable of extracting a sample no larger than 10 mL.[62] The fluid flow rates that can be accommodated using split SFE/GC should be substantially higher than for on-column SFE/GC.

Applications of on-line SFE/GC. The ability of both split and on-column SFE/GC to perform quantitative analysis of trace analytes from complex matrices has been demonstrated for a number of samples ranging from environmental solids to flavor and fragrance

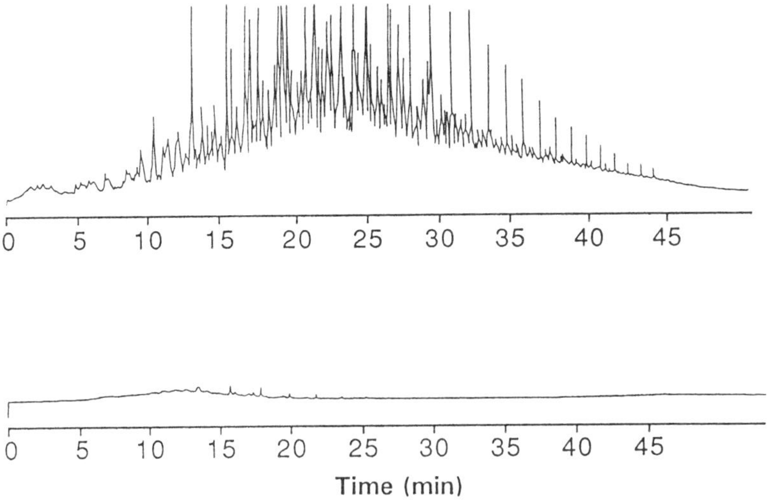

Figure 5.8. Split SFE/GC/FID analysis of a solid hydrocarbon waste (A) using CO_2/0.3% formic acid at 375 atm. The lack of significant peaks in the second extract (B) indicates that the first 30-min extraction was quantitative (reprinted with permission from Ref. 63).

compounds.[39,47,62,63,69-73] Table 5.5 shows a comparison of conventional liquid solvent Soxhlet extraction, off-line SFE, split SFE/GC, and on-column SFE/GC for the determination of PAHs on urban air particulates (NIST Standard Reference Material 1649). Note that even though the SFE/GC analyses were performed independently by two different investigators using different supercritical fluids, and using different SFE/GC approaches, the results of both SFE/GC techniques gave excellent agreement with each other and with the certified values (based on a 48-h Soxhlet extraction). Note also that the conventional method required 1 g of the particulate matter and 3 days for sample extraction and concentration (see Table 5.4). In contrast, split SFE/GC and on-column SFE/GC required only 50 mg and 2 mg of sample, respectively, and only 30 min and 15 min, respectively, for the sample extraction and concentration steps to be completed.

A comparison of the reproducibility obtained using SFE/GC with that obtained using conventional GC injections of liquid solvents is shown in Table 5.6. Split SFE/GC analysis of hydrocarbon standards extracted from alumina is compared to a split autosampler injection of the same quantity of standards in a liquid solvent.[63] On-column SFE/GC of replicate 1-mg samples of basil spice is compared to

Table 5.6. Comparison of GC/FID Peak Area Reproducibility Obtained Using Split SFE/GC, On-column SFE/GC, and Conventional GC Injections

Split SFE/GC *vs* split GC (alkane/aromatic test mixture)			On-column SFE/GC *vs* on-column GC (basil spice)		
	% RSD[a]			% RSD[b]	
Compound	SFE/GC	GC	Compound	SFE/GC	GC
Benzene	3.5	1.2	1,8-Cineole	17	11
Toluene	2.1	1.8	$C_{10}H_{18}O$ isomer	11	10
Octane	4.4	2.2	Estragole	9	11
Decane	1.4	2.5	Eugenol	17	10
Naphthalene	3.8	2.8	$C_{15}H_{24}$ isomer	9	9
Dodecane	3.1	3.3	$C_{15}H_{24}$ isomer	6	10
Pentadecane	1.8	5.2	$C_{15}H_{24}$ isomer	12	10
Phenanthrene	1.5	4.3	$C_{15}H_{24}$ isomer	6	10
Pyrene	2.2	3.4	Beta-selinene	9	15
Chrysene	5.4	3.4	$C_{15}H_{24}$ isomer	8	10

[a]Relative standard deviations (% RSD) for raw peak areas are given for 5 replicate SFE/GC analyses of the standards spiked on alumina, and 5 replicate split autosampler injections (adapted from Ref. 63).

[b]Relative standard deviations (% RSD) for raw peak areas are given for 4 replicate SFE/GC analyses of 4 different 1-mg samples of basil, and for 4 replicate 1-μL on-column injections of a methylene chloride extract of a 1-g sample of basil (adapted from Ref. 62).

conventional 1-μL on-column injections of a methylene chloride extract of basil.[62] As is shown in Table 5.6, both split and on-column SFE/GC yield reproducibilities similar to conventional liquid solvent injections. It should also be noted that the deviations shown for the on-column SFE/GC analysis of basil include the effects of sample inhomogeneity in the 1-mg samples as well as any irreproducibility in the SFE/GC technique itself.

Although on-line SFE/GC is in the early stages of development, a fairly broad range of applications has been reported, as demonstrated in Table 5.7. While many of the SFE/GC analyses are qualitative, an increasing number of quantitative results are being reported, including the analysis of certified standard reference materials,[39,63,71] the recovery of spikes, and the demonstration of

Table 5.7. Representative Applications of On-line SFE/GC

Sample matrix	Analytes	Refs.
Environmental solids		
Air and exhaust particulates	PAHs, alkanes, O-PACs, S-PACs	39,62,63,67,73
Wood smoke particulates	Guaiacol and syringol derivatives	73
Cigarette smoke particulates	Phenols, nicotine, N-heterocycles	39
Soil and sediments	PCBs, PAHs, fuel hydrocarbons	39,45,71
Sorbent resins		
Tenax:		
Vehicle exhaust, spikes	Alkanes, benzenes, PAHs, PCBs	47,66
Polyurethane foam (PUF):		
Air pollutants (exhaust, roofing tar, wood smoke)	Alkanes, benzenes, PAHs, phenols, S-PACs, guaiacol and syringol derivatives	57,62
Cigarette smoke volatiles	Phenols, N-heterocycles	57
Coal gasifier wastewater	Phenols, N-heterocycles, hydrocarbons	71
Charcoal	Hydrocarbons	44
Silica, alumina (columns)	Fuel hydrocarbons	63,69,70
Food products		
Chewing gum	Terpenes, menthol flavors	72
Citrus fruit peels	Terpenes, aldehydes, alcohols	62,72
Tobacco	Nicotine, menthol	73
Spices (*e.g.*, basil, mint, sage, rosemary, cinnamon, chili powder)	Terpenes, aromatic and aliphatic alcohols, esters sesquiterpenes, oxy-sesquiterpenes, cinnamaldehyde, coumarin	62,72,73
Miscellaneous		
Coal	Sesquiterpenes, biological markers	73
Treated utility pole	PAHs, O-, S-, N-PACs	73
Conifer tree and bush needles	Terpenes, camphor, bornyl acetate	72
Aromatic cedar wood	Cedrene, cedrol	72
Eucalyptus tree leaves	Terpenes, oxy-terpenes, sesquiterpenes	52

quantitative extraction by performing multiple extractions of a single sample.[39,57,62,63,69-73]

The development and application of on-line SFE/GC techniques is in its infancy, and any definitive statement of the potential abilities of various approaches to SFE/GC coupling would be premature. However, of the SFE/GC techniques presently available, approaches utilizing the GC column for collection and focusing of the extracted analytes appear to have significant advantages over techniques utilizing cold-trapping units external to the GC, particularly for more volatile analytes and if quantitative results are desired. On-column SFE/GC has advantages when maximum sensitivity from small samples is desired, while split SFE/GC appears to be more useful for larger and more concentrated samples. Split SFE/GC also has greater potential for the analysis of samples containing large amounts of water and samples requiring the use of polar modifiers to obtain quantitative extraction.[63,71]

The potential for performing class-selective SFE/GC by sequentially extracting samples using different pressures (or fluids) has also been demonstrated by the fractionation of PAHs and alkanes.[67,73] The ability to perform SFE/GC (or more properly, SFC/GC) analysis of different classes of analytes that are selectively eluted from sorbents has been demonstrated by the class-selective determinations of saturate, unsaturate, and aromatic fractions from fuel samples.[63] Future development of such class-selective SFE/GC techniques could be extremely useful since the potential exists to perform quantitative analysis of selected compound classes including extraction, class-fractionation, and chromatographic analysis in a total time of less than 1 h per sample.

5.5 On-line (Coupled) SFE/SFC

A number of unique devices have been developed to perform on-line SFE where the technique is coupled to SFC. In the on-line mode, SFE requires a source of compressed extraction fluid, a high pressure extraction cartridge or cell, and frequently a focusing device for concentrating the extracted solutes prior to SFC.

SFE/SFC configurations. On-line SFE/SFC equipment can range from relatively simple devices to complex arrangements involving multiple pumps, valves, and ovens. A simple approach to conducting SFE on-line in a static extraction mode is shown in

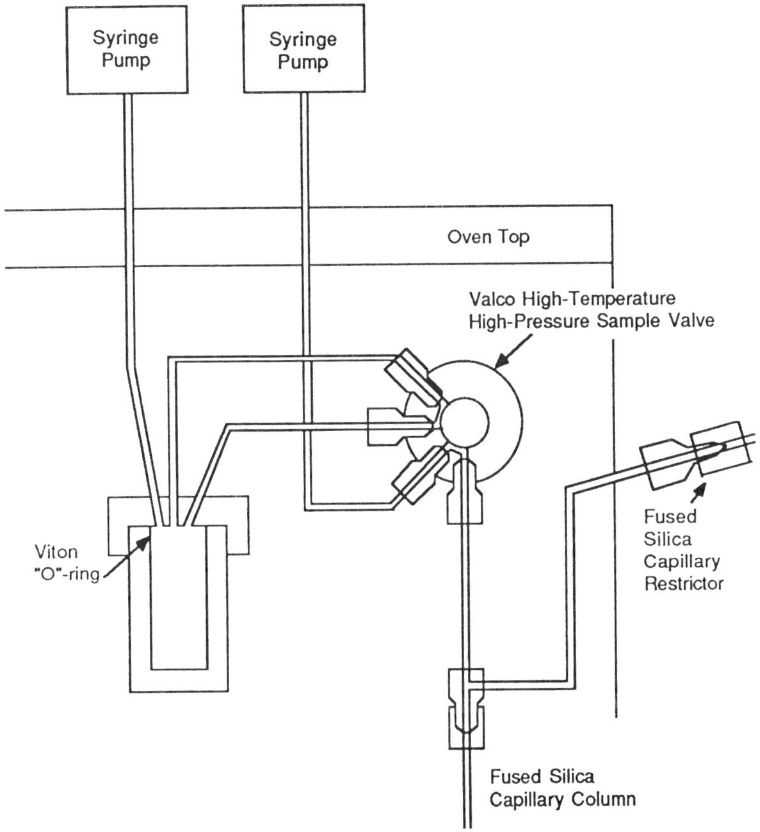

Figure 5.9. Schematic diagram of a supercritical fluid extractor/injector (reprinted with permission from Ref. 74).

Figure 5.9 where two pumps are employed: one for extracting the sample from the cell and the other for performing SFC. The sample cell in this case was a 1-cm^3 stainless steel vial sealed by a Viton "o-ring" closure. A high temperature sample valve was arranged downstream from the sample cell to permit the solute-laden extraction fluid to be introduced into the fused silica open tubular column equipped with a splitter assembly. In this arrangement, there was no focusing of the extract before SFC.

This early attempt[74] at coupling SFE with SFC deserves some further comment. The choice of material for closure seals can be critical for SFE and for the lifetime or replacement of system components. For example, Viton o-rings will eventually blister when exposed repeatedly to high pressure CO_2,[75] and alternative materials such as nitrile-containing polymers and Teflon will result in longer service lifetimes. Similar considerations apply in the case of switching valves where seal components can be adversely affected by high temperatures. Interestingly, the extraction device shown in Figure 5.9 appears to be adequately temperature controlled with respect to the transport lines between the extractor module and the chromatography column. Improper temperature control can result in extract precipitation in the lines and valving, and can lead to failure in the SFE/SFC module. A possible problem in this regard exists when the split restrictor vent line is positioned outside the oven. Precipitation of the vented extract can occur in this narrow capillary restrictor, leading to erratic analytical results.

A dynamic extraction scheme for coupling SFE with SFC is illustrated in Figure 5.10 where the concept of an accumulator trap is utilized to focus the sample prior to SFC. Once again, a high pressure pump is utilized for the SFE, followed by the extraction cell, which is coupled to a back pressure restrictor inserted into a low dead-volume accumulator tee. During extraction, the tee is vented to atmosphere and the extract is concentrated (or accumulated) in the tee. After extraction, the valve is switched and supercritical fluid is introduced through the side arm of the tee to transfer the sample to the chromatographic column. The transfer tubing between the accumulator trap and chromatographic column is critical to obtaining good chromatographic performance. With a section of uncoated fused silica tubing as the transfer line, the analytes are unretained during transfer until they reach the analytical column where they are concentrated in the column stationary phase due to the difference in migration rates of the extracted solutes in the two sections of tubing (phase ratio focusing). Focusing between the transfer tubing and column can be improved by operating the transfer region at a temperature that is sufficient to lower the density of the supercritical fluid. This results in a reduction of the extract solubility in the supercritical fluid mobile phase and greater concentration in the stationary phase.

More sophisticated approaches for conducting SFE/SFC are possible. On-off and multiport switching valves can be arranged to permit simultaneous extraction or venting of the extraction cell and accumulator, while supplying supercritical fluid to the chromatographic column. Extracts can be focused prior to chromatography by a cryocooling tee inserted in the flow path prior to the column. Subambient temperatures can be attained at the tee

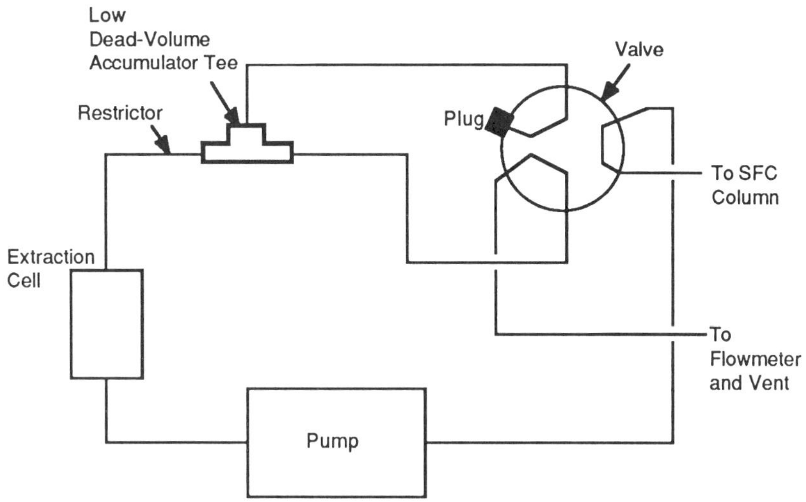

Figure 5.10. Schematic diagram of a combined supercritical fluid extractor/chromatograph utilizing a retention gap to concentrate the extract before chromatography.

by Joule-Thompson expansion of CO_2 through a micrometering valve.[46] Defocusing of the collected solutes is achieved by cessation of the external CO_2 flow and by increasing the temperature of the chromatographic oven in which the tee is located. In this arrangement, purging of the extractor cartridge can be accomplished during SFC analysis by opening a needle purge valve when a multiport valve is in the "column" position.

For on-line SFE/SFC static extractions, the sample to be extracted is held under pressure in a cell for a finite length of time before introduction into the supercritical fluid chromatograph. Aside from the ability to precisely measure fundamental parameters, such as partition coefficients of solutes in supercritical fluid media, the static method seems to have few advantages to offer the analyst. Extraction in this case is very dependent on the mass transfer kinetics of the solutes from the sample matrix. Wheeler and McNally[40] have shown that the static mode can even take days to effect complete extraction.

Recycle extraction schemes offer some interesting possibilities for SFE/SFC. One reported mode of operation[76] involved the pressurization of a recycle loop by pumping extraction fluid into the recycle loop until the desired extraction pressure had been attained for a given temperature. The recycle loop of the system was then sealed and a second pump in-line with the loop was used to circulate a fixed volume of fluid (determined by the dimensions of the recycle loop) through the sample in an extraction cell. In this mode, intermittent sampling *via* an injection valve can be utilized to characterize the extract in the circulating fluid phase. Accurate quantitative data from such a device can only be achieved by judiciously adjusting the size of the recirculation loop relative to the anticipated level of extractables from the sample matrix for a given extraction pressure and temperature. Continuous recirculation of the extraction fluid through a sample, with subsequent precipitation at a lower temperature and pressure, is another possible recycle mode.

A novel approach to SFE/SFC is the use of a thermal modulator between the extraction cell and the chromatographic column.[77] The modulator is usually made from a short segment of a fused-silica open tubular column, which is coated on the outside with electrically conductive paint so that it can be heated rapidly by a pulse of electric current. As the supercritical fluid stream from the extraction cell flows through the modulator, solutes partition into the stationary phase in the usual manner according to the partition coefficient. When an electric pulse is applied, the solutes either desorb as a concentration pulse or absorb as a vacancy pulse because of the reduction in density, depending on the normal operating temperature of the modulator tube. The amplitude of the signal generated is proportional to the concentration of analytes flowing through the system. By performing a series of pulses and applying multiplex or correlation chromatography, selective and sensitive analysis of target analytes can be achieved. The advantage of this approach is that trapping or accumulation of the sample before the chromatographic step is not necessary, thus eliminating the most troublesome step in on-line SFE/SFC. The major limitation to this approach is that it is currently restricted to nonprogrammed SFC. While only preliminary results have been obtained using this technique, future application appears to be quite promising.

Extraction cell configuration. Extraction cells for SFE/SFC have tended to range between 150 μL and 50 mL in total volume in order to avoid overloading the open tubular and microbore columns that are commonly used in SFC. A particularly attractive source for these devices has been the guard cartridges which are utilized in high performance LC; however, care should be taken to ensure that these cells are properly rated for high pressures. Figure 5.11 illustrates common configurations which are commercially available. The use of these tubular extractors requires that attention be paid to such factors as the total amount of extraction fluid that passes

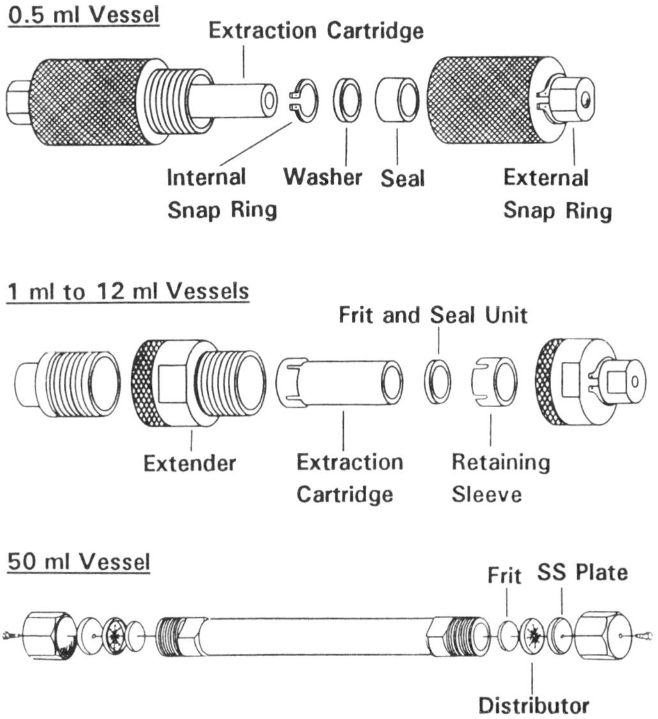

Figure 5.11. Typical extraction cells used in analytical supercritical fluid extraction.

through the cell in order to assure reproducible extraction. One can calculate the linear velocity (u) through the extraction cell using

$$u = \frac{F_a \rho_c}{A_c \rho_a} \tag{5.9}$$

where F_a is the flow rate under ambient conditions, ρ_c is the fluid density within the cell, ρ_a is the fluid density under ambient conditions, and A_c is the cross-sectional area of the extractor cell. Hence, with knowledge of the cell geometry and u, one can estimate the total number of column volumes of fluid/unit of time that pass through the extractor. The degree of sample contact with the extraction fluid is critical, and this can be increased by keeping the volume of the extractor cell small with respect to the sample volume. In fact, coupled extractor cells as small as 85 μL have been reported in the literature.[78]

Solute trapping after extraction. Trapping techniques for combined SFE/SFC are still evolving, but recent research has identified factors which are critical for optimizing the technique. For example, the effect of trapping temperature on the recovery of analytes can be understood by invoking the mass transfer coefficient[35] defined in Equation 5.5, and substituting the partial pressure for the concentration of the solute. This indicates that the difference in partial pressures is the driving force for focusing the solute. Hence, if the solute's partial pressure at the fluid/solid interface is less than its partial pressure in the supercritical fluid, then deposition will take place. Since solute vapor pressure is a function of temperature, lower temperatures should lead to increased rates of solute deposition. This effect is nicely illustrated in Figure 5.12 where the higher trap temperature leads to discrimination in the recovery of the more volatile *n*-alkane analytes.[45] Cryofocusing temperature also has a pronounced effect on peak widths in subsequent chromatography.[79] For instance, too high cryofocusing temperature will lead to peak broadening of the more volatile analytes in the extracted mixture. Conversely, too low cryotrapping temperature may lead to plugging in the restrictor leading from the extraction cell.

An alternative focusing method to thermal-based trapping is the use of a sorbent for concentrating the extracted solutes. In this case, the high surface area of the sorbent combined with the adsorptivity of the solute at the fluid/solid interface creates a favorable situation for concentrating the extract after SFE. When utilizing this method, it is important to assure that breakthrough of the analytes from the sorbent trap does not occur. King[80] has shown that the solute breakthrough volume varies with pressure in the presence of a supercritical fluid; therefore, one must select a low enough pressure or temperature to limit the mobility of the solute on the sorbent trap. Desorption from the trapping medium can be accomplished by increasing the temperature and/or by using the supercritical fluid to desorb the collected sample. The surface areas and porosities of trapping sorbents can be altered by exposure to supercritical fluids,[81] and this may ultimately limit their service lifetimes.

The use of adsorbents after SFE can also aid in the fractionation and cleanup of complex extracts that frequently result from using SFE for complex natural product mixtures. The combination of adsorbent columns with SFE is an emerging area of research, and limited work has been reported to date. Nevertheless, Campbell and Lee[61] have shown that complex petroleum- and coal-derived mixtures can be fractionated using siliceous adsorbents prior to capillary gas

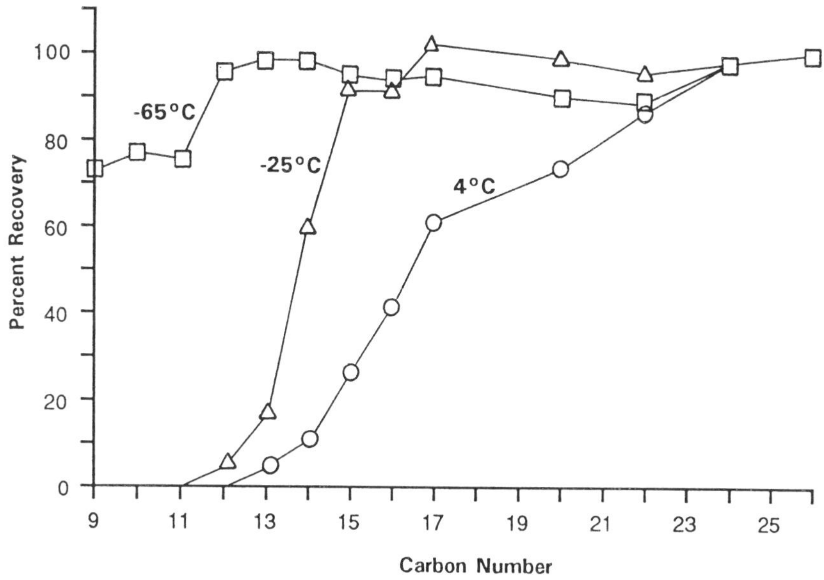

Figure 5.12. Recoveries of extracted *n*-alkanes as a function of carbon number at different cryotrapping temperatures (reprinted with permission from Ref. 45).

chromatographic characterization. Specific adsorbents, such as ion exchange resins, offer the possibility to isolate specific compounds free from interfering coextractives. This principle has been recently demonstrated by Schaeffer *et al.*[82] by successfully capturing an alkaloid on a cation exchange resin from a supercritical CO_2/ethanol fluid phase. The use of different types of supercritical fluids can also influence the desorption of analytes from adsorptive matrices.[80] Several investigators have shown that supercritical nitrous oxide is more effective than supercritical CO_2 in removing solutes from adsorbents and active solids.[39,41,83]

Liquid sample extraction. The SFE devices described above are largely designed for the extraction of solid samples. SFE of liquid samples presents a different challenge to the analyst due to the small, but finite, solubility of water in most supercritical fluid media, and the possibility of phase inversion effects as the extraction fluid is compressed. A simple method has been reported by Hedrick and Taylor[84] for the quantitative extraction of a phosphonate herbicide from aqueous media using a tubular extraction cell. In this method, the fluid is bubbled through a partially-filled cell, collected in the headspace, and routed out of the cell through an exit tube which projects above the liquid miniscus in the extraction cell. The addition of salts and acids to the aqueous medium promoted extraction of the solute into the supercritical fluid phase, thereby reducing the extraction time from several hours to minutes. More elegant devices based on phase switching principles have been developed for affecting SFE of aqueous samples.[85,86] One of the reported methods[85] is based on the use of a phase separator designed for liquid-liquid extraction to partition the dissolved analyte from water to a compressed CO_2 phase, while the other[86] involves adsorption of the target analyte on a hydrophobic sorbent, subsequent removal of the interfering hydrophilic components by a liquid solvent, drying, and desorption into supercritical CO_2.

5.6 Optimization of Analytical SFE

Analytical SFE is currently an evolving technique in which many experimental parameters and problems are still under investigation. Optimization of SFE is, perhaps, the key area of most concern to the analyst since it impacts on the accuracy and precision that can be obtained with this technique. Some of the experimental factors which impact SFE have already been discussed, including the pressure and temperature of the extracting fluid. Concern must also be paid to the time period of extraction, fluid flow rate through the extraction cell, sample matrix effects, homogeneity of the sample, and system contamination.

The optimum extraction time is dependent on the experimental pressure and temperature as well as on the flow rate of the fluid through the extraction cell. For unknown samples, the extraction time can best be found by experimentally conducting successive extractions to determine the completeness of extraction. An example of this procedure is depicted in Figure 5.13 in which two successive extractions were conducted on an aquifer sediment sample with supercritical CO_2 under the listed conditions. In this particular case, a 3-min extraction was not sufficient to complete the extraction of the supercritical CO_2-soluble components.

The use of a nondestructive detector in tandem with SFE/SFC can also aid the analyst in determining the extent of extraction. For example, diverting the extraction fluid through a UV detector prior to chromatographic analysis would allow the extent of extraction to be monitored, provided the compounds of interest absorb light. A more elaborate technique involving the use of radioactive tracers[40,55]

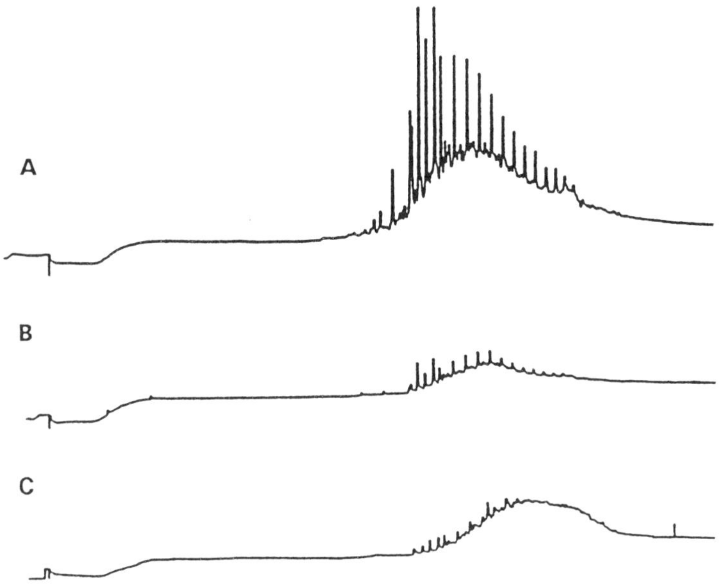

Figure 5.13. On-line SFE/SFC analysis of an aquifer sediment sample for two subsequent extractions: (A) 1st extraction, (B) 2nd extraction, and (C) blank. SFE conditions: 3 min; CO_2; 200 atm; 60°C; cryofocused extract. SFC conditions: 15-m x 50-μm i.d. open tubular column; poly(50% *n*-octyl)methylsiloxane stationary phase; CO_2; 100°C; pressure program from 50 to 300 atm at 3 atm min^{-1}; FID.

has also been employed for measuring the completeness of extraction from soil matrices. Knowledge of a principal component's solubility in the supercritical fluid can also assist the analyst in choosing the proper extraction time, since the percent recovery is a function of the distribution coefficient of the compound, the phase ratio in the extraction cell, and the number of extractions (column volumes) utilized.[87] Optimum extraction will, of course, be realized when the chosen cell volume is evenly filled with sample and there is sufficient contact between the extraction fluid and the sample.

The sample matrix can also exert a profound effect on SFE. Such factors as particle size and shape, surface area and porosity, moisture content of the sample, and level of extractable matter will impact on the analytical results. In general, a reduction in particle size will improve the extraction efficiency[46,88] and lead to a reduction in extraction time; however, deleterious effects may also occur, such as large pressure drops in the extraction cell and plugging of the extractor cell due to sample compaction. Samples having large surface areas and correspondingly high porosities may inhibit

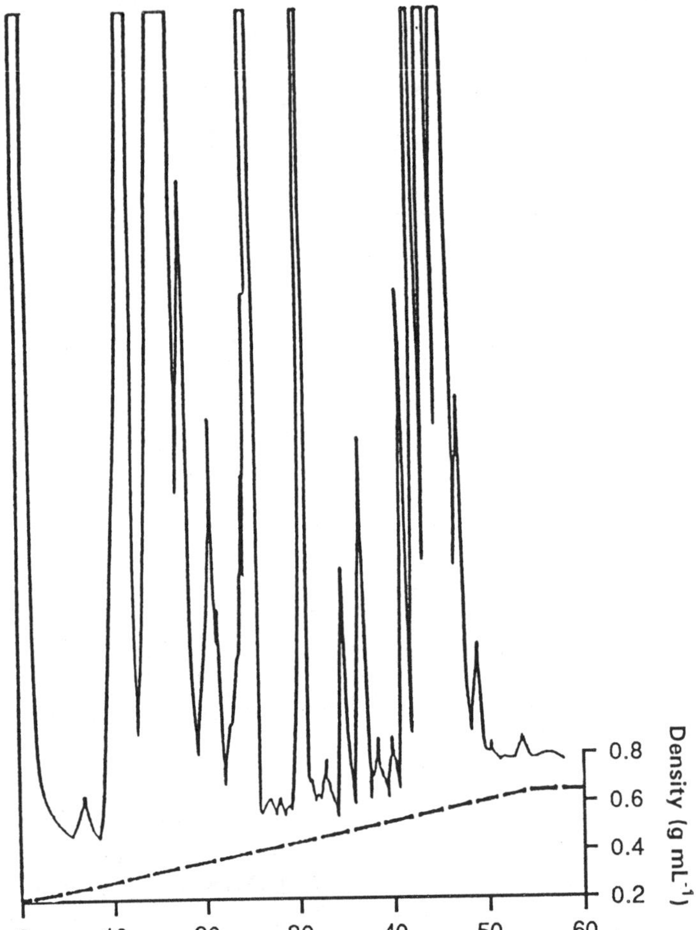

Figure 5.14. Supercritical fluid chromatogram of a supercritical fluid extract (dissolved in chloroform) of spiked poultry liver (15 ppm Bendiocarb). Conditions: 12-m x 100-μm i.d. open tubular column; poly(5% phenyl)methylsiiloxane stationary phase; CO_2; 107°C; program from 0.25 g mL^{-1} to 0.65 g mL^{-1} at 0.008 g mL^{-1} min^{-1}; FID (reprinted with permission from Ref. 95).

extraction, particularly if the sample has a heterogeneous surface with active sites. In this case, higher extraction temperatures have been shown to improve the recovery of analytes, such as pesticides from soil matrices.[40,89] Alternatively, the addition of organic modifiers to the extraction fluid may also aid in the recovery of the analytes from specific soils.[90,91] Moist samples can also present a particularly difficult problem when using supercritical CO_2 as the extracting fluid. For the extraction of nonpolar to moderately polar solutes in matrices containing high moisture levels, it is often necessary to remove most of the water by some technique such as oven drying,[12] microwave drying, or freeze drying.[92] Alternatively, the sample can be mixed with an adsorbent such as anhydrous sodium sulfate which can adsorb water during the extraction. In certain cases, the presence of moisture in a sample matrix can result in better recoveries of trace levels of polar analytes due to enhanced solubility in the supercritical fluid phase[93] and reduction of matrix-analyte interactions.[94]

Sample matrices containing a high level of extractables can present additional problems for the analyst. The presence of a high level of coextractives from the SFE can render subsequent chromatographic analysis useless without additional class-fractionation steps. This is particularly true for lipid-rich sample matrices where such components show a propensity for partitioning into such fluids as supercritical CO_2. An example of this phenomenon is illustrated in Figure 5.14 where a poultry liver sample has been extracted with supercritical CO_2 and subsequently chromatographed using SFC. In such samples, the ability to discern a particular target analyte among the coextracted peaks is often difficult. This problem can be partially overcome by using chromatographic detectors having a specificity for the analyte of interest. For example, King[96] has shown that chlorinated pesticides in the presence of coextracted lipid matter can be selectively monitored using an electron capture detector, while Schneiderman *et al.*[48] has employed an LC electrochemical detector for detecting substances such as vitamins in infant formulations.

Specific problems exist in the field of analytical SFE which must be addressed in the future if the technique is to expand into other areas of application. Handling of the extraction cells should be done with caution since even fingerprints on the surface of the cell can produce a "chromatographic fingerprint" when the extract is analyzed by SFC. Figure 5.15 shows the results obtained when an extraction cell was handled before SFE/SFC. The peaks appearing in the profile are probably lipids from the analyst's hands. This problem is particularly enhanced when using extraction cells that have internal parts, such as ferrules or fritted disks, that must be reassembled before the commencement of extraction. Caution should also be exerted when using detergents for leak testing or cleaning of extraction cells, since it has been demonstrated that surfactants can be readily solubilized in supercritical fluid media.[97]

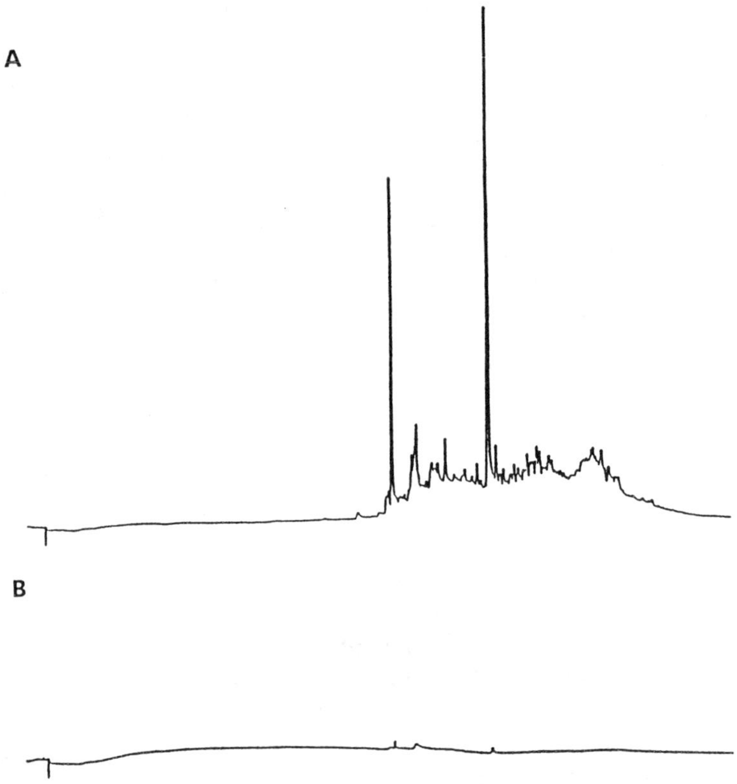

Figure 5.15. SFE/SFC profile (A) after and (B) before exposure of an extraction cell to human fingerprints. SFE conditions: 3 min; CO_2; 200 atm; 60°C; cryofocused extract. SFC conditions: 15-m x 50-μm i.d. open tubular column; poly(50% *n*-octyl)methylsiloxane stationary phase; CO_2; 100°C; pressure program from 50 atm to 300 atm at 3 atm min^{-1} after a 1-min initial isobaric period; FID.

Another vexing problem in analytical SFE is the purity level of the extraction fluid. High purity gases currently available have been purified for use in gas chromatographic trace analysis or SFC. Early analytical SFE studies revealed the need for even higher levels of gas purity because of the concentrating effect of the SFE step. This problem has been attacked by several methods: the use of adsorbent traps prior to SFE, choosing an alternative source for the extraction fluid, and by modifying the detection scheme during the chromatographic analysis. The use of activated alumina has found particular favor among many analysts for rendering CO_2 free from

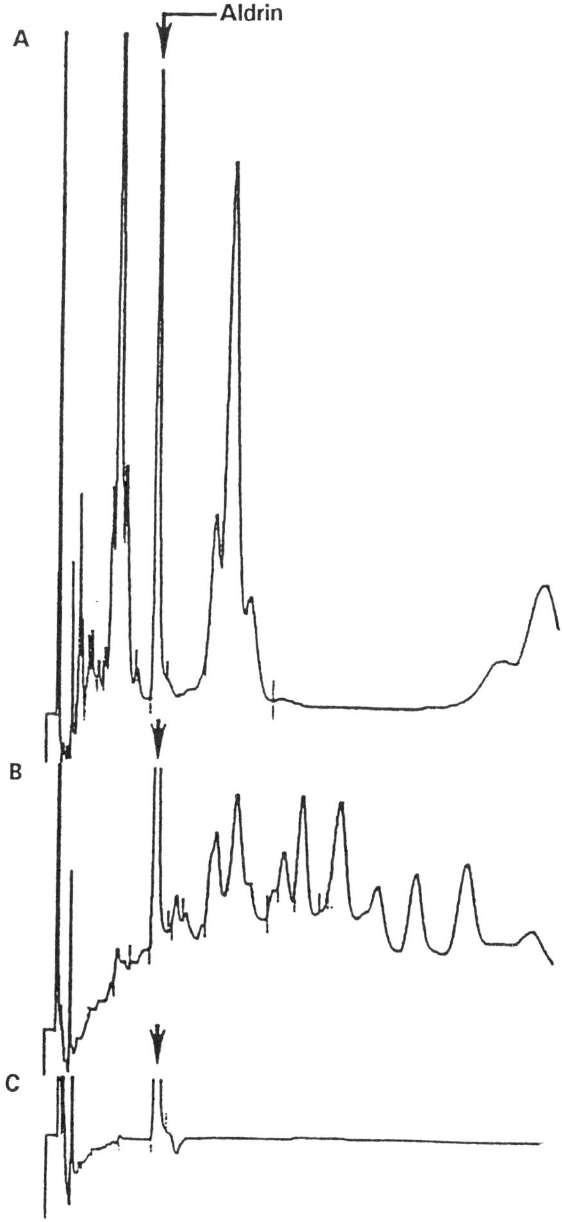

Figure 5.16. Gas chromatographic detection (ECD) of impurities in CO_2 used for SFE showing (A) Teflon contamination, (B) welding-grade CO_2, and (C) SFC-grade CO_2. SFE conditions: ~45 min; ~220 mL CO_2 (liquid); 160 atm; 40°C.

FID-detectable impurities. Food grade CO_2, derived from fermentation sources, also appears to be a good choice for minimizing contamination in the extraction fluid.[68] The use of highly sensitive element-specific detectors, such as the electron capture detector (ECD), in trace analysis creates an additional criterion in gas purity. As is shown in Figure 5.16, there is a considerable difference in the ECD-detectable impurity levels between a tank of welding grade CO_2 and the highly purified SFC-grade product. The introduction of specific materials such as elastomeric-based seals and o-rings into the flow path of the extraction fluid can result in a significant extraction of ECD-sensitive impurities into the compressed fluid. Chlorofluorohydrocarbons utilized in the cleaning of compressed gas cylinders have also been implicated as a source of impurities in fluids used for SFE.

Currently, analytical scale SFE has certain limitations that must be addressed if the technique is to find widespread acceptance in the analytical community. Certain laboratory environments (*i.e.*, regulatory laboratories) prepare samples in a parallel mode, suggesting that the development of extractor modules which allow the processing of multiple samples should be pursued. Such devices may eventually involve highly automated stream switching techniques to allow improved mating of SFE with the various forms of chromatography. The problem of optimal sample size should also be studied, since current equipment has been miniaturized in order to be consistent with the small sample size requirements of microcolumn chromatographic methods. Unfortunately, such small samples may not be representative of the overall sample and, hence, bias the analytical result. This problem is particularly amplified in trace analysis where the detection limits of many chromatographic assays are seriously challenged. The occurrence of small leaks in a microextraction system can be much more problemmatic than in larger extraction equipment and must be eliminated to obtain consistently reproducible results.

Despite the above reservations, very respectable levels of detection and precision have been achieved using analytical SFE. As tabulated in Tables 5.3 and 5.7, a wide spectrum of sample types are amenable to analytical SFE. Detectable amounts of analytes range from mg to pg quantities in such diverse matrices as urban dust, basil, and rat chow. There is a wide range in the recorded precision levels for the cited examples that is partially dependent on the nature of the analyte, the matrix being extracted, and the concentration level of the analyte. In many cases, the RSDs of SFE are as good as those of the chromatographic method used for the quantitation. The whole field of analytical SFE will undoubtedly be improved in the near future as the technique is modified and practiced by additional analytical chemists.

References

1. E. Stahl and W. Schilz, *Chem. Ing. Tech.* **50**, 535 (1978).
2. K.A. Larson and M.L. King, *Biotech. Prog.* **2**, 73 (1986).
3. R.D. Smith, H.R. Udseth, B.W. Wright, and C.R. Yonker, *Sep. Sci. Tech.* **22**, 1065 (1987).
4. E. Stahl and E. Willing, *Pharm. Ind.* **42**, 1136 (1980).
5. G.M. Schneider, *Fluid Phase Equilib.* **10**, 141 (1983).
6. M.A. McHugh and V.J. Krukonis, *Supercritical Fluid Extraction* (Butterworths, Boston, 1986), Chap. 5.
7. J.C. Giddings, M.N. Myers, L. McLaren, and R.A. Keller, *Science* **162**, 67 (1968).
8. J.W. King, *J. Am. Oil Chem. Soc.* **61**, 689 (1984).
9. M. Gitterman and I. Procaccia, *J. Chem. Phys.* **78**, 2648 (1983).
10. J.W. King, *J. Chromatogr. Sci.* **27**, 355 (1989).
11. L.M. Bowman, Jr., Ph.D. Thesis, University of Utah, 1976.
12. J.W. King, J.H. Johnson, and J.P. Friedrich, *J. Agric. Food Chem.* **37**, 951 (1989).
13. E. Stahl, K.W. Quirin, A. Glatz, D. Gerard, and G. Rau, *Ber. Bunsenges. Phys. Chem.* **88**, 900 (1984).
14. E.H. Chimowitz and K.J. Pennisi, *AIChE J.* **32**, 1665 (1986).
15. E. Stahl, W. Schilz, E. Schutz, and E. Willing, *Angew. Chem. Int. Ed. Engl.* **17**, 731 (1978).
16. J.A. Hyatt, *J. Org. Chem.* **49**, 5097 (1984).
17. D.K. Dandge, J.P. Heller, and K.V. Wilson, *Ind. Eng. Chem. Prod. Res. Dev.* **24**, 162 (1985).
18. J.W. King and J.P. Friedrich, *Institute of Food Technologists Meeting*, Chicago, IL, 1987, Abst. No. 202.
19. W.G. Schultz and J.M. Randall, *Food Tech.* **24**, 94 (1970).
20. J.M. Wong and K.P. Johnston, *Biotech. Prog.* **2**, 29 (1986).
21. G. Brunner and S. Peter, *Sep. Sci. Tech.* **17**, 199 (1982).
22. R.T. Kurnik and R.C. Reid, *Fluid Phase Equilib.* **8**, 93 (1982).
23. S.S.H. Rizvi, A.L. Benado, J.A. Zollweg, and J.A. Daniels, *Food Tech.* **40**(6), 55 (1986).
24. G.M. Schneider, *Pure Appl. Chem.* **55**, 479 (1983).
25. K.P. Johnston, D.G. Peck, and S. Kim, *Ind. Eng. Chem. Res.* **28**, 1115 (1989).
26. K.D. Bartle, A.A. Clifford, and G.F. Shilstone, *J. Supercrit. Fluids* **2**, 30 (1989).

27. J.M. Prausnitz, in *Phase Equilibria and Fluid Properties in the Chemical Industry*, T.S. Storvick and S.I. Sandler, Eds. (American Chemical Society, Washington, DC, 1977), p. 37.
28. I.R. Krichevskii, Y.V. Tsekhanskaya, and I.I. Sokolinskaya, *Dokl. Akad. Nauk USSR* **196**, 1129 (1971).
29. W.G. Jennings, *J. High Resolut. Chromatogr. Chromatogr. Commun.* **2**, 221 (1979).
30. N.R. Bulley, M. Fattori, and A. Meisen, *J. Am. Oil Chem. Soc.* **61**, 1362 (1984).
31. G. Brunner, in *Supercritical Fluid Technology*, J.M.L. Penninger, M. Radosz, M.A. McHugh, and V.J. Krukonis, Eds. (Elsevier, Amsterdam, 1985), p. 245.
32. J.M. Del Valle and J.M. Aguilera, *J. Food Sci.* **54**, 135 (1989).
33. Y. Ikushima, N. Saito, S. Ito, T. Asano, and T. Goto, *Bull. Chem. Soc. Jpn.* **59**, 3709 (1986).
34. G. Knaff and E.U. Schlunder, *Chem. Eng. Process.* **21**, 193 (1987).
35. J.T. Swanson, M.R. Andersen, and B.E. Richter, *1989 Symposium/Workshop on Supercritical Fluid Chromatography*, Snowbird, UT, June 13-15, 1989.
36. P.G. Debenedetti and R.C. Reid, *AIChE J.* **32**, 2034 (1986).
37. U. Ender and S. Peter, *Chem. Ing. Tech.* **61**, 324 (1989).
38. B.W. Wright, J.L. Fulton, A.J. Kopriva, and R.D. Smith, in *Supercritical Fluid Extraction and Chromatography, Techniques and Applications, ACS Symposium Series 366*, B.A. Charpentier and M.R. Sevenants, Eds. (American Chemical Society, Washington, DC, 1988), p. 84.
39. S.B. Hawthorne and D.J. Miller, *J. Chromatogr.* **403**, 63 (1987).
40. J.R. Wheeler and M.E. McNally, *J. Chromatogr. Sci.* **27**, 534 (1989).
41. S.B. Hawthorne and D.J. Miller, *Anal. Chem.* **59**, 1705 (1987).
42. M. Saito, Y. Yamauchi, K. Inomata, and W. Kottkamp, *J. Chromatogr. Sci.* **27**, 79 (1989).
43. B.W. Wright, C.W. Wright, R.W. Gale, and R.D. Smith, *Anal. Chem.* **59**, 38 (1987).
44. S.A. Liebman, E.J. Levy, S. Lurcott, S. O'Neil, J. Guthrie, T. Ryan, and S. Yocklovich, *J. Chromatogr. Sci.* **27**, 118 (1989).
45. M.R. Anderson, J.T. Swanson, N.L. Porter, and B.E. Richter, *J. Chromatogr. Sci.* **27**, 371 (1989).
46. B.W. Wright, C.W. Wright, and J.S. Fruchter, *Energy and Fuels* **3**, 474 (1989).
47. S.B. Hawthorne and D.J. Miller, *J. Chromatogr. Sci.* **24**, 258 (1986).
48. M.A. Schneiderman, A.K. Sharma, K.R.R. Mahanama, and D.C. Locke, *J. Assoc. Off. Anal. Chem.* **71**, 815 (1988).
49. M.A. Schneiderman, A.K. Sharma, and D.C. Locke, *J. Chromatogr. Sci.* **26**, 458 (1988).
50. K. Sugiyama and M. Saito, *J. Chromatogr.* **442**, 121 (1988).
51. P. Capriel, A. Haisch, and S.U. Khan, *J. Agric. Food Chem.* **34**, 70 (1986).
52. M.M. Schantz and S.N. Chesler, *J. Chromatogr.* **363**, 397 (1986).

53. B.O. Brady, C.P.C. Kao, K.M. Dooley, F.C. Knopf, and R.P. Cambrell, *Ind. Eng. Chem. Res.* **26**, 261 (1987).
54. F.I. Onuska and K.A. Terry, *J. High Resolut. Chromatogr.* **12**, 357 (1989).
55. M.A. Schneiderman, A.K. Sharma, and D.C. Locke, *J. Chromatogr.* **409**, 343 (1987).
56. J.C. Monin, D. Barth, M. Perrut, M. Espitalie, and B. Durand, *Adv. Org. Geochem.* **13**, 1079 (1988).
57. S.B. Hawthorne, M.S. Krieger, and D.J. Miller, *Anal. Chem.* **61**, 736 (1989).
58. J.H. Raymer and E.D. Pellizzari, *Anal. Chem.* **59**, 1043 (1987).
59. Y. Hirata and Y. Oikamoto, *J. Microcol. Sep.* **1**, 46 (1989).
60. S.B. Hawthorne and D.J. Miller, University of North Dakota, unpublished results.
61. R.M. Campbell and M.L. Lee, *Anal. Chem.* **58**, 2247 (1986).
62. S.B. Hawthorne, D.J. Miller, and M.S. Krieger, *J. Chromatogr. Sci.* **27**, 347 (1989).
63. J.M. Levy, R.A. Cavalier, T.N. Bosch, A.F. Rynaski, and W.E. Huhak, *J. Chromatogr. Sci.* **27**, 341 (1989).
64. I.L. Davies, M.W. Raynor, J.P. Kithinji, K.D. Bartle, P.T. Williams, and G.E. Andrews, *Anal. Chem.* **60**, 683A (1988).
65. B.W. Wright, J.L. Fulton, A.J. Kopriva, and R.D. Smith, in *Supercritical Fluid Extraction and Chromatography, Techniques and Applications, ACS Symposium Series 366,* B.A. Charpentier and M.R. Sevenants, Eds. (American Chemical Society, Washington, DC, 1988), p. 44.
66. M.W.F. Nielen, J.T. Sanderson, R.W. Frei, and U.A. Th. Brinkman, *J. Chromatogr.* **474**, 38 (1989).
67. B.W. Wright, S.R. Frye, D.G. McMinn, and R.D. Smith, *Anal. Chem.* **59**, 640 (1987).
68. R.D. Smith, H.R. Udseth, and B.W. Wright, in *Supercritical Fluid Technology*, J.M.L. Penninger, M. Radosz, M.A. McHugh, and V.J. Krukonis, Eds. (Elsevier, New York, 1985), p. 191.
69. J.M. Levy, J.P. Guzowski, and W.E. Huhak, *J. High Resolut. Chromatogr. Chromatogr. Commun.* **10**, 337 (1987).
70. J.M. Levy and J.P. Guzowski, *Fres. Z. Anal. Chem.* **330**, 207 (1988).
71. S.B. Hawthorne, D.J. Miller, and J.J. Langenfeld, *J. Chromatogr. Sci.* **28**, 2 (1990).
72. S.B. Hawthorne, M.S. Krieger, and D.J. Miller, *Anal. Chem.* **60**, 472 (1988).
73. S.B. Hawthorne, D.J. Miller, and M.S. Krieger, *Fres. Z. Anal. Chem.* **330**, 211 (1988).
74. W.P. Jackson, K.E. Markides, and M.L. Lee, *J. High Resolut. Chromatogr. Chromatogr. Commun.* **9**, 213 (1986).
75. F.W. Schremp and G.R. Roberson, *Soc. Petrol. Eng. J.* **15**, 227 (1975).
76. W.G. Engelhart and A.G. Gargus, *Am. Lab.* **20**(2), 30 (1988).
77. S. Mitra and N.K. Wilson, *J. Chromatogr. Sci.*, in press.
78. K.R. Jahn and B. Wenclawiak, *Chromatographia* **26**, 345 (1988).
79. Q.L. Xie, K.E. Markides, and M.L. Lee, *J. Chromatogr. Sci.* **27**, 365 (1989).

80. J.W. King, in *Supercritical Fluids - Chemical and Engineering Principles and Applications, ACS Symposium Series 329*, T.G. Squires and M.E. Paulaitis, Eds. (American Chemical Society, Washington, DC, 1987), p. 150.
81. J.W. King, R.L. Eissler, and J.P. Friedrich, in *Supercritical Fluid Extraction and Chromatography: Techniques and Applications, ACS Symposium Series 366*, B.A. Charpentier and M.R. Sevenants, Eds. (American Chemical Society, Washington, DC, 1988), p. 65.
82. S.T. Schaeffer, L.H. Zalkow, and A.S. Teja, *Ind. Eng. Chem. Res.* **28**, 1017 (1989).
83. N. Alexandrou and J. Pawliszyn, *Anal. Chem.* **61**, 1986 (1989).
84. J. Hedrick and L.T. Taylor, *Anal. Chem.* **61**, 1986 (1989).
85. D. Thiebault, J.-P. Chervet, R.W. VanNoort, G.J. DeJong, U.A. Th. Brinkman, and R.W. Frei, *J. Chromatogr.* **477**, 151 (1989).
86. W.M.A. Niessen, P.J.M. Bergers, U.R. Tjaden, and J. Van Der Greef, *J. Chromatogr.* **454**, 243 (1988).
87. D.G. Peters, J.M. Hayes, and G.M. Hieftje, *Chemical Separations and Measurements* (W.B. Saunders, Philadelphia, PA, 1974), p. 490.
88. J.M. Snyder, J.P. Friedrich, and D.D. Christianson, *J. Am. Oil Chem. Soc.* **61**, 1851 (1984).
89. J.W. King, 1989 Pittsburgh Conference on Analytical Chemistry, Atlanta, GA, March 6-10, 1989, Paper No. 397.
90. J.R. Wheeler and M.E. McNally, *J. Chromatogr.* **447**, 53 (1988).
91. V. Janda, G. Steenbeke, and P. Sandra, *J. Chromatogr.* **479**, 200 (1989).
92. D.P. Noliomu and C.F. Simpson, *Anal. Chim. Acta* **213**, 237 (1988).
93. K. Zosel, *Angew. Chem.* **90**, 748 (1978).
94. P. Hubert and O.G. Vitzhum, in *Extraction with Supercritical Gases*, G.M. Schneider, E. Stahl, and G. Wilke, Eds. (Verlag Chemie, Deerfield Beach, FL, 1980), p. 37.
95. J.E. France and K.J. Voorhees, Colorado School of Mines, unpublished results.
96. J.W. King, *196th ACS National Meeting*, Los Angeles, CA, 1988, Agrochemical Division, Abstr. No. 142.
97. T.L. Chester, *J. Chromatogr.* **299**, 424 (1984).

CHAPTER 6

PRACTICE OF SFC

6.1 Introduction

The understanding and execution of the operational details of SFC are essential for providing accurate and reproducible data. This chapter gives a description of the practice of analytical SFC and includes a number of practical hints and problem-solving suggestions.

Sections in this chapter cover the topics of general system setup and testing, column selection, sample preparation, mobile phase selection (including the preparation and use of mixed mobile phases), sample introduction, and programming methods. Many of the procedures and principles discussed in this chapter are relevant to SFE as well as to SFC. This chapter ends with an extensive tabulated trouble shooting section which includes symptoms, causes, and possible solutions.

6.2 System Setup and Testing

Cleaning of the pump and connecting lines. The hardware that comes in contact with the supercritical fluid mobile phase must be carefully cleaned prior to use to ensure a low and interference-free background signal. The strong solvating properties of supercritical fluids and the high sensitivities of SFC detectors are the main reasons for this requirement.

The following procedures have proven to be effective for cleaning of the system components. New instrumentation should normally have been cleaned before being shipped and need only be further cleaned if it has been contaminated. However, whenever any new tubing, valve, or filter is installed, it should be carefully cleaned before installation.

Components should be rinsed with "LC" or "Spectro" grade solvents in the following order: (a) hexane, (b) dichloromethane, (c) ethanol, and (d) water. These solvents represent a wide range of polarity and can remove nonpolar oils and fatty acids as well as more polar impurities. Finally, the components should be carefully rinsed with high purity water and dried.

Small parts should be disassembled and placed in a glass beaker for cleaning. The glass beaker can then be filled with the solvent and placed in an ultrasonic bath for 5-10 min. Metal tubing can be cleaned by connecting one end to an aspirator (or other vacuum system) and inserting the other end into a container with the solvent. After cleaning, the parts should be dried before reassembling. A clean (oil-free) gas can be used to dry the tubing. Care should be taken not to leave fingerprints on the cleaned surfaces during reassembling. In addition, Teflon tape should be avoided in connections, particularly when using the electron capture detector, since some fluorinated compounds can be extracted in the mobile phase.

Filling of the pump. Common fluids for SFC and SFE are either liquids or gases at atmospheric pressure and ambient temperature. However, they are generally liquids in the supply containers, pump, and throughout the connecting lines to the injector. After exiting the pump, the mobile phase is usually above its critical pressure, and it becomes a supercritical fluid as it enters the oven which is set at a temperature above the critical temperature of the fluid.

Several types of pumps can be used in SFC, as was discussed in detail in Section 4.2. For use with microcolumns, syringe pumps are preferred. When the selected mobile phase is a liquid at room temperature and atmospheric pressure, a syringe pump can easily be filled from a liquid reservoir. If the mobile phase chosen is a gas at atmospheric pressure and room temperature, it can be taken from a pressurized stainless steel or aluminum gas cylinder. An eductor tube in the cylinder makes it possible to transfer the mobile phase as a liquid to the pump. High molecular mass impurities in the mobile phase can, however, be minimized if the mobile phase is transferred as a gas from the cylinder and condensed in the pump. The gas cylinder should be connected to the pump *via* clean reducing unions and stainless steel tubing.

Reciprocating piston pumps continuously pump the liquid eluent from the reservoir to the column. This is an advantage for large-bore columns or for large volume supercritical fluid extractions. To avoid air bubbles in the pumping system, liquid mobile phases (including modifiers) should be degassed before use. The mobile phase can be degassed using an ultrasonic bath, or by purging with an inert gas.

The stainless steel tubing between the mobile phase source and the pump should be as short as possible and have "large" internal diameter (*e.g.*, 0.5 mm). Furthermore, no pressure regulator should be used. Lengthy tubing from the gas cylinder to the pump (*e.g.*, ≥10 m) results in inefficient filling of the syringe pump or poor operation of the reciprocating piston pump, because the pressure drop along the tubing is sufficiently high that the mobile phase has low density when it enters the pump. An adsorption trap (*e.g.*, activated neutral alumina and/or molecular sieve) might be used on-line between the cylinder and the pump to extract trace impurities from the gas. It is, however, recommended to use a mobile phase of sufficient purity that additional purification is not necessary. These impurities should be <100 ppb for SFC and <10 ppb for SFE mobile phases that are used in combination with the FID.

Mobile phases of high compressibility require special filling techniques. Syringe pumps without temperature regulated pump cylinders can be filled using gas cylinders with eductor tubes and a helium head pressure of 1,200-1,500 psi.[1,2] When the syringe pump has a cooling jacket around the pump cylinder,[3] it can be filled with compressible fluids from a gas cylinder with an eductor tube, but without helium head pressure. The vapor pressure of a mobile phase like CO_2 in such a gas cylinder is sufficient to force liquid CO_2 from the cylinder and into the pump. For efficient pump filling, the temperature of the pump cylinder should be below 10°C. An additional advantage of a cooled pump cylinder is that the resulting lower compressibility of the fluid will be important for accurately programming steep pressure or density gradients[3] with ramp rates of 100-150 atm min^{-1}.

Reciprocating piston pumps with cooled pump heads are also used, generally with packed columns, for mobile phases which are gaseous at room temperature and atmospheric pressure. Cooling to temperatures at which the mobile phase is a liquid is recommended to achieve low compressibility of the fluid in the pump heads.[4] It should be noted that care must be taken when mixed mobile phases are used with any cooled equipment. The temperature may influence the miscibility of the modifier in the primary fluid, resulting in phase separation and nonuniform mixtures.

Changing of the mobile phase. If the pump to be used is a syringe type, there are several procedures that can be used to change the mobile phase. For all procedures, the pump is first emptied and the new mobile phase is then used to clean the pump cylinder and tubing. After disconnecting the carrier fluid line from the column oven, the pump is completely filled with the mobile phase, the pump is stopped for at least 1 min, and then the mobile phase is vented through the carrier line for at least 2 min. It is recommended that this procedure be repeated at least twice. This procedure will clean the carrier line and the pump cylinder if they are not extensively contaminated.

An alternative procedure is recommended when switching between mobile phases or modifiers of greatly different polarities. The pump cylinder is filled with 50 mL of the new fluid, and purged. The pump is then refilled with approximately 50 mL of new fluid, and with the purge line connected to a restrictor, the mobile phase is then pressurized to increase its solvating power. The pump is held at maximum pressure for at least 1 min before the mobile phase is allowed to slowly vent through the restrictor. Depending on the mobile phase used, this procedure should be repeated several times. The last step in this procedure is to fill the entire pump cylinder with the mobile phase, pressurize the pump for at least 1 min, and finally allow the pump cylinder to bleed to atmosphere through the restrictor. In some cases, rinsing the pump cylinder with an organic solvent early during cleaning is very helpful.

Still another approach is described by the following sequence of steps: (a) slowly bleed off the fluid through a restrictor, (b) evacuate the cylinder, (c) purge with helium gas to remove any residual impurities, and (d) introduce the new fluid. It is important to note that restriction should always be used to control the purging of the pump cylinder in all of the cleaning procedures. Rapid expulsion of gas from the pump can lead to hazardous conditions, such as oxygen displacement from the room (if the room is small), flying projectiles from the expansion jet, and dislodging of the piston seals.

When changing the mobile phase in a reciprocating piston pump, a restrictor (*e.g.*, a 20-cm x 10-μm i.d. fused silica capillary) should be connected to the pump. To clean the pump, the new mobile phase should be purged through it for 5 min at high pressure. This procedure cleans the pump of previously used solvents. When using modifiers with different compositions, or when switching from one mobile phase to another, there is normally no need for more special pump cleaning procedures, because the pump chambers will be cleaned whenever a new mobile phase is pumped through the pump head.[5]

Regardless of the mobile phase chosen, safety precautions should always be considered. If a nonflammable, inert, and nontoxic mobile phase such as CO_2 is used, the same safety precautions as in GC or LC are necessary. If the selected mobile phase is toxic, corrosive, or flammable, such as *n*-pentane, ammonia, or carbon disulfide, the instrument should be placed under a fume hood. It is recommended that a gas leak detector be installed in the pump housing to monitor the surrounding air if flammable mobile phases are used. The purge lines and column outlet lines of the instrument should be connected to a special device for disposal of hazardous fluids (*e.g.*, ammonia into water and N_2O to exhaust lines). Fittings, pressure lines, and seals should be checked frequently for leaks.

Cleaning of the injection valve. Memory effects or poor reproducibility could result from a dirty injection valve in the SFC instrument. Repeated injections of solvents such as dichloromethane, toluene, methanol, or acetonitrile often help to clean the injector. Here again, it is recommended that solvents of different polarities be used to flush out both polar and nonpolar contaminants from the valve. If this cleaning procedure is not successful, it is necessary to remove the injection valve from the instrument and to take it apart for more careful cleaning. The individual parts of the injector should then be cleaned with a series of solvents in an ultrasonic bath as discussed earlier in this section. Before reassembling the valve injector, a pocket lens can be used to examine the rotor and valve surfaces for scratches. If the rotor is scratched, it should be replaced. Small scratches on the rotor can be filled with sample, possibly leading to broad solvent peaks, memory effects, and poor reproducibility. Caution should be exercised in reassembling the valve to avoid crushing the rotor.

Installation of conventional packed columns. Installing a packed column for SFC is similar to procedures known from LC. To avoid dead volume, a short section of stainless steel or fused silica tubing with an internal diameter of approximately half the diameter of the column is recommended for connecting the column to the injector and the detector. Straight cuts at the ends of the stainless steel tubing and properly installed fittings are necessary in order to minimize the dead volume. These precautions minimize the occurrence of split peaks and excessive band broadening.

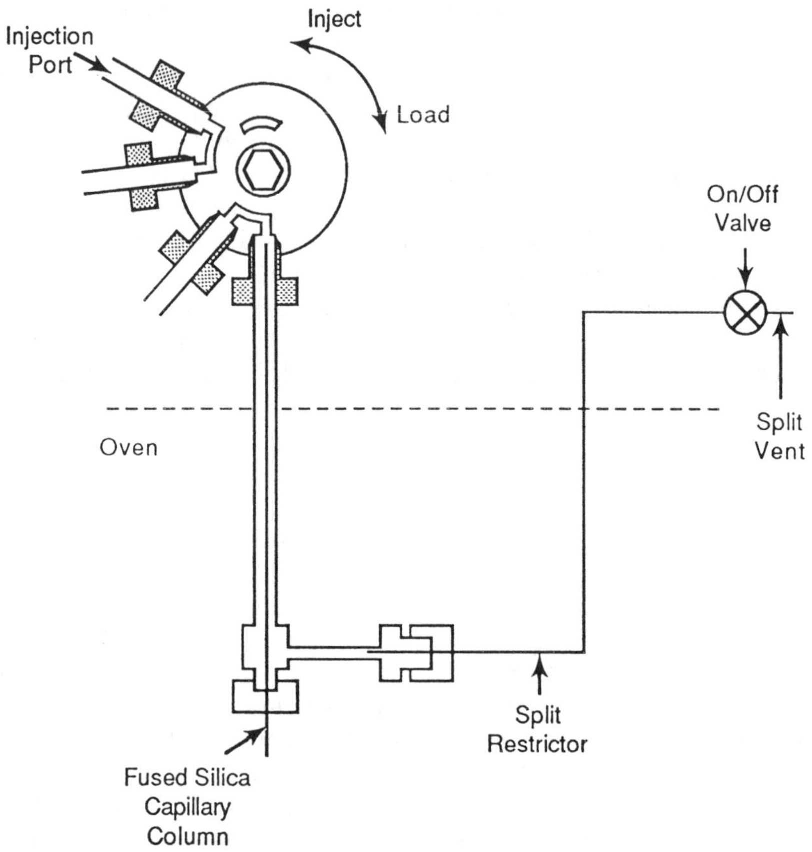

Figure 6.1. Schematic diagram of a split/splitless internal loop injector for open tubular and packed capillary columns (reprinted with permission from Ref. 6).

Installation of open tubular and packed capillary columns. Packed capillary columns are usually adapted with short fused silica capillaries on both ends, and the column installation is similar to open tubular column installation. For split injection, a splitter assembly is connected to the injection valve (Figure 6.1). The column is installed in the splitter T-piece by first removing the splitter nut, sliding the splitter nut and then a graphite ferrule with a slightly larger drilled hole than the outer diameter of the column onto the free end of the column, making a perpendicular score across the column approximately 1 cm below the end of the column with a diamond-, sapphire-, or carbide-tipped scribing tool, and pulling the fused silica straight out on both sides of the score, thereby, leaving a straight cut.

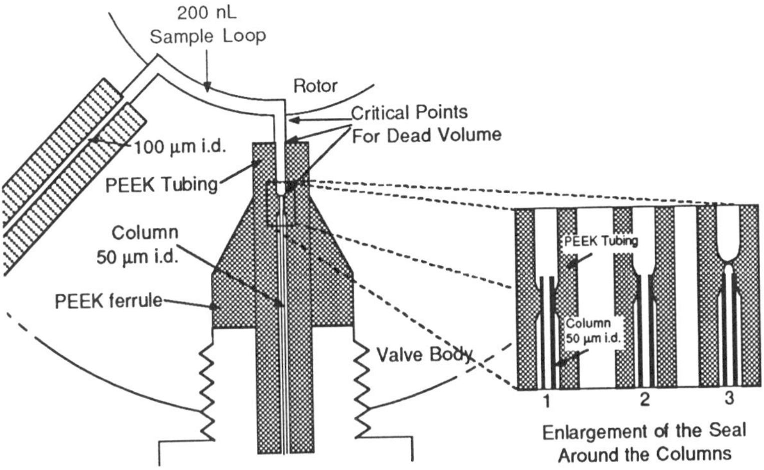

Figure 6.2. Schematic diagram of an internal loop valve injector with direct capillary column connection. Enlargement figure shows (1) too high, (2) optimal, and (3) too low column position (reprinted with permission from Ref. 8).

It is necessary to remove the end of the column which may have become fouled or plugged with graphite shavings from the graphite ferrule. The cut can be checked using a pocket lens. Bad cuts cause dead volume and nonuniform flow, which lead to peak broadening and discrimination. To obtain a smooth surface, the column end can be polished using 1,000-Å diamond powder, followed by careful rinsing.

The column can then be inserted into the splitter assembly. The beginning of the column should be carefully positioned below the rotor.[7] This is easily achieved by threading the column into the splitter assembly until it moves no farther, and then pulling it back to the desired distance.

If the injection mode is direct or timed split, the column should be connected directly to the injection valve using special sleeves, ferrules, unions, butt connectors, and nuts (Figure 6.2). These sleeves and ferrules should be compatible with the supercritical fluid mobile phase (*e.g.*, PEEK material is recommended for work at temperatures <150°C). To install the column, a properly cut end is threaded just through the ferrule. If the column extends beyond the ferrule or sleeve, the rotor of the injection valve can be scored and damaged. If the column end, on the other hand, is located too deep inside the sleeve, column plugging and/or unacceptable dead volume results.

Using a marker or "liquid paper," the desired position of the column at the base of the sleeve tubing can be marked, and the column can be pulled out of the tubing, ferrule, and nut. The tubing and ferrule are then inserted into the injection valve, and the column is pushed into the ferrule and positioned by locating the mark on the column. During tightening of the nut with appropriate wrenches, the mark on the column is carefully positioned at the base of the ferrule.

In the oven, the column should hang on a basket so that no contact with the oven wall occurs. This prevents temperature gradients or hot spots along the column. The outlet end of the column is connected to a restrictor before or after the detector.

The restriction can be made in the column end itself, or in a separate length of capillary tubing which can be connected to the end of the column either before or after the detector. The connection between the column and restrictor tubing must be made without any dead volume. The column end should have a clean, smooth, and perpendicular cut to achieve a good connection. When a butt connector is used for the connection, a sleeve (*i.e.*, a piece of fused silica with a slightly larger internal diameter than the column outer diameter) is used inside the butt connector to minimize the dead volume and to align the column and restrictor capillary. The ends of both the restrictor and the column should have the same outer diameter when using a sleeve. Both ends are pushed into the butt connector and the nuts are carefully tightened.

Startup testing. After installing the column, the mobile phase is allowed to purge through the system. The restrictor end should be heated to at least 150°C before the mobile phase is allowed to pass through the column to compensate for the Joule-Thompson cooling effect.[9,10]

For leak testing and cleaning of the column, a programmed pressure or density gradient should be applied up to the maximum allowable pressure of the column. In some instances, a simultaneous temperature program may be additionally effective. The hold time for the final pressure and temperature should be approximately twice the column dead time. Several repetitive density programmed runs may be necessary to adequately clean the system. When the baseline becomes stable, a pure solvent such as hexane or dichloromethane should be injected in order to determine the approximate dead time.

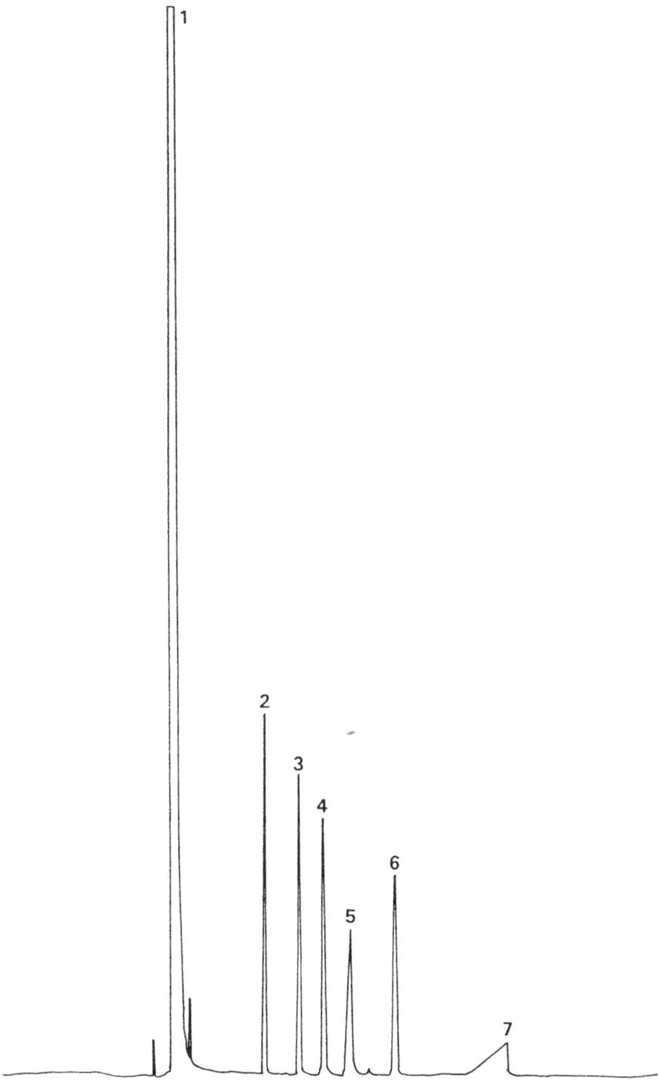

Figure 6.3. SFC chromatogram of an acidic test mixture. Conditions: 10-m x 50-μm i.d. open tubular column, polymethylsiloxane stationary phase (d_f = 0.25-μm); CO_2; 94 atm; 90°C; split injection, 2-3 ng per component; frit restrictor at 325°C, FID at 3.2 x 10^{-11} AFS. Peak identifications: (1) dichloromethane, (2) *n*-decane, (3) 1-octanol, (4) 2,6-dimethylphenol, (5) 1,6-hexanediol, (6) *n*-dodecane, (7) nonanoic acid (reprinted with permission from Ref. 11).

The shape of the solvent peak also gives the first indication as to whether or not the column is correctly installed.

An appropriate test mixture is required to evaluate the performance of the total system. The composition of the test mixture depends on the kind of column and detector which are to be used. When using an FID, an efficiency test mixture could be, for example, a mixture of *n*-alkanes such as *n*-hexadecane, *n*-octadecane, *n*-eicosane, and *n*-docosane, each with a concentration of 100 μg mL^{-1}. If inertness is important, better choices of test solutes are components with different functional groups. Such a polarity test mixture could contain an *n*-alkane, a fatty alcohol, a fatty acid, and a phenylalkane if CO_2 is the mobile phase. A chromatogram of the test mixture shows that the column is properly prepared and installed when sharp peaks are obtained and no peak splitting is observed (see Figure 6.3).

Depending on the inertness of the system, some components can exhibit slight tailing. If nonpolar compounds exhibit tailing, the cause is usually a poor connection of the column to the injector or detector.

Daily performance testing. It is necessary to occasionally monitor the efficiency of the column and to ensure that the restrictor is not plugged. A test mixture as described above, or containing the solutes that are normally analyzed, is recommended for daily performance testing. Checking the efficiency of the column (*i.e.*, the number of theoretical plates) is a good way to follow the useful lifetime of the column.

A comparison of retention data usually indicates the performance of the restrictor with time. Increasing retention times are a sign of plugging of the restrictor with high molecular mass material or particles from the sample or the column (if a packed column is used). Sample solutions should be filtered, and in-line micro screens (2-μm pores) should be installed in the injection valve where sample is injected to avoid these problems. The restrictor for the UV detector should be heated by either placement directly against the lamp housing or by insertion into the oven.

6.3 Sample Preparation

Sample preparation for chromatography, in general, is required to pretreat the sample in such a way as to permit subsequent analysis of the analytes of interest with minimal interference by other components. In the most fortunate cases, this requirement is met by the mere selection of a suitable solvent which is soluble in the mobile phase, effectively solubilizes the sample, and possesses good chromatographic properties. Of equal importance to solvent selection is the concentration of the solutes in the injection solvent. The concentration must be at a level which is compatible with the effective injection volume and the detector sensitivity requirements, and sufficiently dilute to prevent column overloading and subsequent loss of resolution. An injection method that involves solvent elimination might be necessary for trace analysis where, otherwise, the required injection volume would destroy the chromatographic performance of the column. Additionally, in cases where the chemical nature of the solutes of interest results in poor chromatographic efficiency, limited solubility in the mobile phase, or insensitive detection, chemical derivatization can be incorporated in the sample preparation scheme.

These general sample preparation requirements must also be satisfied in the effective practice of SFC. Liquid solvents, such as *n*-alkanes, dichloromethane, carbon disulfide, etc., normally used in GC are also suitable solvents for SFC. Solute concentrations on the order of μg to low mg mL^{-1} are typically used. Sample solutions at these concentrations can be directly injected onto packed column systems, while split injection or solvent elimination injection are normally used to meet the injection volume and sample capacity requirements of open tubular column systems (see Sections 4.4-4.6). In trace analysis, where solute concentrations are on the order of ng mL^{-1}, sample preconcentration is needed. For example, a few mL of a biological fluid must be evaporated or lyophilized and the residue reconstituted in a few microliters of the injection solvent before SFC analysis. Improvements in solvent elimination injection methods (Section 4.6) or sample concentration by SFE (Chapter 5) may minimize the need for labor-intensive sample preconcentration procedures.

It is not mandatory that the injection solvent for SFC be a liquid; supercritical fluids may also be utilized as sample preparation solvents. Many biological sample pretreatments deliver the solutes in a polar solvent which can disturb the phase equilibrium of the supercritical mobile phase or degrade the solute biological activity. Niessen *et al.*[12] have utilized a phase-switching technique in which the solute, dissolved in a polar solvent, is injected onto a precolumn where the polar solvent is removed. The solute is then desorbed in a supercritical mobile phase and analyzed by SFC. Xie *et al.*[13] have demonstrated that the biological activity of a model drug, ouabain, could be retained by using supercritical CO_2 for introducing the sample into the SFC system.

The commonly used supercritical mobile phases (CO_2 and N_2O) are good solvents for relatively nonpolar compounds. The solubilities of polar solutes, especially high molecular mass or biochemically-important compounds with multiple polar functionalities, are poor in supercritical CO_2 and N_2O. Chemical derivatization can be used to "mask" various polar functionalities, thereby improving the solubilities of many compounds. Derivatization techniques common to GC have been incorporated successfully into SFC sample preparation methodology. Chester *et al.* have been extremely successful in the use of trimethylsilyl derivatives for the SFC analysis of oligosaccharides[14] with degrees of polymerization up to 25, inositol triphosphates, and phytic acid.[15] Kuei *et al.*[16] have adapted methylation procedures for the SFC analysis of glycolipids and maltodextrins. Derivatization of free amine functionalities permits the subsequent use of CO_2 as the mobile phase, since the derivative both enhances the solubility of the compound and protects it from reacting with the mobile phase.

In addition to improving the solubilities of polar solutes in the relatively nonpolar fluids, derivatization can also aid in the separation process. Veuthey *et al.*[17] have demonstrated the separation of enantiomeric amino acids by SFC using chloroformate derivatization reagents. In situations where a derivatization reagent also incorporates a heteroatom-containing moiety (*e.g.*, N or P) into the molecule, sensitive and selective thermionic detection can be employed. David and Novotny[18] have used the reaction of *o*-phenylenediamine with α-ketoacids to form nitrogen-containing quinoxalinol derivatives which were analyzed by open tubular column SFC and nitrogen thermionic detection. Sensitive phosphorus thermionic detection was employed for the SFC analysis of hydroxysteroids as their corresponding dimethylthiophosphinic esters.[19]

6.4 Mobile Phase Selection

Common mobile phases. The most commonly used supercritical fluid mobile phase is CO_2 because it is nontoxic, inexpensive, has low critical pressure and temperature, has solvating power greater than hexane, readily accepts a wide variety of modifiers, is residue-free, has UV transparency, and provides no response in the FID. Table 3.1 lists a number of the possible primary mobile phases that could be used in SFC. Historically, a number of these mobile phases have been used by researchers for a variety of applications.[20-24]

Theoretically, one can use any compound that is thermally stable at the supercritical operating conditions. Practically, instrumental and safety limitations narrow the possibilities. Nitrous oxide, which is the second most commonly used mobile phase, exhibits a polarity similar to CO_2 and should provide no response in an FID if pure.[25,26] However, N_2O is an oxidizer, and it is not always a good practical choice for the separation of certain oxidative materials. Supercritical SF_6, which has a very low solvent strength, has been used to effect the separation of nonpolar complex hydrocarbon matrices.[27] However, since SF_6 decomposes in an FID to form HF, proper venting procedures and detector modifications (*i.e.*, gold plating of the FID collector) must be employed. Chlorinated and fluorinated hydrocarbons usually give off some acidic decomposition products under supercritical conditions.[20] Acetonitrile, when used with alumina stationary phases under supercritical conditions, gives off ammonia.[20] Diethyl ether, *n*-pentane, methanol, isopropanol, *n*-heptane, diisopropyl ether, cyclohexane, benzene, tetrahydrofuran, and ethyl acetate are sufficiently stable under supercritical conditions, but are restricted in use to detectors other than the FID.

Selection of mobile phase. Overall, the use of a specific mobile phase in SFC is largely determined by the specific types of compounds that must be resolved and the detector that is to be used. In choosing a supercritical fluid, one may want to consider, in addition to solute-mobile phase reactivity mentioned earlier, the solvent selectivity, availability, cost, and toxicity. Supercritical xenon, for example, has the distinct advantage of being IR transparent, yet it is one of the most expensive supercritical fluid mobile phases that can be purchased.

With respect to detector considerations, CO_2, N_2O, and SF_6, in theory, provide no response in the FID or UV detector. When used with spectrometers (*e.g.*, MS and FTIR), spectral background subtraction may be necessary. With CO_2, it is also possible to use a number of detectors, including electron capture, photoionization, thermionic, sulfur chemiluminescence, ion mobility, atomic emission plasma, light scattering, and fluorescence.

Mobile phase purity is as important in SFC as it is in GC and LC. Commercial suppliers of SFC-grade fluids have made significant improvements in stock purities during the last several years. Residual hydrocarbon impurities in CO_2 were an early problem in SFC, and, in some cases, were so excessive that they prohibited the use of detectors such as the FID. The purity of the supercritical fluid becomes especially important when using selective detectors, such as the electron capture detector, which are inherently very sensitive. Trace levels of halogenated species which can be emitted from Teflon seals, Teflon tape, and from residual Freons which may be present in the supercritical fluid tank, are easily detected by electron capture detectors. Moreover, when coupling SFE to SFC (or GC), any trace level (low ppm) impurities are easily concentrated and can be significant interferences in the analysis.

Selection of modifiers. Mobile phase selectivity and solubilizing power can be modified by the addition of modifiers to the primary supercritical fluid phase. Depending on the situation, if the maximum solvent power of the primary supercritical fluid mobile phase is not high enough to elute the components of interest (*i.e.*, polar or high molecular mass compounds), then higher solubilizing power may be achieved by adding a modifier. Both the solvent power and the selectivity of the fluid mixture are determined by the chemical identity and concentration of the modifier. The modifier effect manifests itself by altering solute retention, improving chromatographic efficiency, and in some cases, altering elution order. For packed column SFC, adding as little as 0.5 mole percent of a modifier has altered capacity factors and separation factors by orders of magnitude.[28]

Table 6.1 lists the modifiers that have been used in SFC with CO_2 as the primary fluid. A number of these modifiers have also been used to a limited extent with other primary fluids, such as N_2O, SF_6, and various Freons. One of the advantages of CO_2 is that, as a supercritical fluid, it is chemically compatible and miscible with a large number of modifiers. The alcohol modifiers, such as methanol, are the most commonly used. However, with most modifiers, FID detection is not possible. Formic acid, when it is pure, is one of the few modifiers that does not give a response in the FID. Modifiers have the ability to significantly enhance supercritical fluid solubilizing power for separating some polar or high molecular mass solutes.

When modifiers are used, it is important to understand the chemical interactions of certain modifiers and supercritical fluids, as well as interactions between modifiers and solutes. For example, primary or secondary amines cannot be used as modifiers in CO_2 because of the formation of urea derivatives and carbonates.[26,29] Modifiers which can easily be oxidized should not be mixed with N_2O, because of its oxidizing nature. The use of

Table 6.1. Modifiers That Have Been Used in SFC Using CO_2 as the Primary Component

Binary	Ternary	Quaternary and higher
Methanol	Methanol + citric acid	Methanol + water + methylamine
Ethanol	Methanol + acetylacetonate	Methanol + water + ethylamine
Isopropanol	Methanol + dioctylamine	Methanol + water + triethylamine
n-Hexanol	Methanol + sodium heptanesulphonate monohydrate	Methanol + dimethyloctylamine + sodium heptanesulphonate monohydrate
Tetrahydrofuran	Methanol + tetramethylammonium hydroxide	Methanol + tributylamine + acetic acid
Acetonitrile	Acetonitrile + isopropanol	Methanol + tributylamine + acetic acid + citric salicylate
Chloroform	Acetonitrile + N-benzoxycarbonylglycyl-L-proline	Acetonitrile + N-benzoxycarbonylglycyl-L-proline + triethylamine
Methylene chloride	Acetonitrile + tetrabutylammonium bromide	
Methyl chloride	*n*-Hexane + methanol	
1,4-Dioxane	*n*-Hexane + ethanol	
2-Methoxyethanol	*n*-Hexane + *n*-propanol	
Dimethyl sulfoxide	*n*-Hexane + isopropanol	
Acetic acid	*n*-Hexane + *n*-butanol	
Formic acid	*n*-Hexane + 2-butanol	
N,N-dimethylacetamide	*n*-Hexane + isobutanol	
Sulfur hexafluoride	*n*-Hexane + *t*-butanol	
Trichlorofluoromethane	*n*-Hexane + tetrahydrofuran	
Carbon disulfide	*n*-Hexane + 1,4-dioxane	
Propylene carbonate	*n*-Hexane + acetone	
Water	*n*-Hexane + ethyl acetate	
n-Propanol	*n*-Hexane + benzene	
n-Butanol	*n*-Hexane + methylene chloride	
n-Pentanol	*n*-Hexane + nitromethane	
n-Heptanol	Formic acid + water	
n-Decanol		
2-Butanol		
2-Pentanol		
t-Butanol		
Methyl-*t*-butyl ether		
Diisopropyl ether		
n-Hexane		
Diethyl ether		
Carbon tetrachloride		
Ethyl acetate		
Nitromethane		

ammonia as a modifier in SF_6 leads to the formation of solid ammonium fluoride in an FID.[26]

In choosing a modifier for a specific set of solutes, the most important factor is solubility. Definitive guidelines for the selection of modifiers are presently nonexistent, but this will change as the technology matures. Suggestions have been made with respect to modifier polarity.[30] However, the performance of each modifier depends to a large extent on the chemical structures of the solutes of interest. In general, formic acid is of interest for polar solutes since an FID can be used. Propylene carbonate is an effective modifier for hydrocarbon matrices, polynuclear aromatic hydrocarbons, isocyanates, polymers, and drugs.[31] Dimethylacetamide can be effective for the separation of basic compounds. Methanol and acetonitrile have been used to solubilize solutes with both acidic and basic moieties. Using a secondary modifier or co-modifier such as trifluoroacetic acid and tetrabutylammonium hydroxide for acidic and basic solutes, respectively, can even further extend the applicable range of SFC to very polar solutes.[31]

6.5 Preparation of Mixed SFC Mobile Phases

Two general approaches have been taken in the preparation of mobile phase mixtures: those involving multiple pump systems and those in which the mixture is prepared prior to addition to the pump. A step-by-step procedure for preparing premixed mobile phases was described by Raynie *et al.*[32] A 300-mL stainless steel mixing vessel, as shown in Figure 6.4, was used. This vessel was carefully cleaned and evacuated to less than 100 torr by connecting it to a vacuum pump. The vessel was then weighed, a previously degassed modifier was introduced through a septum into the vessel, and the vessel was weighed again. From the net weight, the number of moles of modifier could be calculated. With the mixing vessel still on the balance, it was then connected to a liquid CO_2 cylinder and the desired mass of CO_2 was introduced.

The contents of the vessel were carefully mixed and then emptied in a vertical position into the syringe pump cylinder. The empty vessel was weighed again, and this weight was subtracted from the total weight (minus modifier) to give the net weight of CO_2. From this, the mole fraction of modifier in CO_2 in the pump cylinder could be calculated from

$$X = \frac{\text{Moles of modifier}}{\text{Moles of modifier} + \text{Moles of } CO_2} \qquad (6.1)$$

Several comments should be made concerning the mixing of fluids in this way. The mixing vessel contents consist of a gas phase and a liquid phase. The liquid phase is a mixture of CO_2 and liquid modifier. Since the volatility of CO_2 is usually much greater than the volatilities of the organic modifiers, the gas phase is assumed to be pure CO_2 (*i.e.*, the amount of organic modifier in the gas phase is negligible). Since it is of interest to transfer only the liquid phase into the pump, both the pump cylinder and mixing vessel should be maintained at ambient temperature (*i.e.*, do not cool or warm). To help maintain the liquid miscibility, the pump cylinder should not be cooled at any time while using a mixed mobile phase. It is important to know as much about the fluid miscibility properties of the mixtures used in order to avoid phase separation.

The best results are obtained by changing the amount of modifier added for differing concentrations, and always adding a constant amount of CO_2. This procedure has been used for a variety of CO_2 + modifier mixtures in the range of 1-10 mol%. With this procedure, the average relative standard error for the mixing process (defined as the difference between actual and target mol% of the mixture, divided by the target mol%, times 100) is minimal.

Campbell *et al.*[33] used pressure and volume measurements from two pumps and mixed the components by pumping back and forth between the pumps to get mixtures of known concentrations of two gaseous components. However, the experimental error and irreproducibility with such a system is thought to be considerable. A system using three pumps has been used with some success.[34] In this system, pump A was filled with pure CO_2 and pump B with a predetermined amount of modifier. The modifier was transferred to pump A and the resulting mixture was pumped at constant pressure into pump C. This mixture was transferred back into pump A for subsequent use as the chromatographic mobile phase. It was observed that mixing was not optimal using this procedure, probably because of the low temperature around the pump heads. During use of the first 25% of the pump volume, retention times increased, implying decreasing modifier concentration. For the rest of the pump volume, the retention times remained constant.

A computer-controlled two-pump system to deliver on-line modified mobile phases at flow rates compatible with microcolumns (*i.e.*, $\mu L\ min^{-1}$) has been reported,[35] which shows retention time

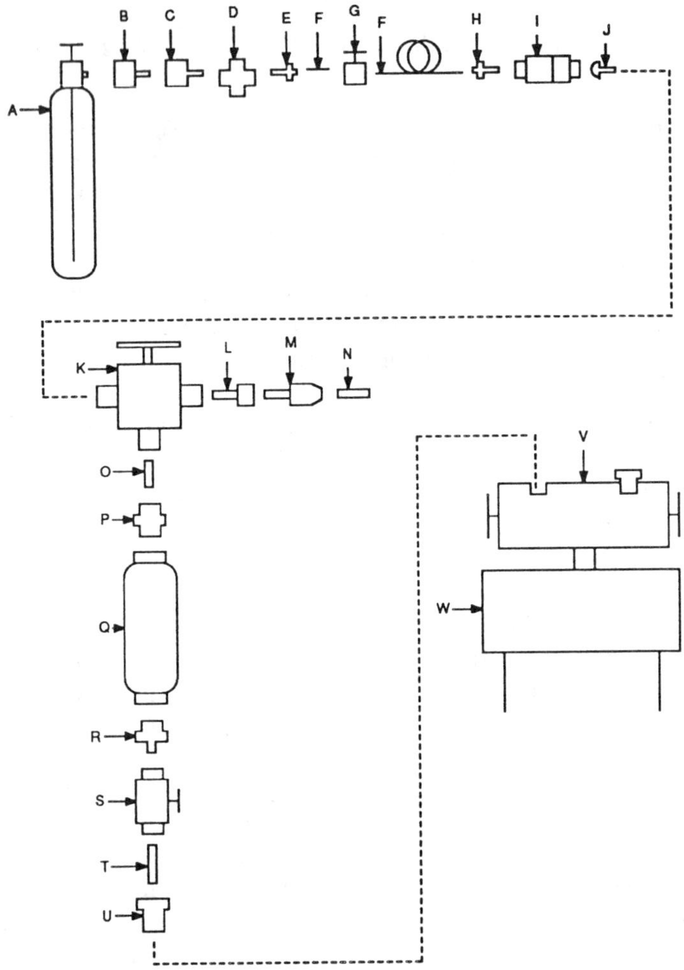

Figure 6.4. Detailed schematic diagram of fluid mixing device components. Except where noted, all fractions are in inches, fittings and tubing are stainless steel, and o-rings, seals, and gaskets are Teflon. (A) Size A (16 liter) SFC grade CO_2 cylinder with eductor tube (Scott Specialty Gases, Plumsteadville, PA); (B) CGA-320 brass fitting; (C) 1/4 FNPT - 1/4 tube (Swagelok (Sw)); (D) 2-μm high flow screen filter (Nupro); (E) 1/4 tube-1/16 Sw; (F) 1/16 ss tubing, 1-2 feet on cylinder side of needle valve, 4-6 feet (coiled) on mixing vessel side of needle valve; (G) Union Bonnet metering valve (Whitey); (H) same as E; (I) full flow quick connect (Sw); (J) 1/4 port connector (Sw); (K) 3-way ball valve (Whitey); (L) 1/4 tube-1/8 FNPT (Sw); (M) syringe adapter (Scott); (N) needle guide, ss tubing, 1/16-o.d. x 0.012-i.d. x 1 1/8; (O) 1/4 ss tubing, 2-3 in.; (P) 1/4 Sw-1/4 MNPT (Sw); (Q) 300 mL sample cylinder, rated to 5000 psi (Scott) filled with a few stainless steel bars to aid mixing; (R) 1/4 MNPT-1/8 Sw (Sw); (S) 2-way ball valve (Whitey); (T) 1/8 high pressure tubing, 1 1/2 in. (High Pressure Equipment (HIP), Lincoln, NE); (U) nut for pump valve V; (V) 3-way, 2-stem high pressure valve (HIP); (W) syringe pump. Syringe: 10 mL gas-tight, Luerlok (#1010, Hamilton, Reno, NV). Needle: 26 ga., 5 in. exposed, "A" point (#1199, Unimetrics, Anaheim, CA) (reprinted with permission from Ref. 32).

reproducibility indicative of reproducible mixing. This system may also be adapted for gradient elution. Another report[5] used two piston pumps, rather than syringe pumps, with back-pressure regulation for on-line mobile phase mixing. A multiple pump system for on-line mixing appears to be the best approach to mobile phase mixing, especially if compositional gradients are reproducibly generated during the chromatographic run.

The estimated critical parameters for the mixed mobile phases can be determined using the recommended procedures of Reid *et al.*[36] These methods have been adapted from those of Chueh and Prausnitz[37] for critical pressure estimations, and from Kreglewski and Kay[38] for critical temperatures.

6.6 Column Selection

Column selection in SFC should be based primarily on the criteria of selectivity, efficiency, inertness, sample capacity, and speed of analysis. The first consideration should be the selection of the stationary phase with the necessary selectivity to accomplish the desired separation. A variety of solid supports and stationary phases are available for packed and open tubular column SFC. In LC, the need for a wide variety of stationary phases with specialized functional groups is lower, mainly due to the ability to achieve considerable solute retention modification with composition gradient elution. In SFC, because composition gradient elution is not widely used and the number of potential fluid mixtures (and polarities) is limited, selective stationary phases play a more significant role.

Both packed and open tubular columns can provide a range of selectivities. In most cases the choice of packed or open tubular columns is mainly dependent on the four criteria of efficiency, inertness, sample capacity, and speed. In general, open tubular columns are superior in terms of efficiency and inertness, while packed columns are preferred when sample capacity and speed are of concern.

Efficiency. It is obvious from Chapter 2 that open tubular columns can provide greater total numbers of theoretical plates than packed columns. This is not to say that packed columns cannot provide sufficient numbers of theoretical plates for many separations. However, if a complex mixture is to be separated and the maximum number of theoretical plates is important, then an open tubular column with the appropriate stationary phase should be selected.

Surface activity. Open tubular columns are identified as being more inert than their currently available packed column counterparts. The deactivation of the large surface areas possessed by porous silica packing materials is a challenging problem. In Chapter 2, innovative approaches to the deactivation of these materials, as well as the use of alternative materials, were discussed. If neat CO_2 is to be used, open tubular columns are currently preferable for polar solutes.

Sample capacity. Open tubular column sample capacity can be increased by increasing the stationary phase film thickness or by increasing the column internal diameter. However, because of the significant loss in efficiency experienced from increasing the column diameter, the film thickness is the only parameter that can be varied for practical reasons. The film thicknesses in open tubular columns are limited to 1 μm or less due to the high viscosities of the coating solutions. Therefore, the largest sample capacity that can be expected is ~100 ng per component. The effect of film thickness on retention is often ignored in practice. However, the column cannot be indiscriminantly changed to one of similar phase and thicker film without considering the overall effect on retention. Otherwise, elution pressures may be shifted too high to be attained by the pump.

Column internal diameter does play a significant role in the sample capacity of a packed column (Chapter 2). The larger the i.d. of the column, the greater is the amount of packing material (and therefore surface area inside), and the greater is the sample capacity. However, larger columns require greater volumetric flow rates, and may require splitting before detection.

Speed of analysis. It was shown in Chapter 2 from theoretical considerations, that packed columns can provide smaller plate heights than open tubular columns of practical dimensions, and that the internal diameter of the packed column does not affect the efficiency or speed of an analysis as it does with an open tubular column. If large numbers of theoretical plates are required, small packed capillaries or, of course, open tubular columns can be used.

Packed column selectivity. Reversed phase LC columns have been shown to not be as universal in SFC as in LC. Other column selectivities are used as often, or more frequently, than the reversed phases. Normal phase alumina and silica particles have proven useful for the analysis of nonpolar hydrocarbons and halogenated hydrocarbons. The elution can be compared to that of normal phase LC; however in LC, the selectivity of these materials is controlled by the degree to which they are hydrated. These materials are strong adsorbents and can prove too adsorptive for CO_2 to solubilize the more polar analytes.

The reversed phase packings modified with C_8 and C_{18} exhibit dipole-dipole interactions from surface silanols in addition to the desired dispersion characteristics. These phases have been successfully applied to the separation of a wide variety of polar and nonpolar compounds. Some size and shape selectivity has also been exhibited.[39]

Cyanopropyl, short chain alkanes (C_1-C_3), phenyl, and nitrophenyl phases can be considered to exhibit dipole-dipole interactions and/or dipole-induced dipole interactions. These phases can be used for a variety of polar or polarizable samples (*i.e.*, aromatics, acids, esters, olefins, and ketones).

Divinylbenzene polymer beads have both polar and nonpolar characteristics. The polarity is often due to catalytic impurities or microporosity, but it should be noted that these materials can be modified with different functional groups to give a variety of properties. Due to the polymeric nature of the particles, relatively large adsorption and long retention times can be experienced, particularly for aromatic compounds. However, as has been noted in Chapter 2, some advantages regarding surface activity, as well as the ability to use these materials in supercritical ammonia, may be important.

Silver nitrate-treated particles have also been used to provide dipole-induced dipole interactions. This type of phase has been used for the class separation of compounds from petroleum samples into aliphatics, olefins, and aromatics.[33]

Amino and diol phases are examples of commercial phases that exhibit hydrogen bonding. The amino material has been used in the analysis of polycyclic aromatic hydrocarbons, drugs, pesticides, and herbicides, while the diol material has been used for steroids and metal chelates in addition to the previous compound types.

Chiral separations require highly selective stationary phases. Most chiral separations in SFC have been studied using packed columns. This important application area is treated in detail in Chapter 7.

As SFC has progressed, the need to separate more complex mixtures has led to the use of pressure or density programming with operating temperatures of up to 200°C. Considering the high solvating of supercritical fluids, this practice prompted the question as to the stability of the conventional LC column packing when operated under these conditions for any length of time. In the final analysis, one can conclude that conventional column activity and selectivity are altered by use at temperatures greater than 100°C.[40] The effects are observed with nonpolar analytes, but the changes are more dramatic with polar analytes. Polar stationary phases (*e.g.*, amino) were found to be more heat sensitive than nonpolar phases (*e.g.*, C_{18} and phenyl).

Open tubular column selectivity. Open tubular columns for SFC are available in several column diameters and with a variety of stationary phases. Typically, 50-μm i.d. columns are used in practice, and larger (100-μm i.d.) or smaller (25-μm i.d.) columns can be used for lower efficiency (higher loadability) or higher efficiency (lower volumetric flow rate) applications, respectively.

The lengths of open tubular columns range from a few meters for rapid separations to usually around 10 meters for practical work. Longer columns (up to ~35 m) can be used for very high resolution applications.

Table 6.2 lists the various open tubular column stationary phases and applications for which these columns have been found to be useful. As previously discussed for packed columns, the methyl and *n*-octyl phases provide mainly dispersive interactions, the phenyl and biphenyl phases are polarizable (and useful for the separation of the more polar compounds), and the Carbowax and cyanopropyl phases are responsible for dipole-dipole type interactions with moderately polar or polarizable solutes.

The liquid crystalline polysiloxane has found considerable use for the separation of isomers according to their shapes. This important selectivity is discussed in some detail in Sections 7.2 and 7.5.

Open tubular column evaluation. To ensure that each step of the open tubular column preparation process was performed correctly, each column should be evaluated using test procedures that define its useful limits and are representative of its intended application. Ideally, the column should be evaluated under SFC conditions with active test compounds that effectively probe efficiency and inertness. A narrow bore capillary column that performs well and exhibits low thermal bleed when evaluated by GC may suffer from considerable stationary phase washout when subjected to the action of supercritical fluid mobile phases. High

Table 6.2. Open Tubular Column Stationary Phases and Applications

Biphenyl or Phenyl	Methyl	Cyanopropyl
Alcohols	Hydrocarbons	Olefins
Polyols	Non-ionic surfactants	Aromatics
Phenols	Peroxides	Halogenated aromatics
Aldehydes	Aromatic and aliphatic amines	PCBs
Ketones	Organophosphates	Esters
Ethers	Organosilane esters	Ethers
Peroxides	Polysiloxanes	Class fractionations
Steroids	Organotin compounds	Nitriles and other polar compounds (low molecular mass)
Carboxylic anhydrides	Halogenated pesticides	Pesticides and metabolites
Esters	Azo compounds	Pharmaceuticals
Amines	Antioxidants	
Imines	Simulated distillation	
Amides	Nitriles	
Imides	Halogenated pesticides	
Nitrogen PAHs	Surfactants	
Nitrogen heterocycles	Oligomers	
Schiff bases	Polymer additives	
Non-ionic surfactants	Pharmaceuticals	
Isocyanates		
Isothiocyanates		
Nitriles		
Nitro compounds		
Nitroso compounds		
Organosilane esters		
Sulfides		
Disulfides		
Sulfoxides		
Sulfones		
Thiols (Mercaptans)		
Ureas		
Halogenated pesticides		
Antioxidants		
PCBs		
High molecular mass polar compounds		
With derivatization:		
Amino acids		
Hydrazines		
Sulfonates		
Polysacharrides		

Carbowax	Octyl	Liquid Crystal
Carboxylic acids	Hydrocarbons	Dioxins
Aromatics	Halogenated hydrocarbons	PAHs
Esters	Steroids	Olefin isomers
Ethers	Simulated distillation	Substituted aromatics
Triglycerides	Dioxins	Diastereomers
Flavors		Steroids
Fragrances		
Essential oils		
Food additives		
Low molecular mass polar compounds		

bleed columns can plug restrictors, foul detectors, and exhibit low efficiencies and changing capacity factors in SFC. Selective bleeding of components of stationary phases comprised of mixed monomers can also result in polarity changes and thus alter the selectivity of the column. Final conditioning of the column with a supercritical fluid mobile phase is conveniently accomplished in the chromatograph just prior to testing by combining pressure and temperature programming over the intended range of column use.

Many analyses in SFC are performed at temperatures where adsorptive interactions are most significant. Elution of polar and high molecular mass solutes is commonly accomplished by programming mobile phase pressure or density instead of column temperature. Testing of the SFC column should be done at lower temperatures (<120°C) to enable probe compounds to adequately reflect surface activity. Molecules with high percentages of active functional groups are recommended as suitable activity probes (*e.g.*, low molecular mass diols, acids, and phenols). Columns intended for analysis of basic materials with compatible mobile phases should include amines as probes as well. Test compounds should be introduced at <10 ng quantities for sensitivity to trace levels of active sites. Efficiency may be determined using peak width measurements from isobaric/isothermal runs, or Trennzahl determinations from programmed runs. While density or pressure programming is more representative of typical analyses, accurate Trennzahl and capacity factor measurements require tight control over linear velocities with the changing diffusion and solvation properties of programmed mobile phases. Figure 6.3 shows a typical test chromatogram of a good open tubular SFC column with run parameters as indicated.

6.7 Practical Injection Considerations

There are some practical aspects of injection in SFC which may not be intuitively obvious based on one's experience in LC or GC. This section will cover some of those considerations to improve the performance and to avoid some of the pitfalls that can occur.

Injector temperature. Andersen[41] showed that the temperature of the injector can have a dramatic effect on performance in SFC. He demonstrated that higher molecular mass materials are more effectively transferred to the column when the injector is warm. This is due to an increase in solubility at higher temperatures. This is shown in Figure 6.5 in which chromatograms are displayed that were obtained by analyzing a polyethylene standard (PE 740).[42] The only difference in conditions between the chromatograms was the increase in injector temperature as indicated on the figure. One would be tempted to say that this sample was not soluble in CO_2 if only chromatogram 6.5C were observed. Obviously, a heated injector has advantages over a cool one for the analysis of some heavy molecular mass materials. This is especially true if the sample must be heated to be solubilized. Care must be exercised, however; the valve must not be hot if one is using a volatile solvent, because the solvent will boil out of the valve and poor reproducibility and sample loss will result. Obviously, higher boiling solvents must be used when the valve is hot.

Injection valve and column connections. Even though "microbore" LC valves are used in SFC, satisfactory results are not achievable using manual actuation as is commonly done in LC. Richter *et al.*[43] showed that one could improve the precision of an SFC system from 11% RSD to 3% RSD by simply changing from a manual actuator to a high-speed pneumatic actuator driven by helium gas. This is true for either packed or open tubular columns.

Chester and Innis[44] showed that the orientation of the valve (*i.e.*, which port is the sample port and which is the waste port) and the addition of a waste port restrictor can improve reproducibility. The addition of the waste port restrictor prevents loss of sample from the valve during valve transit from the load to inject positions.

It is obvious that any dead volumes in connections can lead to inferior performance in SFC, especially if they occur in the injection connections. Francis *et al.*[7] reported the results of studies to determine some practical steps to improve efficiency in split injection with open tubular column SFC. They found that poor performance would result if the splitter was not completely seated in the injection valve. This dead volume can be eliminated by using a gasket cut from PTFE tubing. The internal diameter of the splitter has an effect; a 0.010 inch i.d. splitter gave an efficiency 1.4 times higher than a splitter with 0.020 inch i.d. They also reported that the position of the column inside the splitter can influence the chromatographic performance.[43,45]

In split injection, it is not uncommon to have the splitter plug when analyzing a concentrated or high molecular mass sample. This plugging becomes much more pronounced when the split restrictor

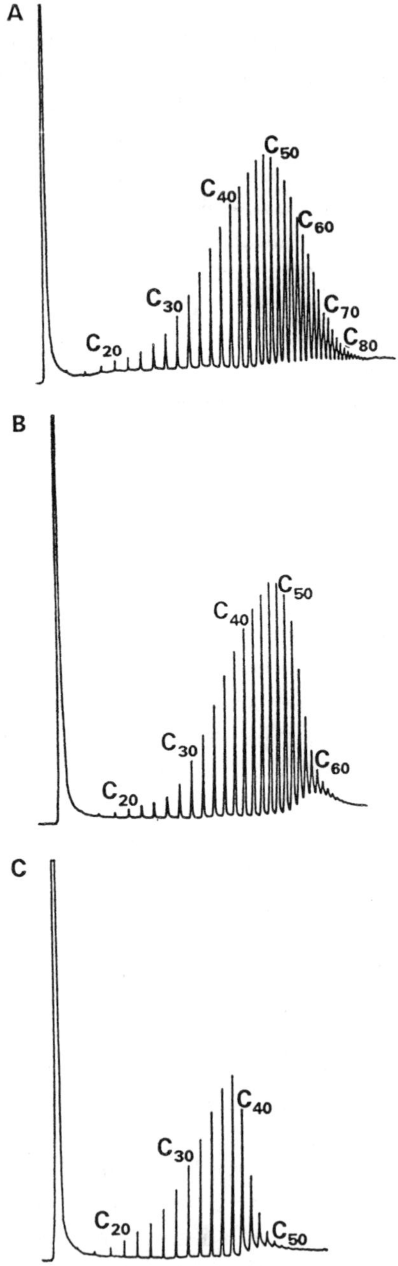

Figure 6.5. SFC chromatograms of polyethylene standard (PE 740) with injection temperatures of (A) 115°C, (B) 80°C, and (C) 50°C. Conditions: 7-m x 50-μm i.d. open tubular column, poly(5% phenyl)methylsiloxane stationary phase (d_f = 0.25 μm); CO_2; 130°C; multi-linear density ramp; timed-split injection; FID (375°C) (reprinted with permission from Ref. 42).

is at room temperature. This can be minimized by having the splitter in a heated zone to help prevent compounds from precipitating.

The inlet tubing connecting the packed column to the injector is also important. Work with typical microbore packed columns has shown that inlet tubing of 0.007 inch i.d. gives excellent results in SFC.[46] It is also vitally important that no dead volume exists in the connections at the injector or at the column inlet. If the stainless steel tubing has a polished cut, it will perform well. However, if the tubing has been cut with a file or is uneven, unsatisfactory results are obtained. These will be manifest as broad solvent peaks and a reduction in chromatographic efficiency.

The tubing which delivers the mobile phase to the injector is important also. A larger than necessary internal diameter tubing on the injection valve allows unnecessary back flow and diffusion of the sample solvent, causing broad solvent peaks, especially when performing timed-split or direct injection. This can be minimized, and in some cases eliminated, by attaching a short piece (5-10 cm) of 0.005-inch i.d. stainless steel tubing on the pump port of the injection valve.[47]

Connecting the column to the injector is also critical for timed-split injection. Currently, fittings made of PEEK or graphite-Vespel work well. Care must be taken to ensure that no dead volume exists. In addition, rotors with 0.5-μL volumes perform better (*i.e.*, more reproducible) than those with 0.2 or 0.06-μL volumes.

Results have been reported which show that under some circumstances, the use of helium headspace pressurized tanks as sources of SFC grade fluids can cause problems in injection.[3] Poor reproducibility of absolute peak areas and retention times can be observed when helium becomes entrained in the CO_2 and is delivered to the pumping system. In contrast to these results, Rosselli *et al.*[2] found little difference in performance when using cylinders of CO_2 with and without helium head-pressure for microbore column SFC.

Injection solvent effects. Several reports have appeared which document that injection performance can be affected by the properties of the sample solvent.[48-50] Berg and Greibrokk[48] have shown that the use of high density chlorinated solvents resulted in solute peak splitting at pressures as low as 106 atm, while toluene and hexane gave no peak splitting below 125 atm. The explanation given was that under high density conditions (*i.e.*, high injector pressure and/or high density solvents) a portion of the solute band moved through the column or precolumn quickly, while the remainder was retained by the stationary phase.

Bohm *et al.*[49] reported that with low split ratios, the retention times of some polycyclic aromatic compounds increased with increasing density of the solvent, while at high split ratios the retention times remained constant. This was explained as resulting from the modification of the column surface at the front of the column by the solvent, or from a change in the split ratio due to changes in the flow dynamics in the split restrictor.

Hirata and Inomata[50] found that the largest volume that could be introduced without peak splitting using their direct injection method (which employed mixing chambers) was dependent on the polarity of the solvent. Injection volumes of 1.0 μL or less can be injected with chloroform, *n*-hexane, ethyl ether, dichloromethane, and carbon tetrachloride; injection volumes of 0.5 μL or less can be made with acetone and tetrahydrofuran; and maximum injection volumes of 0.2 μL can be made with 2-propanol, methanol, acetonitrile, and pyridine. In the mixing chamber approach, the sample solvent is mixed with the CO_2 mobile phase as soon as possible to dilute the solvent and focus the solutes. This prevents peak splitting. However, if the sample solvent is polar, it takes longer to mix with the nonpolar CO_2, and peak splitting is more difficult to eliminate.

It would appear that the injection pressure must be chosen carefully for each sample solvent to avoid split peaks. The density of the sample solvent, and more importantly the sample solvent polarity, affect the chromatographic peak shapes, primarily due to miscibility. There may also be a stationary phase affect, but no documentation of that currently exists in the literature. For the best performance, conditions must be optimized for the samples being studied.

Choice of injection technique. The choice of injection technique depends on the sample being analyzed, the information and performance desired, and the column configuration employed. For packed columns with ≥0.5-mm i.d., direct injection is most widely used. With the internal loop valve injection systems currently installed on most commercially available SFC instrumentation, one can inject 0.06, 0.2, 0.5, or 1.0-μL volumes directly. If solvent venting is used, larger volumes can be injected. Timed-split injection can be successfully used with packed as well as open tubular columns. This gives the advantage of being able to easily change the amount of injected material.

With narrow bore (<0.5-mm i.d.) columns, there are other considerations, since injections of ≥0.5 μL are more challenging. Conventional split injection is easy to implement; however, it is plagued with problems that make good quantitative performance difficult. Discrimination, nonlinearity, changing split ratios, plugged splitters, etc., are some of these problems. If care is taken when making connections, and if internal standard methods are used, good accuracy and precision can, however, be achieved. Split injection cannot be used for trace analysis because of the small amounts injected. Timed-split resolves most of the practical problematic issues of split injection, but it is still limited to injection volumes of <0.5 μL, making it still less than desirable for trace analysis.

For injection of volumes larger than 0.5 μL into open tubular columns, a different approach is needed. A delayed split approach or one of the solvent venting techniques can be employed. These are useful if the components of interest are somewhat less volatile than the solvent; otherwise, loss of analytes will occur. In all cases, reductions in the width of the solvent peak and in peak tailing are obtained by returning the sample valve to the "load" position soon after injection.

6.8 General Operating Parameters and Methods

Selection of the linear velocity. The linear velocity of the mobile phase (u) is an important parameter in chromatography which affects the most important performance factors: efficiency and the speed of analysis. The general van Deemter equation (Equation 1.11) shows that there is an optimum linear velocity (u_{opt}) of the mobile phase for a given chromatographic system at which the height equivalent to a theoretical plate (h) is at a minimum, or the efficiency is at a maximum. The optimum linear velocity of the mobile phase in a 50-μm i.d. open tubular column is 0.1 cm s^{-1} to 0.2 cm s^{-1}. In practice, because of the otherwise long analysis times, a linear velocity is selected which is ~10 times higher than the optimum linear velocity. The linear velocity can be changed by varying the length of the frit restrictor, or by reducing the orifice diameter in a tapered restrictor.

Influence of pressure. An important difference between GC and SFC is the solvating power of the SFC mobile phase. Since solubility is dependent on density, parameters such as temperature and pressure are important for controlling retention in SFC.

An increase in pressure produces an increase in density at constant temperature. The density will influence both the solvating power and the diffusion coefficients. In general, at low densities (*e.g.*, 0.2 g mL^{-1}), a supercritical fluid is more like a gas with low solvating power and relatively high diffusion coefficients. At high densities (*e.g.*, 0.7-1.0 g mL^{-1}), a supercritical fluid is more like a liquid with strong solvating power and low diffusion coefficients. In SFC, density (or pressure) programming is the primary method for developing a separation; it is analogous to temperature programming in GC, and eluent composition programming in LC.[51-53] Both density programming and composition programming in SFC are comparable, because they influence the solvating power of the mobile phase. During a positive density or pressure program, the solvating power increases with increasing density, but at the same time the diffusion coefficients decrease, which results in a loss in efficiency.[54]

Influence of temperature. Another important physical property which influences solvating power, efficiency, and selectivity is the temperature. In supercritical CO_2, at a pressure of 100 atm and at a temperature of 50°C, the density is 0.41 g mL^{-1}. If the temperature is increased while the pressure remains constant, the density and, consequently, the solvating power decreases. Supercritical CO_2 at a pressure of 100 atm and at a temperature of 150°C has a density of only 0.33 g mL^{-1}. This effect means that the solvating power can be increased by performing a negative temperature program.

On the other hand, the mobile phase solvating power can also be increased by increasing the temperature. Compounds are more soluble in solvents as they are warmed as long as the density remains constant. This turns out to be more effective for increasing solubility than by reducing the temperature as described in the previous paragraph. However, care must always be taken to not exceed temperatures at which the analytes are stable.

The temperature also influences the diffusion coefficients of solutes in the supercritical fluid.[55] With increasing temperature, diffusion coefficients increase and higher chromatographic efficiency results. For this and for increased solubilities, the temperature in SFC should be moderately elevated (column temperatures between 80°C and 150°C are recommended). When using density programming to increase the solvating power of the mobile phase, a simultaneously applied temperature program can compensate for the loss of efficiency which results from the decreasing diffusion coefficients with increasing density.

Another physical property which depends on the temperature is the entropy. When strongly ordered stationary phases are used in

SFC, which separate solutes according to size and shape, the temperature affects the selectivity of the stationary phase. Examples of these types of separations are the resolution of polycyclic aromatic hydrocarbons on liquid crystal phases[56] or the separation of optical isomers on chiral phases.[57] These columns have high selectivity at low temperatures. Increasing the temperature provides more energy which results in a loss of order of the stationary phase and causes an increase in entropy and a decrease in selectivity.

6.9 Pressure and Density Programming

Pressure or density programs are most frequently used in SFC; to date, temperature and eluent composition programming are of less importance. All commercially available instruments are capable of pressure programming; for the production of density programs, a microcomputer must be used to calculate density values from pressures and temperatures.

The selection of pressure or density programs in SFC follows the same rules as for selecting temperature programs in GC or eluent composition programs in LC. Analysis times are reduced by using faster ramp rates as is demonstrated in Figure 6.6, in which separations of an *n*-alkane mixture with pressure ramp rates of 35, 100, and 160 atm min^{-1} are compared.[58] Such fast ramp rates allow extremely short analysis times. Pressure ramp rates of most commercially available SFC instruments are limited to 100 atm min^{-1}, whereas the maximum pressure limit is between 425 and 500 atm.

For fast method development, density programming is superior to pressure programming. Of course, there is basically no difference, but prediction of the retention behavior is much easier by density programming. For example, the logarithm of capacity ratios is a linear function of the reciprocal temperature at constant density, while it is a complicated function of temperature at constant pressure (see Section 3.7). At constant density, retention is also related to the molecular mass of the components of a mixture of similar compounds, such as a homologous series

$$\ln k = a + b M \tag{6.2}$$

where a and b are constants and M is the molecular mass of the solute. For polycyclic aromatic hydrocarbons, for example, the molecular mass is related to the number of carbon atoms. Therefore, plots of ln k *vs* carbon number give straight lines (Figure 6.7).[59]

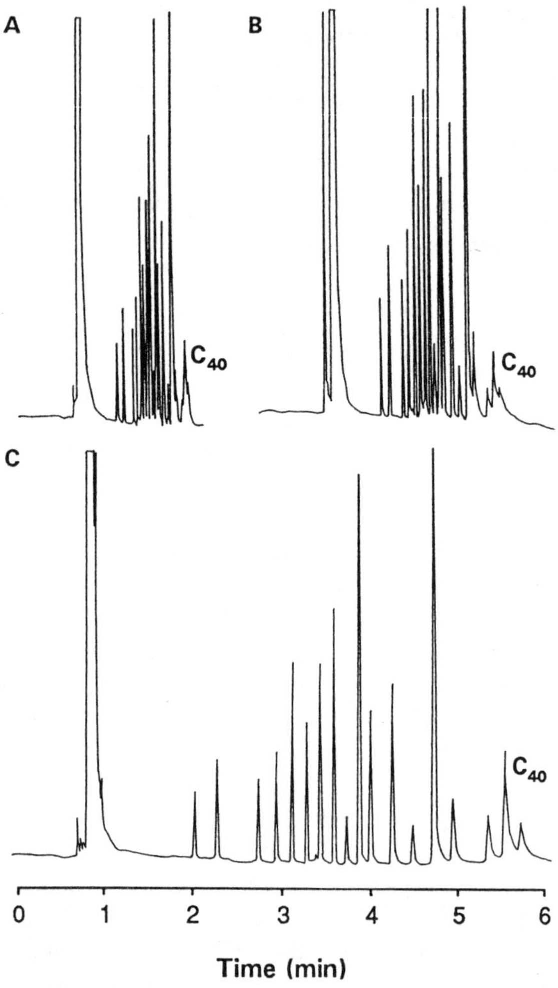

Figure 6.6. SFC chromatograms of an n-alkane mixture with pressure ramp rates of (A) 160, (B) 100, and (C) 35 atm min^{-1}. Conditions: 1.5-m x 25-μm i.d. open tubular column, poly(5% phenyl)methylpolysiloxane stationary phase; CO_2; 100°C (reprinted with permission from Ref. 58).

The translation of a linear pressure program into an optimized density program is shown in Figure 6.8.[54] Although these separations of a polystyrene sample (ave. mol. wt. = 2,000 g mol^{-1}) are no longer state-of-the-art in SFC separations, they clearly demonstrate the different program types applied. With an isothermal (210°C) linear pressure program and n-pentane as the mobile phase (Figure 6.8A), high resolution was obtained in the first part of the

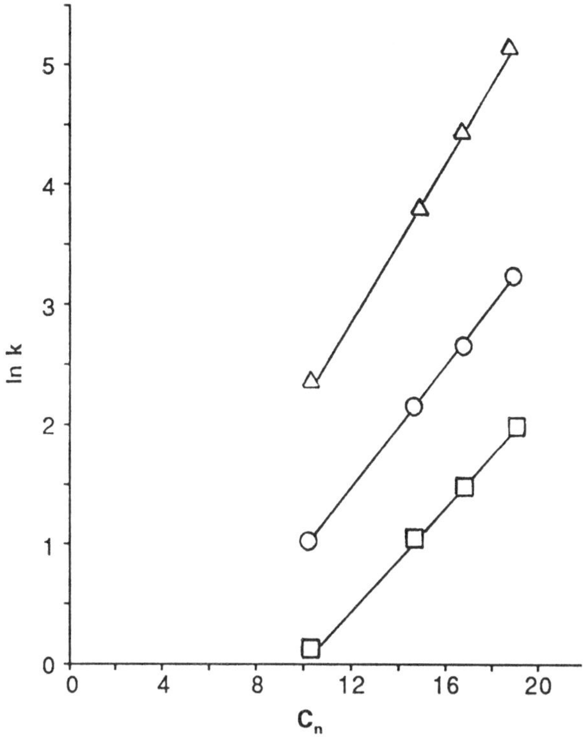

Figure 6.7. Logarithm of capacity ratio as a function of carbon number of polycyclic aromatic hydrocarbons at (△) 110°C, (o) 150°C, and (□) 200°C. Conditions: packed column, silica particles; propane (reprinted with permission from Ref. 59).

chromatogram, but this deteriorated somewhat in the high molecular mass region. Peaks of the higher oligomers were quite broad. These peaks were compressed when the pressure program was substituted by a linear density program (Figure 6.8B). For the elution of compounds in a homologous series during a linear density program, the retention time of an oligomer (t_n) was found to depend on the reciprocal of the degree of oligomerization (n)

$$t_n = t_\infty - \frac{A}{n} \tag{6.3}$$

where t_∞ is the elution time of a hypothetical oligomer with an infinite degree of oligomerization, and A is a constant. Therefore, the distance between two oligomer peaks decreases as the degree of oligomerization increases. From these observations, an asymptotic

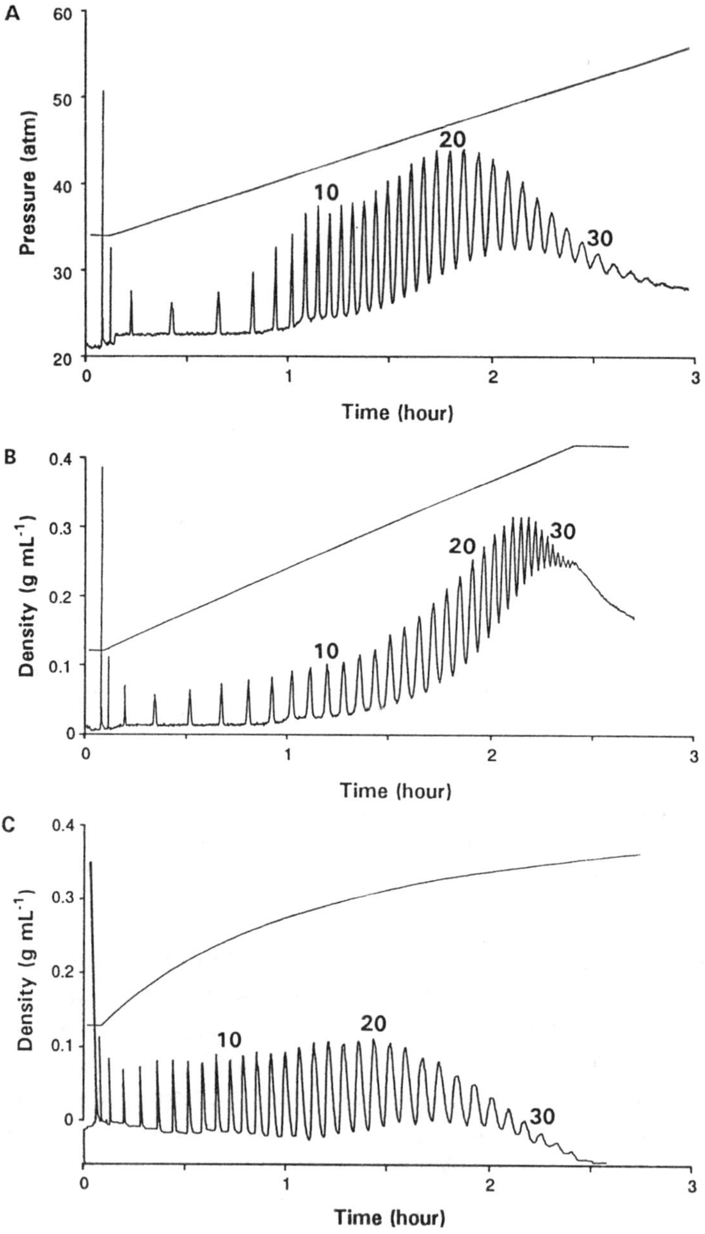

Figure 6.8. SFC chromatograms of a mixture of polystyrene oligomers (PS 2000) by applying (A) a linear pressure program, (B) a linear density program, and (C) an asymptotic density program. Conditions: 10-m x 100-μm i.d. open tubular column, poly(50% phenyl)methylsiloxane; *n*-pentane; 210°C; FID (reprinted with permission from Ref. 54).

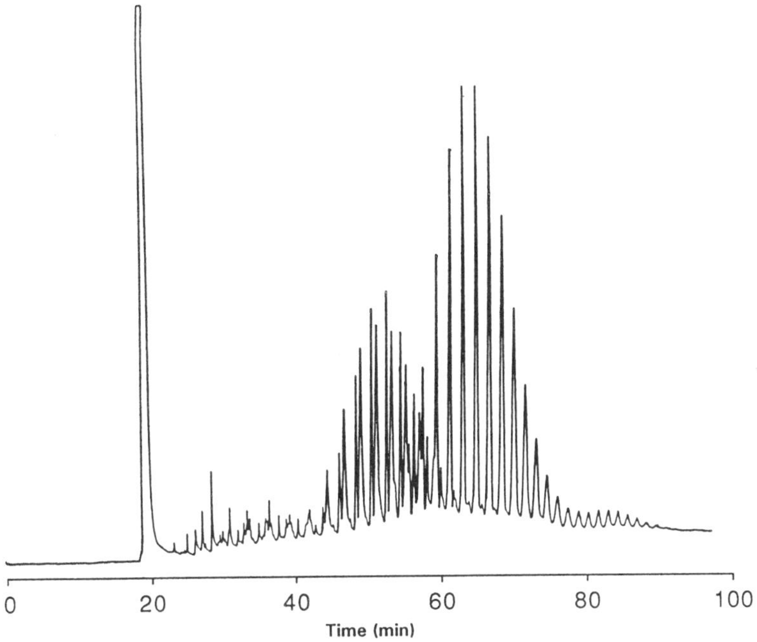

Figure 6.9. SFC chromatogram of a condensation product of stearic acid and polyethylene glycol. Conditions: 20-m x 50-μm i.d. open tubular column, polymethylsiloxane; CO_2; 120°C using a multilinear density program; FID (reprinted with permission from Ref. 60).

density program, *i.e.*, a rapid density increase at first which eventually flattens out, results in almost equal spacing of all oligomer peaks and near baseline resolution (Figure 6.8C). Usually, asymptotic density programs are calculated on the basis of the half rise time of the density and either a density or a time cut-off. A valuable approximation to an asymptotic density program is a multilinear density program which has been applied to obtain the separation of a condensation mixture of stearic acid and polyethylene glycol (Figure 6.9).[60]

At the end of either a pressure or density program, care must be taken during decompression. Especially with open tubular columns, the mobile phase can dissolve into the stationary phase at high densities. A rapid decrease in density could cause physical damage to the stationary phase film by the expanding mobile phase. Therefore, a moderate decompression ramp is recommended.

In packed column SFC, pressure and density programs can be achieved by varying the column outlet restriction by using a needle or a reciprocating valve (Section 4.9). These valves can be controlled electromechanically. The mobile phase flow rate can be held constant or varied during a pressure-programmed run. In SFC using open tubular columns, the column outlet usually is a fixed restrictor having a constant flow resistance (Section 4.8). A pressure/density program is created by increasing the piston speed of the syringe pump, resulting in higher flow rates and linear velocities of the mobile phase. This increase is more pronounced as the analysis temperature is increased. Therefore, separation efficiency in open tubular column SFC systems decreases at higher pressure/density because of two reasons: first, at a higher density, diffusion of the sample molecules is hindered; and second, one moves away from the h-u minimum in the van Deemter plot to higher linear velocities and worse h-values. Therefore, the highest resolution is observed in the first part of a pressure/density programmed separation. The effect of hindered diffusion at higher densities may be overcome in part by adding a simultaneous temperature program (refer to Section 6.10). For constant-flow analyses in open tubular column SFC, new restrictor concepts must be developed (refer to Section 4.9).

6.10 Temperature Selection and Programming

The analysis temperature is one of the main parameters that is chosen before starting a chromatographic run. The temperature is usually chosen to be beyond the critical temperature of the mobile phase. In addition, the thermal stability of the compounds to be analyzed must be considered. Many compounds such as peroxides, azo compounds, explosives, and pharmaceuticals decompose at moderate to high temperatures. If additional peaks from decomposition products are observed at certain analysis temperatures, or if quantification fails, the analysis temperature should be lowered. On the other hand, the analysis temperature should be as high as possible in order to approach the "gas-like" region where diffusion coefficients and separation efficiencies are high (see Section 3.7). Therefore, a mobile phase with a low critical temperature is most desirable. Even though CO_2 has a critical temperature of 31.2°C, a temperature of 100°C is usually the first choice for a separation. For the analysis of thermally stable, nonvolatile components like oligomers, analysis temperatures may be

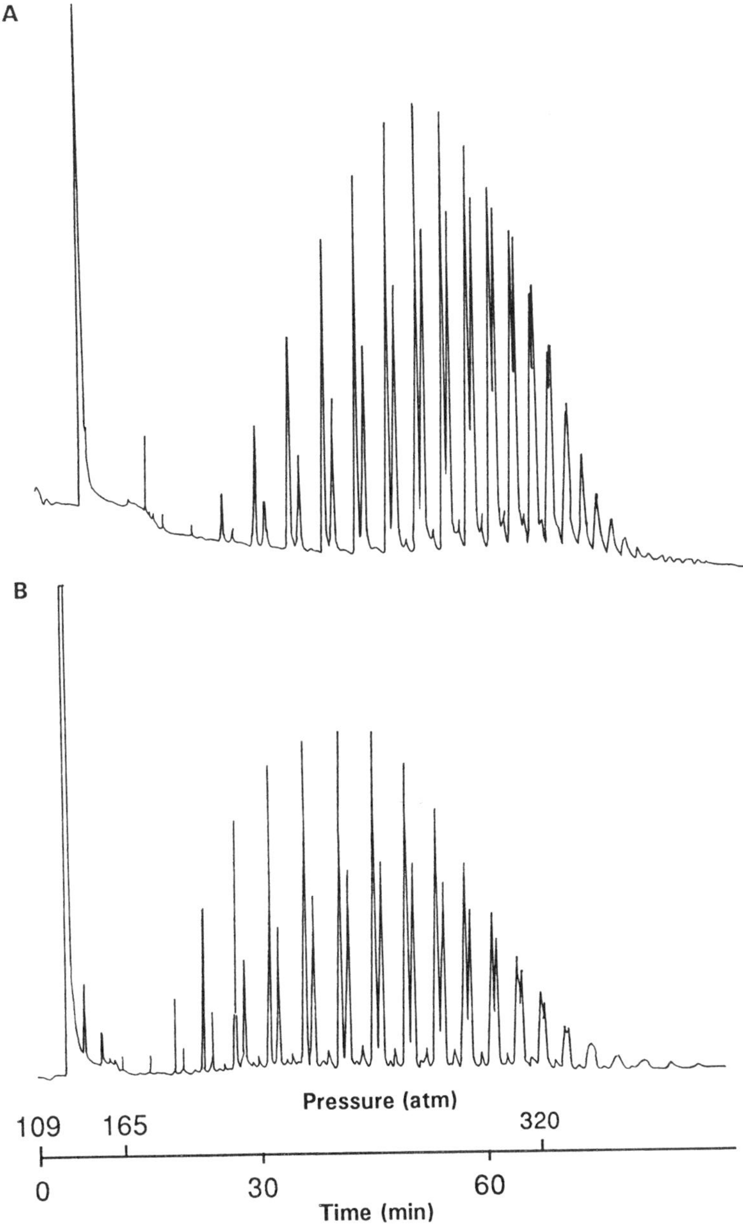

Figure 6.10. Pressure-programmed separation of polyethoxylate oligomers at (A) 120°C and (B) 160°C. Conditions: 25-m x 50-μm i.d. open tubular column, polymethylsiloxane stationary phase (d_f = 0.5-μm); CO_2; pressure program from 109 to 165 atm at 40 atm min^{-1}, 165 to 267 atm at 2 atm min^{-1}, and 267 to 320 atm at 1 atm min^{-1} after a 2-min isobaric period; FID (reprinted with permission from Ref. 61).

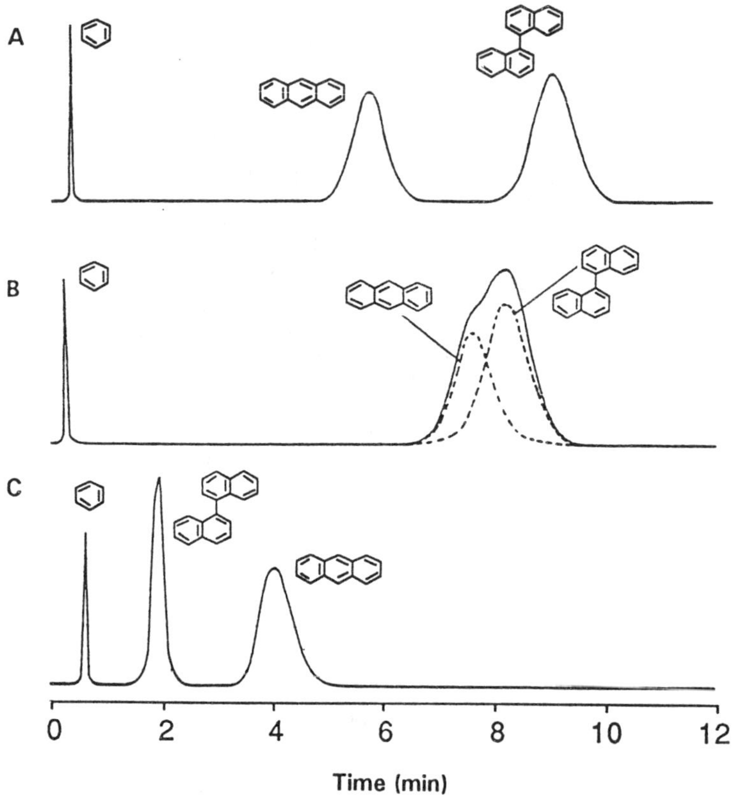

Figure 6.11. Effect of temperature on the separation of anthracene from 1,1′-binaphthyl at (A) 300°C, (B) 250°C, and (C) 200°C. Conditions: 4.25-cm x 6-mm i.d. packed column, alumina particles, 120-140 mesh; *n*-pentane; 38.7 atm; UV (251 nm) (reprinted with permission from Ref. 62).

increased (Figure 6.10).[61] Thermally labile samples may be analyzed at temperatures as low as 40°C; at lower temperatures, reproducibility becomes unacceptable. At constant analysis pressure, predictions of retention behavior at varying temperatures is difficult; at constant density, analysis speed is faster at higher temperatures (see Section 3.7).

If a sample mixture contains components of different chemical structures, varying the temperature may show surprising results. Since selectivity is a function of temperature, it may be tuned by adjusting the temperature. Figure 6.11 shows the coelution of anthracene and 1,1′-binaphthyl at 250°C, but they separate at lower

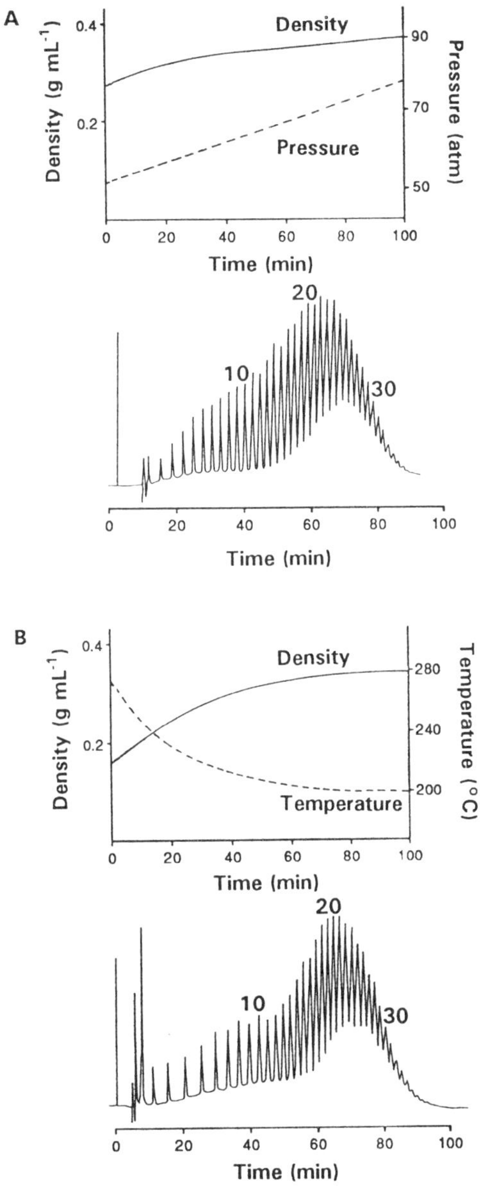

Figure 6.12. Separation of an oligostyrene mixture (PS-2000) by (A) positive pressure program at 210°C and (B) negative temperature program at 60 atm. Conditions: 1-m x 0.25-mm i.d. packed capillary column, C_{18}-particles (Develosil Lop-ODS); *n*-pentane; UV (210 nm) (reprinted with permission from Ref. 63).

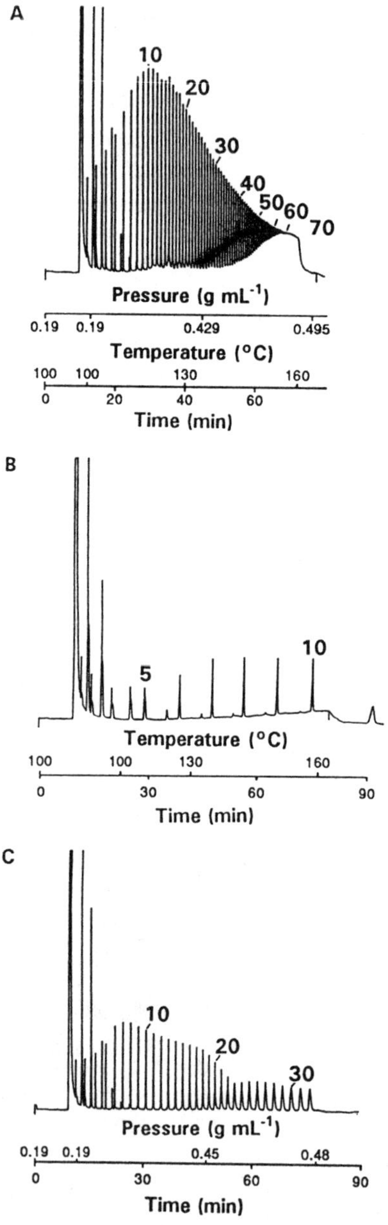

Figure 6.13. Separation of polymethylsiloxane by (A) combined programming of density and temperature, (B) temperature programming at constant density, and (C) density programming at constant temperature. Conditions: 10-m x 50-μm i.d. open tubular column, polymethylsiloxane stationary phase (d_f = 0.25 μm); CO_2 (reprinted with permission from Ref. 64).

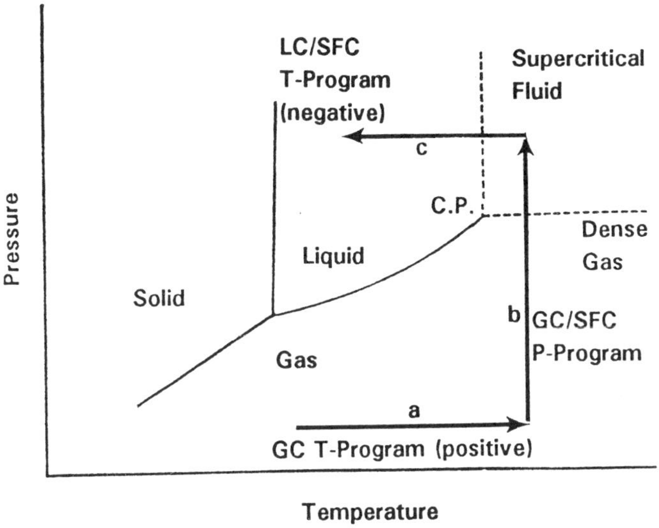

Figure 6.14. Schematic diagram of a programming sequence consisting of a positive temperature program **a** followed by a positive pressure program **b** and by a negative temperature program **c**. C.P. = critical point (reprinted with permission from Ref. 65).

temperature (200°C) as well as at higher temperature (300°C).[62] In fact, their retention order is reversed at 300°C compared to 200°C.

Density programming is achieved either by positive pressure programming (Sections 3.6 and 6.9) or by negative temperature programming. While the separation of a polystyrene 2000 sample on a packed column in Figure 6.12A was performed using a pressure program at isothermal conditions, a similar separation was obtained using an asymptotic negative temperature program at a constant analysis pressure (Figure 6.12B).[63] Since the starting density is lower for the temperature programmed separation, the resolution in the first part of the chromatogram is slightly better.

Density (pressure) programs and temperature programs can be combined either simultaneously or consecutively. Prediction of retention behavior is more difficult when using simultaneous pressure/temperature programming than when using simultaneous density/temperature programming. The successful elution of a wide-ranging polymethylsiloxane sample was achieved by superimposing a temperature program on a density program (Figure 6.13A), while isopycnic (Figure 6.13B) and isothermal conditions (Figure 6.13C) only eluted small portions of the sample.[64]

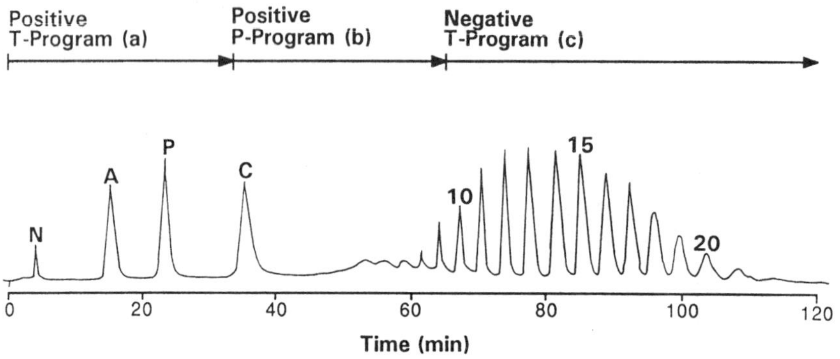

Figure 6.15. SFC chromatogram of a PAC and polystyrene mixture using a consecutive temperature/pressure/temperature program. Conditions: 25-cm x 4.6-mm i.d. packed column, silica particles (10 μm); 5% 1,4-dioxane in n-pentane; UV (254 nm) (reprinted with permission from Ref. 65).

By using a temperature program, the resolution of complex mixtures can be improved. At low densities and low temperatures, the first part of the separation is "stretched," while the peaks of the slowly eluting nonvolatile components are improved at the higher analysis temperatures.

While the differences between the special physical properties of the "gas-like" and "liquid-like" regions in SFC are somewhat leveled by running simultaneous density/temperature programs, they are emphasized using consecutive density/temperature programs. Figure 6.14 schematically describes a temperature/pressure/temperature program which touches the gas region as well as the supercritical and liquid regions.[65] Starting at a subcritical analysis pressure, the analysis temperature is increased from subcritical to supercritical values. At the same time, the density of the mobile phase decreases and elution is effected by an increase in volatility. During this period, the eluent is a gas. At the end of the isobaric temperature program, the most "gas-like" point in the program is reached. Now, the analysis pressure is increased while keeping the temperature constant, resulting in a transition from the gaseous to the supercritical region. During this period, the solubilities of the analytes are enhanced due to the increasing density. To elute the highest molecular mass components in the sample, a negative temperature ramp is applied, further increasing the eluent density.

At the end of the complete program, the mobile phase is a liquid. By applying the program described above, the regions of GC and LC, as well as SFC, are covered. Therefore, this technique has been named "unified" chromatography.[66]

In Figure 6.15, the application of a consecutive pressure/-temperature/pressure program is demonstrated by the analysis of a mixture of four polycyclic aromatic hydrocarbons (naphthalene, anthracene, pyrene, and chrysene) and a polystyrene sample.[65] While naphthalene, anthracene, and pyrene eluted during the positive temperature program when the eluent was a gas, chrysene and the lower styrene oligomers eluted during the pressure program. The largest oligomers did not elute before the density was increased using a negative temperature program.

Most analyses by SFC are accomplished under isothermal conditions. However, the temperature must be chosen carefully. More sophisticated separations may require temperature programming, either simultaneously or consecutively.

6.11 SFC Trouble Shooting

Symptom	Cause(s)	Solution(s)
Pump failure.	Pump electronics overheat, problem with supply current.	Check if fan is operating.
		Check that no electrical wires are caught in fan.
		Switch pump on and off.
Poor pump cylinder refill (syringe pumps).	Low mobile phase cylinder pressure.	Replace mobile phase cylinder and/or include 15-30 min equilibration time when filling.
	Warm pump cylinder.	Cool pump cylinder. Check cooling system.

Symptom	Cause(s)	Solution(s)
Pump unable to reach or keep high pressures.	Leaks.	Check for leaks in entire system.
		Plug outlet flow and see if pumping is high to hold a constant pressure.
		Plug pump cylinder outlets and check for leaks by carefully listening, feel if connections are cold, use soap solution on connections and watch for foam.
No flow/pressure (reciprocating pumps).	No delivery of mobile phase. Bubbles or cavitations.	Degas or vent. Increase prepressure. Check level(s) of mobile phase container(s).
Leaky connections.	Too-hard material in ferrules and/or sleeves with loose fit around tubing or columns.	Graphite or graphitized Vespel ferrules tend to flow and seal well.
	Ferrules not properly seated.	Replace ferrules.
No detector response.	No sample getting into the column.	Check that there is carrier flow at the restrictor outlet.
	Plugged column or restrictor.	Make sure that there is flow through the column. If not, check to see if the restrictor is plugged. If it is, replace it. Check the column to see if it is

Symptom	Cause(s)	Solution(s)
No detector response. (continued)	Plugged column or restrictor. (continued)	plugged. If it is, remove plugged portion of the column or replace the column.
		If packed column, rinse with solvent and/or replace column frits.
	Poor injector alignment.	For rotary valve injector, make certain that the injection valve rotor is properly aligned with the injection port. Make sure that the valve is not plugged. If it is plugged, clean the valve and splitter assembly by sonicating them in an appropriate solvent.
	Injector valve not operating properly. (pneumatic actuator)	Make sure the injection valve pneumatic system is working properly. Check the helium tank pressure and check for leaks to make sure the valve is moving into the inject position.
	Leaks in the system	Check for leaks at all connections.
	Valve malfunction (reciprocating pumps).	Clean valve (sonicate in acetone).

Symptom	Cause(s)	Solution(s)
No detector response. (continued)	Valve malfunction. (continued)	Use filter upstream of the valve.
		Replace valve.
	Different pressures in the pumps (dual syringe pumps).	Adjust pressures, use check valves in pump connection line.
		Compensate for different compressibilities.
		Recalibrate both transducers.
	No mobile phase.	Fill pump.
	Too low solubilities of the analytes in the mobile phase	Change mobile phase or add modifier.
	Sample not arriving at the detector.	Make sure that the column and/or restrictor is positioned correctly in the detector.
	Amount of solutes getting into the detector below the detection limit.	Inject more or improve focusing.
	Detector does not match or give response for the solutes.	Change detector.
	Detector not working properly.	For FID, check to see if the flame is on. Also check that O-ring on detector base

Symptom	Cause(s)	Solution(s)
No detector response. (continued)	Detector not working properly. (continued)	is not leaking and that the detector gas flows are set properly.
		For UV detector, make sure that the light source is functioning. Be sure that appropriate wave-length has been selected for your sample.
		Check the alignment of the flow cell and make sure there is flow through the cell by checking flow out of the restrictor.
		Make sure that the flame is lit.
Broad solvent peak.	Unswept dead volume in the elution flow path.	Make sure there is no unswept dead volume in the butt connections for open tubular columns and at restrictor connection on outlet of packed columns.
		Make sure injector valve is properly aligned.
		Check for flow restrictions.
		Check the connection of the column to the injection valve,

Symptom	Cause(s)	Solution(s)
Broad solvent peak. (continued)	Unswept dead volume in the elution flow path. (continued)	making sure there is no unswept dead volume when direct column connection is used or where the splitter assembly is connected to the valve. Check that line has not clogged with undissolved analyte. In the split injection mode, position the column 1-3 cm from the valve. Install a splitter with a small internal diameter. In timed split or direct injection, make sure there is no dead volume in the carrier inlet connection to the injection valve. Use a 5 to 10-cm x 0.005-inch i.d. tube or a 50-100-μm i.d. fused silica capillary at the carrier inlet connection to the valve. Make sure the tube is cut square and is not rough on the surface. Then make sure the tube is seated properly in the valve. This precaution will lessen possible back-diffusion of solute in this region.

Symptom	Cause(s)	Solution(s)
Broad solvent peak. (continued)	Improper size tubing used to install packed columns.	Use square cut (preferably laser-cut) 0.007-inch i.d. tubing to connect microbore packed columns. Make sure the tubing is seated properly in the valve and column end fitting.
	Excessive solvent volume injected.	In split injection, make sure there is enough flow from the split vent.
		Install a rotor with a smaller volume loop.
		In timed split injection, make sure that the injection duration does not introduce too much sample volume on the column. If it does, change the injection duration to a shorter time.
		If large sample volume must be injected, use injection techniques with solvent elimination.
	Improper mobile phase linear velocity.	Check carrier shut off valve to make sure it is on. If it has a small leak, it may appear that the linear velocity through the column is too slow.

Symptom	Cause(s)	Solution(s)
Broad solvent peak. (continued)	Improper mobile phase linear velocity. (continued)	Make sure the column restrictor gives the proper flow for the column being used.
		Refill pump if it is out of mobile phase.
		Check for blockages in tubing downstream of the pump.
		Make sure the injection rotor is aligned in the flow path.
	Injector rotor scored.	Replace rotor.
	Injector rotor out of round (distorted due to overtightening of the nut or tension adjustment).	Replace injection valve or rotor.
	Post detector restrictor clogged (*e.g.*, UV).	Keep restrictor end in solvent. If plugged, hold a heat source at the end of the restrictor. If this fails, replace restrictor or cut off end.
		Place restrictor in oven or tape against UV lamp housing.
Spiking and/or tailing of trailing edge of the solvent peak.	Dead volume in the connection of the column to the injection valve.	Position column inlet so that connection area is swept by the mobile phase. Adjust

Symptom	Cause(s)	Solution(s)
Spiking and/or tailing of trailing edge of the solvent peak. (continued)	Dead volume in the connection of the column to the injection valve. (continued)	the column in the connection by positioning it at the point where the ferrule sits flush with the inlet end of the column.
	Dead volume in the connection of the restrictor to the outlet of the packed column.	Make sure that the connecting tubing and ferrule are seated properly in the column end fitting.
	Splitter restrictor is plugged or partially plugged.	Trim the end of the splitter restrictor so that flow is reestablished, or replace the restrictor.
	Detector temperature is too low.	Increase detector temperature.
	Restrictor is too fast.	Replace restrictor.
	Poor butt connection.	Remake connection.
	Sample is back-flushed into mobile phase inlet to the injector during injection.	Use 50-100-μm i.d. tubing as inlet tubing to the injector.
	Use of certain injection solvents.	Try another injection solvent (*e.g.*, with CO_2 mobile phase, spiking is not as pronounced with hexane as with ethyl acetate or toluene).

Symptom	Cause(s)	Solution(s)
Split solvent peak.	Poor connection of column to injection valve, or column to restriction capillary.	Remake connections.
Broad humps after solvent peak.	Usually eddy current in the connection between the column and restrictor.	Remake connection.
	Insufficient miscibility of sample solvent in mobile phase and/or slow purge of splitter.	Use smaller volume splitter, faster split flow, or mixed solvents to help miscibility.
Split chromatographic solute peaks.	Unswept dead volume in the chromatographic system.	Check for dead volume in the column connection to the injection valve, and check the butt connectors.
	Column overloading.	Make sure the concentration of the sample is appropriate for the column being used.
		Make sure the split vent has proper flow and is not plugged.
		Be sure a correct injection time is used in timed split injection.
	Multiple sample plugs when injecting large volumes.	Start program at low pressure followed by fast pressurization to regular initial pressure.

Symptom	Cause(s)	Solution(s)
Split chromatographic solute peaks. (continued)	Sample may adsorb to the walls of the precolumn (if used).	Place the precolumn inside the oven.
	Injector rotor scored or worn unevenly.	Replace the rotor.
	Internal dead volume in injector valve.	Eliminate dead volume.
	Excessive solvent volume during injection.	Inject smaller volume.
	Marginal solubility of injection sample solvent in the mobile phase.	Change injection solvent, or mix solvents.
Poor solute peak shape.	System not properly deactivated.	Change to a deactivated column and restrictor, or add modifier to the mobile phase.
	Dead volume.	Check connections.
	Column out of position with respect to injector.	Change column inlet position.
	Too large injection volume or solute concentration.	Inject $\leq$500 nL in open tubular columns and $\leq$5 μL in packed columns. Inject $\leq$400 ng per component in open tubular columns and $\leq$1 μg per component in packed columns.

Symptom	Cause(s)	Solution(s)
Poor solute peak shape. (continued)	Scored injection rotor.	Replace the rotor. If valve body is scored, have it repolished.
	Solute condensation in restrictor.	Increase restrictor temperature or change restrictor type.
Pen does not return to baseline after peak elution (not tailing).	Poor connection between column and restrictor caused by poor cut of open tubular column or overtight connection.	Check and remake connection.
	Solute/solvent diffusion into the mobile phase supply tube or fitting connected to the injection valve.	Use a 0.007-inch i.d. or smaller supply tube; remake fitting.
Overall broad peaks.	Injection solvent not miscible with mobile phase.	Switch to a solvent more miscible with the mobile phase (*e.g.*, use dichloromethane with CO_2, or use mixed solvent).
	Excessive injection volume.	Inject less solvent, or use solvent elimination injection.
	Damaged column.	Break off several inches from the front of the open tubular column, or replace the column.
	Overloaded column.	Change to another stationary phase, higher capacity

Symptom	Cause(s)	Solution(s)
Overall broad peaks. (continued)	Overloaded column. (continued)	column, or inject less sample.
Broad early eluting chromatographic peaks.	Too large volume injected.	Adjust amount injected (*i.e.*, faster time in timed split injection or higher split ratio in split injection).
	Injector rotor scored.	Replace rotor.
	Peaks are eluting during isobaric portion of program.	Inject at a lower density (pressure).
	Injector out of round.	Replace injector.
	Polar compounds run on nonpolar stationary phase.	Select appropriate column.
	Column overloaded.	Dilute sample.
Broad late eluting chromatographic peaks.	Too low detector (restrictor) temperature.	Increase detector temperature. Use direct heat on the restrictor.
No flow from column or split vent.	Split assembly or column is plugged.	Trim off the portion of column that is plugged, or replace column. Then make sure that there is flow from the splitter restrictor. If not, trim off plugged portion of the restrictor or change the splitter restrictor.

Symptom	Cause(s)	Solution(s)
No flow from column or split vent. (continued)	Split assembly or column is plugged. (continued)	Heat restrictor or insert it in a liquid.
		Change to timed split injection. This is recommended when the sample is high molecular mass or reactive. Note that when timed split injection is used, more sample is typically injected as compared to dynamic split and it may be necessary to use a more dilute solution.
		Change solvent used to dissolve sample.
	Sample precipitation in injector or on head of column.	Heat injector and solution.
		Change solvent used to dissolve sample.
	Too much sample injected.	Use larger split ratio by changing split restrictor or using faster injection time in timed split injection.
		Use smaller sample loop in rotor.
	Sample too concentrated.	Dilute sample.

Symptom	Cause(s)	Solution(s)
Elution of only low molecular mass compounds.	Partially plugged or improperly deactivated restrictor.	Change restrictor.
	Too low heat at restrictor.	Increase temperature of restrictor.
	Oven temperature set too low.	Increase oven temperature.
Phantom peaks.	Strongly retained samples on certain stationary phases.	Switch to a less retentive column.
	Retention gap is too short.	Use longer retention gap.
	Dirty mobile phase.	Switch mobile phase cylinder, or use an activated alumina trap between mobile phase supply cylinder and pump.
	Dirty sample preparation equipment.	Wash the equipment with various solvents, or switch to other equipment. Filter sample.
	Dirty connections.	Clean lines and have only one user per cylinder.
	Grease in the pump from the pump seals or valve seals.	Empty and refill the pump 3 or 4 times (25 to 50 mL each time) without letting the pump go to atmospheric pressure between refills.

Symptom	Cause(s)	Solution(s)
Phantom peaks. (continued)	Grease in the system due to contaminated mobile phase.	Replace supply cylinder of mobile phase.
	Poorly conditioned column.	Condition by cycling the density program overnight.
Poor retention characteristics.	Improper restriction.	Check column restrictor to make sure that it has proper flow rate for your selected column. Replace restrictor if necessary.
Too short or no retention of solutes.	Leaky connection at detector end of column.	Make sure there are no leaks at the restrictor butt connection.
	Operating at improper temperature.	Change the operating temperature of the oven.
	Mobile phase flow rate set too high.	Reduce mobile phase flow rate.
	Improper stationary phase for the sample.	Select another column.
	Stationary phase stripped off the column.	Replace column with a column that has immobilized stationary phase.
	Too high solvent strength in the mobile phase.	Reduce polar modifier concentration, add nonpolar modifier, or switch mobile phase.

Symptom	Cause(s)	Solution(s)
Too long retention of solutes.	Too much restriction.	Check that there is proper flow from the column restrictor. If not, trim off part of fixed restrictor or change variable restrictor to increase flow, or replace the restrictor with one that has the proper flow characteristics [*e.g.*, use flow rate of 1-1.5, mL min^{-1} (gaseous) for 50-μm i.d. column at 100-150 atm. Use flow rate of 1.5-2 mL min^{-1} (gaseous) for 100-μm i.d. column at 100-150 atm. Use 40-100 mL min^{-1} flow rate restrictors for microbore packed columns. This is for CO_2 only].
	Partially plugged restrictor.	Check restrictor. Heat restrictor with gas purge.
		Rinse with high density mobile phase overnight.
	Partially plugged union.	Remake connection to union. Use fused silica sleeved low dead-volume butt connectors, PEEK tubing connections, or glass connections.

Symptom	Cause(s)	Solution(s)
Too long retention of solutes. (continued)	Column plugged.	Remove the portion of the open tubular column that is plugged, or replace the column.
		For packed columns, backflush with HPLC pump or remove and sonicate frits.
	Not enough solvent strength in the mobile phase.	Add polar modifier or increase modifier concentration.
	Improper stationary phase selected for the sample being analyzed.	Select new column, providing less interaction with solutes.
	Pressure or density ramp rate has changed.	Check for leaks in system (*e.g.*, observe set point vs. actual during a run to determine if system is pressurizing accurately. If the pump is not keeping up with the set point, check for leaks around the pump seals. If there is a leak, replace pump seals).
	Improper choice of density, pressure, and temperature.	Modification of these three parameters have a large effect on solute retention.

Symptom	Cause(s)	Solution(s)
Irreproducible retention characteristics.	Partial plugging of the column restrictor. This occurs especially when analyzing reactive materials.	Replace the column restrictor. Changing to a different type of restrictor (or a deactivated one) might be advantageous.
	Out of mobile phase.	Refill pump.
	Malfunction of check valves (reciprocating pumps).	Clean or replace valves.
	Pump cylinder warm.	Cool pump cylinder.
Irreproducible retention times when using modifier.	Immiscibility or variation in concentration of modifier in pump cylinder for premixed mobile phases.	Do not cool pump; operate at high pressures, and use long equilibrium time. Avoid temperature gradients and flow dynamics disturbance.
Peaks smaller than expected.	Leaky injector.	Replace seals.
	Incompletely filled sample loop.	Attach a restrictor to the waste port and inject when loop is filled.
	Column plugged or connection loose.	Check seals and/or replace column.
	Restrictor problem.	Replace restriction and/or modify temperature.
Peak area decreases during multiple injections.	Polar material deposits in the restrictor or on the column.	Increase detector temperature, derivatize sample, or inject derivatizing

Symptom	Cause(s)	Solution(s)
Peak area decreases during multiple injections. (continued)	Polar material deposits in the restrictor or on the column. (continued)	agent after each sample analysis.
Inconsistent area counts (poor reproducibility).	Scored rotor in injector, caused by particulates in sample or from column packing blow-back.	Inspect for scoring, replace worn parts, and filter samples. Packed column inlet frit must be tight and in place before application of pressure.
	Injection rotor scored.	Replace rotor.
	Injector head out of round (distorted due to overtightening of the nuts or tension adjustment).	Replace injector head and rotor.
	Injector rotor out of alignment.	Realign rotor with valve.
	Injection valve pressure adjustment nut too tight.	Loosen nut. Replace rotor. Readjust pressure nut. If this does not fix the problem, replace valve.
	Valve inlet tubing from the pump is too large i.d. or not properly seated in the valve.	Replace with small i.d. tubing. Reseat tubing, making sure that it is square and no unswept dead volume.
	Leaks in the column connections.	Check for leaks and tighten any

Symptom	Cause(s)	Solution(s)
Inconsistent area counts (poor reproducibility). (continued)	Leaks in the column connections. (continued)	connections where leaks are found.
	Splitter partially plugged.	Replace splitter restrictor.
	Column not positioned correctly in the injector connection.	In split injection, position column 1-3 cm from the injection valve.
		In timed split, place the column at the end of the ferrule or 1 mm back.
	FID gas flows have changed.	Readjust gas flows.
	Injection valve too hot for solvent being used.	Cool valve or use higher boiling solvent.
	Injection valve not moving all of the way to the inject position.	For timed-split, try longer injection duration.
		Check helium tank pressure. Make sure it is at least 100 psi.
		Check for leaks in the pneumatic system on the injection valve.
	Low solubility of certain components.	Increase temperature of injection valve from ambient to ~50°C and/or change solvent if possible.

Symptom	Cause(s)	Solution(s)
Inconsistent area counts (poor reproducibility). (continued)	Loss of volatile solvent from syringe or valve during injection.	Switch injection solvent, if possible.
	Accumulation of material in restrictor during on-line SFE.	Switch to deactivated restrictor; adjust flow of cooling gas to cryofocusing temperature.
Low FID Sensitivity.	Incorrect flowrates of mobile phase, causing different flame properties.	Optimize make-up gas flow, maybe up to 60 mL min^{-1}.
High background current from FID.	Detector flame was off for period while cycling the SFC pump.	Leave flame on continuously.
Baseline shift when pressurizing (FID).	Contaminated carrier or carrier delivery system. Check baseline rise by ramping to highest pressure while studying the FID signal. If the observed rise stays constant when the pressure is isobaric for 30 min, there is some contamination in the carrier stream.	Make sure right quality carrier is used. If mobile phase trap is used, make sure that there is clean neutral alumina in the carrier fluid trap. Replace lines and clean any parts that may be dirty.
	FID gases not set properly, causing sensitivity change as a function of pressure.	Check and optimize the flame gases. The largest baseline rise will be observed when the hydrogen flow is set too high.

Symptom	Cause(s)	Solution(s)
Baseline shift when pressurizing (FID). (continued)	Carrier flow affects flame.	Check flow rates of FID detector gases.
		Make sure the column restrictor is not protruding into the flame.
	High mobile phase flow rate.	Move restrictor down 1 mm from the flame. If problem persists, try moving the restrictor down further (can be positioned as much as 3-4 mm from flame).
		Turn off make-up gas, and/or make a hotter flame.
	Poor electrical connections.	Check all electrical connections, especially the signal cables. Be sure that external power sources are not causing interference.
Uneven baseline before a peak.	Excessive flooding of sample during injection with inadequate refocusing.	Reduce volume injected. Use a retention gap with the first 30 cm at room temperature.
FID spiking, especially at high pressure.	Restrictor not heated properly. Sample may be overloading restrictor.	Reduce sample load. Increase detector temperature. Make sure to have good heat transfer to restrictor.

Symptom	Cause(s)	Solution(s)
FID spiking, especially at high pressure. (continued)	High molecular mass particles entering flame.	Increase detector temperature.
	Restrictor failure.	Replace the restrictor.
	Column restrictor protruding into the flame.	Reposition the column so that it is just below the base of the flame.
	Material (carryover) coming off column.	Make several solvent injections at high pressure to reduce carryover. Cycle column to clean it, or remove and clean the column by rinsing it with an organic solvent.
	Excessive column flow at high pressure.	Position restrictor down 2-3 mm from the flame jet tip. If the problem persists, increase the restriction.
Excessive negative baseline shift when pressure or density programming (UV).	Large refractive index change due to changing density of the mobile phase.	Thermostat flow cell so that the detection takes place with the mobile phase in a liquid state.
	Improper alignment of the flow cell.	Realign the flow cell so that maximum throughput of light is achieved.
	UV detector heating up.	Cool UV detector cell.

Symptom	Cause(s)	Solution(s)
Excessive negative baseline shift when pressure or density programming (UV). (continued)	Leak in detector cell.	Check connections.
	Gradient in pre-mixed mobile phase cylinder.	Mix thoroughly or add modifier directly into syringe pump.
	Contaminants in mobile phase.	Use more pure mobile phase, or add adsorbent trap.
Excessive baseline rise when pressure or density programming (UV).	Wavelength selected too close to the absorbance of the mobile phase.	Select appropriate wavelength. Change to a modifier with low absorbance.
Spikes in baseline or excessive noise in the baseline (UV).	Helium head space in carrier cylinders result in out-gassing and bubble formation.	Do not use helium head space in mobile phase cylinders.
	UV flow cell contaminated.	Flush the flow cell with solvent.
	Fingerprint or other dirt on flow cell window.	Carefully clean the flow cell window with solvent and non-abrasive, lint-free tissue.
	Leak around the cell windows.	Reassemble.
	Cell window(s) broken.	Replace window(s).
	Bubbles or droplets in the cell.	Check miscibility of mobile phase

Symptom	Cause(s)	Solution(s)
Spikes in baseline or excessive noise in the baseline (UV). (continued)	Bubbles or droplets in the cell. (continued)	components and change pressure or temperature, or use different mobile phase composition.
		For diaphragm pumps, check diaphragm for leaks.
	Silica "fines" trapped in the UV cell when using a packed column.	Clean cell and use new column.
	Leaks in the flow cell connections.	Check all connections for leaks. If a liquid leak detection solution is used, make sure that none of the solution comes in contact with the flow cell window. If the flow cell window is contaminated, clean it with a solvent and a nonabrasive, lint-free tissue.
	Bubble trapped in the flow cell.	Remove the bubble by ramping to a higher pressure while keeping a restriction after the cell.
		Increase pressure, degas eluent, or check for leaks.
	Temperature fluctuations in the flow cell.	Make sure that the flow cell is thermostated below

Symptom	Cause(s)	Solution(s)
Spikes in baseline or excessive noise in the baseline (UV). (continued)	Temperature fluctuations in the flow cell. (continued)	the critical temperature of the carrier.
		Check the insulation around the column outlet to the detector flow cell. Make sure that no hot air from the oven is blowing on the flow cell.
	Stray light reaches the photomultiplier.	Make sure that the flow cell cover is properly positioned and secured.
	Poor electrical connections.	Check all electrical connections, especially the signal cables. Be sure that external power sources do not cause interference.
	Background scans used as reference spectra are collected under conditions at or near the critical point where the fluid is opalescent (this occurs because the flow cell is after the packed column pressure drop).	Make sure the initial pressure is high enough so that the pressure in the flow cell is above the critical pressure.
	Reference spectrum region from a reconstruction is not selected properly.	Select a reference region as close to the peak as possible, preferably on the higher pressure side of a peak (*i.e.*, after

Symptom	Cause(s)	Solution(s)
Spikes in baseline or excessive noise in the baseline (UV). (continued)	Reference spectrum region from a reconstruction is not selected properly. (continued)	the peak). Try to avoid a reference region where the pressure ramp is changing rapidly (*e.g.*, 40-50 atm min^{-1}).
	Column bleed.	Make sure the stationary phase being used is compatible with the mobile phase. Replace the column.
	Mobile phase modifier is not evenly mixed with primary fluid.	Use lower concentration of modifier.
		Increase the temperature of the pump cylinder to improve the miscibility of the modifier in the primary fluid.
		Change to a different and compatible modifier.
	Irreproducibility (run-to-run) in solvent-modified SFC.	Pressure and/or temperature too low for complete mixing of solvents.
Reduced UV signal.	Cell window broken.	Replace window.
	Material deposited on windows.	Clean windows.

Symptom	Cause(s)	Solution(s)
Reduced UV signal. (continued)	Reduced intensity of UV lamp.	Replace lamp.
Poor peak shape for on-line SFE/SFC.	Insufficient solute focusing.	Use longer restrictor from extraction cell. Cool the trap more.
	Too "tight" coolant CO_2 outlet; trap stays pressurized during trapping.	Decrease flow resistance from trap.
	Too large dead volume in trap.	Use smaller trap, or apply solute focusing.
	Adsorptive trap surfaces.	Trap inside deactivated precolumn.
	At the beginning of extraction, column is still pressurized.	Depressurize column before extraction.
Irreproducible SFE/SFC peak areas.	Leaky extraction cell.	Tighten, but frequently need to replace cell.
	Irreproducible cryotrap temperature.	Check cooling fluid flow rate and monitor cryotrap junction temperature.
	Leak in plumbing, or insufficient extraction.	Check for leaks at connections and increase flow rates from reservoir.
	Focusing not sufficient.	Cool the trap more.

Symptom	Cause(s)	Solution(s)
Irreproducible SFE/SFC peak areas. (continued)	SFE restrictor is plugged.	Do not introduce too much water and do not cool the trap too much.
"Phantom" peaks in SFE/SFC.	Adsorption of solutes.	Deactivate all parts.

References

1. H.E. Schwartz, P.J. Barthel, and S.E. Moring, *J. High Resolut. Chromatogr. Chromatogr. Commun.* **10**, 668 (1987).
2. A.C. Rosselli, D.S. Boyer, and R.K. Houck, *J. Chromatogr.* **465**, 11 (1989).
3. N.L. Porter, B.E. Richter, D.J. Bornhop, D.W. Later, and F.H. Beyerlein, *J. High Resolut. Chromatogr. Chromatogr. Commun.* **10**, 477 (1987).
4. D.R. Gere, R. Board, and D. McManigill, *Anal. Chem.* **54**, 736 (1982).
5. A. Giorgetti, N. Pericles, H.M. Widmer, K. Anton, and P. Daetwyler, *J. Chromatogr. Sci.* **27**, 318 (1989).
6. M.L. Lee, B. Xu, E.C. Huang, N.M. Djordjevic, H.-C.K. Chang, and K.E. Markides, *J. Microcol. Sep.* **1**, 7 (1989).
7. E.S. Francis, M.G. Jones, T.J. Shaw, and B.A. Jones, *1989 Symposium/Workshop on Supercritical Fluid Chromatography,* Snowbird, UT, June 13-15, 1989.
8. J.P. Touminen, K.E. Markides, and M.L. Lee, *J. Chromatogr.*, submitted.
9. R.D. Smith, J.L. Fulton, R.C. Petersen, A.J. Kopriva, and B.W. Wright, *Anal. Chem.* **58**, 2057 (1986).
10. S.R. Goates, N.A. Zabriskie, J.K. Simons, and B. Khoobehi, *Anal. Chem.* **59**, 2927 (1987).
11. B.A. Jones, Lee Scientific, unpublished results.
12. W.M.A. Niessen, P.J.M. Bergers, U.R. Tjaden, and J. Van Der Greef, *J. Chromatogr.* **454**, 243 (1988).
13. Q.L. Xie, K.E. Markides, and M.L. Lee, *J. Chromatogr. Sci.* **27**, 365 (1989).
14. T.L. Chester and D.P. Innis, *J. High Resolut. Chromatogr. Chromatogr. Commun.* **9**, 209 (1986).
15. T.L. Chester, J.D. Pinkston, D.P. Innis, and D.J. Bowling, *J. Microcol. Sep.* **1**, 182 (1989).

16. J.C. Kuei, G.R. Her, and V.N. Reinhold, *Anal. Biochem.* **172**, 228 (1988).
17. J.L. Veuthey, M. Caude, and R. Rosset, *Chromatographia* **27**, 105 (1989).
18. P.A. David and M. Novotny, *J. Chromatogr.* **452**, 623 (1988).
19. P.A. David and M. Novotny, *J. Chromatogr.* **461**, 111 (1989).
20. W. Asche, *Chromatographia* **11**, 411 (1973).
21. B.W. Wright, H.T. Kalinoski, and R.D. Smith, *Anal. Chem.* **57**, 2823 (1985).
22. D. Bartmann, *Ber. Bunsenges Phys. Chem.* **76**, 336 (1972).
23. L.G. Randall, in *Ultra High Resolution Chromatography, ACS Symposium Series 250*, S. Ahuja, Ed. (American Chemical Society, Washington, DC, 1984), p. 135.
24. S.T. Sie and G.W.A. Rijnders, *Sep. Sci.* **2**, 729 (1967).
25. J.C. Fjeldsted, R.C. Kong, and M.L. Lee, *J. Chromatogr.* **279**, 449 (1983).
26. R.B. Denyszyn, *Scott Gases Application Note* (1986).
27. H.E. Schwartz and R.G. Brownlee, *J. Chromatogr.* **353**, 77 (1986).
28. J.M. Levy and W.M. Ritchey, *J. High Resolut. Chromatogr. Chromatogr. Commun.* **8**, 503 (1985).
29. J.C. Kuei, K.E. Markides, and M.L. Lee, *J. High Resolut. Chromatogr. Chromatogr. Commun.* **10**, 257 (1987).
30. J.M. Levy and W.M. Ritchey, *J. High Resolut. Chromatogr. Chromatogr. Commun.* **10**, 493 (1987).
31. J.M. Levy, M. Ashraf-Khorassani, A. Rosselli, and D. Boyer, Suprex, unpublished results.
32. D.E. Raynie, S.M. Fields, N.M. Djordjevic, K.E. Markides, and M.L. Lee, *J. High Resolut. Chromatogr. Chromatogr. Commun.* **12**, 493 (1989).
33. R.M. Campbell, N.M. Djordjevic, K.E. Markides, and M.L. Lee, *Anal. Chem.* **60**, 356 (1988).
34. S. Hoffman and T. Greibrokk, *J. Microcol. Sep.* **1**, 35 (1989).
35. C. Borra, G. Mapelli, F. Andreolini, A. Trisciani, and S. Trestianu, *Proc. Tenth Int. Symp. Cap. Chromatogr.*, P. Sandra and G. Redant, Eds. (Huethig, Heidelberg, 1989), p. 1205.
36. R.C. Reid, J.M. Prausnitz, and B.E. Poling, *The Properties of Gases and Liquids*, 4th ed. (McGraw-Hill, New York, 1987).
37. P.L. Chueh and J.M. Prausnitz, *AIChE J.* **13**, 1099 (1967).
38. A. Kreglewski and W.B. Kay, *J. Phys. Chem.* **73**, 3359 (1969).
39. K.M. Payne, B.J. Tarbet, J.S. Bradshaw, K.E. Markides, and M.L. Lee, *Anal. Chem.* **62**, 1379 (1990).
40. C.W. Saunders and L.T. Taylor, *Chromatographia* **28**, 253 (1989).
41. M.R. Andersen, *LC/GC* **6**, 566 (1988).

42. A. Rynaski, *1989 Symposium/Workshop on Supercritical Fluid Chromatography*, Snowbird, UT, June 13-15, 1989.
43. B.E. Richter, D.E. Knowles, M.R. Andersen, N.L. Porter, E.R. Campbell, and D.W. Later, *J. High Resolut. Chromatogr. Chromatogr. Commun.* **11**, 29 (1988).
44. T.L. Chester and D.P. Innis, *J. Microcol. Sep.* **1**, 230 (1989).
45. E.S. Francis, Lee Scientific, unpublished results.
46. J.T. Swanson, Lee Scientific, unpublished results.
47. B.E. Richter, *1988 Workshop on Supercritical Fluid Chromatography*, Park City, UT, Jan. 12-14, 1988.
48. B.E. Berg and T. Greibrokk, *J. High Resolut. Chromatogr. Chromatogr. Commun.* **12**, 322 (1989).
49. M. Bohm, F. Umland, and B. Wenclawiak, *J. Chromatogr.* **465**, 17 (1989).
50. Y. Hirata and K. Inomata, *J. Microcol. Sep.* **1**, 242 (1989).
51. F.P. Schmitz and E. Klesper, *J. Chromatogr.* **388**, 3 (1987).
52. C.R. Yonker and R.D. Smith, *Anal. Chem.* **59**, 727 (1987).
53. A. Wilsch and G.M. Schneider, *J. Chromatogr.* **357**, 139 (1986).
54. J.C. Fjeldsted, W.P. Jackson, P.A. Peaden, and M.L. Lee, *J. Chromatogr. Sci.* **21**, 222 (1983).
55. M. Novotny, W. Bertsch, and A. Zlatkis, *J. Chromatogr.* **61**, 17 (1971).
56. H.-C.K. Chang, K.E. Markides, J.S. Bradshaw, and M.L. Lee, *J. Chromatogr. Sci.* **26**, 280 (1988).
57. P. Macaudiere, A. Tambute, M. Caude, R. Rosset, M.A. Alembik, and I.W. Wainer, *J. Chromatogr.* **371**, 177 (1986).
58. R.D. Smith, E.G. Chapman, and B.W. Wright, *Anal. Chem.* **57**, 2829 (1985).
59. D. Leyendecker, *PhD Thesis*, Aachen, 1986.
60. D. Leyendecker, D. Leyendecker, and F. Höfler, *GIT Fachz. Lab.* **33**, 465 (1989).
61. B.E. Richter, Lee Scientific, unpublished results.
62. S.T. Sie and G.W.A. Rijnders, *Sep. Sci.* **2**, 755 (1967).
63. Y. Hirata, F. Nakata, and S. Murata, *Chromatographia* **23**, 663 (1987).
64. D.W. Later, E.R. Campbell, and B.E. Richter, *J. High Resolut. Chromatogr. Chromatogr. Commun.* **11**, 65 (1988).
65. B. Gemmel, F.P. Schmitz, and E. Klesper, *J. High Resolut. Chromatogr. Chromatogr. Commun.* **11**, 901 (1988).
66. D. Ishii and T. Takeuchi, *J. Chromatogr. Sci.* **27**, 71 (1989).

CHAPTER 7

APPLICATIONS

7.1 Introduction

The true test of the usefulness of an analytical technique is its range of applicability in solving real-world problems. SFC and SFE have always been able to attract interest, owing some to the intrigue stimulated by the use of the word "supercritical," which is often taken to infer extraordinary ability. Practical analytical applications of the techniques have really only appeared in the last several years, but they are expected to grow considerably in the future.

It would be impossible to cover in detail all applications of SFC and SFE in this chapter. Instead, selected examples of applications have been grouped into different categories to cover the main principles and techniques. It is hoped that the reader will gain insight into the fundamental principles governing the practices of SFC and SFE, as well as details concerning the analysis of specific samples.

7.2 Fossil Fuels and Hydrocarbons

The increasing demand for petroleum and the dwindling supply of this natural resource have encouraged the petroleum industry to process higher boiling crude fractions and to consider alternate sources, such as tar sands, shale oil, and coal. With this greater diversity of fuels have come the need for new and better methods for hydrocarbon analysis, particularly for heavy hydrocarbons. SFC offers new opportunities for hydrocarbon analysis that arise from the availability of the FID, and from the possibility of achieving high resolution of heavy hydrocarbons that exceed the analysis range of GC.

Hydrocarbon type analysis. Hydrocarbon type analysis primarily refers to the separation and quantitative measurement of saturates, olefins, aromatics, and polar components. Fractions may be further grouped into (a) straight-chain, branched, and cyclic paraffins; (b) olefins according to the number of unsaturated groups, or aliphatic *vs* alicyclic structures; and (c) aromatics according to the number of rings. Polar compounds include heavy condensed aromatic hydrocarbons and compounds containing sulfur, nitrogen, and oxygen heteroatoms.

Before SFC was available, most methods for hydrocarbon type analysis were based on LC with packed columns operating in the adsorption mode. Many of the procedures practiced today in the refinery laboratory employ open columns and labor-intensive routines, including the collection of fractions and their gravimetric work-up. These techniques have not benefitted from the tremendous advances in resolving power, instrumentation, and automation that have occurred in the last 15 years in LC, simply because of the lack of a suitable detector for easy and accurate quantitation. SFC, in combination with the FID, has provided a solution to this problem, since the detector, widely used in GC, has been shown to have a nearly uniform response for hydrocarbons.

The standard method for the analysis of light distillate products (naphtha, jet fuel, and diesel oil) is ASTM D1319, a fluorescence indicator adsorption (FIA) method. Saturates, olefins, and aromatics are separated in an open column, but remain in the column as separate bands which fluoresce in the presence of admixed dyes after stimulation by UV light. Using this procedure and other LC versions, the saturates can be adequately separated from the aromatics, whereas the separation of olefins from aromatics is marginal.

In applying SFC to this separation, the use of high fluid densities, temperatures close to the critical point, and 1-mm i.d. columns packed with small-particle silica, tends to minimize the volatility effects that cause band broadening. However, it was found that with CO_2, the separation of saturates from olefins was incomplete because of the polarizability of CO_2 compared to hexane or perfluorinated hydrocarbons used as solvents in LC. In other words, the gain in quantitation was offset by a loss in resolution.

Similar remedies that have improved the resolution of saturates and olefins by LC have been introduced into the SFC technique. Silver-modified silica was used to retain olefins by complex formation.[1] However, difficulties have been encountered in the production of silver-modified columns of predictable and stable separation behavior. The use of sulfur hexafluoride as mobile phase, which is reasonably compatible with the FID, is analogous to the use of perfluorinated hydrocarbons in LC, in that olefins are better resolved from saturates.[2,3] With unmodified columns, the separation of olefins was marginal. A gasoline analysis could be completed in 20 min, but the method did not apply to higher boiling fuels. Incorporation of a silver-loaded, strong cation exchange silica column, in addition to another silica column, and valves for column switching and backflushing, as well as the use of a mixed SF_6/CO_2 solvent, led to an analysis method that achieved adequate separation of all types in 50 min (Figure 4.52), and it was also applicable to middle distillates.[3] However, the instrumentation was somewhat complex, and SF_6 is an environmentally objectionable fluid. Also, certain polar compounds are reactive with silver, and can be modified or permanently retained. While polar compounds are generally only present in significant amounts in higher boiling fractions, the recent trend of adding oxygenates and polar additives to gasoline has changed this situation.

A new SFC method for hydrocarbon type analysis has recently been developed that retains the simplicity of a single silica column operated with CO_2 as the mobile phase. This method[4] employs ultraviolet absorption and flame ionization detectors in series, where the UV detector acts as an olefin-selective detector and the FID acts as a general mass sensitive detector for all three hydrocarbon types. It has been shown that by operating the UV detector at 190 nm, olefins of the types present in gasoline can be measured not only with high sensitivity but with a uniform detector response that permits quantitation of the olefin fraction.

This new method eliminates the need for the separation of olefins from saturates. As a result, the chromatographic equipment and methodology becomes extremely simple, in that a single silica column, isopycnic conditions, and only forward elution conditions are used. Samples are injected neat. The method is independent of sample size since calculations for quantitation are based solely on peak ratio measurements. Dual UV/FID chromatograms of a gasoline sample are shown in Figure 7.1. The olefin quantitation by the SFC method correlates well to results by the standard FIA method for a wide range of commercial gasolines.

That volatility effects, in addition to polarity differences, are significant in SFC, even at temperatures close to T_c, is demonstrated by the observation that with increasing pressure, the elution order of alkyl substituted aromatics can change. Merely altering the stationary phase/mobile phase pair can shift the volatility/polarity balance. Polar stationary phases paired with nonpolar mobile phases lead to better hydrocarbon type separations.[5]

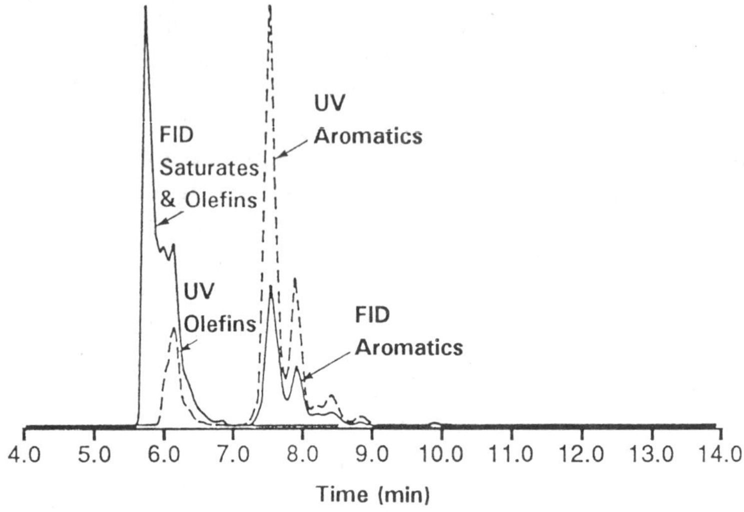

Figure 7.1. SFC chromatogram of standard gasoline using dual UV/FID detection. Conditions: 50-cm x 1-mm i.d. packed column, silica particles; CO_2; 40°C; 200 atm (reprinted with permission from Ref. 4).

While no better packing than activated silica has been found for the separation of saturates from aromatics, silica has proven to be a poor LC packing material for aromatic ring number separations because alkyl substitution, depending on type, can cause both early and late elution compared to the unsubstituted aromatic compound. Because of significant volatility contributions in SFC, band broadening due to alkyl substitution is absent. Only late elution occurs, and generally on the basis of carbon number. For example, the first monoaromatic compound following benzene is toluene, which is then followed by xylene, *etc.* The same is true for the naphthalenes. The volatility effect leads to a cleaner type separation in SFC, at least in the low boiling fuel range, because of the limited alkyl substitution possibilities. The ever increasing band broadening for higher boiling fractions because of lower diffusivity at higher density makes SFC an unsuitable technique for hydrocarbon type separations of heavy fractions, unless means can be devised to further focus the increasingly complex band.

To extend the analysis to compounds which are present in small but not insignificant levels in fossil fuels, SFC has been coupled to MS with selected ion detection. Both microbore and open tubular columns were used for the analysis and characterization of the complex hydrocarbon mixtures.[6]

Simulated distillation. Standardized GC methods (*i.e.*, simulated distillation) exist for the measurement of boiling point distributions of petroleum fractions boiling to 1050°F at atmospheric pressure. The need to support heavy hydrocarbon conversion processes has been the driving force to extend simulated distillation to higher temperatures. Specialized GC equipment, incorporating thin-film, highly stable open tubular columns, on-column injection, and temperature programming to 430°C, has been commercially introduced to extend boiling point measurements to 1400°F and beyond.[7,8] These high temperatures, however, expose analytes to possible thermal degradation.[9]

The promise of SFC in performing simulated distillations under mild, nondegradative conditions and in extending the analysis to even higher boiling materials generated considerable research activity early on. Both with open tubular and packed columns, high efficiency separations of species up to and exceeding 100 carbons were achieved under temperature conditions that did not exceed 150°C. Short open tubular columns coated with nonpolar siloxane polymers, or short microbore columns containing 5-10 μm C_{18} bonded silica, were used. Simulated distillation was attempted on whole crudes, crude residues, waxes, and the whole range of distillation products.

A significant advantage was found in using a poly(50% *n*-octyl)methylsiloxane stationary phase in similated distillation by open tubular column SFC.[10] The correlation of the elution of aromatic compounds with boiling point was considerably better by SFC than by the conventional GC method. This is especially important in the analysis of petroleum products that have significant aromatic compositions.

However, a number of obstacles were encountered that prevented broad application of the technique. Low temperatures in the injector caused precipitation of wax components and memory effects due to sample carry-over. Some heavy fractions proved to be insoluble in CO_2, even at high densities.

Recent advances have helped solve some of these problems and have given SFC simulated distillation new impetus. Direct injection and other automated injection modes, such as timed-split, have eliminated or greatly reduced discrimination. The availability of heated injectors has largely eliminated the precipitation of high molecular mass components. Use of higher boiling solvents, such as *n*-decane, for initially dissolving heavy samples, has been helpful. Low viscosity samples can often be injected neat. The availability of commercial SFC equipment that can be programmed beyond the traditional 400 atm pressure should extend the analyzable boiling range even further.

Coulombe and Duquette[11] found FID response factors to vary with hydrocarbon types. Variations were small between saturates/aromatics, but were more significant with heterocyclic components. These variations may not be intrinsic to the SFC technique, unlike those due to flow changes during pressure programming. The latter problem is minor in open tubular column SFC, but is more severe when using packed columns because of the broader CO_2 flow range involved. It appears that for accurate boiling point distribution measurements, response corrections as a function of elution must be developed and incorporated into the test.

Polycyclic aromatic compounds. The analysis of polycyclic aromatic compounds (PACs), widely found in fossil fuels and combustion products, are of interest because of their alleged carcinogenic properties. They occur commonly as isomers in complex samples and extend over a wide range of molecular masses. PACs have been analyzed since the earliest days of SFC. In 1976, Gouw and Jentoft reported near complete resolution of benzo[*a*]pyrene and benzo[*e*]pyrene from automobile exhaust.[12,13] PACs have also been isolated and measured by SFC from refinery solid wastes, coal tar pitch, lignite tar, creosote oil, soot, heavy distillates, and vacuum residues from cat-cracking processes.

Open tubular column SFC, with its inherent advantages of low pressure drop and high resolution has been applied to the separation of PACs with more than 10 aromatic rings.[14] While for real samples, analysis is generally preceded by a series of separation and concentration steps to yield a PAC-rich sample, the high selectivity of SFC for PACs often permits a simplified sample preparation procedure. The identification and measurement is further aided by the availability of sensitive and selective detection devices, including multi-wavelength UV, scanning fluorescence, and MS.

Open tubular columns containing bonded liquid crystalline polysiloxane stationary phases have achieved PAC resolution superior to that obtained by GC with the same phases because of the more ordered structure of the stationary phase at the low temperatures used in SFC (Figure 7.2). Increased molecular ordering at high pressures contribute further to improved shape selectivity. Alkanes could be eluted as a group ahead of the PACs with these columns as well.[15]

Packed columns for PAC analysis have contained a range of packings from silica to chemically modified silica, including bonded C_{18}, NH_2, and others. A column packed with C_{18} material achieved

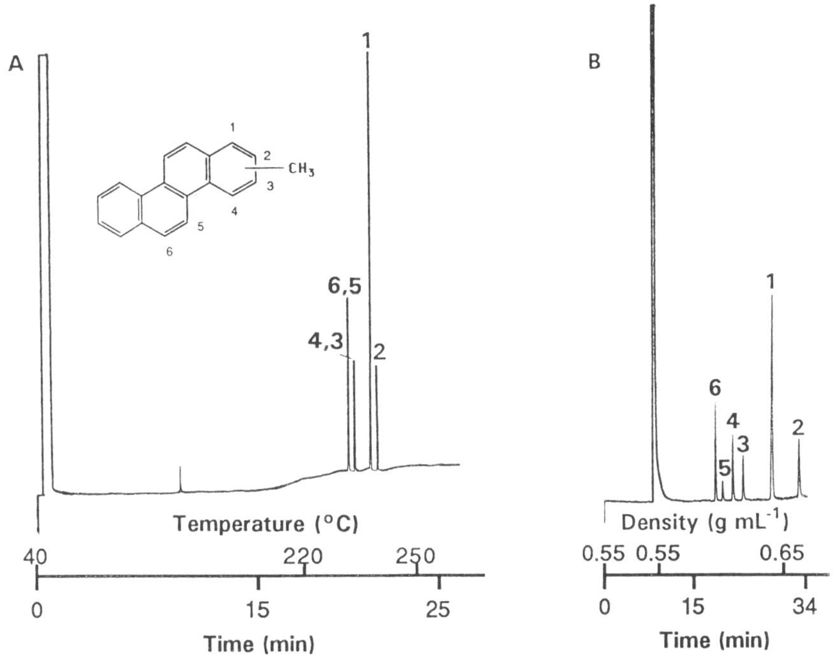

Figure 7.2. Chromatograms of six methylchrysene isomers separated on the same liquid crystalline stationary phase by (A) GC and (B) SFC. GC conditions: 15-m x 200-μm i.d. open tubular column; 0.15-μm film thickness; temperature program from 40°C to 200°C at 10°C min^{-1}, then 4°C min^{-1} to 240°C. SFC conditions: 10-m x 50-μm i.d. open tubular column; 0.15-μm film thickness; CO_2; 100°C; density program from 0.55 g mL^{-1} to 0.70 g mL^{-1} at 0.005 g mL^{-1} min^{-1} after a 10-min isopycnic period. Peak numbers represent the positions of methyl substitution on the chrysene structure (reprinted with permission from Ref. 15).

a higher selectivity compared to silica, but the separation took longer. Alkyl substituted PACs were separated from their parent compounds. The amino-bonded columns separated PACs according to the number of aromatic rings, with the alkylated species eluting near their respective parent compounds.

Juvancz *et al.*[16] reported an on-line two-dimensional SFC system for a packed capillary column and an open tubular column separation of complex PAC mixtures. The packed capillary in the first dimension, being an aminosilane-bonded silica column, provided a rapid chemical class separation, while the open tubular column in the second dimension, being a liquid crystalline polysiloxane column, provided high resolution of closely related isomers. Multiple injection in the first dimension followed by trace enrichment cryofocusing before the second dimension also allowed analysis of trace level compounds in complex PAC matrices.

Jinno *et al.*[17] reported that PAC retention was mainly controlled by molecular size and dipole moment in the case of silica columns or columns that have a high residual level of silanol groups. Equations were derived to correlate and predict PAC retention. The molecular descriptors that correlated well with retention were a correlation factor based on the number of double bonds and the number of primary and secondary carbon atoms, the molecular length-to-breadth ratio, the molecular connectivity, and the van der Waal's volume. Furthermore, a relationship based on polarizability was derived to predict the elution of a specific PAC as a function of the pressure of the mobile phase at constant temperature. This equation was successfully applied in the prediction of the elution of 12 PACs.[18]

7.3 Agrichemicals

Agrichemical applications are often challenging for any analytical method because of the complexities of the sample matrices, the detection levels required, and the varieties of the compounds involved. Issues such as food contamination and environmental safety from pesticide, herbicide, and fungicide usage are of paramount concern. Once the environmental fate of a pesticide and its resulting metabolites have been determined, methods allowing quantitative measurement of these residues in soil, water, plant, and animal tissues must be developed. In many cases, trace amounts of the parent or its metabolite in these complex matrices must be detected at part per billion (ppb) or part per trillion (ppt) levels.

Typical sample analysis involves an initial extraction of the compounds of interest into a liquid phase from the solid or liquid sample matrix, several sample clean-up steps, concentration, possible derivatization, and final analysis by GC (or LC for thermally labile compounds). Method development times can range from months to a year, resulting in extraction and analysis procedures which may require several labor intensive days. Degradation or chemical modification of the desired compounds is also of concern.

In evaluating new methodologies for agrichemicals, issues such as ease of use on a routine basis, commercial availability of equipment, ruggedness, reproducibility, and accuracy must be considered.[19] In addition, with the new pesticides being introduced into the market and the growing number of sample matrices to be examined, technologies that can be applied to multiple compounds or compound classes are essential. The current regulatory environment is insistent on advances in multi-residue analyses.

In this section, applications of analytical supercritical fluid chromatography and extraction are presented which demonstrate both qualitative and quantitative analyses relating to the agricultural industry. Carbamate pesticides, sulfonylurea herbicides, naturally occurring pyrethrin insecticides, and organophosphorus insecticides represent a range of agrichemical compounds to which SFC has successfully been applied. Many of these compounds are thermally labile, rendering GC inappropriate, and they have poor UV absorbances, making LC difficult for trace level analyses. Using SFC with both packed and open tubular columns, these compounds can be separated and detected using a variety of selective and sensitive detectors.

Detection using an FID. The most commonly used detector, the FID, has been extensively used with SFC for detecting pesticides.[20-25] For relatively simple samples, short open tubular columns and rapid pressure programming achieved separations of several carbamate pesticides present in a standard mixture in less than 2 min,[20,22] as shown in Figure 7.3. Open tubular columns were ideal for this application since pressure differences along the length of the column were minimal. This approach can greatly reduce analysis times and lead to lower detection limits by sharpening the chromatographic peaks, thus compensating for the smaller injection volumes and limited column loading capacities.

For more complex matrices, longer open tubular columns have provided the necessary separation efficiencies.[21,26] Several sulfonylurea herbicides extracted from a ground water sample were analyzed using SFC with an FID.[27] These chemically similar herbicides were well resolved from background interferences. Detection of these sulfonylurea herbicides, representing 5 ppb in a 1-liter ground water sample, was easily achieved.

Linuron and diuron herbicides, extracted from wheat and soil matrices, were also analyzed using SFC with an FID.[25] Density programming allowed good resolution of linuron from the more complex wheat grain background (see Figure 7.4).

Naturally occurring pyrethrin insecticides, extracted from the *Chrysanthemum cineraniaefolium* flower, are structurally similar and are difficult to resolve chromatographically. Typically, GC analysis has been used; however, thermal degradation of two components was observed. No degradation incurred under mild SFC conditions, and good separation and detection of an extract containing six compounds was obtained.

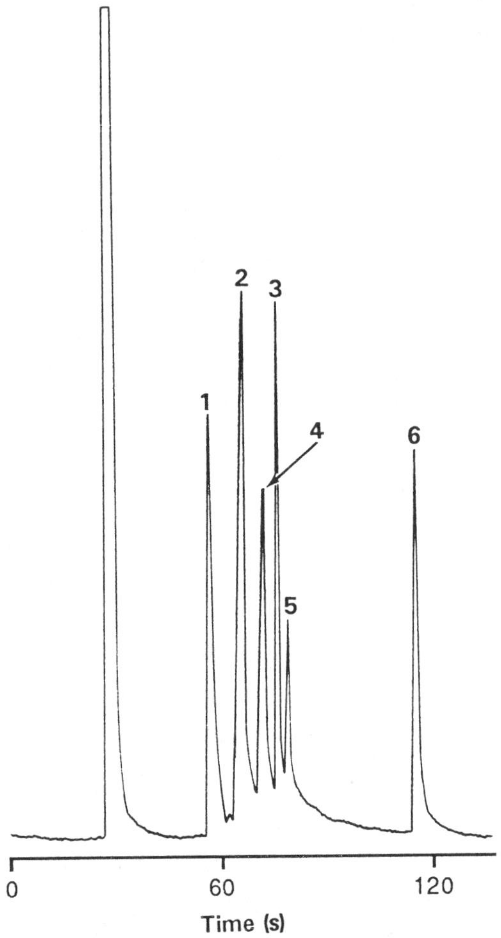

Figure 7.3. SFC chromatogram of carbamate and acid pesticides. Conditions: 1.5-m x 25-μm i.d. open tubular column, poly(5% phenyl)methylsiloxane stationary phase (d_f = 0.15 μm); CO_2; 100°C; pressure program from 135 atm to 350 atm at 50 atm min^{-1}; 60-nL injection volume (1:80 split ratio); FID (350°C). Peak identifications: (1) propoxur, (2) dicamba, (3) 2,4-D, (4) carbaryl, (5) silvex, (6) phenmedipham (reprinted with permission from Ref. 22).

Packed column SFC with an FID has also been applied to pesticide analysis. Linuron and diuron, extracted from soil and wheat grain using SFE, were separated from various background impurities using a microbore packed column.[25] The larger injection volumes accommodated by the microbore packed columns afforded detection limits more appropriate for trace residue analyses. Subnanogram quantities of linuron were detected, representing 10 ppb levels of the pesticide in 1-g soil samples.

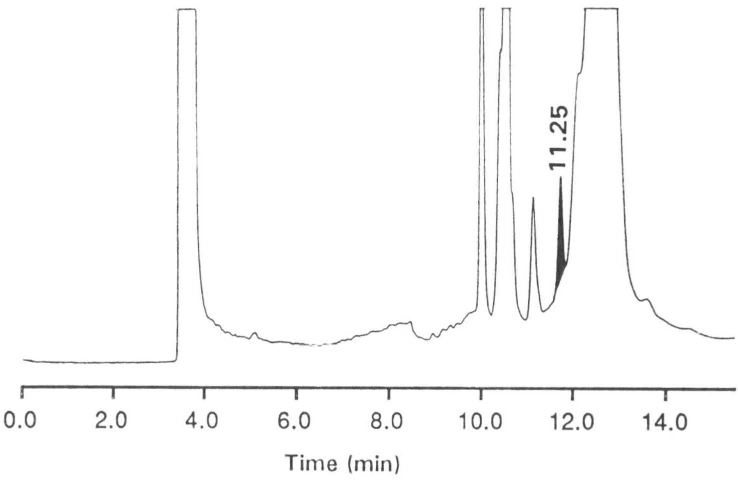

Figure 7.4. SFC chromatogram after SFE of linuron from wheat grain. SFE conditions: 1 g wheat grain + 200 μL ethanol; CO_2; 120°C; 10 min; 0.7 g mL^{-1}. SFC conditions: 10-m x 100-μm i.d. open tubular column, poly(25% cyanopropyl)-methylsiloxane (d_f = 0.5 μm); CO_2; 110°C; density program from 0.2 g mL^{-1} to 0.6 g mL^{-1} at 0.12 g mL^{-1} min^{-1} after a 6-min isopycnic period; FID (400°C). Peak identification: 50 ppm linuron (reprinted with permission from Ref. 25).

UV absorbance detection. UV absorbance detectors have been easily interfaced to both open tubular and packed column SFC, allowing the use of mobile phases or mobile phase modifiers not possible when using the FID.[23,28-31] Packed column SFC was used to analyze sulfonylurea herbicides, as well as several additional pesticides and their precursors on a conventional silica column.[23,24,31] The addition of a small amount of organic modifier to the CO_2 was required to prevent severe band broadening, to reduce retention, and to sharpen the chromatographic peaks, thereby improving the detection limits.

Recently, a UV photodiode array detector was interfaced to open tubular column SFC, allowing sensitive detection of several herbicides and pesticides.[32] Comparable sensitivities to single wavelength detection were obtained. With multichannel capability, UV detection can be used to provide additional resolution of peaks not resolved chromatographically.

Element selective detection (NPD, FPD, and ECD). Many agrichemicals contain nitrogen, phosphorus, sulfur, or halogens, making elemental specific detectors quite advantageous in reducing background interferences present in the matrix. The nitrogen-phosphorus detector (NPD) has been used in the analysis of a

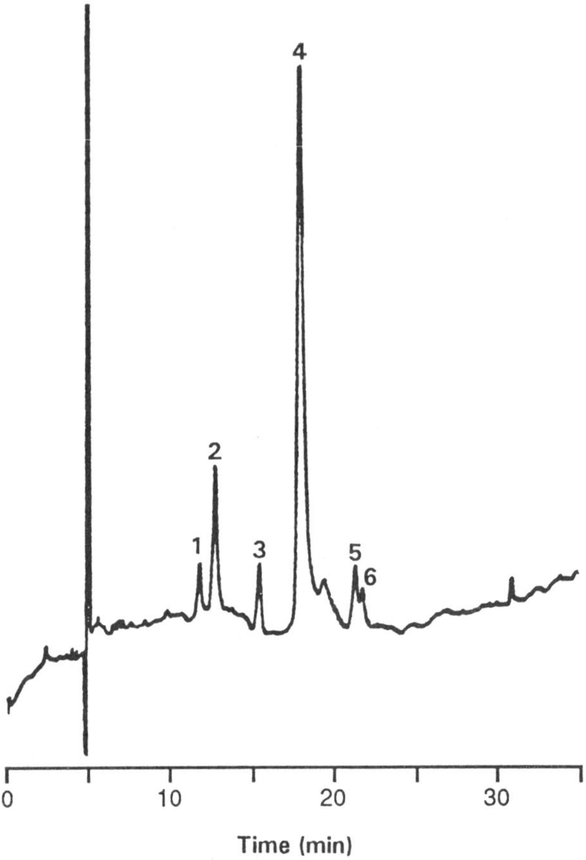

Figure 7.5. SFC chromatogram of carbamate pesticides in parsley. Conditions: 10-m x 50-μm i.d. open tubular column, polymethylsiloxane (d_f = 0.25 μm); CO_2; 100°C; multilinear density program from 0.2 g mL^{-1} to 0.75 g mL^{-1} over 30 min; 200-nL injection (1:10 split ratio); NPD (325°C). Peak identifications: (1) aldicarb, (2) methomyl, (3) mesurol, (4) oxamyl, (5) carbofuran, (6) carbaryl (reprinted with permission from Ref. 26).

standard mixture containing 10 ppm each of aldicarb, methomyl, diflubenzuron, and phenmedipham. Carbamates extracted from a parsley sample and analyzed using an NPD are shown in Figure 7.5.[26] Detection of 1.6 ppb was reported.

The flame photometric detector (FPD) is specific for phosphorus and sulfur-containing compounds. With the dual flame design, larger samples could be injected without fear of extinguishing the flame.[33] Quenching due to co-elution of other organic compounds in the sample was also eliminated. Optimization of the flame gases has led to similar detector sensitivities as experienced in GC.[34] However, with CO_2 as the mobile phase, there is an emission at the wavelength

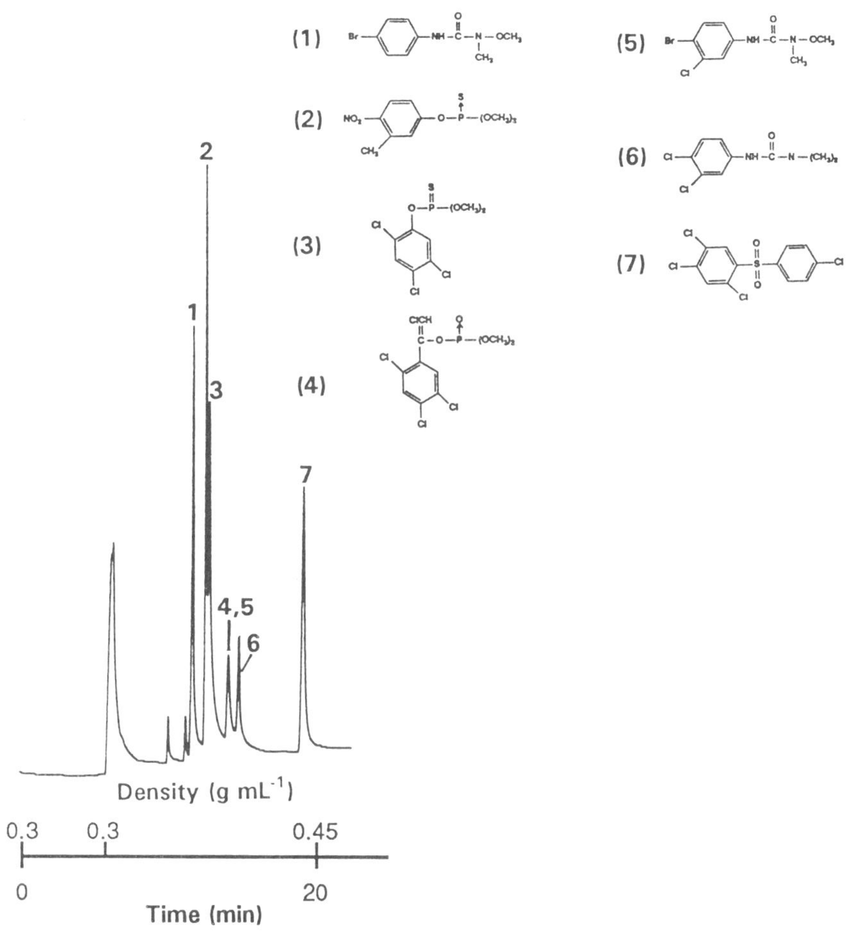

Figure 7.6. SFC chromatogram of thermally labile pesticides. Conditions: 6-m x 50-μm i.d. open tubular column, poly(5% phenyl)methylsiloxane stationary phase (d_f = 0.25 μm); CO_2; 100°C; density program from 0.3 g mL^{-1} to 0.4 g mL^{-1} at 0.01 g mL^{-1} min^{-1} after a 5-min isopycnic period, ECD (350°C). Peak identifications: (1) metobromuron, (2) fenitrothion, (3) fenchlorphos, (4) chlorbromuron, (5) tetrachlorvinphos, (6) diuron, (7) tetradifon (reprinted with permission from Ref. 36).

used to monitor sulfur. With density or pressure programming, the intensity of the emission increases. Therefore, at the present time, the FPD is more useful for monitoring of phosphorous-containing compounds.

The electron capture detector (ECD) has been successfully used with open tubular column SFC in the analysis of a triazole fungicide metabolite[35] and labile pesticides.[36] The sensitivity of this detector to O_2 required passing the CO_2 through an oxygen trap[36]. In addition, elevated detector temperatures were necessary to counteract the cooling of the effluent during mobile phase decompression. An approximate detection limit of 35 ng mL^{-1} of the metabolite in solution was obtained. A number of thermally labile pesticides were also analyzed, and their separation is shown in Figure 7.6. Several urea, phosphorothioate, and phosphate derivatives were successfully chromatographed. It was found that most of these pesticides decomposed at analysis temperatures above 120°C. Detection of captafol and captan was not successful using the ECD. Since these compounds have been previously chromatographed *via* SFC with an FID, failure to detect these with an ECD may be due to the fact that these analytes were not sufficiently vaporized in the detector cavity. Further optimization of detector design and conditions should facilitate the use of this detector in SFC.

FTIR detection. The IR detector can provide structural information about eluting compounds as well as chromatographic detection. When coupled with SFC, IR detection can be accomplished off-line *via* direct deposition onto a solid substrate, or on-line with a flow cell.[37-39] IR detection is very amenable for use with both packed and open tubular columns, and can be useful in resolving peaks of interest from co-eluting impurities. Several pesticides have been analyzed using SFC with on-line IR detection.

Monitoring wavelengths where CO_2 does not absorb allows the easiest and most sensitive detection of eluting compounds. A standard mixture of a nitroaromatic herbicide and a nitrosamine were separated using microbore packed column SFC with pressure programming.[40] With FTIR detection, a reconstructed chromatogram was easily generated.

In the IR regions where CO_2 absorbs, spectral baseline subtraction methods must be applied. A standard pesticide mixture containing aldicarb, methomyl, captan, and phenmedipham was separated and detected on-line.[41] The Gram-Schmidt reconstructed chromatogram generated from the FTIR data and modified to correct for the IR absorbance of CO_2 during the density program indicated that these components were well resolved and easily measured.

IR spectra were used to distinguish the identities of six pyrethrin pesticides present in a sample extract.[41,42] Past methods for analyzing these pesticides required GC separations, often relying on pattern recognition between the generated chromatograms and those reported in the literature for compound identification. Commercial standards were not available. SFC/FTIR has been used to separate and confirm the identities of similar materials using a density program. The mild SFC conditions eliminated the degradation observed during the GC analysis of two of the pyrethrins.

Ion mobility detection (IMD). 2,4-Dichlorophenoxyacetic acid (2,4-D) is an organochlorine herbicide. Analysis of soil extracts for this herbicide has usually included sample clean-up, derivatization, and GC with an ECD. Replacing the ECD with an ion mobility detector (IMD) eliminated the need for sample clean-up, and when SFC was coupled to the IMD, both sample clean-up and derivatization were no longer required.[43] Figure 7.7 illustrates the selective detection of 2,4-D over many other matrix components. This represents a major savings in time as well as a reduced probability for sample loss.

SFC/MS. The MS can serve as a universal detector for pesticide analysis. In the selected ion monitoring mode, the MS does not monitor most interferences present in the sample extracts, thereby allowing the elimination of many of the sample clean-up steps normally required. Thus, sample preparation and analysis times can be greatly reduced.

MS detection for packed and open tubular column SFC has been used to detect acid, carbamate, and organophosphorus pesticides.[44-50] Various interfaces have been proposed to accommodate both open tubular and packed column flow rates, including a high flow rate interface developed by Kalinoski and Smith.[44] With this interface, an 8-component pesticide mixture was separated within 8 min using a microbore column and 2% 2-propanol in CO_2 with pressure programming. The modified CO_2 mobile phase enhanced the chromatographic peak shapes and improved the detection limits. In a cherry extract, the average detection limits for all 8 pesticides were estimated to be 100 ppb, using total ion current monitoring (Figure 7.8).

Other interfaces designed to accommodate packed column SFC include a moving belt interface with a modified thermospray source for deposition,[48] and a typical thermospray source in the filament-on mode.[51] A simple direct interface between open tubular column SFC and MS required no instrument modifications. Using a simple open tubular column direct interface,[52] triazine metabolites were analyzed with good sensitivity.

Quantitative analysis. Aspects of sample introduction, especially in open tubular column SFC, are continually being addressed.[53,54] Such factors as injection valve design, column

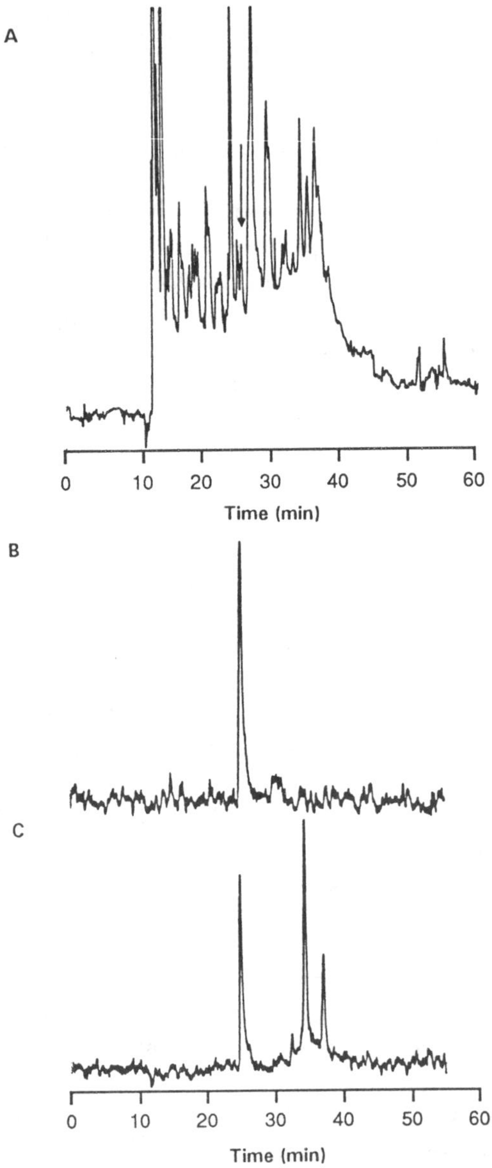

Figure 7.7. SFC chromatogram of 2,4-D in a soil extract using ion mobility (A) nonselective detection (drift time monitored between 10 and 25 ns), (B) selective detection of a 2,4-D standard (drift time monitored between 15.0 and 15.5 ms), and (C) selective detection of 2,4-D in soil. Conditions: 20-m x 100-μm i.d. open tubular column, polymethylsiloxane (d_f = 0.4 μm); CO_2; 100°C; pressure program from 120 atm to 280 atm at 4 atm min^{-1}; 60-nL direct injection; IMD, 600 mL min^{-1} air drift gas, 200 mL min^{-1} air makeup gas, 150°C, -250 V electric field, 7.0-cm drift length, ambient pressure (reprinted with permission from Ref. 43).

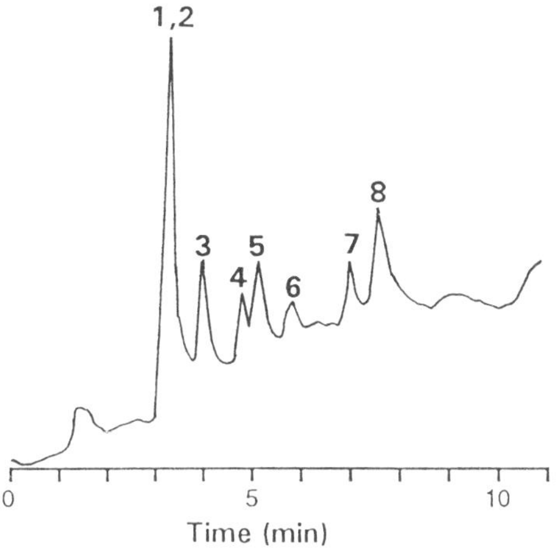

Figure 7.8. SFC chromatogram (total ion) of an extract of cherries spiked with 20-30 ppm each of organophosphorus pesticides. Conditions: 1.0-m x 1.0-mm i.d. packed column, amino-bonded silica particles (APS-Hypersil), 5-μm particle size; 2% 2-propanol in CO_2; 75°C; 100-150 μL min^{-1}; 200-nL injection; 2-propanol CIMS, scan from 190-500 daltons. Peak identifications: (1) chlorpyrifos methyl, (2) chlorpyrifos, (3) iodofenphos, (4) leptophos, (5) methidation, (6) tetrachloroinphos, (7) phosmet, (8) famphur (reprinted with permission from Ref. 44).

position relative to the injector, use of split or direct injection mode, injection loop size and configuration, and injection temperature have been shown to influence chromatographic results.

Chromatographic reproducibility in terms of retention time, peak area, and peak height has been studied in detail with respect to routine analysis of agricultural samples.[23,27,32] Typical reproducibilities using open tubular column SFC for a series of herbicide standards were between 2 and 3% RSD based on raw peak areas and peak heights.[27] This data represents 5 repetitive injections of each standard prepared at 3 concentrations between 10 and 100 μg mL^{-1}. Direct injection using an injection loop size of 1 μL gave a retention time reproducibility of 0.1% RSD. Five repetitive injections of high concentration standards (100 μg mL^{-1}) yielded RSD values typically <2% for both raw peak areas and heights. No internal standards were used.

The peak area reproducibility reported for the analysis of pesticides and herbicides using open tubular column SFC with photodiode array UV detection was 1.2% RSD.[32] The retention time reproducibility was 0.5% RSD.

The detection limits for open tubular column SFC with an FID are typically on the order of low to sub-nanograms. However, limited sample volume loading on the open tubular column is an area in which improvements are necessary. In this regard, SFE offers great potential as a sample introduction method for SFC,[53,55] and is expected to greatly enhance the use of SFC for trace and ultra-trace level analyses in the agricultural industry.[54]

The detection limits for packed column SFC/UV were investigated compared to currently utilized LC/UV methods for analyzing sulfonylurea herbicides.[23] These detection limits were comparable, if not better in some cases, due to the extended UV window of CO_2. The larger columns could accommodate larger sample sizes in order to improve sensitivity, as long as the column efficiency was appropriate for the desired separation.

Structure-retention studies. Studies have been initiated to enable empirical predictions of retention characteristics based on the chemical structure and functional group variances within a group of agricultural compounds.[24,31] Such correlations would facilitate SFC method development for new compounds and metabolites. Packed column SFC results, using silica and C_{18} columns, were compared with results obtained using two different open tubular columns.[24] Trends were found relating molecular structures of the compounds of interest, the functional groups present, and the regiospecificity of these functional groups. Later studies examined the effects of organic modifiers added to the CO_2 mobile phase on retention characteristics of the same group of agricultural compounds.[31] Again, trends relating to molecular structure, functional group, and regiospecificity were found. Modifiers dramatically affected retention order, and mobile phases could be tailored to achieve the desired separation. The silica stationary phase used in this study could mimic some polar matrices from which pesticides or metabolites might be extracted. Understanding modifier effects on retention could be very useful in designing initial SFE experiments.

SFE. SFE has been applied to agrichemicals for qualitative analysis, as a sample introduction method, and for quantitative analysis of trace components.[25,53,55,56] As a sample introduction method, part or all of the extracted materials are introduced onto the chromatographic column for analysis and identification. Pesticides extracted from parsley,[26,55] herbicides (linuron and diuron) extracted from soil and wheat grain,[25] and a herbicide extracted from an enzyme cell culture[56] are a few examples. These qualitative experiments are useful for determining the presence of a component in a matrix, or in evaluating the potential application of SFE to a particular analysis.

A recent article by King[57] discussed four parameters useful in understanding solute behavior in a supercritical fluid, and in designing extraction experiments in SFE. Understanding the solubility of the compound of interest is useful for optimizing conditions for selectively extracting the desired material. The desired components could easily be enriched during the extraction of organochlorine pesticides from porcine fat by precise pressure control.

The quantitative evaluation of extraction is easily conducted using radiolabeled tracers. The radioactivity present in the extract can be measured using liquid scintillation counting, and the remaining matrix can be combusted and counted to determine the mass balance. The collection efficiency can also be monitored in this way.

The extraction of strongly adsorbed pesticide residues is often difficult or impossible without degrading the compounds by harsh treatments such as reaction with strong acid or base, pyrolysis, or even high temperature distillation. The use of supercritical methanol as an extraction solvent for removing ^{14}C-labeled pesticide residues from soil and plant tissues was compared to results obtained by high temperature distillation.[58] After 2 h of extracting with supercritical methanol at 150 atm and 250°C, extraction efficiencies were >95% (±5% RSD) for pesticides from plant tissues. Soil extraction efficiencies were not as high, but in most cases they were >60%. Efficiencies obtained with supercritical methanol for both matrices were significantly higher than those obtained using high temperature distillation for the pesticides examined. However, the possibility of chemical reaction with the methanol was suspected for one compound.

The quantitative evaluation of extraction parameters has been conducted using ^{14}C-labeled herbicides, linuron and diuron.[25,56] The extraction effluent was collected, and recoveries were determined by liquid scintillation counting. The effects of temperature and fluid density, each independently controlled, as well as the effect of mobile phase modifier (type and amount) were evaluated. The optimized extraction conditions were compared to conventional methodologies previously used. SFE for almost 1 h produced comparable or better extraction efficiencies than the conventional procedures, which required up to 3 days of labor intensive effort (see Table 7.1). For the compounds examined, modifying the CO_2 with small amounts of organic solvents, such as methanol or ethanol, was necessary for quantitative recoveries.

SFE has been used for extraction of sulfonylurea from soil, again using radiolabeled ^{14}C compounds and liquid scintillation counting.[56,59] Several soil types were examined, representing

Table 7.1. Comparison of Recoveries of Diuron and Linuron from Soil by SFE and Classical Extraction[a]

Compound	Method	Time	Modifier volume	Percent recovery
Diuron	SFE	35 min	200 μL MeOH	99
	Classical	3 days	>1 L solvent	96
Linuron	SFE	50 min	500 μL EtOH	95
	Classical	3 days	>1 L solvent	82

[a]Taken from Ref. 25.

different organic content and pH. Effects of density, temperature, modifier type, and modifier amount on extraction efficiency were examined for chlorsulfuron, the active agent in Glean® herbicide. Once the optimized extraction conditions were achieved, the results were compared to those from classical extraction procedures. Freshly spiked samples, as well as soil samples spiked and aged up to 2.5 months, were analyzed. In all cases, the results obtained using SFE were comparable to the classical methods in terms of extraction recovery, with major savings in time and labor. Sample collection, extraction, and LC/UV analysis could be conducted within 1.5 h, compared with at least a full day for sample preparation and extraction using conventional methods. The extraction reproducibility, based on 6 repetitive extraction experiments, was ±5% RSD.

To demonstrate the multi-residue application of SFE to the class of sulfonylurea herbicides in soil, five sulfonylureas were simultaneously extracted using SFE, collected by an LC guard column inside the chromatographic oven, and analyzed by reversed phase LC/UV[60] (Figure 7.9). The extraction recoveries, however, were not reported.

7.4 Explosives

The use of GC for the analysis of military explosives was first employed for the determination of 2,4,6-trinitrotoluene (TNT) in castable explosives containing TNT and $Ba(NO_3)_2$, 1,3,5-trinitro-1,3,5-triazacyclohexane (RDX), 1,3,5,7-tetranitro-1,3,5,7-tetraaza-cyclooctane (HMX), or NH_4NO_3.[61] Although GC analysis was found

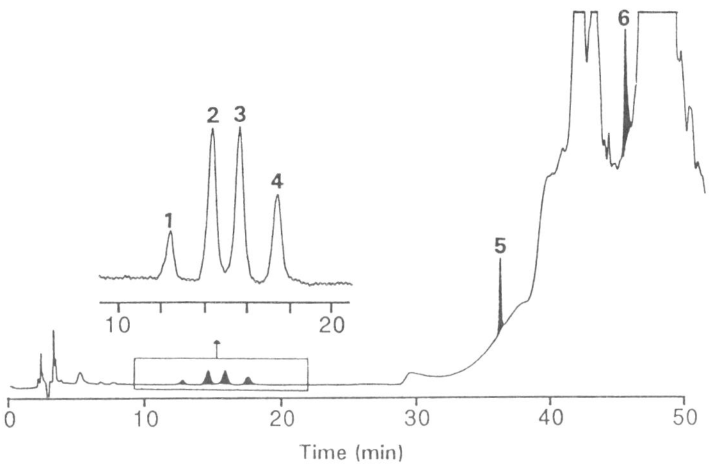

Figure 7.9. SFE/LC chromatogram of several herbicides (5 of which are sulfonylureas) in soil. SFE conditions: 1 g dry soil + 250 μL water; CO_2; 60°C; 0.86 g mL^{-1}; 20-min extraction; collection on 2-cm x 4.6-mm i.d. guard column, C_{18} silica particles. LC conditions: 25-cm x 4.6-mm i.d. packed column, phenyl-bonded silica particles (Zorbax); water/acetonitrile mobile phase (pH 3 with phosphoric acid); 1.5 mL min^{-1} flow rate; UV (200 nm, 15-nm band width). Peak identifications: (1) thiameturon methyl (Harmony®), (2) metsulfuron methyl (Ally®), (3) sulfometuron methyl (Oust®), (4) chlorsulfuron (Glean®), (5) chlorimuron ethyl (Classic®), (6) hexythiazox (Savey®) (reprinted with permission from Ref. 60).

to be quantitative for TNT, analysis for RDX or HMX gave only decomposition products. Other explosives that cannot be analyzed by GC are HNBB (hexanitrobibenzyl), tetryl, and PETN (pentaerythritol tetranitrate). Therefore, GC analysis of composite mixtures of explosives gives only an approximate composition.[62]

Douse[63] has shown that by using SFC, the previously mentioned nonvolatile and/or thermally labile explosives could be chromatographed. He also demonstrated that these explosives, along with TNT, could be separated from a complex mixture without tedious sample clean-up and with no permanent damage to the column. Figure 7.10 is an analysis of seven high explosives by SFC using an FID. Griest[65] has shown a separation of high explosives using packed-column SFC and UV detection.

Since TNT is listed on the "priority pollutant" list (U.S. EPA Appendix VIII compounds), the analysis of it and its residue products is of great interest. As mentioned, TNT is analyzed readily with SFC using conventional chromatographic detectors. Figure 7.11 shows the analysis of TNT and its residue products on a packed column with an FID.[66] However, to detect TNT and other explosives at trace

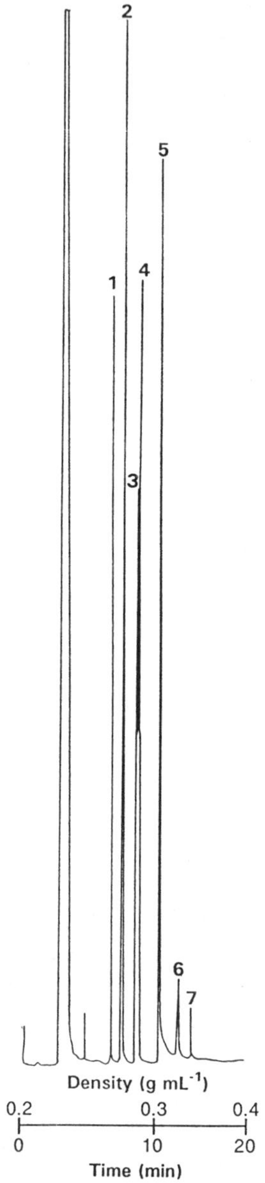

Figure 7.10. SFC chromatogram of high explosives. Conditions: 4-m x 50-μm i.d. open tubular column, poly(5% phenyl, 1% vinyl)methylsiloxane (d_f = 0.25 μm); CO_2; 100°C; density program from 0.2 g mL^{-1} to 0.4 g mL^{-1} at 0.2 g mL^{-1} min^{-1} after a 4-min isopycnic period; FID. Peak identifications: (1) 2,6-dinitrotoluene, (2) 2,4-dinitrotoluene, (3) 2,4,6-trinitrotoluene, (4) 1,3,5-trinitrobenzene, (5) picric acid, (6) 1,3,5-trinitro-1,3,5-triazacyclohexane, (7) 1,3,5,7-tetranitro-1,3,5,7-tetrazacyclooctane (reprinted with permission from Ref. 64).

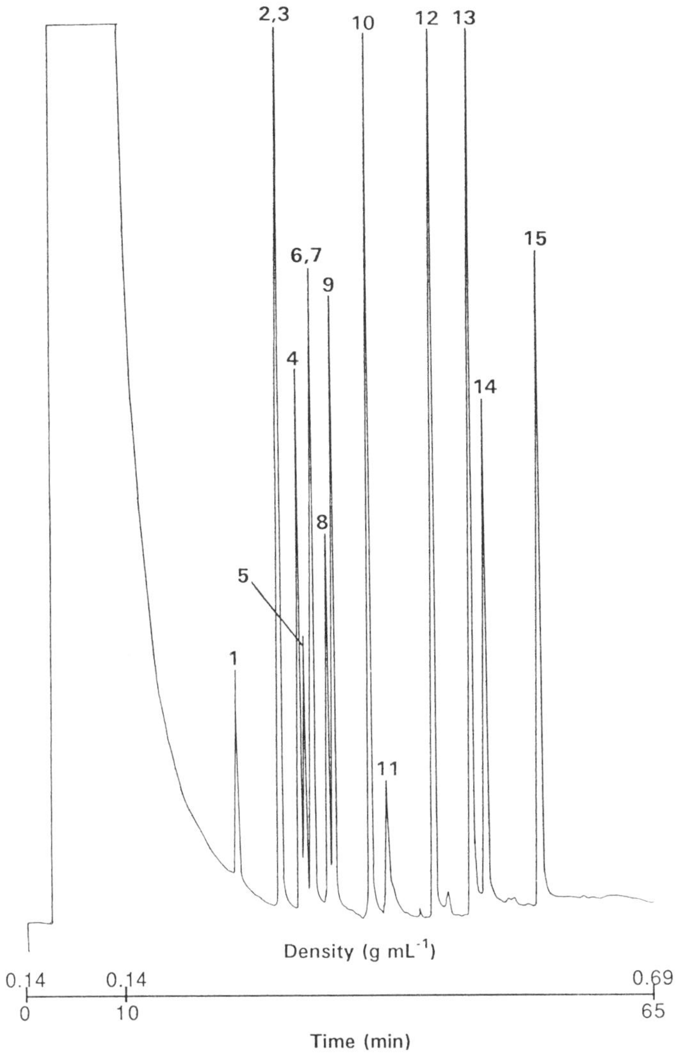

Figure 7.11. SFC chromatogram of an explosive and PAH standard mixture. Conditions: 25-cm x 250-μm i.d. packed column, poly(50% octyl)methylsiloxane bonded silica, 5-μm particles; CO_2; 100°C; density program from 0.14 g mL^{-1} to 0.64 g mL^{-1} at 0.01 g mL^{-1} min^{-1} after a 10-min isopycnic period. Peak identifications: (1) naphthalene, (2) 2,6-dinitrotoluene, (3) dibenzofuran, (4) 2,4-dinitrotoluene, (5) N-nitrosodiphenylamine, (6) 2,4,6-trinitrotoluene, (7) 1,3,5-trinitrobenzene, (8) biphenyl, (9) 2-nitronaphthalene, (10) phenanthrene, (11) 2-nitrophenol, (12) pyrene, (13) benz[*a*]anthracene, (14) 1-nitropyrene, (15) benzo[*a*]pyrene (reprinted with permission from Ref. 66).

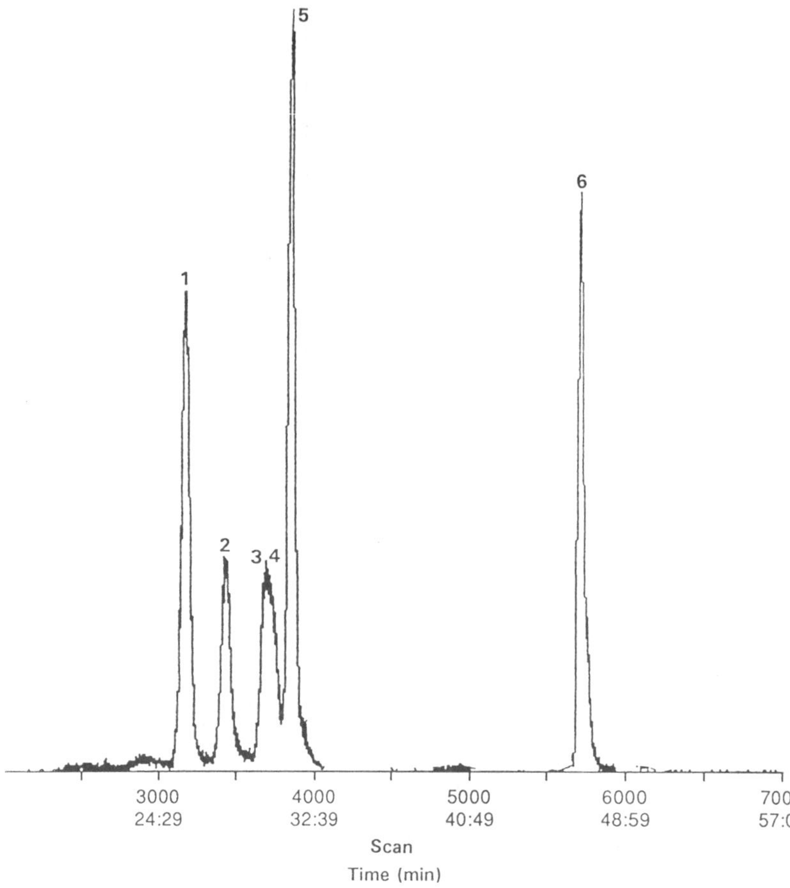

Figure 7.12. SFC/MS total ion chromatogram of TNT and related compounds. Conditions: 10-m x 50-μm i.d. open tubular column, poly(5% phenyl)methylsiloxane (d_f = 0.15 μm); CO_2; 100°C; density program from 0.3 g mL^{-1} to 0.75 g mL^{-1} at 0.0075 g mL^{-1} min^{-1} after a 10-min isopycnic period; 0.01 ng each by direct injection; methane-NICI-MS (selected ion monitoring). Peak identifications: (1) 2,6-dinitrotoluene, (2) 2,4-dinitrotoluene, (3) 4-nitrophenol, (4) 1,3,5-trinitrotoluene, (5) 2,4,6-trinitrobenzene, (6) 1-nitropyrene.

levels, mass spectrometry is the most powerful on-line detector. Minimum detectable quantities of 1-5 ng for full scan operation and below 100 pg for selected-ion-monitoring after electron impact ionization are common. With negative ion chemical ionization (NICI), one hundred-fold improvements in detection limits can be obtained. Figure 7.12 shows the NICI SFC/MS analysis of 10 pg of TNT and other related compounds.[67]

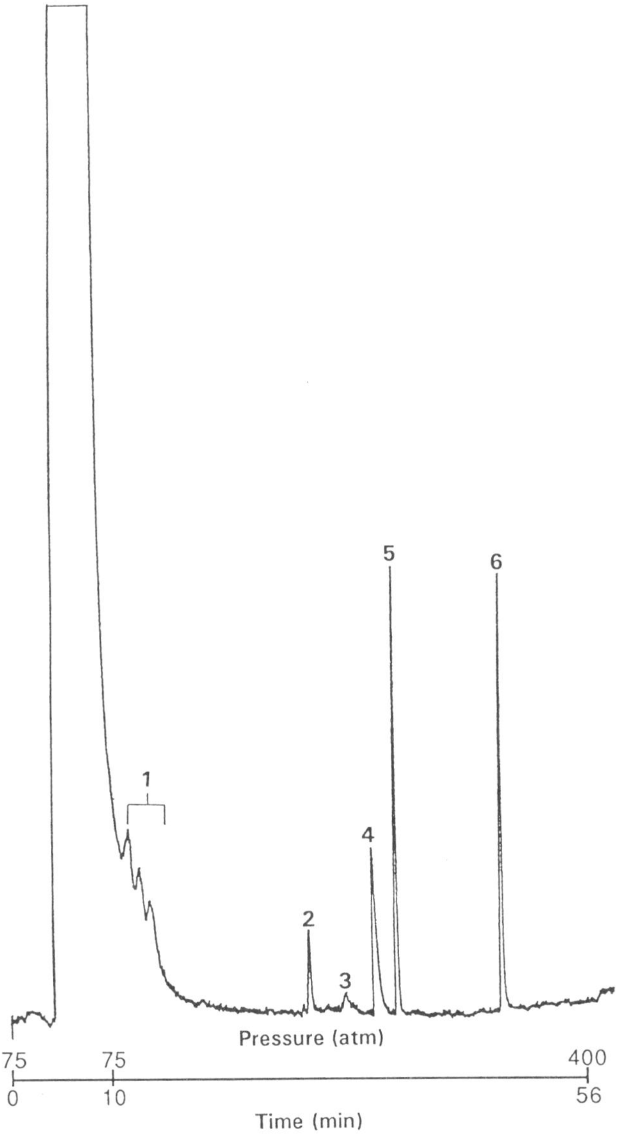

Figure 7.13. SFC chromatogram of a double-base propellant. Conditions: 25-cm x 250-μm i.d. packed column, poly(50% octyl)methylsiloxane bonded silica, 5-μm particles; CO_2; 120°C; pressure program from 75 atm to 400 atm at 7 atm min^{-1} after a 10-min isobaric period; FID. Peak identifications: (1) candelilla wax, (2) 2-nitrodiphenylamine, (3) resorcinol, (4) nitroglycerin, (5) triacetin, (6) di-*n*-propyl adipate (reprinted with permission from Ref. 66).

Propellant analysis is another important application of SFC. The determination of stabilizers in the propellant is important for the prediction of the safe "shelf life" of the propellant. A major stabilizer used is diphenylamine, which is capable of combining with the nitrogen oxides liberated during decomposition of the propellant. This leads to production of nitro- and nitroso- derivatives of diphenylamine. In routine GC analysis, these derivatives cannot be analyzed because they decompose to diphenylamine. Also nitroglycerin, the major component of most propellants, cannot readily be determined by GC without some thermal degradation.

Recent articles by Ashraf-Khorassani and Taylor[68,69] have shown that SFC analysis of propellants and stabilizers does not lead to decomposition of the nitro- or nitroso-derivatives of diphenylamine, but can produce complete separation of these derivatives. Also, chromatographic analysis of nitroglycerin was achieved with no decomposition. Figure 7.13 shows the analysis of a propellant formulation using a packed capillary column.[66]

7.5 Drugs and Pharmaceuticals

The use of very potent drugs in low dosage and the development of novel drug delivery systems require employment of the most sophisticated analytical methods. In the area of drug analysis, SFC may be regarded as a complementary technique to the more traditional techniques such as immunoassays, LC, GC, and coupled chromatography/mass spectrometry.[70] This is partially due to the ready interfacing of SFC with mass spectrometry, thus providing a viable alternative to GC/MS and LC/MS for drug analysis. In addition, SFC may be advantageously applied in the analysis of thermally labile drugs, which are not as ideally analyzed by GC. Prior to the late seventies, drug analysis was performed in most clinical laboratories as part of clinical chemistry under the general title of toxicology. These analyses, primarily involved the identification and quantitation of both therapeutic and illicit drugs. The results were interpreted by the clinicians to check for a patient's compliance with drug therapy, to account for possible toxic side effects, and to verify the presence of toxic or illicit drugs as a cause of clinical disorders.

In the late seventies, the specialty of "therapeutic drug monitoring" (TDM) emerged as a result of increasing evidence from clinical pharmacology that, for selected groups of drugs, there existed a significant correlation of clinical response with plasma concentrations. Pippenger[71] proposed the rationale for TDM as (a) checking for noncompliance, (b) adjusting the dosage for individual variations in drug metabolism, (c) identifying altered drug utilization due to a disease state, (d) compensating for a changing physiological state, and (e) establishing the best concentration range for optimum therapeutic response. At about the same time, the advent of LC and reliable and fast immunoassays, such as the enzyme multiplied immunoassay technique (EMIT) and fluorescence polarization immunoassay (FPIA), rendered clinical drug monitoring feasible in a timely manner of 10 min to several hours. The specificity of these methodologies are such that reliable measurement may be achieved in the presence of structurally similar metabolites and other drugs, such as in serum samples from patients on multiple drug therapy.

Using these technologies, the following groups of drugs are now routinely monitored: (a) antiasthmatics (*e.g.*, theophylline), (b) antiepileptics (*e.g.*, phenytoin, valproic acid, and others), (c) antiarrhythmics (*e.g.*, procainamide and others, including flecainide), (d) antimicrobials, (e) antidepressants, (f) anticancer agents, and (g) immunosuppressants such as cyclosporine. Of particular interest is the analysis of the last group in ensuring the success of organ transplantation. With recent introduction of newer immunosuppressants such as macrolide, the clinical laboratory will possibly be involved in its clinical monitoring. In summary, TDM, as practiced currently by the clinical laboratory, involves rational drug measurement with short turn-around time, possibly followed by interactive interpretation with the requesting clinicians.

While TDM has become a most active clinical specialty, the epidemic abuse of illicit drugs has propelled the toxicology area of the clinical drug laboratory into controversy regarding its reliability of substance abuse testing. The number of laboratories engaged in drug testing has grown significantly, due not only to increasing numbers of drug-related crimes, but also to workplace testing and screening. A recent publication[72] verified that most drug labs are indeed able to adequately identify the presence of five illicit drug groups: amphetamines, cannabinoids, cocaine, opiates, and phencyclidine, as recommended by the National Institute of Drug Abuse.[73]

Potential applications of SFC in substance abuse testing have been demonstrated. Crowther and Henion[74] reported on the use of a polar packed column in SFC with mass spectrometric detection for the analysis of cocaine. Fast analysis was achieved at high linear velocity without significant loss in efficiency. The high efficiency also enabled higher sensitivity than was possible by LC/MS. In addition, it was shown how SFC/MS is preferred over GC/MS for the analysis of thermally unstable polar drugs.

Tetrahydrocannabinol and six common metabolites were resolved on a nonpolar open tubular column, but a more selective column was recommended for baseline separation of all metabolites.[75] The analysis of a human urine extract spiked with 9-carboxy THC was also demonstrated. Acid extracts of heroin samples for fingerprint information have been analyzed using an efficient and inert open tubular column with universal flame ionization detection.[76] In addition, a series of opiates were successfully analyzed using open tubular column SFC with ion mobility detection without the need for derivatization.[77]

With ample technology already available for TDM and toxicology, the role of SFC, as pointed out previously, may be characterized as complementary in providing added selectivity to already established techniques. Since SFC has had limited application in clinical analysis, a preliminary study of the analysis of phenobarbital by open tubular column SFC provided an examination of the technical feasibility and results compared to an established routine clinical fluorescence polarization immunoassay.[78] Phenobarbital was chosen because of its therapeutic application as an antiepileptic, as well as its potential abuse resulting in overdose toxicological cases. As such, its monitoring may be regarded clinically as part of TDM and/or toxicology. Figure 7.14 shows the SFC chromatograms of extracts of a drug-free serum and a patient serum. Precision studies showed that within-run and day-to-day mean phenobarbital concentrations were 25.8 (n = 5) and 23.8 (n = 13) mg L^{-1}, respectively, compared to 25.8 mg L^{-1} estimated by fluorescence polarization immunoassay. The corresponding coefficients of variation were 6.9 and 12.9%. The precision must be improved to the clinically accepted limits of <5% and <10% for within-run and day-to-day coefficients of variation, respectively.

Therapeutic antibiotic drugs. These compounds are often of high molecular mass with multiple polar functional groups on each molecule. The absence of chromophores makes it difficult to perform trace analysis in complex biological matrices. Underivatized erythromycin A has been analyzed using open tubular column SFC/FID with CO_2 as mobile phase at 40°C[79] and by using a packed aminosilane column with 8% methanol in CO_2 and CIMS detection.[80]

Prostaglandins. The possibility of analyzing underivatized prostaglandins by open tubular column SFC using neat CO_2 as mobile phase and the universal FID[81] for detection was recognized

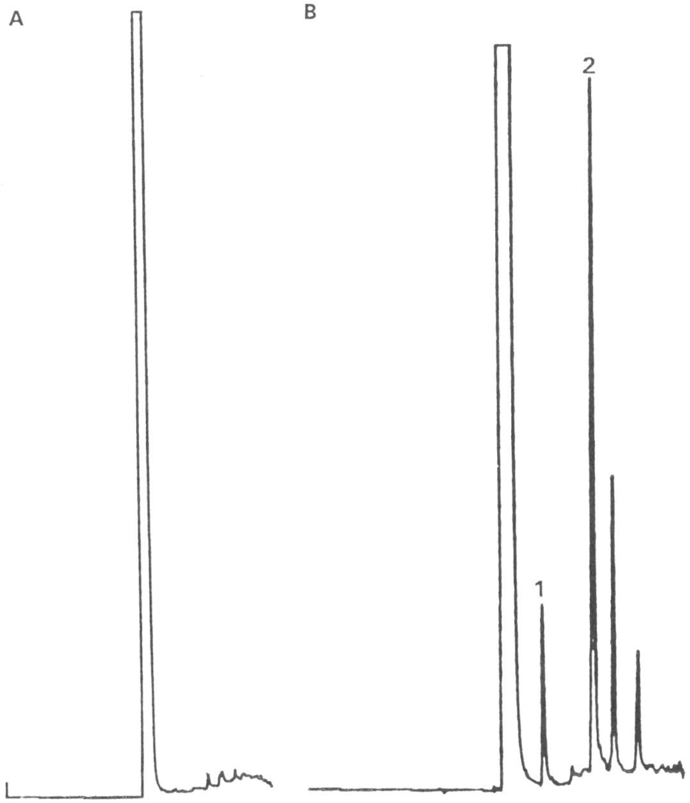

Figure 7.14. SFC chromatograms of (A) a drug-free serum extract and (B) a serum extract of a patient medicated with phenobarbital. Conditions: 10-m x 50-μm i.d. open tubular column, polymethylsiloxane stationary phase; CO_2; 120°C; density program from 0.25 g mL^{-1} to 0.75 g mL^{-1} at 0.06 g mL^{-1} min^{-1}. Peak identifications: (1) internal standard, (2) phenobarbital (reprinted with permission from Ref. 78).

early. In biological samples, prostaglandins are present only in trace amounts, making it preferable to use solvent elimination sample introduction methods. Underivatized prostaglandins have been analyzed from water solutions by Koski *et al.*[82] using SFE injection (Figure 7.15).

Basic drugs. Berger and Deye[83] demonstrated the use of tetrabutylammonium hydroxide in methanol as an additive to supercritical Freon 23 mobile phase for successful elution of promazine and thioridazine on a packed column. Benzodiazepines can be eluted with pure CO_2 if a well deactivated and short column is used as demonstrated by Koski,[84] while 5-13% methanol modifier

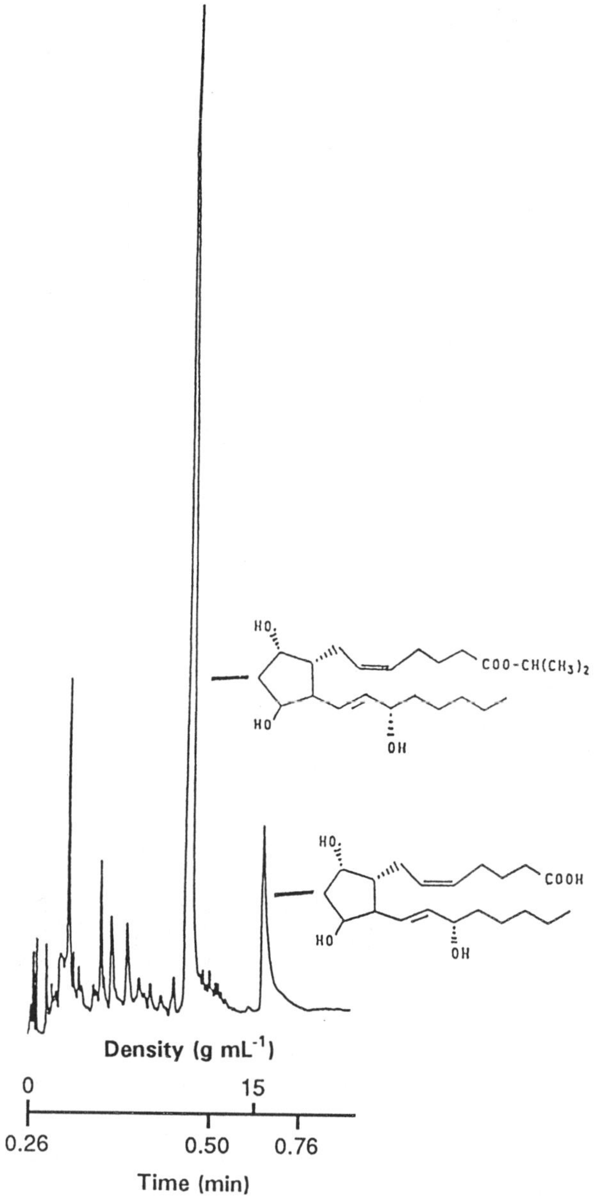

Figure 7.15. SFE/SFC chromatogram of an aqueous solution of prostaglandins. SFE conditions: 15 μL (13.3 ppm) aqueous prostaglandin loaded on 20 mg XAD-2; 3-min N_2 purge followed by 100-μL *n*-hexane rinse; CO_2; 0.80 g mL^{-1}; 45°C; 45 min. SFC conditions: 6.5-m x 50-μm i.d. open tubular column, poly(oligo-ethylene oxide)methylsiloxane stationary phase (d_f = 0.15 μm); CO_2; 100°C; FID (reprinted with permission from Ref. 82).

in CO_2 was recommended for packed column analysis.[85] Medium polarity stationary phases were used for both column types.

For basic drugs of higher polarity and primary amine functionality, it is no longer possible to use CO_2 as the mobile phase. An initial feasibility study of the use of ammonia as mobile phase with open tubular columns gave rapid elution of a primary amine even when injected as a hydrochloride salt.[86]

Steroids. The steroid hormones are members of a class of numerous and extremely similar molecules with widely divergent physiological effects. Mixtures of these components are difficult to separate, which can be compounded by the biological matrix in which they are found. A recent report described the use of SFC with selective thermionic detection[87] and FTIR[88] for the analysis of a number of steroid hormones. The described instrumentation and methodology provide sensitive and specific detection for following the metabolism of these compounds.

An alternative detection approach was introduced for several steroids based on thiophosphinic ester derivatization.[87] The thiophosphinic group, when reacted with the steroyl hydroxyl groups, introduces specificity for phosphorus selective detection. This procedure provides femtogram detectabilities that allow detection and profiling of the trace level steroids found in plasma and urine samples.

While FTIR is not as sensitive as thermionic detection, it can provide important structural information for the identification of these compounds. The efficacies of LC and SFC coupled to on-line FTIR detection have been compared for the analysis of steroids.[88] The chromatography was comparable with respect to both separation efficiency and usefulness of FTIR spectra. Open tubular column analysis of underivatized anabolic steroids, dexamethasone and betamethasone, the adrenocortical prednisone steroids, and the cortical steroids and hydrocortisone was demonstrated.[75] The SFC separation could be performed at a low enough temperature to avoid epimerization between the betamethasone and dexamethasone forms.

The resolution of steroid isomers was improved significantly using a liquid crystalline polysiloxane stationary phase[89] when compared to results obtained using other stationary phases. Figure 7.16 shows the separation of androstane diols and triols where the retention mechanism is related to solute molecular shape rather than to properties such as volatility and polarity.

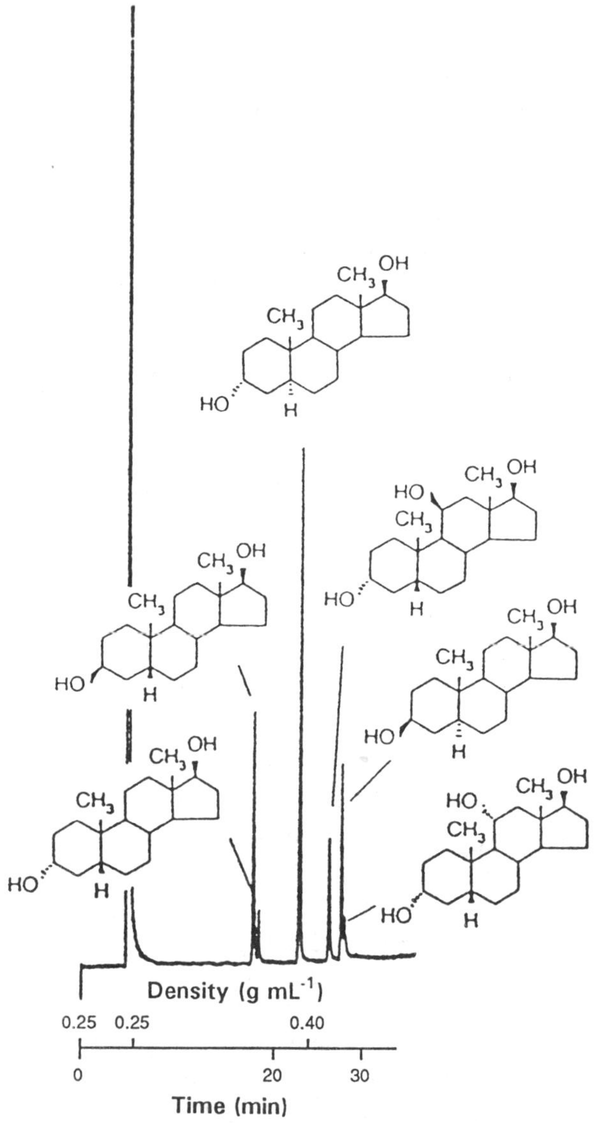

Figure 7.16. SFC chromatogram of four androstane diols and two androstane triols separated on a liquid crystal column. Conditions: 6-m x 50-μm i.d. open tubular column (d_f = 0.15 μm); CO_2; 120°C; density program from 0.25 g mL^{-1} to 0.50 g mL^{-1} at 0.0075 g mL^{-1} min^{-1} after a 5-min isopycnic period; FID (reprinted with permission from Ref. 89).

7.6 Polymers

Of the many fields in which SFC has attracted considerable attention, the field of polymer analysis is one of the most important. The interest in SFC is focused on several different aspects:

(a) Determination of low molecular mass components in polymers, such as residual monomers and solvents; oligomers; and additives like stabilizers, plasticizers, and surface treatments. For the determination of these compounds in a high-polymer matrix, the combination of SFE and SFC is very effective.

(b) Separation of homologous series (*i.e.*, oligomers) of low molecular mass. For such samples, the determination of the molecular mass distribution is of interest.

(c) Assay of parallel homologous series. Such information may be of special interest for industrial purposes (as "fingerprint" information).

(d) Purity tests for chemicals such as monomers and reactive oligomers and prepolymers with respect to purity, batch-to-batch reproducibility, fingerprint information, etc.

Polymer separations can be subdivided into separations in which the members of the homologous series are separated one by one (especially with oligomers), and into those in which the polymer chains are so long that it is not possible to separate the homologs (*i.e.* the peaks of the individual homologous members overlap to give one more or less broad peak for the whole sample. This peak reflects the molecular mass distribution of the polymer sample). The latter yields a chromatogram which resembles one obtained from a size exclusion separation.

In true size exclusion chromatography, there is no interaction between the analytes and the surface of the stationary phase, and according to the separation mechanism, the largest molecules are the first to appear at the end of the column and the smallest molecules are the last to leave the column. This separation technique has only rarely been used with supercritical mobile phases (*e.g.*, for polyvinyl pyrrolidone and polystyrene).[90] Polystyrene samples have also been chromatographed by SFC based on a retention mechanism (*i.e.*, based on interaction between sample molecules and the stationary phase). For the latter case, molecular mass ranges of the polystyrene samples reached 600,000 and 1,200,000, respectively. As examples of the separation of polystyrene standards having narrow molecular mass distributions, Figure 7.17 and 7.18 show examples of both the retention and size exclusion types.

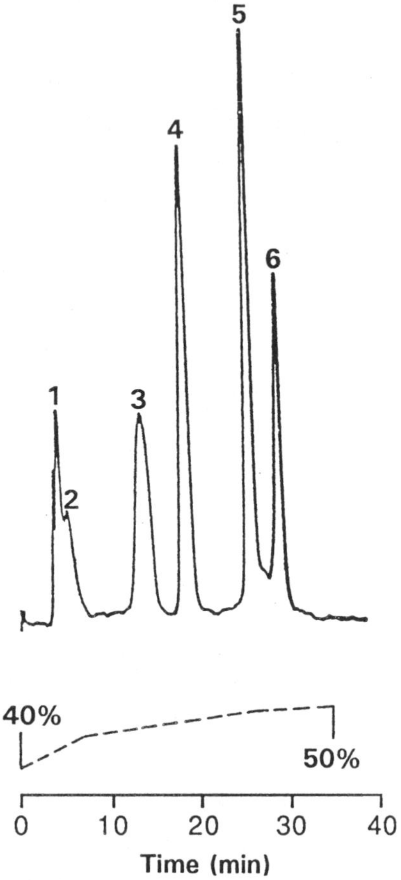

Figure 7.17. SFC chromatogram of polystyrene standards using composition gradient elution. Conditions: 250-mm x 4.6-mm i.d. packed column, 10-μm silica particles (LiChrosorb); 1,4 dioxane gradient (40-50%) in pentane; 250°C; 185 atm; 1 mL min^{-1} flow rate (liquid); UV (254 nm). Peak identifications: polystyrene standards with molecular masses of (1) 3,600, (2) 10,000, (3) 53,700, (4) 93,000, (5) 254,000, (6) 600,000 (reprinted with permission from Ref. 94).

In the early days of SFC, Giddings and coworkers[91-93] examined whether or not different polymers could be transported through a chromatographic column by means of various supercritical fluids (*i.e.*, "migration tests"). From the same lab, a study originated on the use of supercritical media in field flow fractionation of polystyrene.[96]

In the remaining part of this section, "oligomer" separations will be reviewed; this implies that discussion will be restricted to such samples for which the sample components are composed of repeating units. This section does not aim for completeness, but rather gives

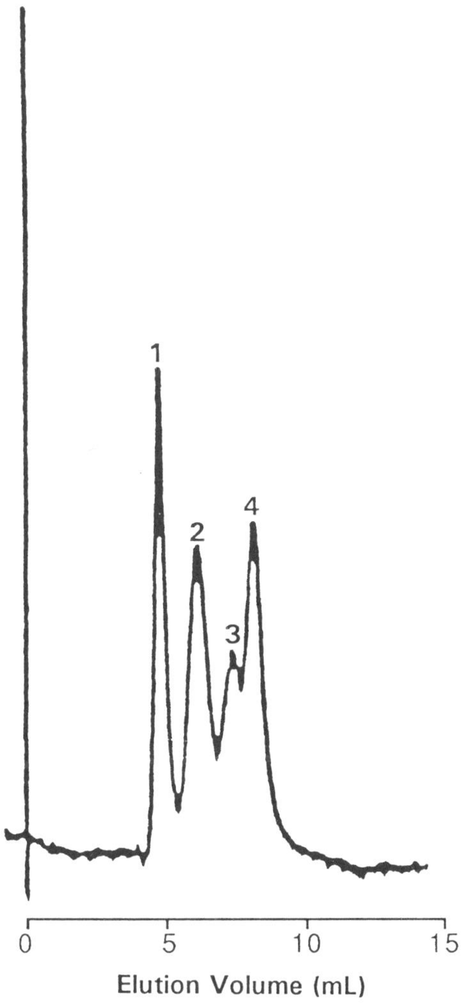

Figure 7.18. SFC chromatogram of polystyrene standards using size exclusion chromatography. Conditions: two 250-mm x 4.6-mm i.d. packed columns, pore glass of 200/400 mesh (37-75 μm), 8.6-nm mean pore size; dichloromethane; 244°C; 317 atm; 0.63 mL min^{-1} flow rate (liquid); UV (266 nm). Peak identifications: polystyrene standards with molecular masses of (1) 110,000, (2) 17,000, (3) 4,000, (4) benzene (reprinted with permission from Ref. 95).

some exemplary separations of the different oligomer types separated to date, with special focus on recent results. A short reference to biopolymers is included at the end of this section.

Hydrocarbon oligomers. Different technical alkane mixtures (*i.e.*, waxes) have been separated, as *e.g.*, a Fischer-Tropsch wax which contained alkanes from *ca.* C_{20} to $>C_{100}$.[97] Such a wax can be referred to as poly(methylene) since it contains both odd- and even-numbered carbon chains. More frequently, however, separations of ethylene oligomers containing only even-numbered chains have been reported. Separations of these oligomers have been used for the calibration of simulated distillation analysis[9,98] (see Section 7.2). Additionally, a separation of α-olefins has been achieved by Knowles.[99] Such alkane separations are normally within the domain of GC (including high-temperature GC). SFC cannot compete with GC for the separation of low molecular mass alkanes, but SFC offers the ability to separate analytes of high, as well as those of low, molecular mass. Furthermore, SFC exerts less thermal stress on a sample than GC does. In all, SFC is an attractive technique for hydrocarbon samples, especially if compounds of low or moderate thermal stability are present.

Other hydrocarbon-type homologs which have been separated using SFC are the oligomers made from the α-olefin, 1-decene,[100] and the butadiene oligomers.[49,101]

Vinyl arene oligomers. This class of compounds is comprised of oligomers with an aromatic side group in the repetitive unit, such as polystyrene/styrene oligomers. Styrene oligomers have been the most intensively studied oligomers by SFC (approximately 50 papers involve styrene oligomers). These oligomers were chosen for the first pressure programmed separation reported in the literature.[102] Furthermore, styrene oligomers have been used for testing the applicability of multiple gradients.[103,104] Finally, use was made of styrene oligomers for evaluating the dependence of retention on temperature, pressure, and density of the mobile phase, and on the pore diameter[105] of the stationary phase. For minimizing the time requirements for such fundamental studies, styrene oligomers were replaced by aromatic compounds which were shown to resemble them with respect to their retention behavior.[106]

Like the other vinyl arene oligomers, styrene oligomers have been studied primarily using packed columns. Gradient techniques which have been used for the separation of these oligomers include pressure/density gradients, composition gradients, and temperature gradients. With both pressure and composition gradients, up to *ca.* 70 homologs could be separated,[107,108] which corresponds to a molecular mass range of 7,300. With a proper combination of mobile and stationary phases, the separation of stereoisomers was possible.[109]

Other compounds belonging to the vinyl arene class which have been separated by SFC are oligomers from vinylnaphthalene (1- and

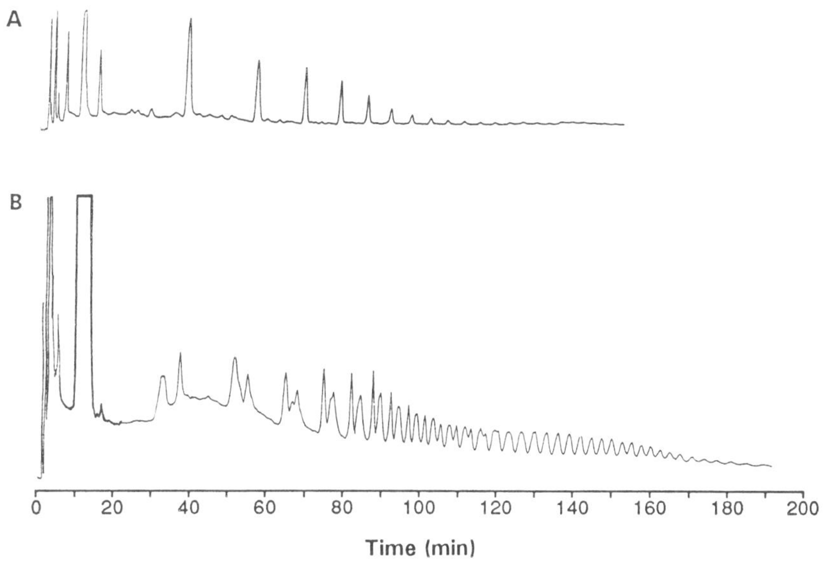

Figure 7.19. SFC chromatograms of 2-vinyl naphthalene oligomers using composition gradient elution for (A) radical-initiated oligomerization and (B) anionic-initiated oligomerization. Conditions: 250-mm x 4.6-mm i.d. packed column, 10-μm silica particles (LiChrosorb); 1,4-dioxane gradient (5-50% (v/v) in pentane, concave gradient over 160 min; 270°C, 1 mL min^{-1} flow rate (liquid); initial pressure of 49 atm in (A) and 51 atm in (B); UV (278 nm) (reprinted with permission from Ref. 115).

2-isomers), vinylbiphenyl, N-vinylcarbazole, and vinylpyridines. Vinylnaphthalene and biphenyl oligomers were used to evaluate the effects of simultaneous programming of composition and temperature,[110-112] and vinylcarbazole oligomers were separated successfully with a simultaneous composition/pressure gradient.[106,110,113] Furthermore, 2-vinylnaphthalene oligomers have been used for comparing composition programming with packed columns and density programming with open tubular columns. As a result, it was found that the efficiency of the separation was much higher with density programming on open tubular columns, but the molecular weight range eluted was higher with the packed column composition programming technique.[114] Actually, up to 50 homologs in a 1-vinylnaphthalene oligomer sample could be separated, which corresponds to a molecular mass range of over 7,700.[115]

With 2-vinylnaphthalene, the occurrence of side reactions during anion-initiated oligomerization was proved by SFC separations. Since parallel homologous series were not observed with radical-initiated oligomers, it was concluded that the conditions for the anionic preparation had not been optimized;[115] separations of radical- and

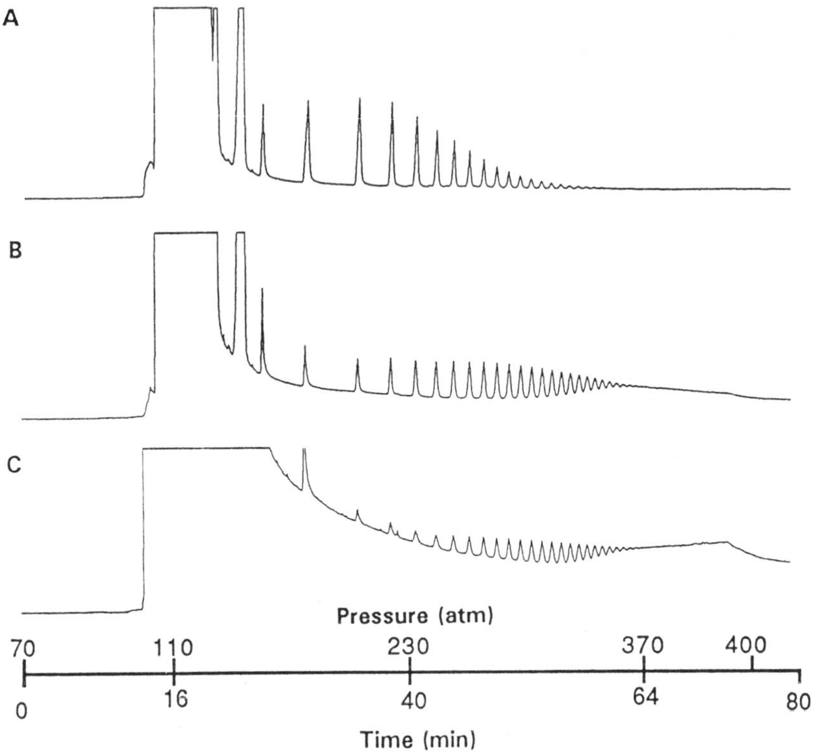

Figure 7.20. SFC chromatograms of TBDMS derivatives of acrylic acid oligomers with average molecular masses of (A) 900, (B) 2,000, and (C) 10,000. Conditions: 10-m x 50-μm i.d. open tubular column, polymethylsiloxane (d_f = 0.2 μm); CO_2; 120°C; pressure program from 70 atm to 80 atm at 1 atm min^{-1} and from 80 atm to 400 atm at 5 atm min^{-1}, 1.5 cm s^{-1} linear velocity; FID (400°C) (reprinted with permission from Ref. 118).

anion-initiated products are shown in Figure 7.19. Since this time, the reasons for the side reactions have been found, the preparation was optimized, and SFC was used for "monitoring" the preparation.[108] Finally, 2-vinylpyridine oligomers which had been difficult to separate on silica[116] have been successfully chromatographed on a polystyrene stationary phase.[117]

Acrylic acid and acrylate oligomers. Separations of these polar oligomers have only recently been reported. *t*-Butyldimethylsilyl (TBDMS) derivatives of acrylic acid oligomers, which contain one carboxylic acid side group per repetitive unit, were successfully chromatographed by Pinkston[118] on an open tubular column coated with polymethylsiloxane by means of neat CO_2 and

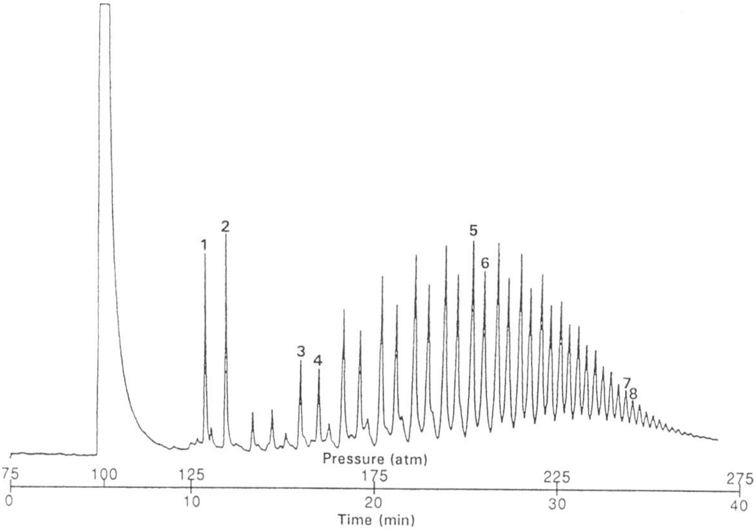

Figure 7.21. SFC chromatogram of ethoxylated alcohols (Neodol 45-7T) with straight-chain C_{14} and C_{15} alcohols spiked as retention standards. Conditions: 10-m x 50-μm i.d. open tubular column, polymethylsiloxane stationary phase (d_f = 0.2 μm); CO_2; 120°C; pressure program from 100 atm to 275 atm at 5 atm min^{-1} after a 5-min isobaric period; FID (300°C). Peak identifications: (1) *n*-C_{14} alcohol, (2) *n*-C_{15} alcohol, (3) $C_{14}E_2$, (4) $C_{15}E_2$, (5) $C_{14}E_7$, (6) $C_{15}E_7$, (7) $C_{14}E_{15}$, (8) $C_{15}E_{15}$ (reprinted with permission from Ref. 122).

pressure programming (Figure 7.20). Separation was achieved for some twenty homologs; higher molecular mass species remained unresolved. Other combinations of mobile and stationary phases may improve the separation. Methylmethacrylate oligomers, as obtained from radical oligomerization, were chromatographed with a composition gradient on a column packed with a silica stationary phase.[119]

Oligoethers. Oligomers with the general repeating unit, -O-R-, are very readily separated by means of supercritical CO_2. While there are only single reports mentioning the separations of oxymethylene (R = CH_2)[120] and oxyphenylene (R = C_6H_4)[121] oligomers, separations of oxyethylene (R = CH_2-CH_2) oligomers are reported in more than 40 publications. In the majority of cases, the separations of the poly(ethylene oxide) or poly(ethylene glycol) samples have been carried out by means of open tubular columns using pressure/density programming with a CO_2 mobile phase. The polyethylene oxides, which are frequently used as nonionic

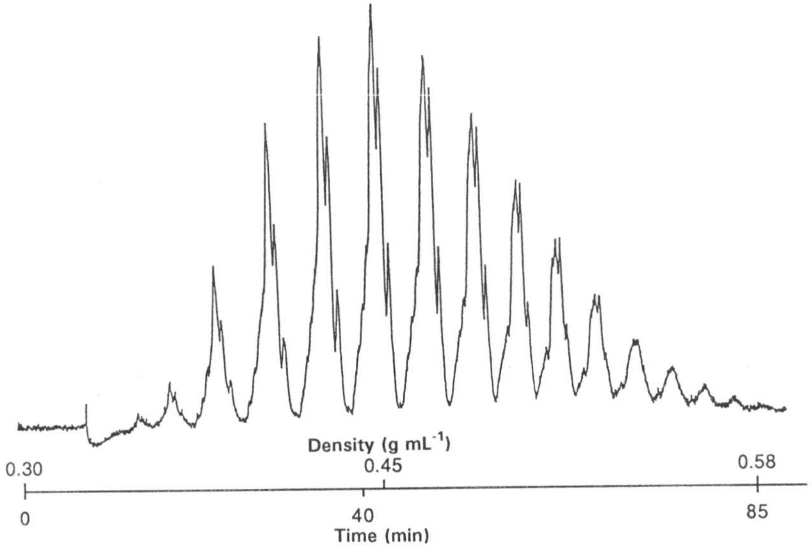

Figure 7.22. SFC chromatogram of thioethoxylates. Conditions: 6-m x 50-μm i.d. open tubular column, poly(5% phenyl)methylsiloxane (d_f = 0.25 μm); CO_2; 100°C; density program from 0.30 g mL^{-1} to 0.45 g mL^{-1} at 0.004 g mL^{-1} min^{-1} after a 5-min isopycnic period, then to 0.58 g mL^{-1} at 0.003 g mL^{-1} min^{-1}; SCD (375°C) (reprinted with permission from Ref. 123).

detergents/surfactants, are produced *via* ethoxylation of fatty alcohols. When such alcohols are technical grade products, they may contain more than only one hydroxyl component; this leads to the formation of parallel series. The SFC chromatogram of such a product is shown in Figure 7.21,[122] where the two main series, originating from ethoxylation of tetradecanol and pentadecanol, are seen along with smaller additional series.

The analysis of C_{12}-ethoxylate thiols by capillary SFC with sulfur chemiluminescence detection (SCD) is demonstrated in Figure 7.22. Ethoxylated alcohols or thiols are widely used as nonionic surfactants. Determination of the distribution of oligomers and numbers of ethylene oxide (EO) units are important because they affect the chemical properties and quality performance of the surfactant. SFC has been shown to be an effective method for separating EO oligomers in ethoxylated alcohols using flame ionization detection.[124] Because the SCD responds only to the amount of sulfur present in each oligomer, the average EO unit can be quantified directly from the molar sulfur content obtained from the SCD. By integrating each broad peak which represents one

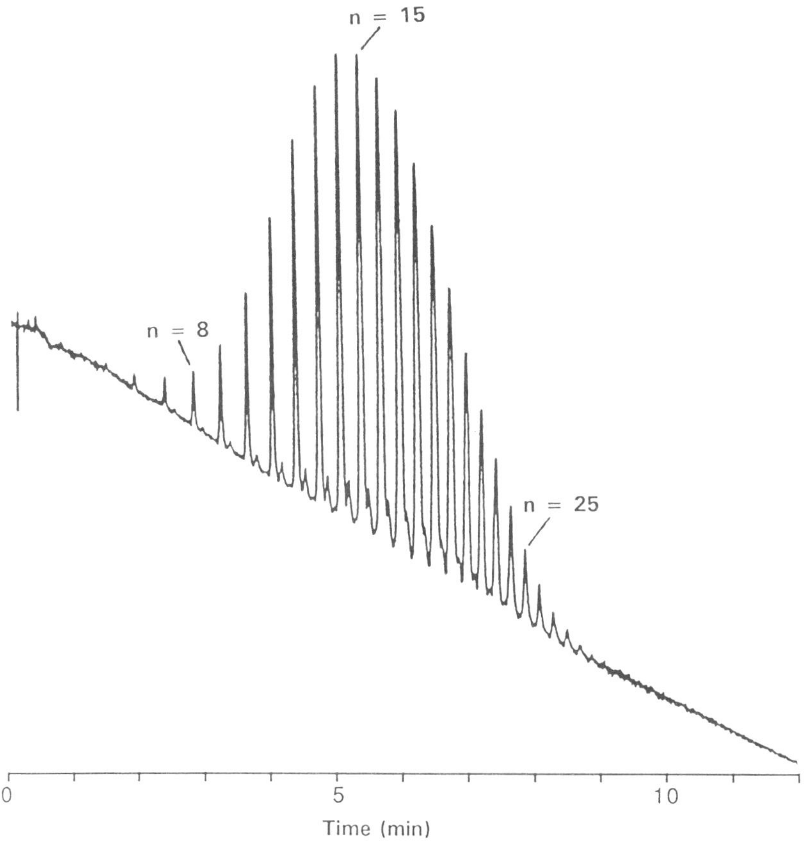

Figure 7.23. SFC chromatogram of polyethylene glycol with an octylphenol-ether end group (Triton X-165). Conditions: 100-mm x 2-mm i.d. packed column, C_{18} bonded silica particles (Nucleosil), 3-μm particle size; 7% (v/v) methanol in CO_2; 170°C; pressure program from 130 atm to 375 atm in 12 min, 2.15 mL min^{-1} flow rate; UV (278 nm) (reprinted with permission from Ref. 130).

oligomer (Figure 7.22), the calculated average EO unit was 6.8, which is in good agreement with the number provided by the manufacturer.[123]

Other products separated by SFC have been obtained by esterification of polyglycols with fatty acids. The separation of such a product, obtained by condensation of stearic acid and poly(ethylene glycol), is shown in Figure 6.10.[125] The two main series in the chromatogram can be assigned as chains bearing one or two stearate ends, respectively. A similar distribution has been reported for a polymer additive composed of esterified poly(ethylene glycol).[126] Additional industrial samples based on poly(ethylene oxide), including

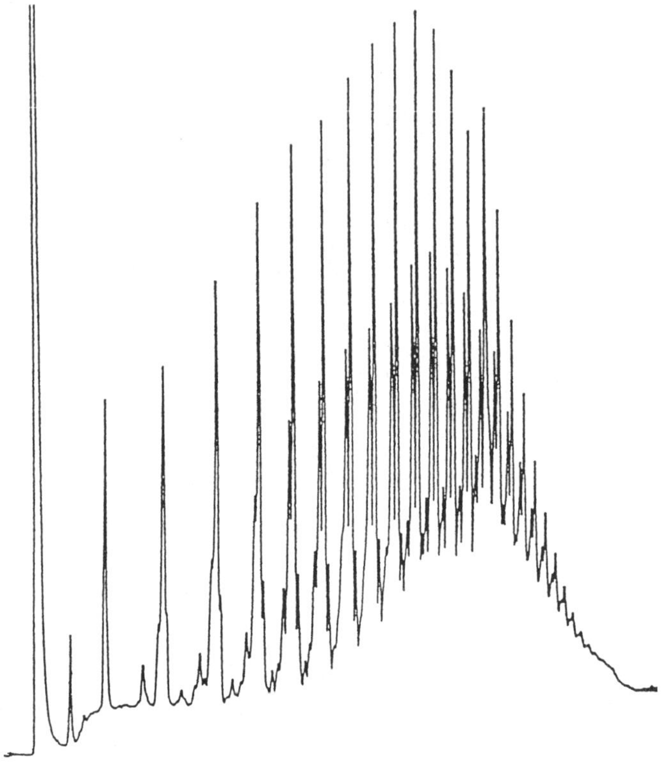

Figure 7.24. SFC chromatogram of a copolymer prepared from equimolar amounts of ethylene oxide and propylene oxide. Conditions: 3-m x 50-μm i.d. open tubular column, poly(30% biphenyl)methylsiloxane (d_f = 0.25 μm); CO_2; 100°C; density program from 0.25 g mL^{-1} to 0.76 g mL^{-1} in 60 min; FID (375°C) (reprinted with permission from Ref. 99).

an ethoxylated amine,[127] and their separation by SFC have been described.[127-129]

In some cases, separations have been performed using packed columns. Due to the polar natures of the samples, nonpolar stationary phases are better suited than are polar ones. Thus, successful separations were obtained on polystyrene[117] and on a C_{18}-silica material.[130] Especially the latter material (3-μm particles) allowed rapid high resolution separations, as exemplified by Figure 7.23.

Like oligomers from ethylene oxide, the corresponding products from propylene oxide have been separated by SFC, using both open tubular and packed columns. A special challenge is the separation of copolymers from ethylene oxide and propylene oxide, because a variety of products containing different amounts of the two monomers can be expected. Even this can be accomplished by SFC as is shown in Figure 7.24 for a product from equimolar amounts of ethylene oxide and propylene oxide.[99]

Condensation type oligomers. Polysiloxanes are a third group of oligomers which have been frequently chromatographed by means of SFC. Most of these applications are concerned with the separation of polymethylsiloxanes and poly(phenyl)methylsiloxanes. Resolution of up to 80 homologs has been achieved;[126] in Figure 7.25, polymethylsiloxane and poly(phenyl)methylsiloxane oligomers have been chromatographed up to a molecular mass of 8,000, corresponding to *ca.* 60 homologs.[127] Siloxane oligomer separations have been used for testing injection techniques[131] and simultaneous density and temperature programming.[132,133] In virtually all cases reported so far, siloxane oligomers have been chromatographed using open tubular columns.

A second type of oligocondensate which should be amenable to SFC analysis are aliphatic oligoesters, since these compounds are quite soluble in different solvents. It appears, however, that caprolactone oligomers have been the only oligoesters separated so far.[112,117,119] According to the rather polar nature of these oligomers, a nonpolar stationary phase was found to yield better separations than a polar phase.

Epoxy prepolymers, formed by reaction of bisphenol A with epichlorohydrin, have also been separated *via* SFC.[117,130,133] In one case, comparison was made between similar products of different manufacturers.[117] Prepolymers of phenol-formaldehyde resins have been separated by means of temperature programmed elution (decreasing T), and molecular mass data have been calculated from the chromatograms of prepolymers of the novolac type.[134]

Miscellaneous. A variety of additional oligomeric products have been analyzed by SFC. Different perfluorinated materials, such as tetrafluoro ethylene oligomers,[100] a fluorinated surfactant,[100] and a "functionalized perfluoropolyether"[135] were chromatographed by means of a pressure gradient. In addition, different end-group or side-group reactive oligomers should be mentioned, such as ethoxy acrylate oligomers,[131,136] allyl ethoxylate oligomers[118] and a novolac-type product with epoxy side groups.[109] Other products of technical interest are, *e.g.*, different types of waxes and hydrocarbon mixtures.

A special analytical problem is the identification and determination of polymer additives. Such additives may be single components, or they may be composed of a homologous series

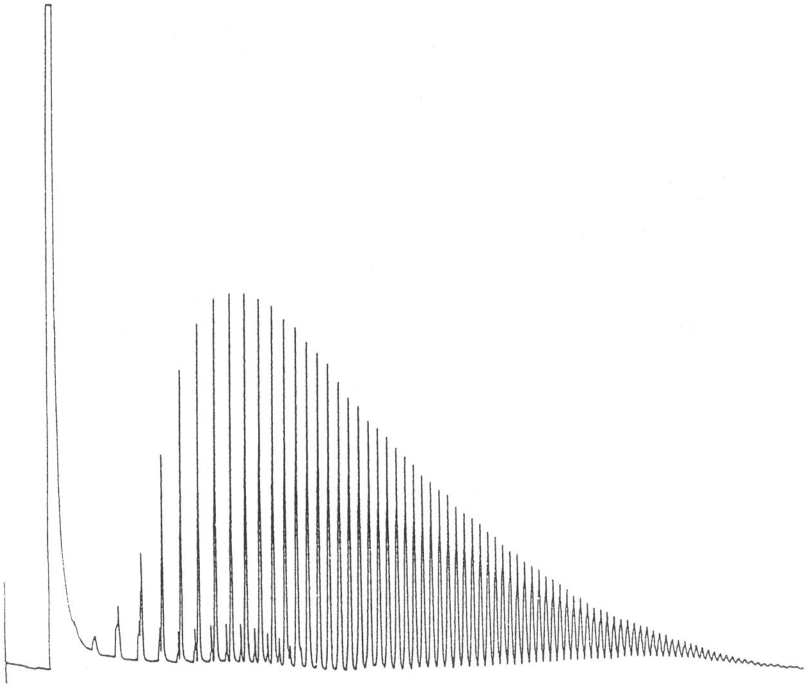

Figure 7.25. SFC chromatogram of dimethylsiloxane oligomers. Conditions: 20-m x 50-μm i.d. open tubular column, polymethylsiloxane stationary phase; CO_2; 120°C; asymptotic density program; FID (reprinted with permission from Ref. 126).

themselves. In an increasing number of publications, separations of polymer additives have been mentioned.[126,129,130,137-142] For the analysis of additives, SFE is of great advantage.[137]

As mentioned above, SFC can be a useful analytical technique for checking lot-to-lot reproducibility of polymer manufacturing. This has been demonstrated[133] for two "polyglycol" lots from the same manufacturer. One of these lots did not perform correctly for the user. The reason for this was found in the difference in the molecular mass distributions of the two products. One study reported the use of SFC for process control monitoring of an epoxy resin manufacturing process.[130]

Preparative separations of oligomer species have also been reported. In one case, the separation of an oligosiloxane (DC-710) which is used as a stationary phase in GC was optimized on an analytical scale[143] and then carried out on a preparative scale.[144] The separated compounds were investigated with respect to their properties as GC stationary phases.

Biopolymers. The analysis of biopolymers by SFC is described in detail in Sections 7.8 and 7.9. A brief discussion is included in this section, however, for completeness. Biopolymers can be distinguished as being composed of a homologous series or being molecularly homogeneous themselves. While, *e.g.*, polysaccharides belong to the former type, polypeptides and proteins are typical of the latter type.

One problem with saccharide oligomers is their high polarity, which is due to the large number of hydroxyl groups attached to the repetitive units. It has been demonstrated that after derivatization, resolution was considerably higher, and the paired peaks could be assigned to the two possible anomers.[145-147] Another example of a homologous series of biopolymers are ubiquinones, which have been separated on a packed column.[148]

Success in the separation of polypeptides strongly depends on their dissolution behavior, which in turn depends very strongly on the amino acid composition and distribution within the molecules. In the very beginning of SFC, glycine, a dipeptide, and a tripeptide were separated in supercritical ammonia.[92] Recently, cyclic peptides, cyclosporin[149] and valinomycin[150,151] (with amide and ester linkages), were chromatographed by SFC. Finally, the separation of amino acids as their PTH derivatives[152] should be mentioned. More detailed information on the separation of oligomers and polymers, including the discussion of general principles, is contained in a recent review article.[153]

7.7 Lipids

The separation of fatty acids, and mono-, di-, and triglycerides has been extensively studied with open tubular column SFC, primarily due to the nonpolar nature of supercritical CO_2, the most widely used mobile phase for SFC. Supercritical CO_2 is an excellent solvent for fats, oils, fatty acids, and most of their derivatives. As a result, triglycerides and their components are often used to demonstrate improvements in SFE and open tubular column SFC techniques.

SFC and high temperature GC. Packed column SFC of mono-, di-, and triglycerides compares favorably with open tubular column SFC for quantitative analysis of easy-to-separate analytes. Schwartz *et al.*[154] used an interesting combination of modifiers (0.3% formic acid and 0.15-0.18% water) for FID detection. For the most difficult separations, open tubular column SFC generally provides the best resolution, although there is some disagreement on this point.[155,156] Sandra *et al.*[156] compared separations using high temperature open tubular column GC and open tubular column SFC of a variety of lipid samples including milk chocolate, palm kernel oil, and several mixtures consisting of mono-, di-, and triglycerides. In most cases, GC compared favorably with open tubular column SFC, although SFC was favored for (a) certain compounds with molecular mass above approximately 1,400, (b) phospholipids, (c) mixtures that require the use of selective phases like cyanopropyl or polyethylene glycol (with temperature limitations of 250-280°C), and (d) samples for which fraction collection is desired. As an example, fatty acid hydroperoxides decompose well below the temperatures required to volatilize these thermally sensitive analytes. As a result, either SFC or LC must be used to separate the positional isomers of unsaturated fatty acid hydroperoxides.[157] Additional work is needed in this area, since separations achieved using open tubular column SFC are not much better than those currently possible using LC.

Derivatization. Derivatization improves open tubular column SFC of mono- and diglycerides, and fatty acids in general, when using certain stationary phases,[158,159] although it is not needed in as many cases as with GC. Mixtures of fatty acid methyl esters from sunflower, peanut, linseed, flax, and black currant oils, and highly unsaturated fatty acid methyl esters isolated from fish oil (C18:4, C20:5, and C22:6) were separated on a packed column containing 5-μm Hypersil silica. Geiser *et al.*[160] were able to circumvent the usual requirement to derivatize the fatty acids before their analysis on packed silica columns by hydrating the column. The result was a significant improvement in the chromatography.

Sample introduction and SFE. SFE has been used both as an injection technique and as a sample preparation technique for the SFC of lipids and fatty acids. Hawthorne and Miller developed an "almost" on-column injection technique in conjunction with solvent venting which greatly facilitated analysis of dilute solutions.[161] A sample of bayberry wax containing di- and triglycerides was injected and then separated on an open tubular column. Berg and Greibrokk[162] used a similar venting technique to analyze dilute solutions of mono-, di-, and triglycerides. Gmür *et al.*[163] extracted and analyzed several high fat food samples using combined SFE/SFC. Samples of fresh and rancid butter, fresh and rancid cheese, raw and refined beef fat, plus raw and refined rapeseed oil were compared. The same group extracted methylketones and free fatty acids from

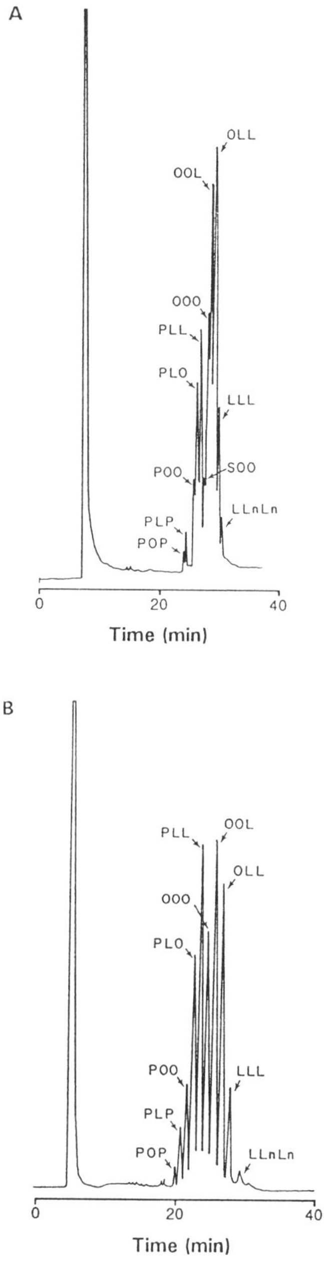

Figure 7.26. SFC chromatograms of soybean oil using (A) polymethylsiloxane and (B) Carbowax 20M stationary phases. Conditions: 10-m x 50-μm i.d. open tubular column (d_f = 0.25 μm); CO_2; 200°C; density program from 0.25 g mL^{-1} to 0.45 g mL^{-1} at 0.005 g mL^{-1} min^{-1}; FID (375°C) (reprinted with permission from Ref. 26).

dairy products.[164] The authors suggested that the techniques developed could be used to monitor cheese ripening, flavor defects in butter, and rancidity in dairy products in general. SFE of triglycerides from soybeans, followed by high temperature GC, was used to examine the effects of extraction pressure on the mix of triglycerides obtained from soybeans;[165] this information is quite useful for the commercial extraction of soybeans and for the optimization of conditions for the analytical extraction of lipids from food samples.

Schomburg and Roeder[166] demonstrated some effective focusing procedures designed to overcome sampling techniques that could result in poor peak shapes. Fatty acid methyl esters were separated on a polymethylsiloxane open tubular column. Fields and Lee[167] used triglycerides as one of their example analyte samples to demonstrate the effect of density and temperature programming on column efficiency.

Column selection. For separations of fatty acids or their derivatives (*e.g.*, fatty acid methyl esters) based on the number of carbon atoms, nonpolar stationary phases, [*e.g.*, polymethylsiloxanes or poly(phenyl)methylsiloxanes] are sufficient. However, separations of triglycerides or fatty acids of the same carbon number, but with a different number of double bonds, are usually better accomplished on more polar stationary phases, such as the cyanopropyl-containing or the polyethylene glycol stationary phases. An excellent example of the selectivity provided by the more polar stationary phases was reported by Richter *et al.*[26] (Figure 7.26), in which soybean oil triglycerides were separated on both polymethylsiloxane and Carbowax 20M columns. Soybean triglycerides contain nearly 90% C_{18} fatty acids, from stearic to linolenic acid. The separation on the Carbowax 20M column was substantially better than that on the polymethylsiloxane column. They were able to achieve very good separations of closely related triglycerides, such as OLL and LLL. This information was applied in the analysis of a variety of samples, including jojoba oil, oil of evening primrose (Figure 7.27), palm oil, beef tallow, corn oil, a chocolate chip extract, and some novel fat substitutes.[157] Saturated fatty acids of various lengths, as well as mono-, di-, and triglycerides can be easily separated on polymethylsiloxanes[168-171] and on similar phases,[24,167,172] while separations of cis/trans isomers of unsaturated fatty acids of the same carbon length can be better accomplished with a cyanopropyl column. The time required for the separation of unsaturated fatty acids is usually longer[81,173] than that required for similar saturated compounds, particularly if the separation of cis/trans isomers is

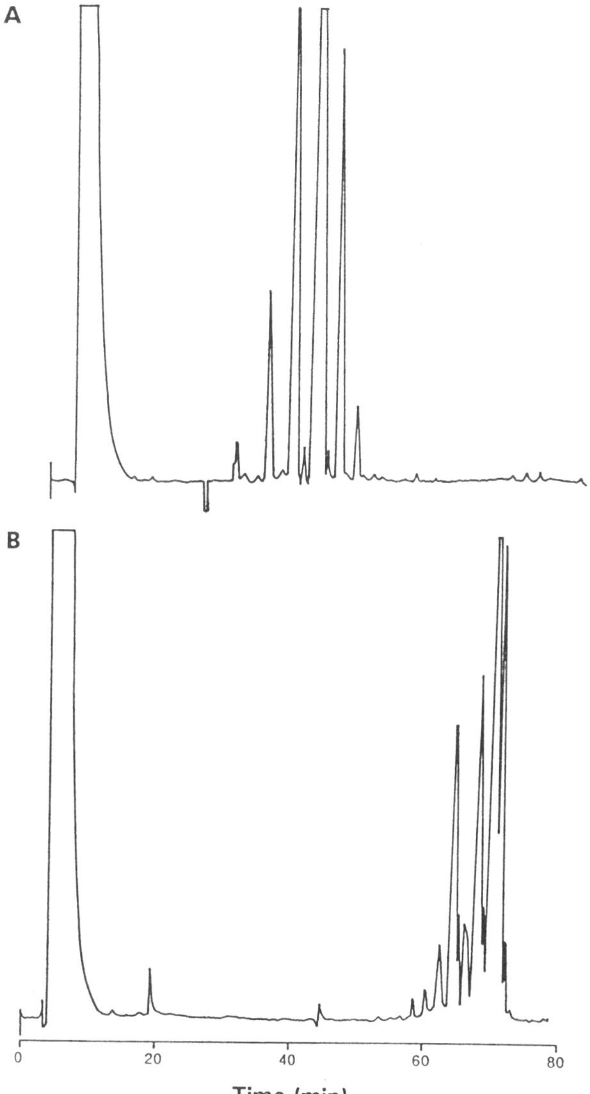

Figure 7.27. SFC chromatograms of (A) jojoba oil and (B) oil of evening primrose. Conditions: 8-m x 50-μm i.d. open tubular column, poly(25% cyanopropyl)methylsiloxane (d_f = 0.25 μm); CO_2; 100°C; density program from 0.35 g mL^{-1} to 0.55 g mL^{-1} at 0.0025 g mL^{-1} min^{-1} (reprinted with permission from Ref. 157).

desired. White and Houck[100,174,175] separated a mixture of mono-, di-, and triglycerides on a poly(50% phenyl)methylsiloxane stationary phase, but used a poly(25% cyanopropyl)methylsiloxane stationary phase to separate tristearin, triolein, trilinolein, and trilinolenin. Similarly, Sandra[176] used a poly(25% cyanopropyl, 25% phenyl)-methylsiloxane stationary phase to separate tripalmitin, tristearin, triolein, trilinolein, and trilinolenin.

The triglycerides that compose fish oil, rapeseed oil, soybean oil, and butterfat were also separated;[177] those of butterfat were identified using MS.[178] Since the majority of the fatty acids in soybean and rapeseed oil are C_{18} with only modest amounts of palmitic acid (<12%), the triglyceride patterns were much more simple than those of butterfat and fish oil.[179] Separation of the triglycerides in soybean and rapeseed oil on the basis of the number of double bonds was not attained. Proot *et al.*[179] separated a series of triglyceride standards, palm kernel oil, and milk chocolate triglycerides. Palm kernel oil contains predominantly saturated fatty acids (C_8-C_{18}), with about 15% of the total being oleic and linoleic acid. A cocoa butter separation was also reported,[180] which was similar to that reported by Proot *et al.*[179] Greibrokk[181] separated a series of saturated fatty acids (C_6-C_{31}) on an open tubular column containing a poly(30% biphenyl)methylsiloxane stationary phase.

Microbial lipids. Separations of lipids from some interesting microbial sources have been reported.[182,183] Glycerol tetraethers were extracted from methanogenic thermophilic bacteria. The archaebacteria were isolated from 2,000 m below the surface near a hydrothermal vent in the Gulf of California. The lipids differed in the number of cyclopentane rings present in the isopranyl side chains. Separation had been difficult to achieve with LC, TLC, or GC, and direct analysis was only possible with open tubular column SFC. Some of these lipids were unique to specific archaebacteria.[183] Triglycerides from fungi have also been examined. Separation of the triglycerides isolated from fungi was improved with the addition of either hexane or methanol to the supercritical CO_2 mobile phase.[184] Separation of various fatty acid esters, which included the ester of gamma-linolenic acid, from the neutral lipids isolated from fungi was also reported using open tubular column SFC.[184]

Lipids as test analytes. Many separations of fatty acids, fatty acid derivatives, and mono-, di-, and triglycerides are contained in reports about new detectors, stationary phases, and injection techniques, *etc.* In 1970, Cashaw *et al.* reported a new micro absorption detector[185] for SFC. Analytes examined included fatty acids, triglycerides, and lecithin, a phospholipid. As improvements in FIDs were made, mono-, di-, and triglycerides were often used to demonstrate the new instrumentation.[122] Several investigators reported the adaptation of MS for use with open tubular column SFC, and have selected fatty acids, fatty acid methyl esters, or triglycerides as example analytes.[51,186-188] Hawthorne and Miller[97] separated bayberry wax components (free carboxylic acids and several di- and triglycerides consisting of combinations of myristic and palmitic acid) with open tubular column SFC. They identified the components using a quadrupole MS with only slight modifications to the normal GC/MS procedures. Ion mobility spectrometry has been examined as a detection system for open tubular column SFC, and methyl esters of saturated fatty acids were selected as the example analytes. IR spectrometry has also shown considerable promise as a detector for open tubular column SFC; several investigators have reported the separation and identification of lipid-derived analytes, *e.g.*, Crisco oil and shortening,[189] free fatty acids,[190] and fatty acid methyl esters.[191] Light scattering detectors have been developed for both packed[192-194] and open tubular column SFC;[195] investigators have used both triglycerides[192,193,195] and fatty acids[194] to demonstrate their performance with SFC. Ogden and McNair[196] separated a series of fatty acids from coconut oil on a cyanoethyl-containing polymethylsiloxane stationary phase.

Restrictor development, like other advances in the SFC field, require evaluations with test analytes; an integral pressure restrictor developed by Guthrie and Schwartz[197] was evaluated using triglycerides. Similarly, Green and Bertsch[198] used fatty acids to examine the effectiveness of their new restrictor for open tubular column SFC.

7.8 Carbohydrates

During the last two decades, there has been an increasing interest in the details of carbohydrate structure. This interest has not been driven by first principles, but by a trail of research that focuses on the structure-function relationships of glycan residues conjugated to lipid and protein components, an area of study more recently referred to as "glycobiology."[199] These glycoconjugates, detected in plant and animal tissues of all types, are associated with countless cellular processes that have aroused and engaged the interest of many biological researchers. From a teleological standpoint, these hybrid molecules provide new domains of specificity and functionality that are the prerequisites to higher evolution. Our structural understanding of these materials, however, remains primitive when contrasted to the analytical expertise we enjoy with proteins and nucleic acid biopolymers. The complication, of course, resides with oligosaccharide characterization.

Perhaps the best studied glycoconjugates are those with proteins. In glycoprotein biosynthesis, the unique interplay of anabolic and catabolic events produces one of the most varied biopolymers currently known. The initial hybrid structure is the result of a single glycan, -[$(GlcN)_2$-$(man)_9$-$(Glc)_3$], attached "en bloc" to the glycosylation products in the nascent protein. The protein may be adorned with multiple copies, although it is not mandatory that every sequon be so covered. These newly linked residues are then modified by a series of integrated enzyme activities that first degrade and then rebuild a new glycan in a series of steps termed processing.[200,201] Herein lies one of the reasons for molecular complexity, for this processing is not a direct translational one analogous to protein synthesis, but the product of many post-translational events that provide numerous opportunities to alter and express the vagaries of the cell. Thus, not only does one observe different glycoproteins from the same cell, but the same protein may give rise to a family of glycoproteins, and even an identical asparagine residue may have a different glycan chain. This resultant molecular heterogeneity serves to create a series of subsets which are frequently termed "glycoforms."

A second reason for the difficulty in oligosaccharide structural analysis lies within the individual saccharide moieties themselves, as shown in Figure 7.28. Each of these features may contribute yet another set of structures, any one of which identifies a unique glycoform. As an example, a hexose residue could be one of three common types (mannose, galactose, or glucose) and may exist in either a pyran or furan ring form. The intervening linkage contributes another set of variables, where the C_1-hemiacetal oxygen can be positioned axial (α-anomer) or equatorial (β-anomer) to the ring and be joined to one of four (C_2, C_3, C_4, or C_6) positions to the adjacent monomer. And, as a final "coupe de grace" these structures frequently occur with multiple branches where the primary structural array is best characterized as a bush. A modification in any single feature represents a different glycoform, which must be determined for an accurate delineation of structure. This large number of structural iterations makes it easy to understand why isolation procedures are not routine and why a carbohydrate sequencer is a most difficult instrument to envisage. If the complexity of structure is not enough to deter the analyst, then the

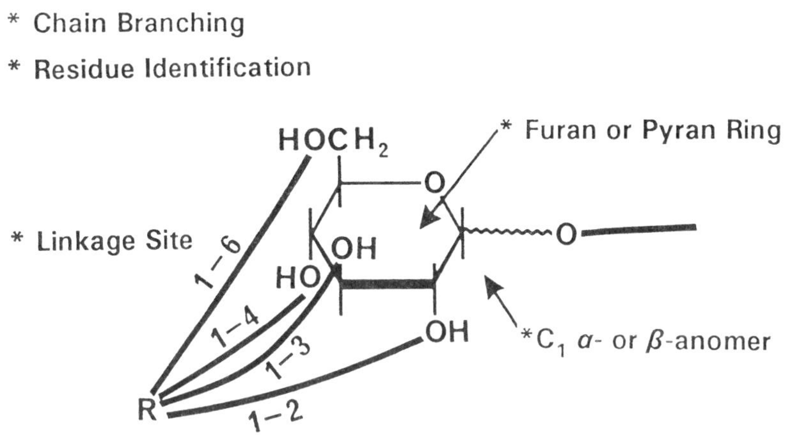

Figure 7.28. Schematic diagram of a carbohydrate conjugate.

heat and acid lability, the large variations in linkage stability, and the absence of photosensitive groups should.

Advantages of SFC. SFC, as a separation technique, brings a number of advantages that make it most appropriate for oligosaccharide characterization. Foremost of these are high analyte diffusivity, which provides excellent column efficiency; relatively low temperature operation, which eliminates thermal degradation of high molecular mass or polar materials; and, the ease of eluant detection with the FID. These features were demonstrated very early by Chester and Innis[145] in a paper that surprised and excited many carbohydrate researchers. Using trimethylsilyl derivatization to make the compounds less polar, CO_2 as a mobile phase, and open tubular columns,[202,203] they demonstrated the separation of a series of glucose oligomers up to degree of polymerization (DP) = 18 (Figure 7.29). The remarkable column resolution was apparent from the baseline separation of each incrementing monomer, and demonstrated again within each oligomer by their α- and β-anomeric separation. A second feature of this report that demanded attention was the simplicity of universal detection by flame ionization. The molecular mass of the higher oligomers exceeded 6,000, a mass range well above anything demonstrated by GC. Certainly, one could argue that the majority of this mass was attributed to the derivatizing agent and that the chemistry of blocking was an encumbrance. However, silylation serves only to induce solvent miscibility with the less polar CO_2; other more polar mobile phases could be considered. Furthermore, derivatization can provide a number of advantages, as

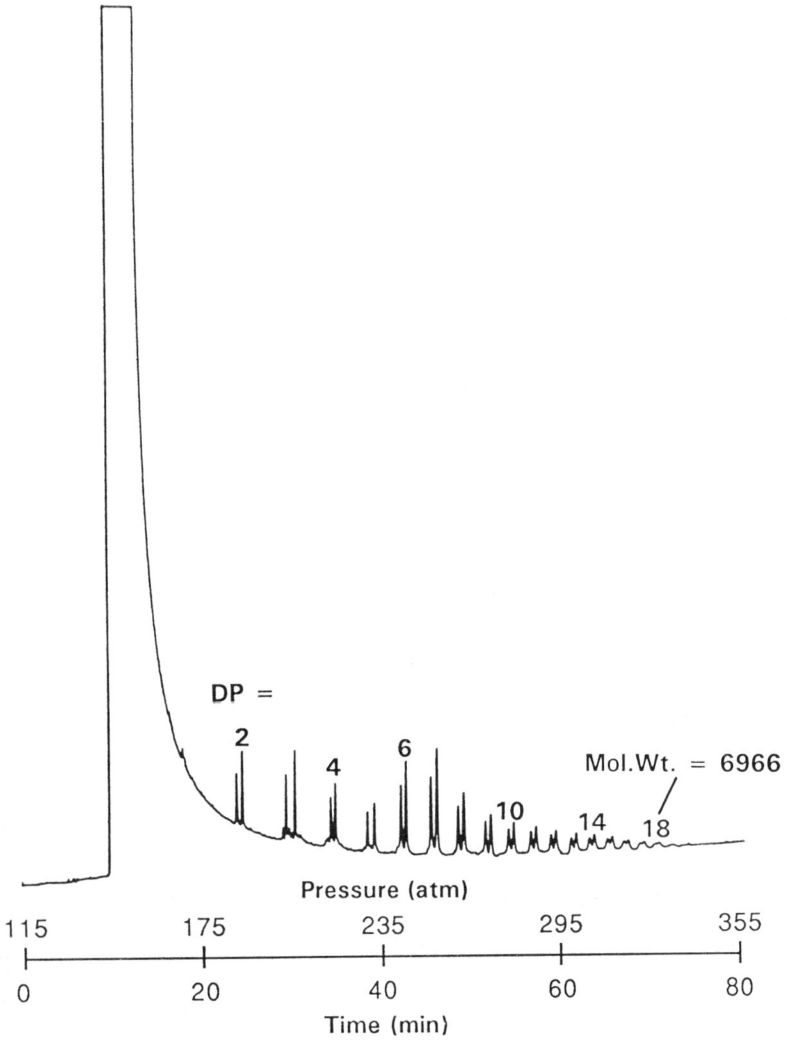

Figure 7.29. SFC chromatogram of silylated Maltrin 100. Conditions: 10-m x 50-μm i.d. open tubular column, poly(5% phenyl)methylsiloxane stationary phase; CO_2; 89°C; FID (reprinted with permission from Ref. 145).

have been observed over the years with GC. The same oligosaccharide sample can be prepared as the methyl or acetyl derivative, and chromatographed on polymethylsiloxane or more polar cyanopropyl stationary phases[204] with comparable separation efficiencies (Figure 7.30). Again, excellent column resolution and anomer separation can be observed, and for mass spectral detection following SFC, these lower molecular mass products have advantages.

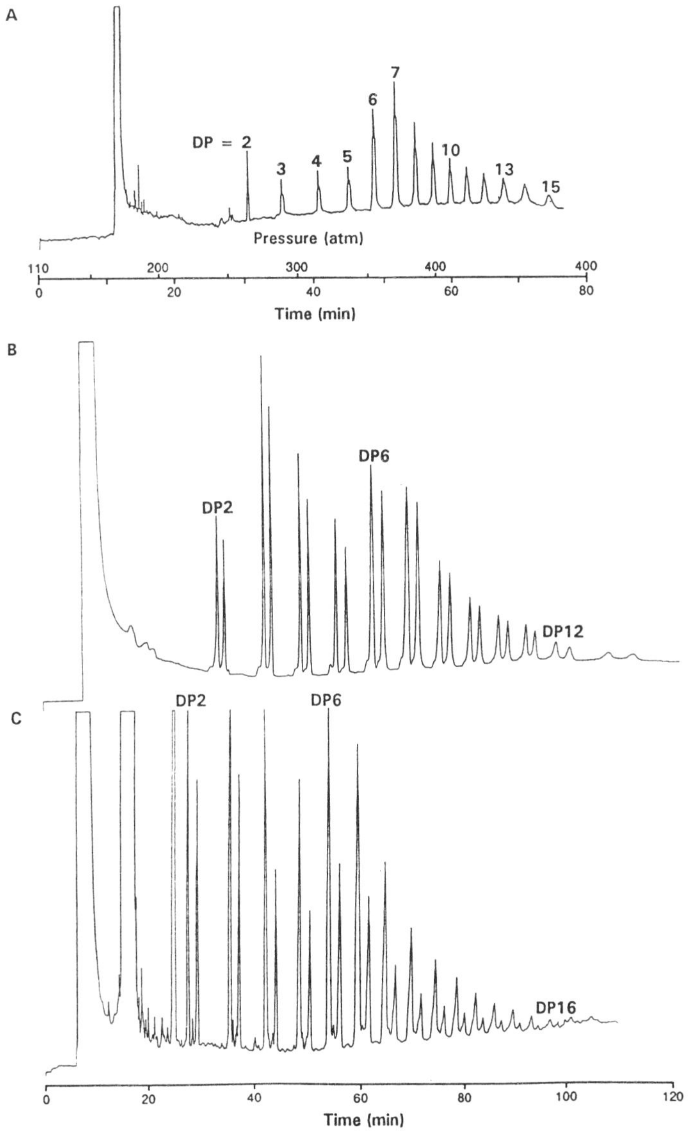

Figure 7.30. SFC chromatograms of derivatized polyglucose samples: (A) permethylated sample on poly(5% phenyl)methylsiloxane stationary phase, (B) peracetylated sample, and (C) permethylated sample on poly(50% cyanopropyl)methylsiloxane stationary phase. Conditions: 10-m x 50-μm i.d. open tubular column; CO_2; 120°C in (A); oven temperature program in (B) and (C) from 120°C to 180°C at 1°C min^{-1} after a 30-min isothermal period; pressure program in (A) from 110 atm to 400 atm at 5 atm min^{-1} after a 5-min isobaric period, and in (B) and (C) from 115 atm to 415 atm at 3 atm min^{-1} after a 10-min isobaric period; FID (reprinted with permission from Ref. 204).

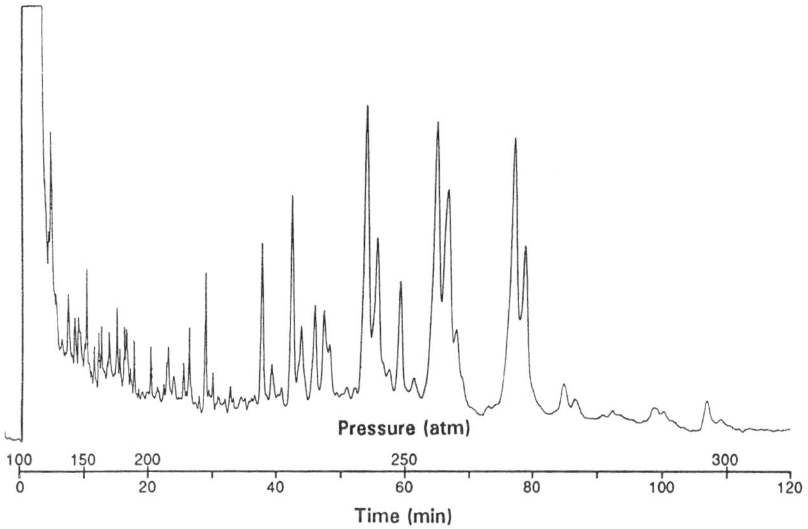

Figure 7.31. SFC chromatogram of permethylated extract of sheep urine. Conditions: 10-m x 50-μm i.d. open tubular column, poly(5% phenyl)methylsiloxane stationary phase; CO_2; 130°C; pressure program from 100 atm to 150 atm at 6 atm min^{-1}, to 170 atm at 5 atm min^{-1}, to 190 atm at 4 atm min^{-1}, to 210 atm at 3 atm min^{-1}, to 230 atm at 2 atm min^{-1}, and to 350 atm at 1 atm min^{-1} after an initial 5-min isobaric period; FID (reprinted with permission from Ref. 204).

Applications. SFC has been applied to a number of problems involving the characterization of oligosaccharides. As an example, in a study of locoweed toxicity in sheep where a plant alkaloid inhibits glycoprotein catabolism, the urine accumulates high concentrations of undegraded glycans. Urine extracts of these materials can be derivatized and profiles obtained using SFC/FID to determine the degree and type of toxicity. Figure 7.31 is an example representing locoweed toxicity in sheep with peaks of high mannose glycans (Man_n-GN, where n = 3 to 6). The paired peaks for each incrementing oligomer are attributed to the alpha and beta anomers of the reducing 2-acetamido-2-deoxyglucose residue.

The characterization and quantitation of glycolipid materials represents another application well suited to SFC/FID. These amphoteric glycoconjugates are important constituents in nervous tissue and are primarily detected in cell membranes. Glycolipids are composed of a long chain base (sphingoid), a single fatty acid, and one or more carbohydrate residues (Figure 7.32). The sphingoid moiety, when conjugated with a fatty acid, is called a ceramide, and

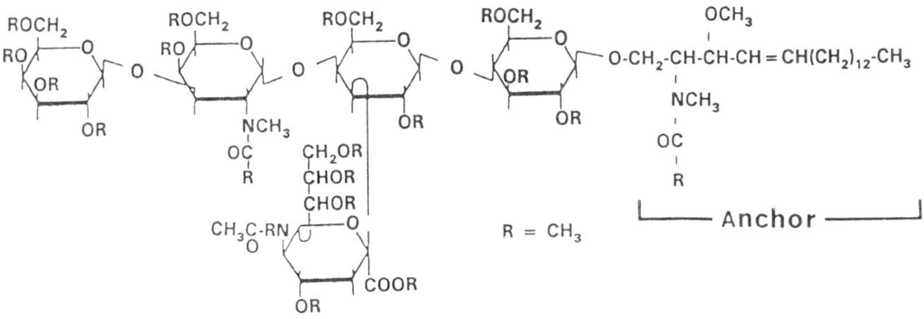

Figure 7.32. Structure of a glycolipid, G_{M1}.

it is this lipophilic portion of the molecule that lies implanted within the membrane and accounts for much of the observed microheterogeneity (*i.e.*, alkyl chain modifications of unsaturation, hydroxylation, and length). The carbohydrate portion of the molecule provides a means for classification,[205] and its interaction with exterior cellular components are again the focus of much biological interest.[206-208]

Kuei *et al.*[209] have evaluated SFC for the characterization of glycolipid samples following permethylation. In this report, several neutral and acidic glycolipid samples and two tissue mixtures were investigated. One sample of considerable complexity was an extract from brain tissue known to consist of gangliosides (neuraminic acid-containing). This material was permethylated and chromatographed on an open tubular column (Figure 7.33). Three major materials were detected, each paired with a component of equal intensity. From a comparison of purified standard materials, using SFC and MS, it was concluded that the detected SFC peaks were split due to a difference in the alkyl chain of the ceramide moiety.

Another important naturally occurring glycolipid class includes those derived from gram-negative bacteria, called endotoxins. These amphoteric lipopolysaccharide molecules vary considerably from species to species and have the general composition of a terminal fatty acid esterified 2-amino-2-deoxyglucose residue conjugated to hydrophilic oligosaccharides. The lipophilic residue is called Lipid A (Figure 7.34), and accounts for all of the tissue toxicity. Lipid A serves as the membrane anchor for the outer polysaccharide coat, and its structural determination has perplexed many scientists for decades. Recent studies of the Lipid A moiety obtained from *Salmonella typhimurium* C30/C21 has indicated the anchor to be a homologous mixture differing in the number of attached fatty acid residues,[210-213] and this heterogeneity appears to be the case with other species.[214-216] In *Salmonella minnesota* Re595, a heptaacyl

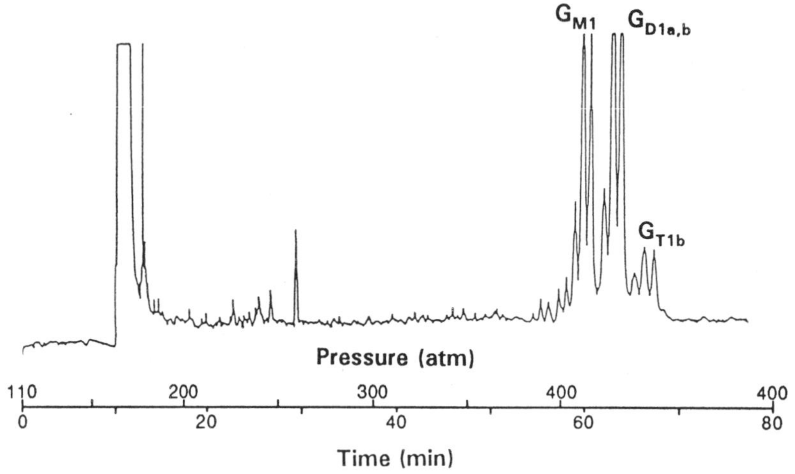

Figure 7.33. SFC chromatogram of a brain extract consisting of the gangliosides, G_{M1}, $G_{D1a,b}$, and G_{T1b}. Conditions: 10-m x 50-μm i.d. open tubular column, poly(5% phenyl)methylsiloxane stationary phase; CO_2; 120°C; pressure program from 110 atm to 400 atm at 5 atm min^{-1} after a 5-min isobaric period; FID (reprinted with permission from Ref. 209).

monophosphoryl Lipid A structure has been characterized as a phosphorylated 2-amino-2-deoxy-*D*-glucose disaccharide, linked β(1→6)[217] with an (R)-3-hydroxytetradecanoyl group at the 2,3-position (GlcN I), and the 2′,3′-position (GlcN II). Additionally, the hydroxyl group on three of these residues (at positions 2, 2′, and 3′) are esterified with palmityl, lauryl, and myristyl groups, respectively.[218]

Lipid A can be obtained from bacterial lipopolysaccharides by hydrolysis and partial purification of the complex mixture by TLC. Extraction of each spot, followed by FAB-MS analysis, have indicated a homologous series of Lipid A components which vary in the degree of acylation.[219] In support of this study, the mixture was also evaluated by SFC following trimethylsilylation (Figure 7.35).[204] It is probable that the later eluting SFC peaks in this chromatogram represent a higher degree of acylation, and the major peaks are comparable to those observed by TLC. In support of this fact is the observation that the SFC/FID peak intensity corresponds very closely to the individual intensities detected by TLC and FAB-MS.[219] Although much work must be undertaken to make full use of SFC for endotoxin studies, these preliminary results would suggest considerable improvement over TLC.

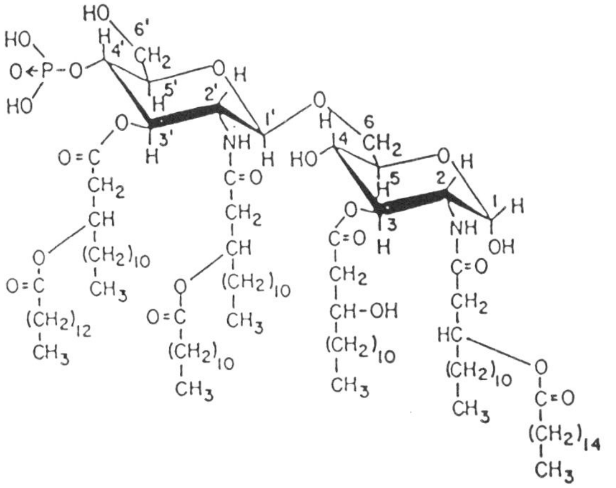

Figure 7.34. Structure of Lipid A.

A most active and difficult area of glycoconjugate investigation involves the relationship of glycan structure to the functional roles exhibited by glycoproteins. The glycan residues are known to serve as recognition determinants in host/pathogen interactions, in protein targeting, and in cell/cell interactions. Many recent reviews have focused on these developments.[199,220-222] As a consequence of the considerable diversity of structure (see above), the need for improved specificity and sensitivity are fundamental. The glycan residues can be released from proteins by either chemical or enzymatic techniques, and this heterogeneous mixture can be resolved by SFC. The advantages of SFC analysis for other oligosaccharides are equally important to the study of these N-linked glycans. As a demonstration of this approach, a glycoprotein was treated with an endoglycosidase that cleaves the glycan residues from the protein at the penultimate residue adjacent to the protein backbone. These materials were directly trimethylsilylated and analyzed using CO_2 as the mobile phase (Figure 7.36). The structure and composition of this fraction has been previously studied by FAB-MS using a permethylated derivative. These data provided a series of ions (incrementing by 204 amu and indicative of a permethylated monomer residue) which can be related to their molecular masses. The three major and three minor peaks can thus be identified as (Man_n-GlcNAc, n = 4 to 9), with partial anomer resolution of the peaks A, B, and C. These problems have recently been approached by specific derivatization at the reducing end and analysis by SFC/MS using NICI for enhanced sensitivity.[223,224] Using this approach,

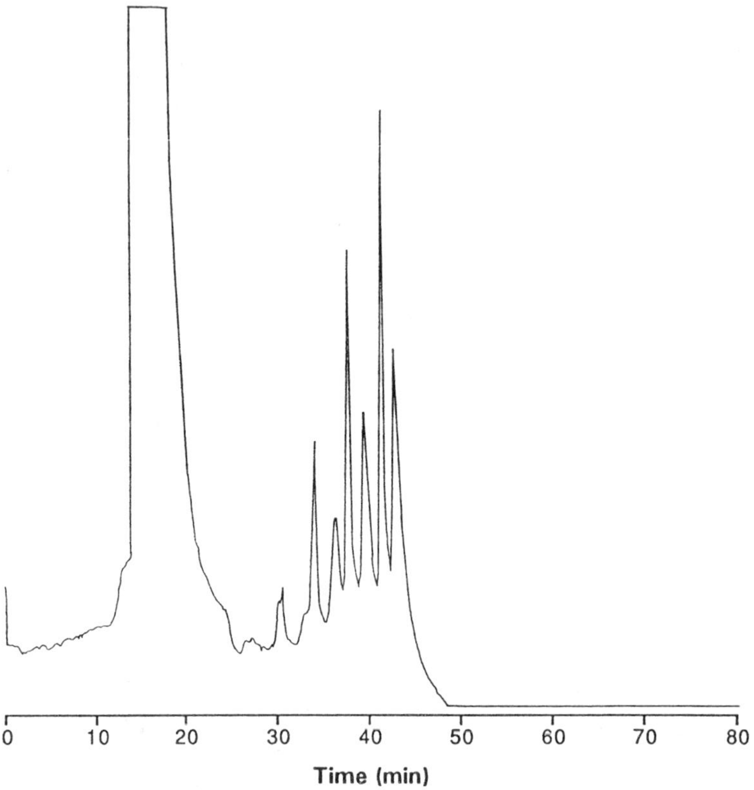

Figure 7.35. SFC chromatogram of *Salmonella Minnesota* Re595 Lipid A prepared by trimethylsilylation. Conditions: 10-m x 50-μm i.d. open tubular column, poly(30% biphenyl)methylsiloxane stationary phase; CO_2; 90°C; pressure program from 150 atm to 400 atm at 2 atm min^{-1} after a 10-min isobaric period; FID (reprinted with permission from Ref. 204).

preliminary data have been presented, indicating femtomole sensitivities.

Detection by MS. As mentioned above, a major feature that SFC brings to sample characterization is the ease of universal detection by flame ionization. For greater structural understanding, detection by MS is frequently considered, and the simplicity of removing a gaseous mobile phase, *vis-a-vis* an aqueous mobile phase, makes SFC an inherently simpler chromatographic instrument to interface with MS when compared to LC. Consistent progress in SFC/MS interfacing has been reported,[225] although there have been complications in the design of an "ideal" SFC/MS interface for polar molecules (*e.g.*, allowing high mass transmission while operating at

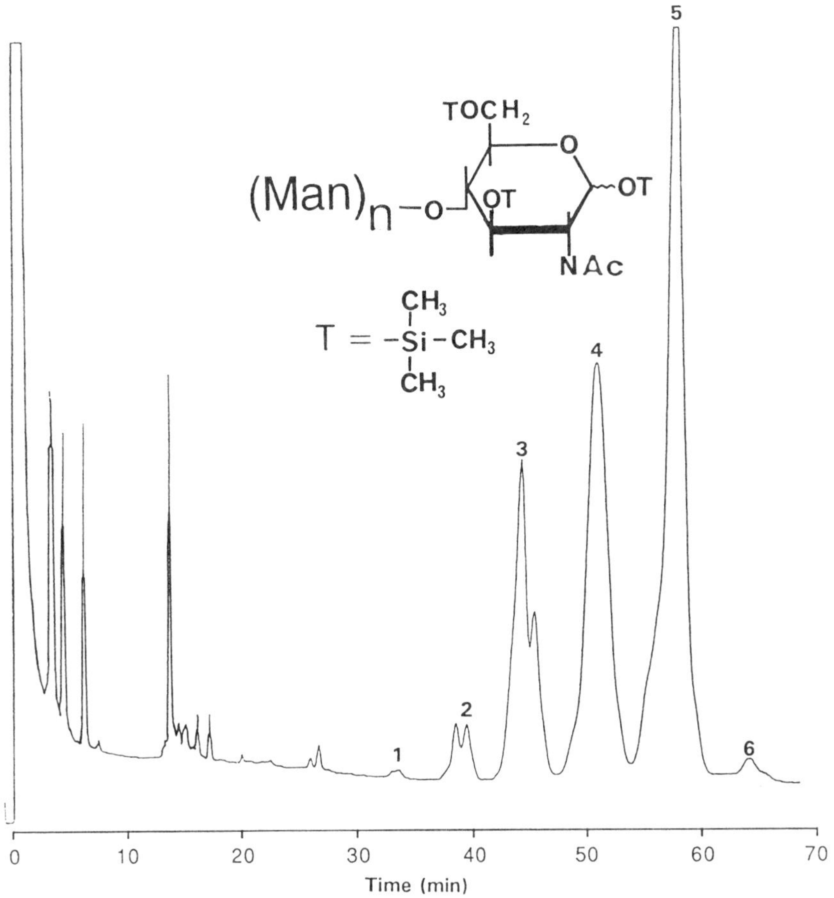

Figure 7.36. SFC chromatogram of a series of trimethylsilyated high mannose glycans isolated from plant lectin by enzyme treatment. Conditions: 10-m x 50-μm i.d. open tubular column, poly(5% phenyl)methylsiloxane stationary phase; CO_2; 130°C; pressure program as in Figure 7.31; FID. Peak identifications: (1) $(Man)_4$GNAc, (2) $(Man)_5$GNAc, (3) $(Man)_6$GNAc, (4) $(Man)_7$GNAc, (5) $(Man)_8$GNAc, and (6) $(Man)_9$GNAc (reprinted with permission from Ref. 204).

ambient temperatures).[226] With increasing interest in SFC, it is highly probable that these problems will be soon resolved. The large volume of gas necessarily delivered to an ion source during MS interfacing can be used to advantage by serving as a reagent gas for CI. Pinkston *et al.* have reported the design of an interface that allowed transmission and mass measurement of silylated polyglucose samples up to DP = 7 using ammonia CI with a quadrupole instrument.[227] Later, using the same materials and a high performance magnetic sector instrument, detection up to DP=10 for permethylated [m/z 2104, $(M + NH_4)^+$], and DP=15 for trimethylsilylated materials [m/z 5850, $(M + NH_4)^+$] were reported (Figure 4.48).[147] In addition, the latter group investigated detection limits, and reported picogram detectability for a permethylated β-cyclodextrin sample by single ion monitoring of the $(M + NH_4)^+$ ion. These data suggest that under appropriate conditions, oligosaccharide separation, high mass transmission, and sensitive detection are obtainable with contemporary SFC/MS techniques.

Polysaccharides, unconjugated to protein or lipid materials, are of great structural interest. This is especially the case in plants and microorganisms where numerous examples exist, showing that these materials possess biological activity. The surface polysaccharides of *Rhizobium* and *Agrobacterium* have received much attention because of their possible involvement in cell-signaling of specific plant cells involved in nitrogen fixation.[228] These surface polysaccharides include extracellular and capsular anionic polysaccharides, lipopolysaccharides, and cyclic glycans. The latter material from these species consists of a cyclic β-1,2-glycan backbone with 17-24 glucose residues.[229] Cyclic glycans have recently been detected in a *Bradyrhizobium* species.[230] These materials were analyzed by FAB-MS which indicated the fraction to consist of two separate glycans of 11 and 12 residues. Support for this finding was obtained by sample permethylation followed by SFC and SFC/MS. Two major components were separated by SFC, providing ions at m/z 2262 and 2466 [$(M + NH_4)^+$] when analyzed by ammonia CI. These data supported the FAB-MS data and clearly indicated two cyclic glycans composed of 11 and 12 residues. Smaller satellite peaks were also detected with 10 and 13 residues.

7.9 Biomolecules

The application of SFC to biological problems has had a surprisingly slow start. This is in direct contrast to the demonstrated benefits and inherent potential that supercritical fluids could bring to the study of biological systems, both in extraction and chromatography. The reasons for this are not clear, but part of the answer may be that alternative chromatographic techniques have reached a high degree of sophistication and are applied with adequate success to many biological problems.[231] In this environment, the acceptance of any new technology must clearly demonstrate unique advantages, not subtle improvements. Although this is easy to do for selected compound classes, biological applications are not extensive.

To consider what types of biological samples are best suited for SFC, it may be best to reflect on the limitations imposed by GC and LC analysis. For GC, these may include (a) samples exhibiting thermal lability and (b) samples of high molecular mass. Whereas for LC, these could include (a) limitations in detection and (b) poor sample resolution. With the great diversity of biological samples, it is impossible to generalize about specific applications.

Many of the earlier reports on supercritical fluids were initiated and sustained by the perceived extraction advantages. Much of that effort was, by necessity, devoted to instrument development, and these endeavors had an obvious focus on lipophilic materials, *e.g.*, oils, waxes, and lipids. The compatibility of these materials with the more convenient and less expensive mobile phases, such as CO_2, confined the applications to nonpolar analytes. Derivatization has extended these techniques to more polar samples, but this has not received widespread attention. Even less attention has been focused on the use of more polar mobile phases, probably due to the limited number of fluids available with physical properties matching analyte requirements and current instrumental capabilities. In the SFC reports summarized below, we have tried to include classes of biomolecules that have fallen outside of the topics already covered. These include amino acids and peptides, and the general category of bacterial products.

Amino acids and peptides. Amino acids and peptides have received scant attention in SFC. This may be attributed, in large part, to the plethora of other very sensitive and routine techniques which are available. Additionally, these materials are quite polar and require derivatization for chromatography when using nonpolar CO_2 as mobile phase. The less polar cyclic peptides, valinomycin[46] and cyclosporin,[149] have been analyzed by SFC/MS. Both of these molecules are composed exclusively of amino acids with aliphatic side chains, and they were chromatographed without derivatization. Leu-enkephalin, prepared as a permethylated derivative, has also been analyzed by SFC/MS.[232] Given the number of reports of applications, mainly for nonpolar peptides, it does not appear that SFC applications will progress further without sample derivatization or the introduction of more polar mobile phases. Two recent reports have described the SFC of amino acids. Most recently, the separation of up to 24 different PTH-amino acid derivatives was reported using

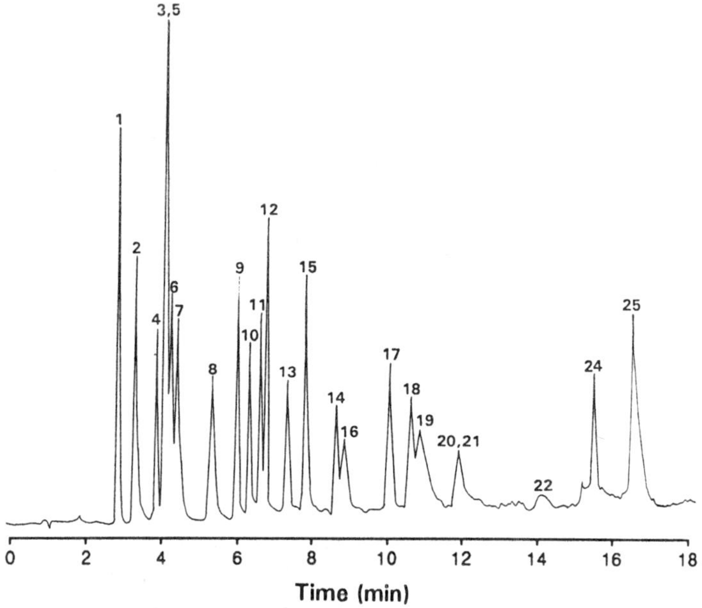

Figure 7.37. SFC chromatogram of PTH (phenylthiohydantoin) derivatives of 24 amino acids. Conditions: 250-mm x 4.6-mm i.d. packed column, cyanopropyl bonded silica, 5-μm particles (Zorbax); 0.001 M tetramethylammonium hydroxide in methanol gradient in CO_2; 40°C; 235 atm outlet pressure; 4.5 mL min^{-1} flow rate for 12 min, then decreased to 3 mL min^{-1}; FID. Peak identifications: (1) proline, (2) leucine, (3) valine, (4) isoleucine, (5) norleucine, (6) α-amino-isobutyric acid, (7) norvaline, (8) α-aminobutyric acid, (9) alanine, (10) methionine, (11) phenylalanine, (12) glycine (13) 4-hydroxyproline, (14) methionine sulfone, (15) threonine, (16) serine, (17) glutamine, (18) tyrosine, (19) tryptophan, (20) lysine, (21) asparagine, (22) histidine, (23) glutamic acid, (24) cysteic acid, (25) aspartic acid (reprinted with permission from Ref. 152).

a packed column with CO_2 as the mobile phase, modified with tetramethylammonium hydroxide and methanol (Figure 7.37).[152] The analysis of these PTH derivatives resolved all but two pairs of amino acids. SFC analysis of the PTH-labeled products of the Edman degradation may have advantages over LC because of the use of more specific and sensitive detectors, and the increased speed of analysis. SFC has also been applied to the resolution of amino acid enantiomers.[152,233-235] Thermal stability is not a requirement of columns in SFC for stereochemical separations.

Microbial products. One unique advantage of SFC is its ability to resolve a great diversity of structures in the same run which normally would require multiple chromatographic approaches. In this regard, the characterization of the various components produced by bacteria, fungi, and other microbes is an important application of SFC.

One class of bacterial components that has been studied by SFC is the tricothecene mycotoxins. These are thermally labile, polar, and difficult to analyze by GC. The SFC studies of these materials were reported in two papers using open tubular column SFC with MS detection.[236,237] The earlier report[236] demonstrated separations of a variety of mycotoxins ranging in molecular mass from 296 to 532. The very large roridin A and E structures were not resolved, but they could be identified by single ion plots. It was also demonstrated that the same components could be analyzed in less than 5 min with little loss in column resolution by rapid programming. The MDQ was approximately 1 picogram with this system using ammonia CI.

In the more recent of the two studies,[237] negative ion CI during SFC/MS was investigated. It was observed that fragmentation varied with the restrictor and source temperature; the higher temperatures contributed greater fragmentation. Ionization also varied from structure to structure under electron capture conditions, which could be explained by considering nucleophilic attack by OH^- or attachment by Cl^-. Details were not provided relative to the sensitivity of the method.

Another class of bacterial components of considerable interest are the ubiquinone structures isolated from the causative agent of "Legionnaire's Disease," *Legionnella pneumophila*. Ubiquinones are a class of isoprenoid quinones, which includes Coenzyme Q-12 (a major link in the electron transport chain), which are ubiquitous in biological tissues. Excellent separations of these materials have been reported using ODS reversed-phase columns with methanol-doped CO_2 as mobile phase (Figure 7.38).[148] The differences in ubiquinone content of different *Legionnella* species has facilitated their differentiation into three separate classes.[238]

Additional studies have described extraction (SFE) of various components from insects. In the past, MS of cell surface components has been used to classify species of various bacteria. It would be interesting to attempt this type of rapid characterization of microbial species by SFE/SFC.

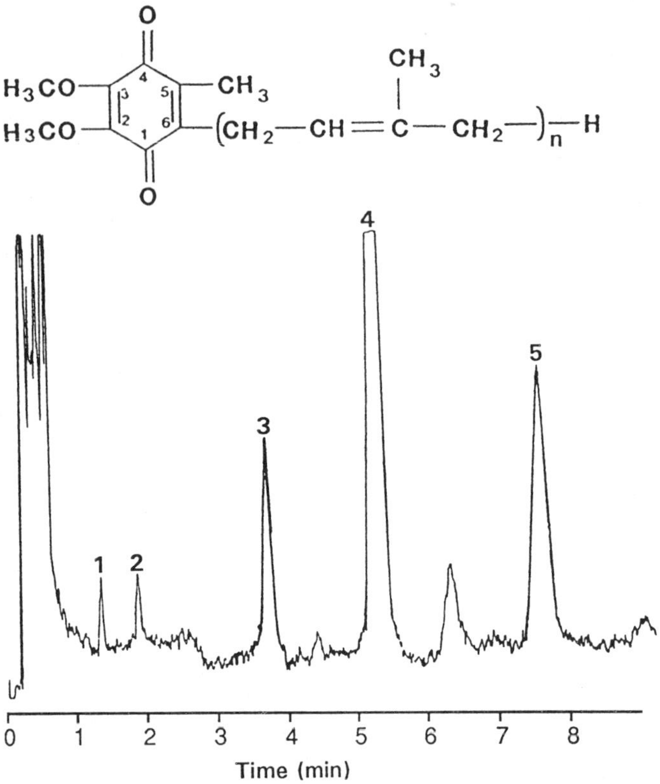

Figure 7.38. SFC chromatogram of ubiquinones (Q-n) from *Legionella pneumophila*. Conditions: 10-cm x 4.6-mm i.d. packed column, C_{18}-bonded silica, 3-μm particles; 1.5% methanol modified CO_2; 40°C; 232 atm average column pressure (69 atm pressure drop); 0.87 g mL^{-1} average mobile phase density, FID. Peak identifications of ubiquinones: (1) Q-8, (2) Q-9, (3) Q-11, (4) Q-12, (5) Q-13 (reprinted with permission from Ref. 148).

7.10 Industrial Chemicals

The great diversity in chemical and physical properties and the need for quality control and assurance have stimulated the development of a large number of analytical and chromatographic methods for industrial chemicals. SFC and SFE emerge as viable alternatives to existing methods with a bonus in economical and ecological performance.[239] Neat and modified supercritical fluids are increasingly being used in analytical and large scale separation processes. Selected applications of SFC for industrial chemicals are discussed in this section.

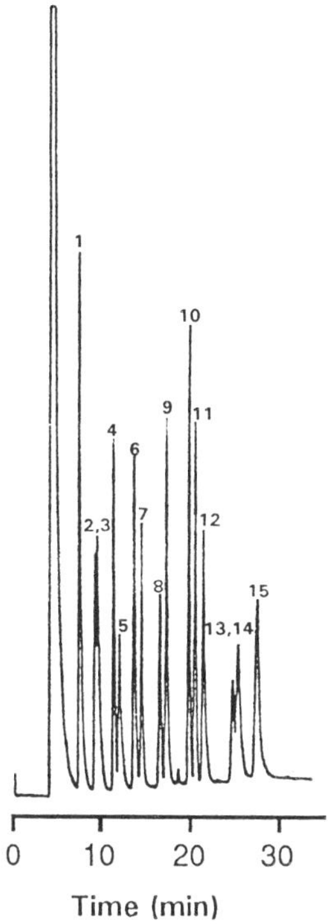

Figure 7.39. SFC chromatogram of aromatic acids. Conditions: 10-m x 50-μm i.d. open tubular column, poly(oligoethylene oxide)methylsiloxane stationary phase; CO_2; 100°C; density program from 0.05 g mL^{-1} to 0.76 g mL^{-1} at 0.015 g mL^{-1} min^{-1} after a 5-min isopycnic period; FID. Peak identifications: (1) benzoic acid, (2) cinnamic acid, (3) 4-chlorobenzoic acid, (4) 3,5-dimethoxybenzoic acid, (5) chlorocinnamic acid, (6) vanillic acid, (7) 3,4-dimethoxycinnamic acid, (8) syringic-acetic acid, (9) ferrulic acid, (10) sinapic acid, (11) *p*-hydroxyphenyl-propionic acid, (12) *p*-hydroxyphenyl-acetic acid, (13) *p*-hydroxyphenyl-acetic acid, (15) *p*-hydroxybenzoic acid, (14) *cis-p*-hydroxycinnamic acid, and (15) *trans-p*-hydroxycinnamic acid (reprinted with permission from Ref. 126).

Carboxylic acids. Carboxylic acids are important natural products and intermediates for a wide variety of compounds. Their highly polar nature and tendency to form hydrogen bonds usually require derivatization in GC techniques. Neat supercritical CO_2 and N_2O have sufficient solvent strength to dissolve and transport the C_1 to C_{30} free acids[170] in concentrations typically associated with open tubular column SFC. Figure 7.39 shows the separation of fifteen aromatic acids from benzoic to trans-*p*-hydroxycinnamic acid.[126]

The elution sequences on identical stationary phases are very similar in GC and SFC: homologous retention on apolar phases, incremental increase in retention according to unsaturation on medium polar phases, and very selective separation of positional and geometrical isomers on cyanopropyl stationary phases.[81] Quantitative analysis can be successfully performed on medium polar, thick film stationary phases such as Carbowax 20M or oligoethylene oxide-substituted polysiloxane stationary phases. The solubilities of polar analytes in these types of stationary phases appear to be much higher than in apolar stationary phases, thus allowing the injection of higher sample amounts without overloading of the column.[240]

Polar compounds such as di- and tricarbonic acids must be analyzed using neat CO_2 on a well deactivated open tubular column, or solubilized with an entrainer such as methanol for packed column SFC. An alternative would be to esterify the acidic groups.

In packed column SFC, highly deactivated polysiloxane phases and water/formic acid modifiers[241] or special polymeric materials, such as divinylbenzene packings,[242] allow the separation and detection of free fatty acids with flame ionization detection. Benzoic acid was eluted on a regular 10-μm C_{18}-modified silica column with only 0.16% methanol as modifier in CO_2 without peak distortion.[243] Salicylic acid, 3-bromobenzoic acid, naphthoic acid, and a series of acidic phenols showed no tailing effects on a 3-μm silica column when CO_2 and methanol with 0.15% citric acid as modifier was used.[130]

Analysis of amines. Amines differ strongly in their basicities and, therefore, reactivities with silanol surface groups or acidic mobile phases. An extensive study of the reactivity of CO_2 with amines led to the conclusion that amines with pK_b values lower than ~9 react with supercritical CO_2.[244,245] The formation of insoluble salts and decomposition products was observed for some compounds, while immiscibility and variable solubility were observed for others.

With respect to chromatographic behavior, several guidelines have been deduced from systematic experiments.[246] Aliphatic tertiary amines such as triethylamine, tributylamine, trihexylamine, and trioctylamine can be eluted from a polymethylsiloxane stationary phase with good peak shapes using CO_2 as mobile phase. To analyze strongly basic amines, two approaches are feasible: (a) conversion to trifluoroacetyl or other acetyl derivatives[247] (see Figure 7.40) or (b) direct analysis of the primary alkyl amines by employing inert mobile phases such as SF_6[248] or N_2O and very well deactivated columns.[247] Unless mixed with modifiers, both fluids exhibit low solvent strength and are limited to low molecular mass and monofunctional compounds.[249]

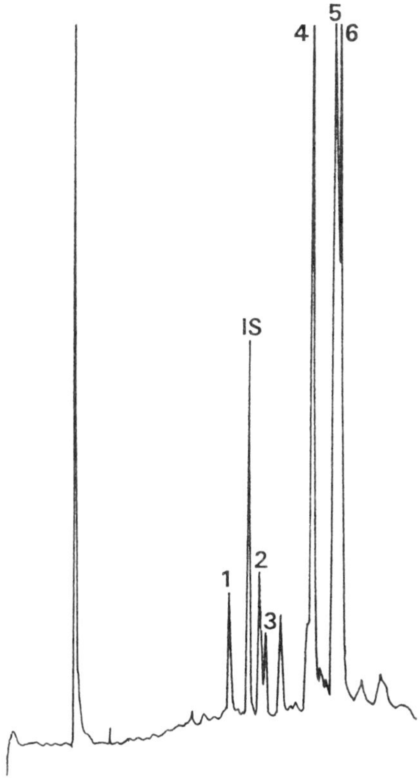

Figure 7.40. SFC chromatogram of primary aliphatic amines as their acyl derivatives. Conditions: 20-m x 100-μm i.d. open tubular column, polymethylsiloxane stationary phase (d_f = 0.2 μm); CO_2; 100°C; pressure program from 100 atm to 200 atm at 2 atm min^{-1}; NPD. Peak identifications: (1) C_{16}-NH_2, (2) $C_{18:u}$-NH (unsaturated), (3) C_{18}-NH_2, (4) C_{16}-NH-$(CH_2)_3$-NH_2, (5) $C_{18:u}$-NH-$(CH_2)_3$-NH_2 (unsaturated), (6) C_{18}-NH-$(CH_2)_3$-NH_2, and C_{17}-NH_2 as internal standard (reprinted with permission from Ref. 247).

Aromatic amines and heterocyclic, basic nitrogen containing products are amenable to packed column SFC. Narcotine, thebaine, codeine, cryptopine, and morphine were separated in less than 2.5 min on an aminopropyl silica with CO_2 and polar amine modifiers.[250] While methyl- and ethylamine modifiers reduce solute capacity factors and influence selectivity on the aminopropyl- and silica columns for these compounds, enhanced resolution was observed in 11-min chromatograms on silica with a CO_2/methanol/methylamine/-water mobile phase. Reaction of CO_2 with the amino group of the modified silica and/or methylamine and water is possible, but no verified reports on the permanent polarity changes or reaction products are currently available.

Analysis of isocyanates. Isocyanates are reactive, and the thermal labilities of many isocyanate derived products makes their analysis difficult. The types of isocyanates used as bulk chemicals for polyurethane manufacturing range from low molecular mass to medium molecular mass polymers, and include monomeric and polymerized derivatives such as ureas, dimers and trimers, blocked isocyanates, and prepolymers of isocyanates and polyols. The various raw materials often contain oligomers and prepolymers in addition to monomers. This makes analysis with LC and TLC, which require derivatization, difficult. Gel permeation chromatography (GPC) provides the only alternative to open tubular column SFC for analyzing oligomeric and prepolymeric isocyanates directly. However, the moisture content and impurities in the mobile phase can create potential problems, and the low resolution of GPC does not provide much detail[251] about polymeric distributions and isomeric compositions.

Open tubular column SFC[252] can overcome the above problems, provided the mobile phase, instrument components, and solvents are kept meticulously free of water and alcohols. Typical chromatograms of 4,4′-diphenylmethane diisocyanate samples are shown in Figure 7.41. Columns with 100-μm i.d. and film thicknesses of 0.25 to 0.5 μm were used to optimize sample capacity for the inherent minimum detection limits of the flame ionization detector. Short column lengths were used to decrease the analysis time.

The selectivities of stationary phases such as polymethylsiloxane, poly(30% biphenyl)methylsiloxane, and poly(50% *n*-octyl)methylsiloxane are very similar for low molecular mass diisocyanates like 2,4-toluene-, isophorone-, 1,6-hexane-, 2,2,4-trimethylhexane-, 1,6-*p*-phenylene-, and *m*-xylylene-diisocyanates for which the elution order remains the same, except for 1,6-hexane-diisocyanate which is less retained on the *n*-octyl phase *vs* the methyl phase. A temperature of 100°C appears to be optimal for maximizing efficiency while avoiding decomposition of labile compounds such as the carbodiimides.

Monomeric and dimeric diisocyanates can be effectively extracted with sub- and supercritical CO_2 from polyisocyanate biurets.[253] Co-extraction and countercurrent extraction principles can be employed. The extracts are precipitated in a cascade of depressurized steps or in a vessel at reduced pressure while recycling the CO_2.

Analysis of alcohols. Alcohols, polyols, and phenols are large scale raw materials for a large number of industrial processes, commercial formulations, and other products. In chromatography, these compounds exhibit a strong and notorious tendency to interact with adsorptive sites in the column.

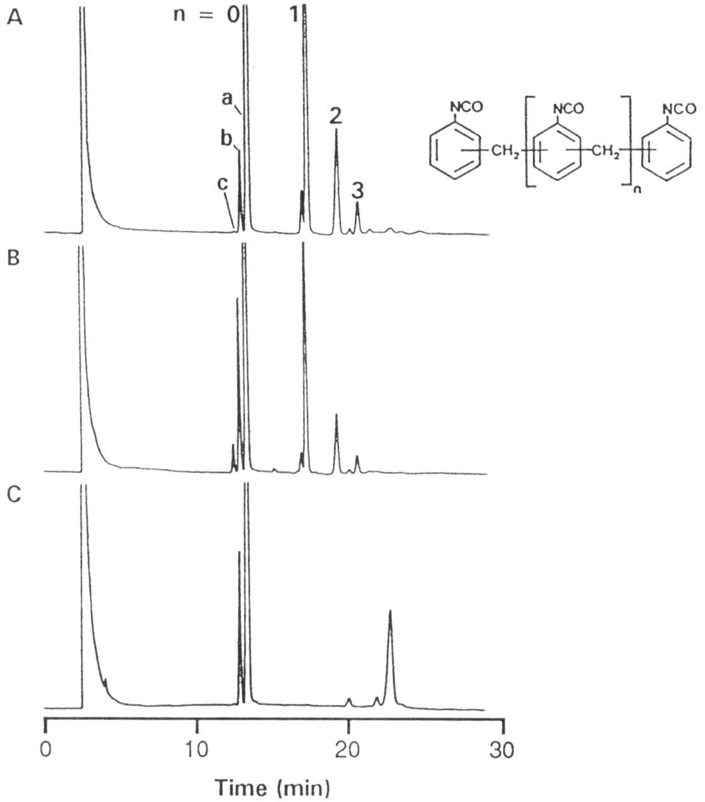

Figure 7.41. SFC chromatograms of liquid diphenylmethane diisocyanates (MDI). The first grouping of peaks (a-c) corresponds to the MDI in the isomars polymeric structure n = 0; the numbered peaks refer to the oligomer series of n = 1-3. (A) Technical-grade MDI, (B) Desmodur VL, (C) Isonate 143L. Conditions: 5-m x 100-μm i.d. open tubular column, polymethylsiloxane stationary phase (d_f = 0.5 μm); CO_2; 100°C; density program from 0.2 g mL^{-1} to 0.7 g mL^{-1} at 0.03 g mL^{-1} min^{-1} after a 5-min isopycnic period; FID. Peak identifications: (a) 4,4′-MDI, (b) 2,4′-MDI, (c) 2,2′-MDI (reprinted with permission from Ref. 252).

The Lewis base properties of supercritical CO_2 pair well with the prominent Lewis acid and hydrogen bonding forces of alcohols and phenols. In a CO_2 environment, the hydroxy groups of the alcohols and phenols are not available for hydrogen bonding with the free silanol groups in the column. As a result, adsorptive interactions (tailing) are reduced and compounds with acidic OH groups elute with undistorted peak shapes.[254]

7.11 Foods and Flavors

The chemical complexities of foods and flavors, coupled with their tendency to react and degrade at elevated temperatures, are considerations that must be taken into account when selecting the analytical methods for their analysis. Preparative scale supercritical fluid extraction has been used for two decades to achieve clean, solvent free, food extracts. Two of the most well-known preparative scale applications of SFE are decaffeination of coffee and extraction of hops for the brewing industry. Present applications of SFE in the food industry are listed in Table 7.2.[255,256]

Table 7.2. Applications of SFE in the Food Industry[a]

Purpose	Material
Decaffeination	Coffee
	Tea
Deodorization	Vegetable oils
	Animal fats
	Brewer's yeast
Aroma recovery	Spices
	Hops
	Tobacco dust
	Coffee
	Wine
	Apples
	Pears
Oil recovery	Soya beans
	Jojoba beans
	Palm kernels
	Cotton seed
	Coconuts and olives
	Peanuts
	Rice
	Avocado
	Cocoa
Fractionation	Cod liver oil
	Fatty acid esters
	Glyceride mixtures
	Lecithin

[a]Taken from Refs. 255 and 256.

There are three main objectives that motivate the analysis of food products: the determination of the natural chemical composition of the products, the determination of added ingredients to enhance the quality of the foods, and the determination of food contaminants. It is important to know the amounts of essential amino acids, fatty acids, triglycerides, and vitamins, because of the nutritional point of view. Plant material may be contaminated by pesticides and herbicides; meat and milk products may contain hormones or other drug residues that were used to treat the animal; and food may be contaminated by microbial toxins. Thus, the requirements for the analytical method can vary significantly depending on the matrix to be analyzed and the objectives in mind.

Vitamins. Vitamins are compounds that are essential in minute quantities to nutrition. They are divided into two groups: water soluble vitamins (B complex and C) and lipid soluble vitamins (A, D, E, and K). The water soluble vitamins act as coenzymes or as precursors for coenzymes in the regulation of metabolic processes. Lipid soluble vitamins are involved in the regulation of calcium and phosphorous metabolism (D), blood coagulation (K), vision (A), and oxidation (E). The lack of vitamins can cause severe diseases like rickets, beriberi, pellegra, and scurvy. Lipid soluble vitamins (A, D, and K) have proven to be difficult to analyze, because of the subtle structural differences between isomers, and thermal and oxidative instability.[257,258] SFC has proven to be a good analytical method to analyze both water soluble[259-261] and lipid soluble vitamins.[259,262,263]

Gere *et al.*[264,265] have used a column packed with 5-μm particles of octadecyl-bonded silica to separate Vitamins A, E, and D, using a methanol/CO_2 mobile phase gradient (0.5% for 7.5 min, to 1.0% in 2.5 min). However, addition of modifiers is not necessary for the elution of lipid-soluble vitamins. White *et al.*[260] reported the separation of a vitamin mixture by SFC using a crosslinked Carbowax 20M open tubular column. Excellent reproducibility of peak areas and retention times was achieved.

A 100% enrichment of Vitamin E from wheat germ powder by SFE/preparative SFC was reported by Saito *et al.*[266] The wheat germ powder was extracted with supercritical CO_2 and the resulting oil was fractionated by SFC on a silica gel column.

Matsumoto *et al.*[259] separated water-soluble and lipid-soluble vitamins on an 80-cm x 0.53-mm i.d. fused silica column packed with 30-50-μm particles of C_{18} silica. Vitamins were eluted with a 1:1(v/v) mixture of hexane and ethanol at 260-280°C, depending on the critical temperature of the mobile phase. The stability of the vitamins at such high temperatures was not tested. Detection was achieved by UV absorption in series with MS. The mobile phase vapor served as the CI reagent gas.

SFE has been utilized in combination with previously established LC methods to quantify vitamin K_1 in powdered infant formulas.[257] For samples spiked with K_1 (between 0.25 and 2.0 μg of K_1 per gram of powder), the recovery was about 95%. Although this work was done using electrochemical detection following LC, other workers have demonstrated good resolution of the K vitamins by open tubular column SFC.[260]

Open tubular columns and CO_2 as the mobile phase have provided separations of a variety of fat soluble vitamins, including various isoforms of A, D, E, and K (Figure 7.42).[260] The K_1 and K_3 isoforms are easily separated, but the three D forms, provitamin D, D_2, and D_3, are not as easily resolved (Figure 7.43). MS detection following packed capillary column SFC has been demonstrated for water and fat soluble vitamins.[259] A combined instrumental approach using SFC/MS, UV detection, and single ion monitoring has been successfully applied to determine vitamins A and D_3. A similar analysis of a series of water soluble vitamins, indicates interesting resolution for vitamins C, B_6, nicotinic acid, and nicotinamide. The compounds were not as well resolved as those by open tubular column SFC, but the analysis required less than 10 min. An interesting feature of these results is that the conditions for separation of both classes of compounds are remarkably similar.

Tocopherols. An unusual application of SFE and SFC for wheat germ tocopherols has been reported.[266] SFE was used to extract the lipid components of the wheat germ, and this was followed by preparative SFC to increase for evaluation the tocopherol concentration in the lipid extract. Multichannel UV detection allowed independent monitoring of lipid and tocopherol elution in order to optimize this difficult separation. While the tocopherol content of wheat germ is typically 0.05%, the product of this procedure was approximately 20% tocopherols. The purification of α-tocopherol has been difficult,[266] and an application of these procedures could prove helpful. Semipreparative scale SFE/SFC was used to concentrate α- and β-tocopherol from wheat germ.[259,266]

Nitrosamines. Nitrosamines are carcinogenic compounds that are formed when amines and nitrites react. Amines are ubiquitous in most foods; nitrites are added to preserve the foods. After the discovery of the health hazards caused by these compounds, several methods have been used for their analysis. The volatile nitrosoamines have been analyzed using GC, and the nonvolatile ones have been

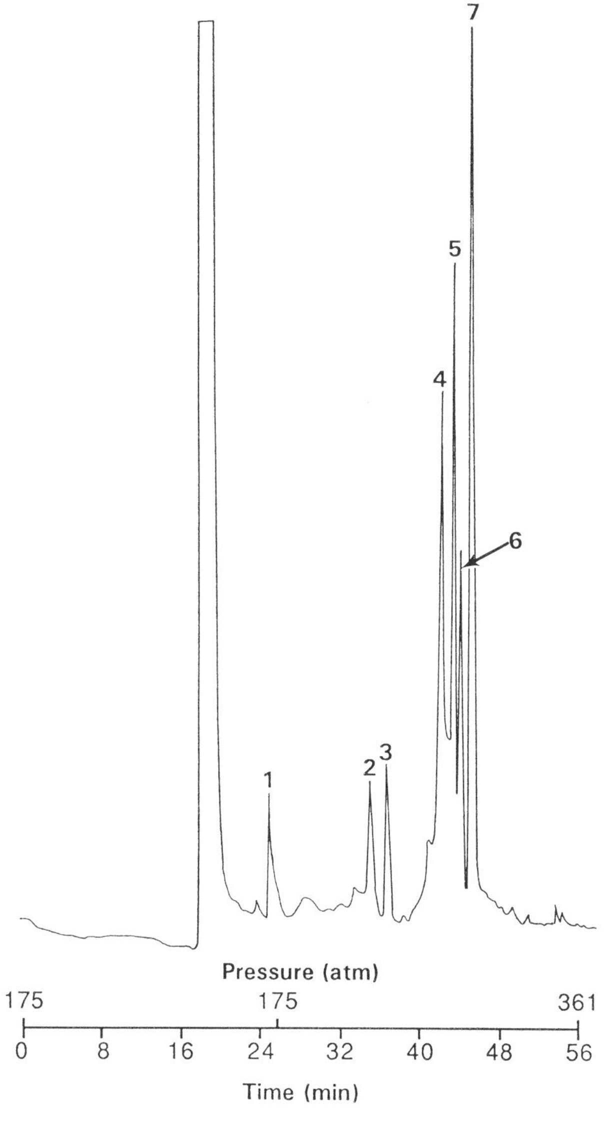

Figure 7.42. SFC chromatogram of seven lipid soluble vitamins. Conditions: 10-m x 100-μm i.d. open tubular column, polyethylene glycol stationary phase (d_f = 0.1 μm); CO_2; 150°C; pressure program from 175 atm to 350 atm at 4.17 atm min^{-1} after an initial 25-min isobaric period; FID. Peak identifications: (1) vitamin K_3, (2) vitamin A acetate, (3) vitamin E acetate, (4) vitamin K, (5) vitamin A, (6) vitamin E, (7) vitamin D_3 (reprinted with permission from Ref. 260).

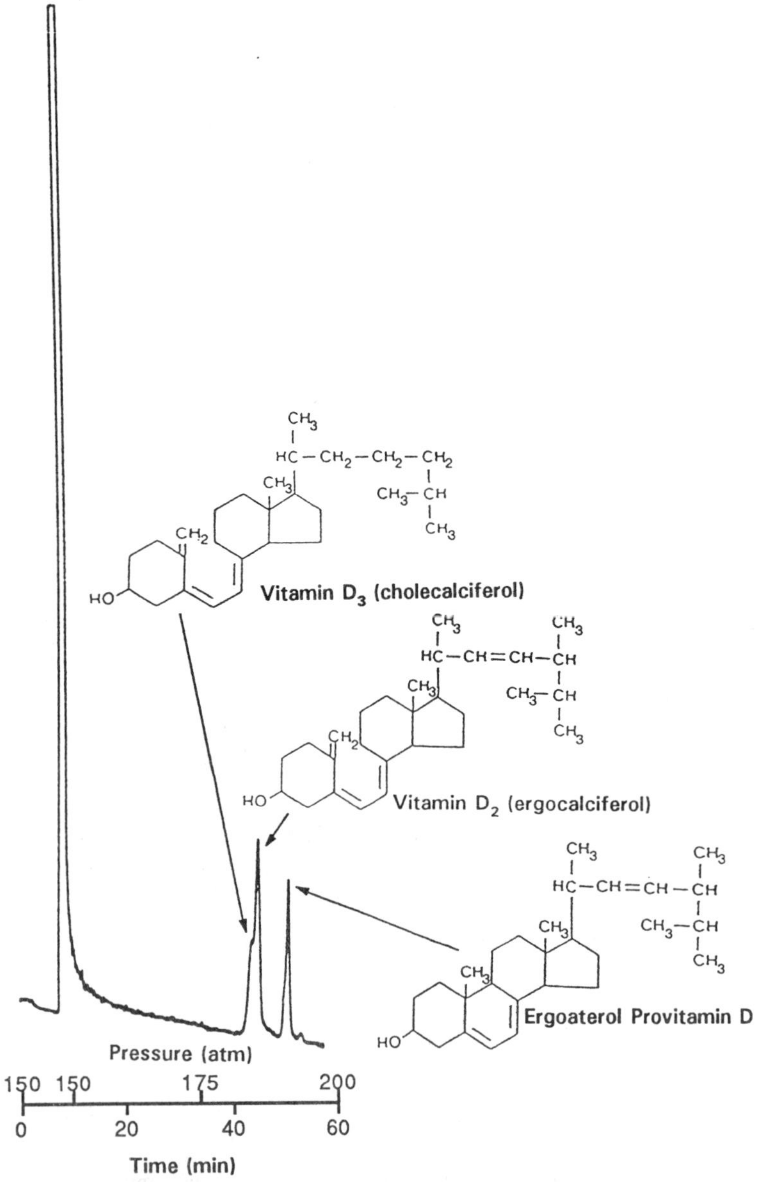

Figure 7.43. SFC chromatogram of three forms of vitamin D. Conditions: column as in Figure 7.42; CO_2; 140°C; pressure program from 150 atm to 200 atm at 1 atm min^{-1} after a 10-min isobaric period; FID (reprinted with permission from Ref. 260).

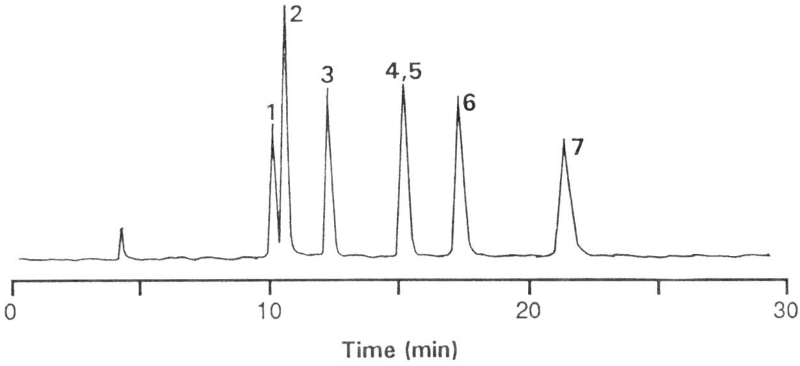

Figure 7.44. SFC chromatogram of nitrosamines. Conditions: 25-cm x 1.35-mm i.d. packed column, C_{18} bonded silica particles (CPTm Spher); 0.9% methanol in CO_2; 46°C; density program from 0.27 g mL^{-1} to 0.40 g mL^{-1} at 0.04 g mL^{-1} min^{-1} after a 4-min isopycnic period; FID. Peak identifications: (1) nitrosodimethylamine, (2) nitrosodiethylamine, (3) nitrosodipropylamine, (4) nitrosodibutylamine, (5) nitrosomorpholine, (6) nitrosopiperidine, (7) nitrosopyrolidine (reprinted with permission from Ref. 268).

analyzed using LC.[267] The use of selective detectors with SFC makes it possible to analyze complex matrices without extensive cleanup and prefractionation steps. Thus, the easily degradable nitrosoamines can be analyzed quantitatively (Figure 7.44).[268]

Milk products. Milk products are subject to proteolysis and lipolysis by microbial action. Oxidation of triglycerides changes the organoleptic properties of milk products. SFE/SFC can be used to analyze the composition of fresh and rancid cheese. The total amount of free fatty acids, due to the breakdown of triglycerides, is higher in the rancid product than in fresh cheese. Also the amounts of the volatile 2-methylketones, which give the fresh cheese odor, are decreased in the rancid product. In this application, steam distillation is not an acceptable preparation method, because some of the triglyceride bonds break at elevated temperatures.[164]

Spices and essential oils. Spices are used to enhance the flavors of foods. The qualities of spices can be reduced by insect infestation, microbial contamination, and environmental factors. Fumigation of spices with methyl bromide and ethylene oxide prevents further insect and microbial contamination, but introduces new toxic contaminants into the material. Irradiation provides a residue-free sterilization method, but qualitative homogeneity is not guaranteed. Spice extracts, obtained by steam distillation or

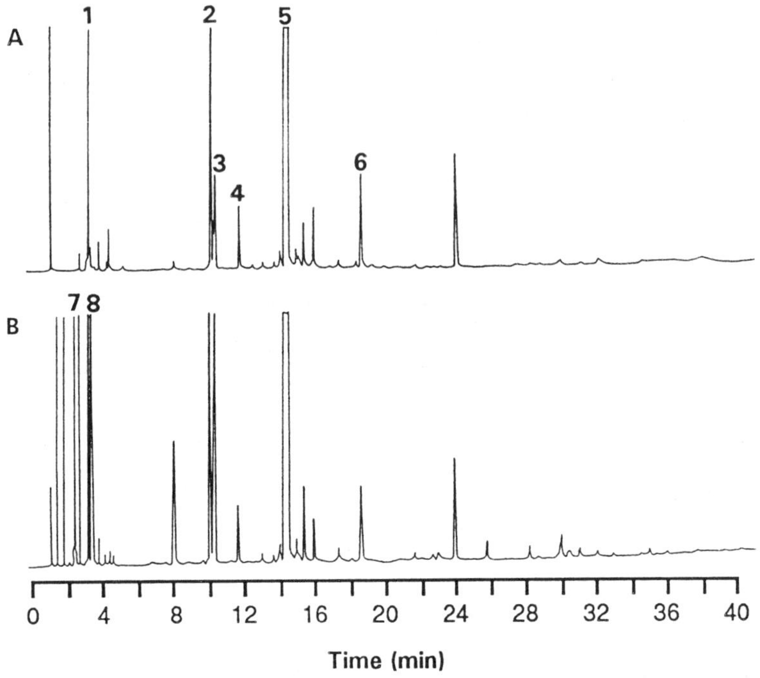

Figure 7.45. GC chromatograms of cardamom oil extracted with (A) liquid hexane and (B) supercritical CO_2. GC conditions: open tubular column, polyethylene glycol stationary phase; temperature program from 80°C to 140°C at 4°C min^{-1} and to 280°C at 10°C min^{-1} after a 5-min isothermal period. Peak identifications: (1) limonen, (2) linalool, (3) linalylacetate, (4) terpinen-4-ol, (5) terpinylacetate, (6) geraniol, (7) sabinen, (8) 1.8-cineol (reprinted with permission from Ref. 270).

supercritical fluid extraction, are easy to use, are easily standardized, have a long shelf life, and are microbiologically safe.[269] In steam distillation, the spice material is exposed to relatively high temperatures which can cause artifact formation, especially due to the hydrolytic influence of the water which is present even in dried spice material. Supercritical fluids offer the possibility for gentle and selective extraction that is not possible by, *e.g.*, steam distillation. This can be seen in Figure 7.45, where cardamom oil has been extracted with hexane and supercritical CO_2. In the hexane extract, some of the linalylacetate has decomposed to linalool.[270] While sonication is not as harsh as higher temperature extractions, some differences can be seen in the chromatographic patterns from SFE and liquid sonication of rosemary (Figure 7.46.).[271]

Essential oils from citrus fruits contain terpenes, sesquiterpenes, oxygenated compounds, and nonvolatile residues. The terpenes

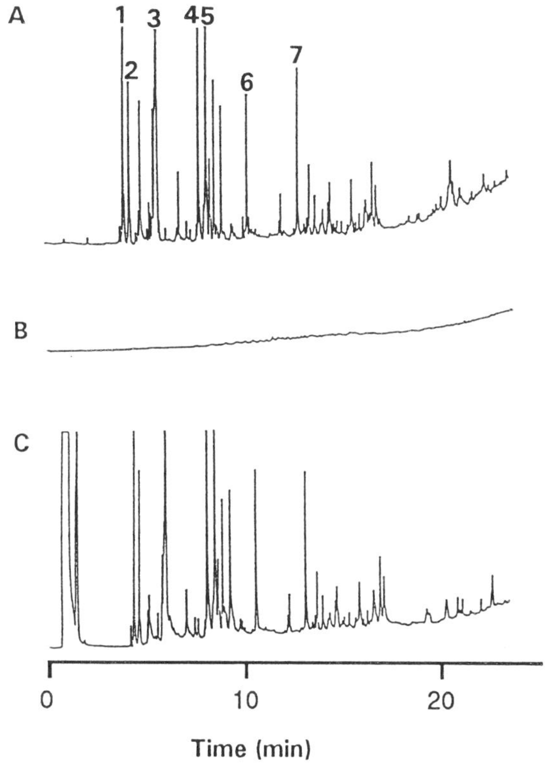

Figure 7.46. GC chromatograms of rosemary extracts using (A) on-line SFE/GC, (B) on-line SFE/GC (second extraction), and (C) GC on-column injection of a methylene chloride extract. SFE conditions: 0.5 μg; 45°C; 10 min; GC conditions: 30-m x 320-μm i.d. open tubular column, poly(5% phenyl)methylsiloxane stationary phase; temperature program from 40°C to 70°C at 30°C min^{-1} and to 320°C at 8°C min^{-1}. Peak identifications: (1) α-pinene, (2) camphene, (3) 1,8-cineole, (4) camphor, (5) bornyl acetate, (7) humulene (reprinted with permission from Ref. 271).

contribute little to the flavor or fragrance of the oil. Since the terpenes are mostly unsaturated compounds, they can decompose from heat, light, and oxygen to compounds with undesirable flavors and odors. To improve the quality of the oil, limonene and other terpenes can be reduced.[272] Figure 7.47 shows capillary gas chromatograms of a lemon oil before removal of the terpenes with supercritical CO_2.[273] The ratio of terpenes to oxygenated compounds decreased considerably. Extraction recoveries of nonvolatile, thermally labile sesquiterpenes from marjoram are better with

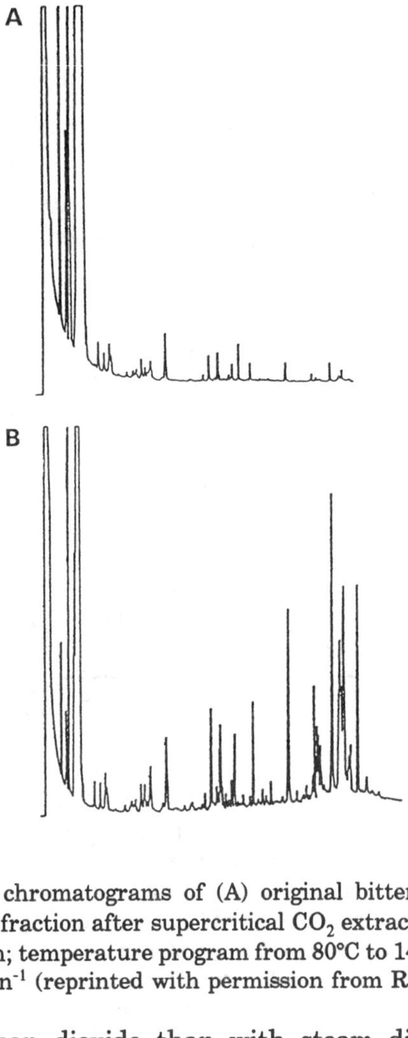

Figure 7.47. GC chromatograms of (A) original bitter lemon oil and (B) concentrated flavor fraction after supercritical CO_2 extraction. GC conditions: open tubular column; temperature program from 80°C to 140°C at 4°C min^{-1} and to 280°C at 10°C min^{-1} (reprinted with permission from Ref. 273).

supercritical carbon dioxide than with steam distillation (Figure 7.48).[270] The major components of citrus fruit oils, terpenes and sesquiterpenes, have been analyzed by several research groups with SFE/GC,[271] SFE/SFC,[274] and SFC/FTIR (Figure 7.49).[275-277] The advantage of FTIR over MS is the possibility to distinguish isomers.[275] The terpenes, *d*- and *l*-carvone, are easily separated by organoleptic methods, because the *d*-form tastes like caraway and the *l*-form tastes like spearmint. In most cases the chiral separation of isomers is not that simple. Optically active terpenes have been separated by LC, but packed and open tubular column SFC provide faster analysis (Figure 7.50).[278,279]

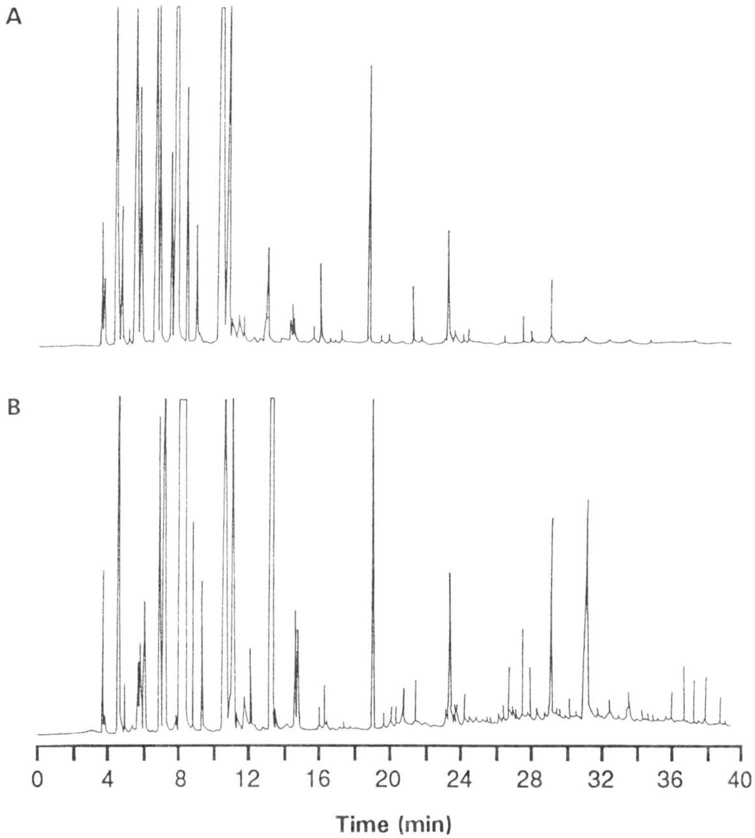

Figure 7.48. GC chromatograms of marjoram oil after extraction with (A) liquid hexane and (B) supercritical CO_2. GC conditions: open tubular column, polymethylsiloxane stationary phase; temperature program from 80°C to 140°C at 4°C min^{-1} and to 280°C at 10°C min^{-1} (reprinted with permission from Ref. 270).

Bitter components. Selective extraction with supercritical fluids has made it possible to use wormwood as a flavoring agent.[280] Absinthine and artabsine are sesquiterpene lactones, and they are the major bitter compounds in wormwood (*Artemisia absinthium L.*). The use of untreated wormwood in alcoholic preparations is prohibited because it contains the toxic compound, thujone. Because the solubility of thujone in supercritical carbon dioxide differs considerably from the solubilities of absinthine and artabsine, SFE can be used to remove the toxic material without the loss of the bitter constituents.[280]

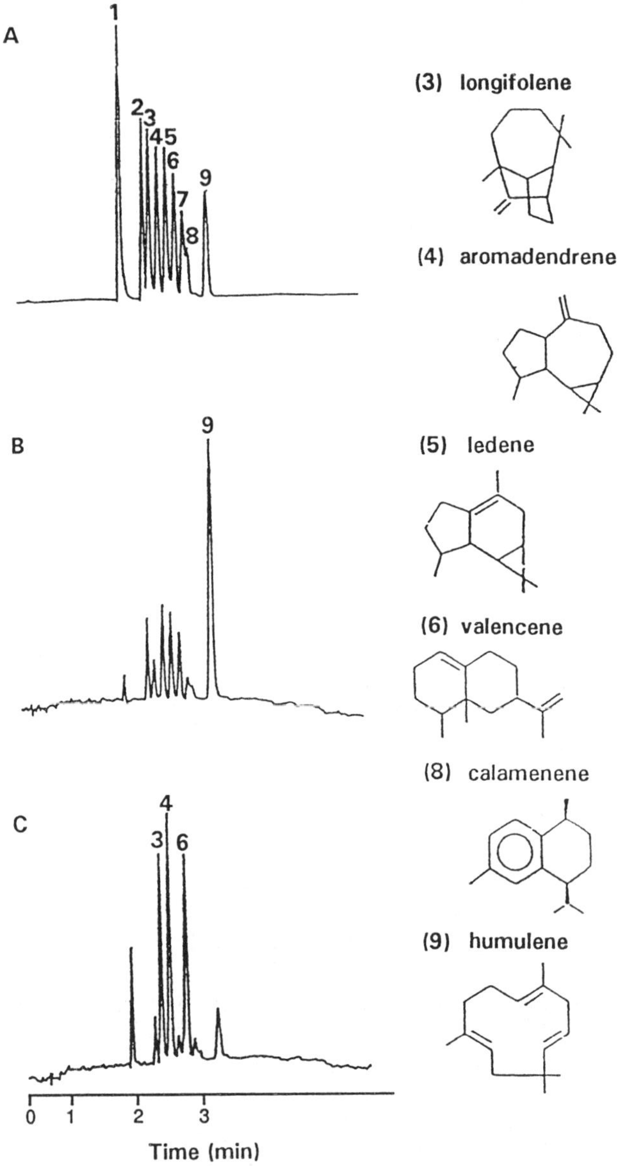

Figure 7.49. SFC/FTIR chromatograms of a mixture of eight sesquiterpene hydrocarbons: (A) Gram-Schmidt total IR, (B) 957-987 cm^{-1} spectral window, and (C) 849-895 cm^{-1} spectral window. Conditions: two 250-mm x 4.6-mm i.d. packed columns in series, Nucleosil 100 and Spherisorb 5-μm particles; CO_2; 40°C; 130 atm. Peak identifications: (1) hexane, (2) longicyclene, (3) longifolene, (4) aromadendrene, (5) ledene, (6) valencene, (7) *trans*-calamenene, (8) *cis*-calamenene, (9) humulene (reprinted with permission from Ref. 275).

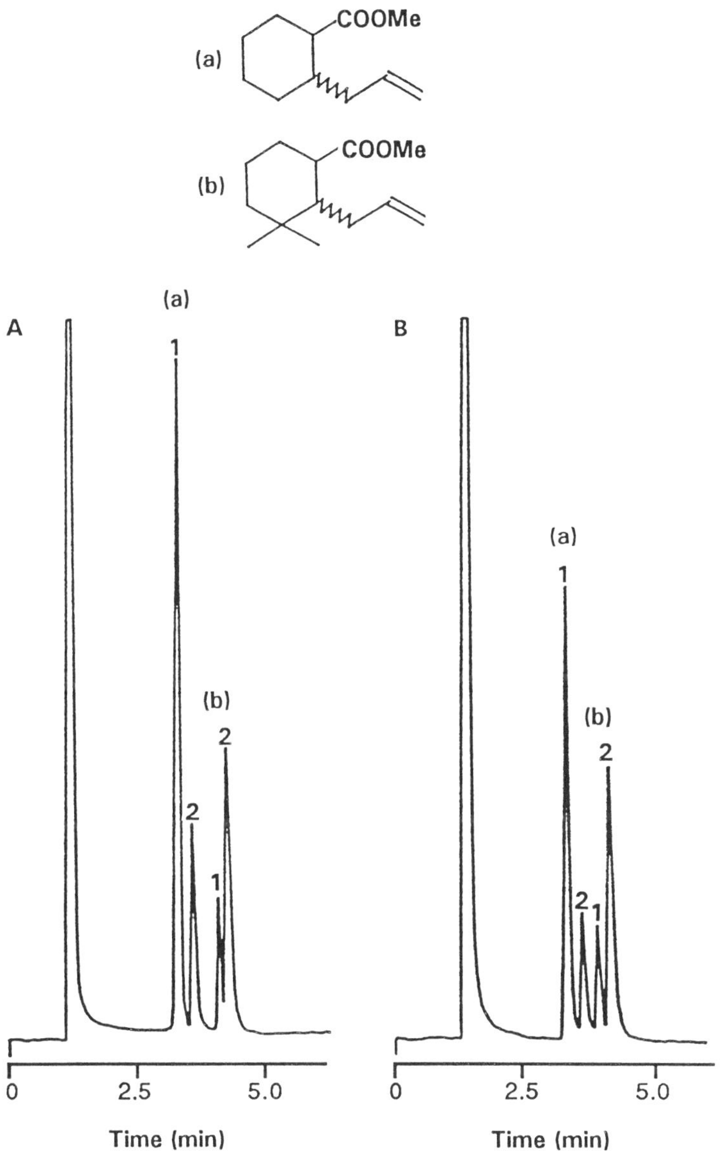

Figure 7.50. SFC chromatograms of synthetic monoterpenes at (A) 40°C and 157 atm, and (B) 25°C and 110 atm. Conditions: 250-mm x 4-mm i.d. packed column, C_8-bonded silica particles (5 μm). Peak identifications: (1) cis, (2) trans (reprinted with permission from Ref. 279).

The bitter taste of beer is due to the α-acids (humulones) and β-acids (lupulones). Methylene chloride has been used to extract hops, but the removal of the solvent from the extract to the minimum allowable concentration (2.2%) is difficult without changing the flavor of the extract.[269] Extracts obtained with supercritical fluid extraction contain 50% humulones and 30% lupulones. During brewing, the α-acids are isomerized to iso-α-acids. Due to similarities in the structures of the humulones and lupulones, their separation has been difficult. Open tubular column SFC provides sufficient resolution for determining the quality of hop extracts (Figure 7.51).[281]

7.12 Natural Products

SFC has become an increasingly popular separation technique for the analysis of natural products. This class of compounds includes a large variety of biological compounds, some of which are discussed separately in other sections of this book. This section addresses work published on the analysis of alkaloids, bacterial lipids, terpenoids, pigments, some naturally occurring hormones, and esters. Various applications of packed and open tubular column SFC with different modes of detection are reviewed. Despite impressive progress, SFC analysis of natural products (especially polar compounds) should be greatly expanded with further development.[282] Many applications require high densities and the addition of modifiers to increase the solvating power of CO_2. Because of the high critical temperatures of polar fluids, and their enhanced chemical reactivities, analysis of very polar natural products is presently better done by other techniques; however, the high efficiencies and the flexibility of detection make SFC an attractive alternative method whenever it is applicable.

Alkaloids. Berry *et al.*[48,51,283] investigated a wide variety of natural products, including indol and ergot alkaloids and steroids. Impressive results (Figure 7.52) were obtained for the separation of monoterpene indole alkaloids isolated from Catharanthus leaves.[283] SFC was also applied in conjunction with thermospray (filament-on mode) and EI ionization MS. Thermospray ionization generated molecular mass information on the eluting components; however, peak intensities did not correspond to the actual ratios of the alkaloids in the mixture.

Aminosilane packed columns with UV detection proved equally suitable for the analysis of ergot alkaloids of the clavine type.[48] The

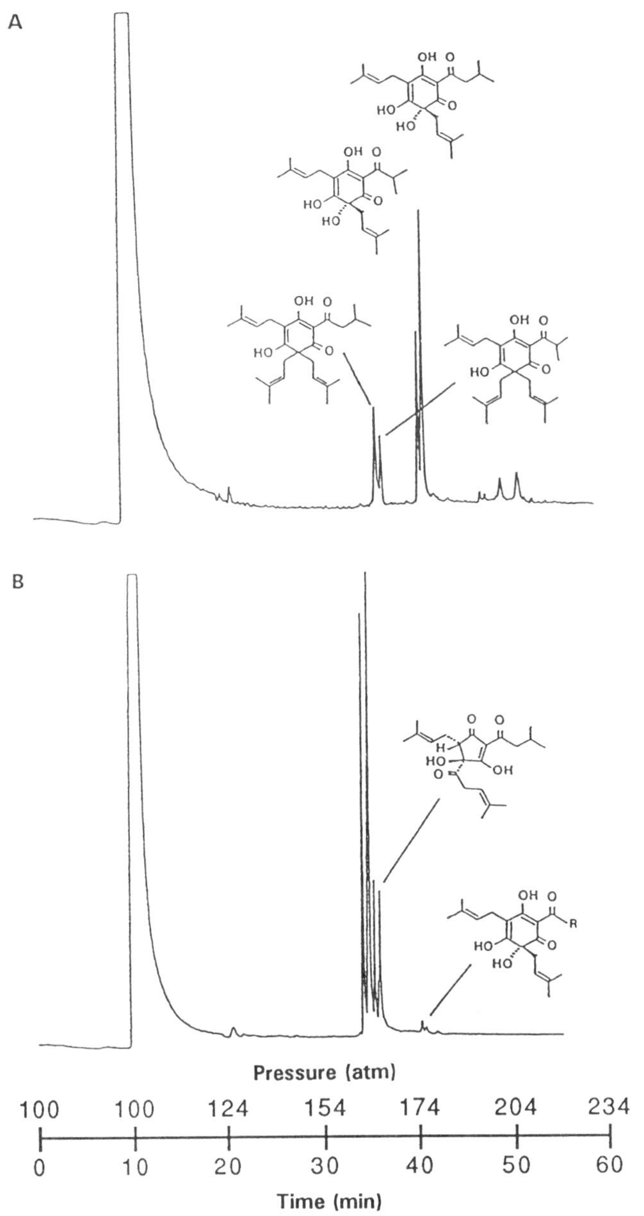

Figure 7.51. SFC chromatograms of (A) hop extract and (B) beer bittering iso-α-acids. Conditions: 10-m x 50-μm i.d., poly(15% oligoethylene oxide)methylsiloxane (d_f = 0.15 μm); CO_2; 80°C; pressure program from 100 atm to 300 atm at 3 atm min^{-1} after a 12-min isobaric period (reprinted with permission from Ref. 281).

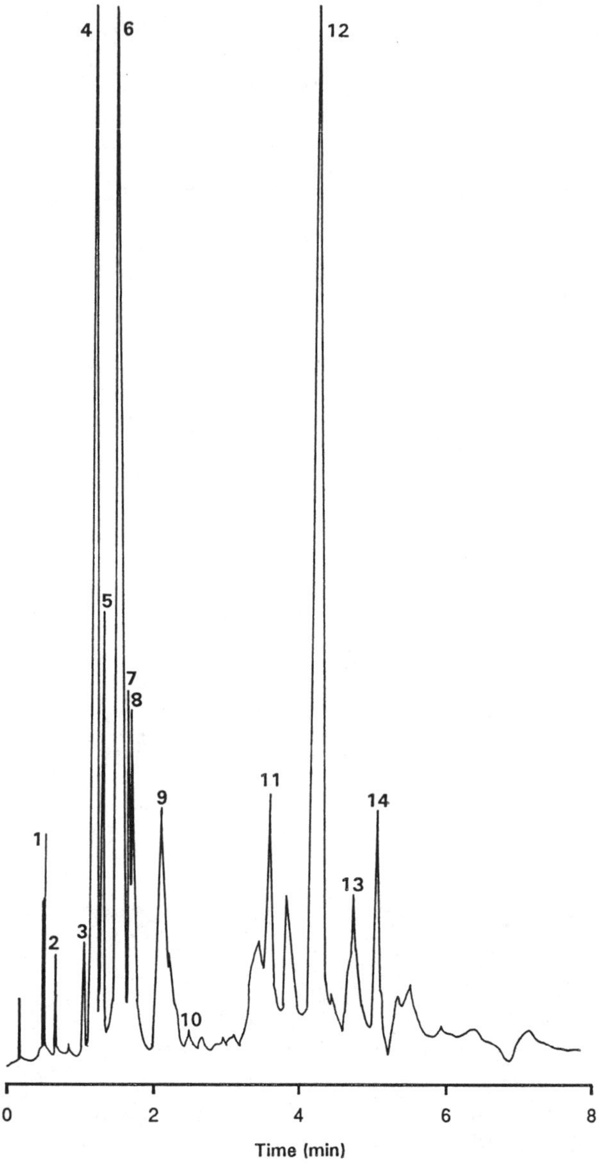

Figure 7.52. SFC chromatogram of *Catharanthus* leaf alkaloids. Conditions: 100-mm x 4.5-mm i.d. packed column, 5-μm NH_2 bonded silica particles (Spherisorb); 5 to 15% gradient of methanol in CO_2; 65°C; UV (290 nm). Peak identifications: (1) lochnerinine, (2) horhammericine, (3) deacetoxyvindoline, (4) catharanthine, (5) vindorosine, (6) vindoline, (7) ajmalicine, (8) vindolinine, (9) epivindolinine, (10) deacetylvindoline, (11) teurosine, (12) 3,4-anhydrovinblastine, (13) perivine, (14) vinblastine (reprinted with permission from Ref. 283).

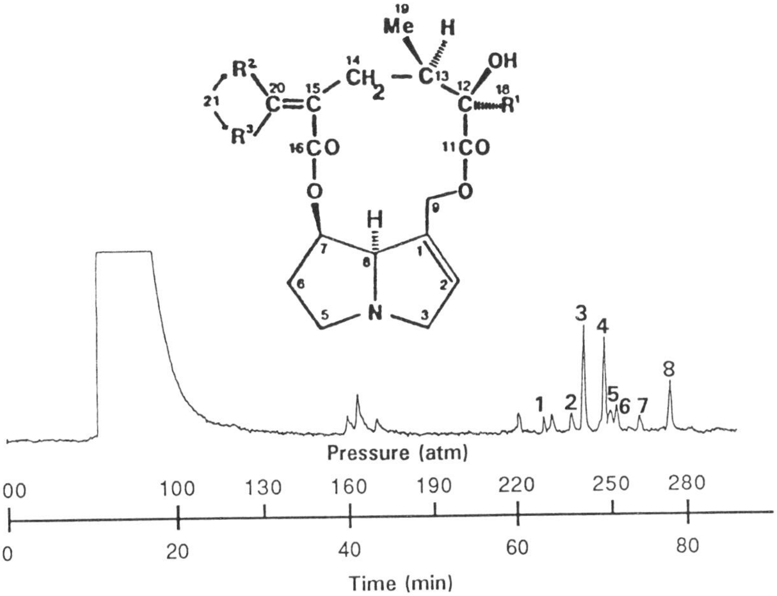

Figure 7.53. SFC chromatogram of pyrolizidine alkaloids from *Senicio anonymus*. Conditions: 10-m x 50-μm i.d. open tubular column, poly(30% biphenyl)methylsiloxane stationary phase (d_f = 0.25 μm); CO_2; 130°C; pressure program from 100 atm to 280 atm at 3 atm min^{-1} after a 20-min isobaric period; FID. Peak identifications: (1) senecionine (R^1 = CH_3, R^2 = H, R^3 = CH_3), (2) integerrimine (R^1 = CH_3, R^2 = H, R^3 = H), (3) senkirkine (R^1 = CH_3, R^2 = H, R^3 = CH_3, N-oxide), (4) neosenkirkine (R^1 = CH_3, R^2 = CH_3, R^3 = H, N-oxide), (5) retrosine (R^1 = CH_2OH, R^2 = H, R^3 = CH_3), (6) otosenine (R^1 = CH_3, R^2 = H, R^3 = CH_3, N-oxide, epoxide), (7) hydroxysenkirkine (R^1 = CH_2OH, R^2 = H, R^3 = CH_3, N-oxide), (8) ananamine (R^1 = CH_3, R^2 = CH_2OH, R^3 = H, N-oxide) (reprinted with permission from Ref. 284).

components of a crude extract from *Claviceps purpurea* were completely separated using a methanol/CO_2 mobile phase gradient. Detection by MS with a thermospray source produced CI spectra if the mobile phase contained a modifier, and EI spectra in the absence of the modifier.[51] Two ergot alkaloids of the peptide type (bromocryptine and ergocryptine) were also separated on the aminosilane column using 20% methoxyethanol as modifier and single wavelength UV detection.

Pyrrolizidine alkaloids are found in certain flowering plants, most of which are toxic. They can be grouped into macrocyclic diesters of the retronecine and otonecine types. If the N-oxides are reduced prior to analysis, no CO_2 modifier is necessary. The open tubular column SFC analysis of an extract from a common plant, *Senicio anonymus*,[284] is shown in Figure 7.53. Senecionine and integerrimine, which differ only in the orientation of the methyl group attached to C_{20}, as well as the cis/trans pair, senkirkine and neosenkirkine, are completely resolved. This analysis is an excellent example of a separation which is better handled by SFC than by any other method.

Because of the present awareness of drug abuse, the analysis of opium alkaloids by methods other than LC has experienced a revival. Most SFC applications describe the separation of standard solutions on columns packed with aminosilane silica. Analysis times are often faster than in LC; however, with a few exceptions, quantitative results have not been reported. Crowther and Henion[285] examined the retention of codeine on a 5-μm Spherisorb-NH_2 column at different modifier (methanol) concentrations. The MDQ was dependent on the type of compound, and was found to be in the low nanogram range for positive ion CIMS. Similar results were obtained by Carroud *et al.*,[193] who separated narcotine and morphine on a Spherisorb-NH_2 column with CO_2/methanol as mobile phase, using a light scattering detector.

Janicot *et al.*[250] have reported the separation of seven opium alkaloids by packed column SFC using a diode array UV detector. The authors examined the influence of water and amines as modifiers on the retention of the alkaloids on bare silica and on aminosilane silica packed columns. Bare silica was preferred as stationary phase if the narcotine/papaverine pair was to be resolved, whereas narcotine, thebaine, codeine, cryptopine, and morphine could be separated within 2 min on the aminosilane silica using CO_2/methanol/triethylamine/water [87.62 : 11.80 : 0.36 : 0.22 (w/w)] as mobile phase. The procedure was successfully applied to the analysis of a poppy straw extract which contained four opium alkaloids. Standard deviations for their quantitative analysis ranged from 2.9% for thebaine to 9.0% for codeine.

Bacterial lipids. Unlike the glycerol bound fatty acids, which are the structural elements of most biological membranes, a select group of organisms known as archaebacteria, contain isopranylglycerol ether lipids.[286] These compounds are so unusual that they serve as chemotaxonomic markers for these organisms. The study of their occurrence and transformations in organic sediments has greatly advanced the understanding of geochemical processes.

The analysis of archaebacterial ether lipids by GC is not possible because of their high molecular masses. LC is not sensitive enough, since these compounds lack chromophores. Open tubular column SFC with FID or MS offers a viable alternative.

If ether lipids are released into a sediment from dead organisms, the polar head groups which are attached to the free hydroxyl group(s) of glycerol, are removed fairly rapidly by hydrolysis. Since living bacteria contain mostly polar lipids, free ether lipids can be related to past methanogenic activity. If a sediment extract is hydrolyzed prior to SFC, and an increase in ether content is observed, this usually indicates ongoing microbial activity in the sediment.[287] Additional support for an active bacterial population can be demonstrated by the presence of squalene analogs, which are recognized as neutral lipids of archaebacteria. The organisms produce these isoprenoids with various degrees of unsaturation. Once discharged into the sediment, these compounds are gradually hydrogenated to the fully saturated isoprenoids.

The SFC analysis of a standard containing cyclopentyl substituted tetraethers and other lipids is shown in Figure 7.54.[288] These compounds were separated on a poly(50% phenyl)methylsiloxane coated open tubular column, using CO_2 as mobile phase.[182] The ethers, which were isolated from the thermophilic archaebacteria, *Thermoplasma* and *Sulfolobus*, are difficult to separate by chromatographic methods other than SFC.[289]

In contrast to the isopranoid-branched ethers of the archaebacteria, *Thermodesulfotobacterium cummune*, an anaerobic thermophile, produces normal, iso, and anteiso branched alkylglycerol ethers.[290] The complexity in the distribution of these membrane lipid ethers is reflected in the SFC chromatograms obtained.

One of the most unusual membrane lipid components was isolated from *Thermomicrobium roseum*, a thermophilic eubacterium.[291] This organism lacks glycerol derived lipids, but instead possesses a series of straight chain and internal methyl-branched 1,2-diols as membrane lipids. The diols are acylated with fatty acid residues at C-2, and they contain polar head groups at the terminal OH-group.

Ubiquinones or coenzymes Q are a class of isoprenoid quinones found in prokaryotic and eukaryotic cells. They function as hydrogen carriers in the electron transport chain of the inner mitochondrial membrane. Several ubiquinones are known, differing only by the number of isoprenoid units contained in their side chains. Because of their diversity, the ubiquinone distribution has been used as a chemotaxonomic marker for bacteria.[292] The separation of ubiquinones, isolated from *Legionnella pneumophila* using a C_{18} reverse phase column with CO_2 and 1.5% methanol as mobile phase, was reported by Gere.[148] A variable wavelength UV detector was used to monitor the eluting compounds. The chromatographic profile

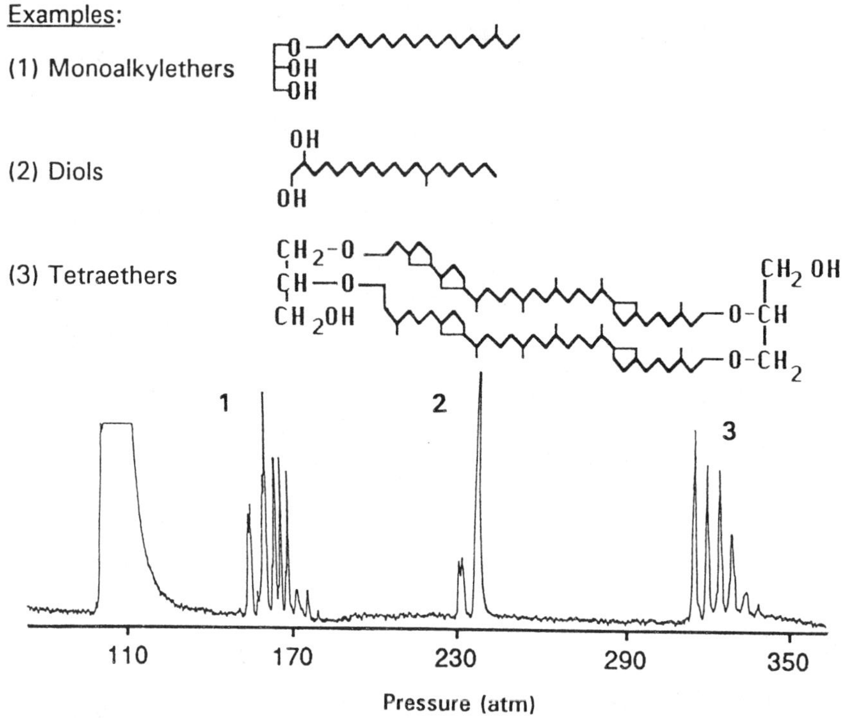

Figure 7.54. SFC chromatogram of bacterial lipid standards. Conditions: 10-m x 50-μm i.d. open tubular column, polymethylsiloxane (d_f = 0.25 μm); CO_2; 100°C; pressure program from 100 atm to 400 atm at 3 atm min^{-1}; FID (reprinted with permission from Ref. 288).

contained peaks for ubiquinone Q-8 through Q-13, with Q-14 missing. The pattern allowed further classification into three distinct groups.[293]

Terpenoids. Terpenes are a diverse group of compounds, which are derived from isoprene precursors. They are found in most higher plants and are largely responsible for characteristic odors. Terpenes are composed of two (monoterpenes) to eight isoprene units (carotenoids) varying in the number of double bonds and cyclic structural components. The analysis of the lower terpenoids is conveniently done by open tubular column GC; however, triterpenes and carotenoids can undergo rearrangement reactions when exposed to high injection temperatures. In the absence of a universal LC detector, SFC offers a real advantage.

The separation of sesquiterpenes by SFC with FTIR detection was demonstrated by Morin *et al.*[275,277,294] An important benefit of this

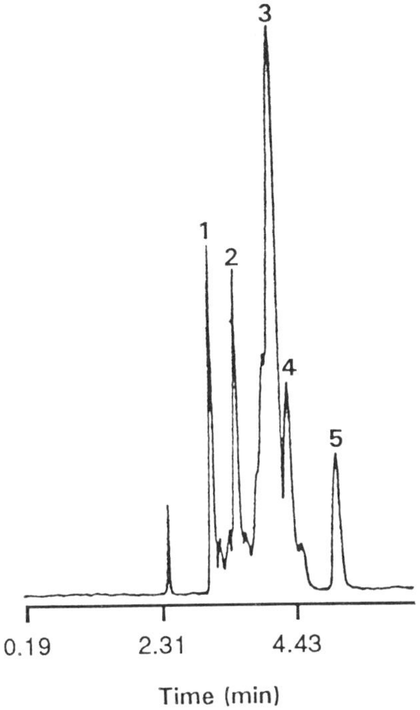

Figure 7.55. SFC chromatogram of sesquiterpenes extracted from Copahu oil. Conditions: two 250-mm x 4.5-mm i.d. packed columns in series, 5-μm silica particles (Nucleosil 400); CO_2; 40°C; 186 atm; FTIR. Peak identifications: (1) α-copaene, (2) trans-α-bergamotene, (3) cartophyllene, (4) β-bisabolene, (5) humulene (reprinted with permission from Ref. 277).

technique is the identification of sample components by their IR spectra. A sesquiterpene standard containing 9 different components could be separated using CO_2 isocratically at 130 atm and 40°C on two bare silica columns in series. The separation was done within only five minutes. The Gram-Schmidt reconstruction of the sesquiterpene fraction of Copahu oil is shown in Figure 7.55. The compounds were identified by comparison of their IR spectra with those published in the literature. In addition to substance identification, chromatographic reconstruction from specific absorption bands allows the grouping of sample components with respect to characteristic structural features, such as trans double bonds or geminal dimethyl substitution.[277] Partially separated components may be resolved by this technique if there are differences in their IR spectra.

SFC has become increasingly important as a monitoring device for solute extractions with supercritical fluids. On-line coupling of an extraction process to SFC instrumentation was used to examine the completeness of extraction of the terpene fraction from caraway seeds.[295] A similar system interfaced to open tubular column GC was used to extract biological markers from complex matrices.[296]

Sesquiterpenoids of the trichothecene group[297] are mycotoxins with presently over sixty different compounds known. Trichothecenes are produced by various species of fungi which are found in agricultural products. A common structural feature of these mycotoxins is the 12,13-epoxytrichothecene ring system. Smith *et al.*[236] have analyzed mixtures of trichothecenes by SFC/FID on a 15-m SE-54 open tubular column with CO_2 as mobile phase, using linear density programming. Detection of the compounds was achieved using FID and MS. If a shorter column and rapid pressure programming were used, the analysis times decreased significantly, however, at the expense of chromatographic resolution. The full-scan MS MDQ in the ammonia CI mode was within the picogram range.

The effect of elevated temperature on a heated restrictor interface to an MS was evaluated using negative ion CI spectra of several trichothecene mycotoxins.[237] Optimum temperatures for the efficient mass transfer into the MS were found to be 100°C above the melting point of the sample components. CI spectra were recorded under electron capture, proton abstraction, and chloride addition conditions. Other workers have also pointed out the problems of incomplete mass transfer into the MS, especially for relatively polar substances or compounds with high molecular masses.[149,298,299]

Pigments. Carotenoids are tetraterpenoids which occur in all photosynthetic and some nonphotosynthetic organisms. Many carotenoids are distributed widely in these organisms, whereas others are found as characteristic markers for certain types of organisms. The distribution of carotenoids at particular levels is often used to study deposition in geological situations such as eutrophic, lacustrine, or marine environments.

All oxygen free, and several hydroxyl-, keto-, methoxy-, and acetyl-substituted carotenoids are soluble in supercritical CO_2 and are amenable to SFC. The separation of α-carotene and β-carotene by SFC was reported by Giddings *et al.*[300] Gere[265] used a 3-μm particle C_{18} reverse phase column with 1% ethanol/CO_2 as mobile phase to separate lycopene, and α- and β-carotene.

The real advantage of SFC is the high efficiency by which complex mixtures from natural sources can be separated. Using crosslinked poly(cyanopropyl)methylsiloxane and Carbowax open tubular columns, Frew *et al.*[301] obtained good separation of β-carotene, echinenone, canthaxanthin, astacene, and fucoxanthin;

however, the more polar carotenoids did not elute. One of the problems with immobilized polar stationary phases in open tubular column SFC is their instability towards higher concentrations of polar modifiers. Rather than adding modifier, the authors employed a short column operated at high flow rates. Detection of the compounds was achieved by MS in the methane CI mode. Adequate sensitivity was obtained for 50-100 ng of component, injected on a 100-μm i.d. open tubular column. The SFC column was coupled to the MS through the existing GC/MS interface. Thermal degradation effects could be minimized by the relatively low interface and source block temperature of 100-120°C. Under these conditions the methane CI mass spectra showed minimal fragmentation, even for thermally labile carotenoids.

Porphyrins are compounds derived from chlorophyll and are found in sediments as desoxophylloerythroetioporphyrin and mesoetioporphyrin. Fossil porphyrins are chelated with nickel or vanadium. After demineralization, they can be separated by SFC with CO_2 as mobile phase[302] as shown in Figure 7.56. Vanadyl porphyrins were analyzed on poly(30% biphenyl)methylsiloxane coated open tubular columns using 8.5 mole percent isopropanol/-CO_2.[303]

Hormones. Probably no other class of natural products has been studied by SFC more than the steroids and steroid hormones.[48,75,87,289,304-308] CO_2 has been the most commonly used mobile phase, although nitrous oxide has been reported for the analysis of cholesterol.[307] Packed column SFC of steroid hormones with 2-methoxyethanol, isopropanol, and chloroform as CO_2 modifiers was reported by Randall *et al.*[308] Later *et al.*[75] demonstrated the SFC/FID analysis of prednisolone from equine urine samples using a polymethylsiloxane open tubular column with CO_2 as mobile phase. In a similar application, David and Novotny[87] reported an extraordinary sensitivity of 120 fg s^{-1} for the phosphorus thermionic ionization detection of the dithiophosphinic ester of pregnanediol by open tubular column SFC.

Ecdysteroids belong to a class of moulting hormones which play an important role in insect development. The separation of these compounds was reported by Morgan *et al.*[309] The authors used a 10-cm x 4.6-mm i.d. column packed with 5-μm Hypersil and CO_2/methanol (4:1) as mobile phase, at a pressure of 300 atm and 80°C. Detection was accomplished with a single wavelength UV detector (235 nm) in series with an MS. Raynor *et al.*[310] separated ecdysterols by open tubular column SFC on a poly(50% cyanopropyl)methylsiloxane column with CO_2; however, ecdysteroids

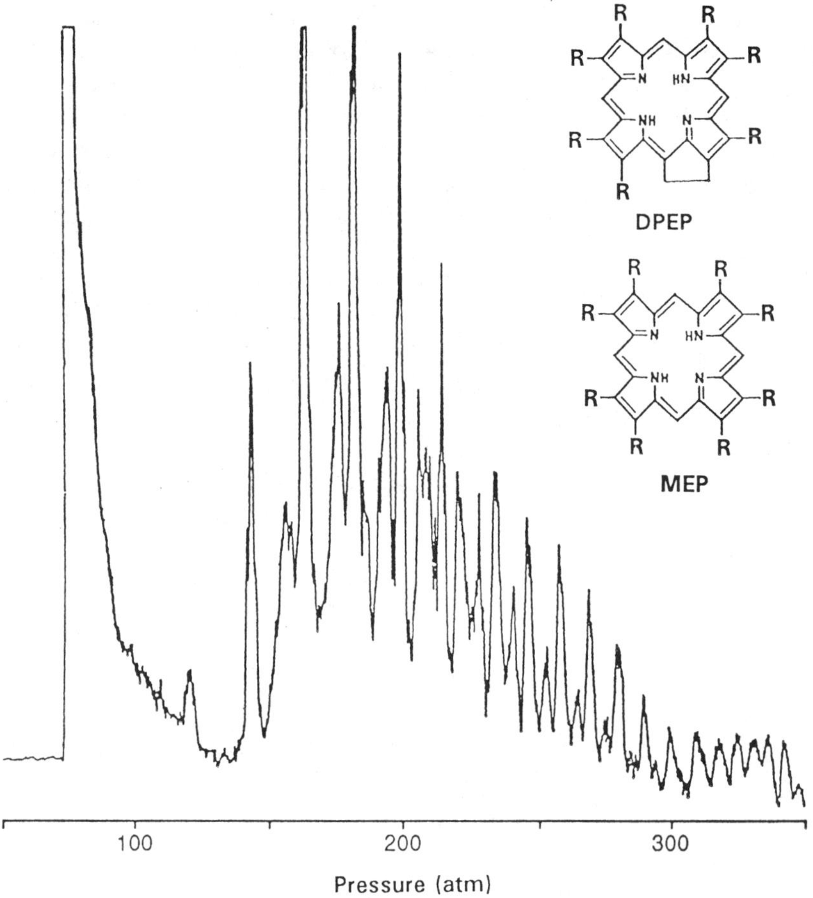

Figure 7.56. SFC chromatogram of demineralized fossil fuel porphyrins. Conditions: 20-m x 75-μm i.d. open tubular column, poly(50% phenyl)methylsiloxane (d_f = 0.25 μm); CO_2; 100°C; pressure program from 125 atm to 350 atm in 50 min after a 10-min isobaric period; FID (reprinted with permission from Ref. 302).

with more than four hydroxyl groups eluted with severe FID spiking. All compounds were successfully separated on a cyanopropyl-bonded silica packed column with 10% methanol/CO_2 at 200 atm and 50°C.

Prostaglandins are cyclopentanoic acids, which are derived from C_{20} polyunsaturated fatty acids. Their wide range of biological functions include effects on blood pressure and blood clotting. Prostaglandins are generally analyzed by open tubular column GC or by LC. SFC offers mild temperature conditions, high efficiency, and high sensitivity without the need for derivatization, which is often used for better volatility or sensitivity. Markides *et al.*[81] reported the separation of a complex mixture of prostaglandins by open tubular column SFC/FID on a poly(50% cyanopropyl)methylsiloxane column using CO_2 as mobile phase.

Naturally occurring esters. Pyrethrins are isolated from the flowers of certain plants and are used as insecticides because of their otherwise low toxicity. SFC offers the advantage of mild operating temperatures for these thermally labile compounds. Wieboldt *et al.*[41,311] have analyzed plant extracts containing a complex mixture of pyrethrins by open tubular column SFC with FTIR detection. The separation was done on a poly(biphenyl)methylsiloxane column, using CO_2 as mobile phase.

7.13 Environmental Pollutants

The category of environmental pollutants covers a wide range of compounds and sample types. The method of analysis must take into account the sample matrix (*i.e.*, air, water, vegetation, soil, etc.) and the desired level of detection, as well as the specific target analyte(s). Therefore, discussions of injection systems, detection systems, column selectivity, and other components of the SFC system are pertinent to the subject of environmental applications. These are discussed in detail in other chapters and will not be repeated here. Furthermore, discussions of agrichemicals and polycyclic aromatic compounds, as well as other classes of compounds which could be environmental pollutants, are included in other sections in this chapter. The topic of supercritical fluid extraction is clearly important in environmental applications; but again, this has been covered in detail in other sections. As a result, the discussion here is limited mainly to the subject of hazardous wastes.

Waste comprises the vast category of undesirable by-products of our industrialized society, things for which we have little or no value. For decades, these materials have simply been dumped and hidden away, but there is now a growing awareness that these materials must be safely reassimilated into our environment.

In 1982, the U.S. Environmental Protection Agency (EPA) issued interim regulations setting permit procedures and operating standards for hazardous waste land disposal facilities. Analytical chemistry plays a critical role in this area because the regulations require disposal facility owners to analyze hazardous wastes and ground water for a broad range of regulated compounds defined in the so called "Appendix VIII and IX" lists. The Appendix VIII and IX

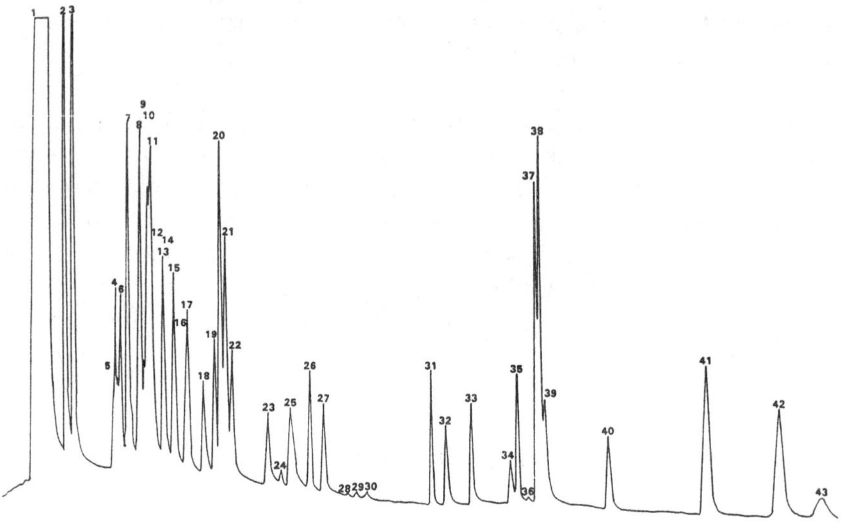

Figure 7.57. SFC chromatogram of selected U.S. EPA Appendix VIII and IX standard compounds. Conditions: 10-m x 100-μm i.d. open tubular column, poly(30% biphenyl)methylsiloxane (d_f = 0.15 μm); CO_2; 100°C; density program from 0.23 g mL^{-1} to 0.70 g mL^{-1} at 0.01 g mL^{-1} min^{-1} after a 2-min isopycnic period; timed split injection (40 nL in 0.1 s at 0°C); FID (350°C); linear velocity range from 0.8 to 4.5 cm s^{-1}. Peak identifications: (1) solvent, (2) *m*-xylene, (3) *o*-xylene, (4) chlorobenzene, (5) phenol, (6) 2-chlorophenol, (7) 2,4-dinitrophenol, (8) benzyl chloride, (9) *o*-cresol, (10) *p*-dichlorobenzene, (11) acetophenone, (12) *p*-cresol, (13) *o*-dichlorobenzene, (14) *m*-cresol, (15) *o*-toluidine, (16) safrole, (17) 2-chloronaphthalene, (18) nitrobenzene, (19) benzotrichloride, (20) naphthalene, (21) 1,2,4-trichlorobenzene, (22) 2,6-dichlorophenol, (23) *p*-chloroaniline, (24) *p*-chloro-*m*-cresol, (25) 1,2,4,5-tetrachlorobenzene, (26) 2,4,6-trichlorophenol, (27) phthalic anhydride, (28) 2,6-dinitrotoluene, (29) 2,3,4,6-tetrachlorophenol, (30) pentachlorobenzene, (31) 4,6-dinitro-*o*-cresol, (32) phenanthrene, (33) anthracene, (34) hexachlorobenzene, (35) *p*-nitroaniline, (36) pentachloronitrobenzene, (37) pentachlorophenol, (38) isodurine, (39) fluoranthene, (40) pyrene, (41) hexachlorophene, (42) 7,12-dimethylbenz[*a*]anthracene, (43) dibenz[*a,h*]anthracene.

lists include several hundred organic and inorganic substances of prime interest to the EPA. The organic compounds cover broad composition, polarity, volatility, and thermal and hydrolytic stability ranges. Additionally, there are broad compositional entries such as coal tar, creosote, cresols, PCBs, aflatoxins, and dioxins, which may contain hundreds of individual components.

The EPA specifies twelve GC and four GC/MS methods for the analysis of these compounds in all types of hazardous wastes. These volatility driven methods are not universally applicable to the diverse

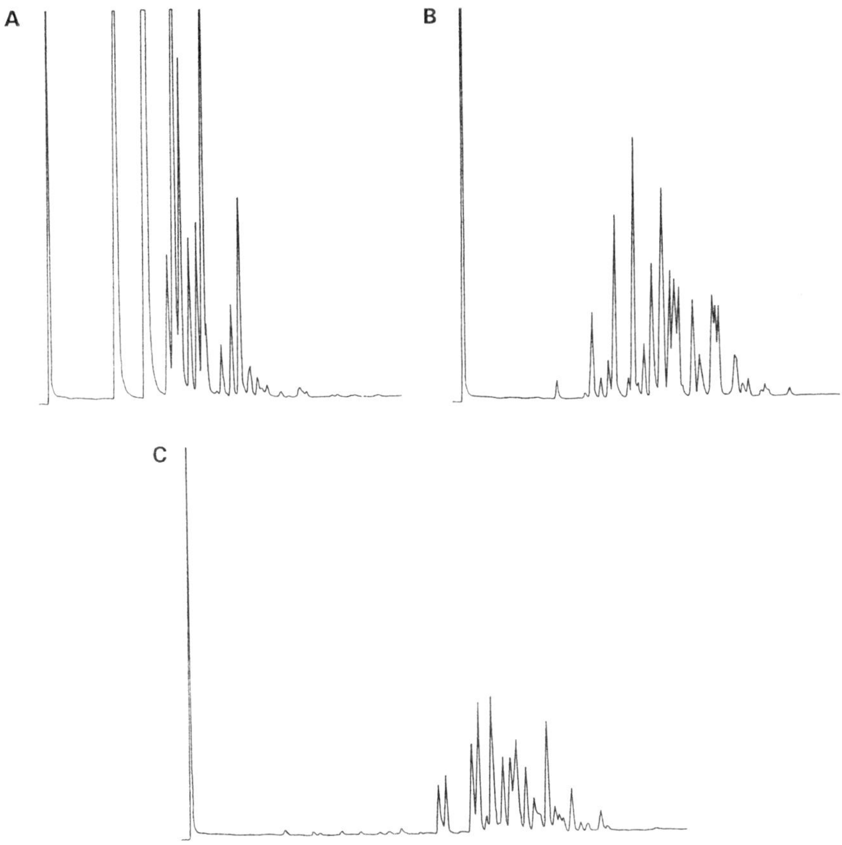

Figure 7.58. SFC chromatograms of (A) Arochlor 1221, (B) Arochlor 1242, and (C) Arochlor 1260. Conditions: see Fig. 7.57.

analyte types in the various matrices encountered in the waste industry. SFC provides a more universal technique for the analysis of such samples.

Retention times and response factors were generated for 270 of the 291 (or 93%) Appendix VIII and IX compounds selected for analysis within a one hour run time using a single open tubular column and CO_2 as mobile phase.[312] SFC is clearly applicable to the broad range of volatile, semivolatile and nonvolatile compound types of interest. Although the retention times of many of the compounds overlap, this is not a serious problem since most wastes in actuality contain a relatively small number of hazardous analytes.

The prime benefit of SFC is that it is flexible enough to encompass the compound class distribution encountered in the area of environmental analysis. To illustrate an application, a mixture of over 40 frequently encountered materials was analyzed using SFC, providing better resolution at the lower retention time portion of the chromatogram (Figure 7.57).[312] than obtained using GC.

Polychlorinated biphenyls (PCBs) are of particular interest because of their occurrence in a broad range of matrices. Chromatograms of Arochlors 1221, 1242, and 1260 are presented in Figure 7.58.[312] Each chromatogram produces a unique pattern enabling the specific Arochlor to be identified. The retention time range corresponds to the molecular mass distribution of the material. The quality of SFC is comparable to that of GC.

7.14 Textile Finishing

Introduction. The textile industry is in the business of converting fiber into fabric, yarn, and other finished goods.[313] In the various steps of the conversion process, a wide range of compounds such as lubricants, anti-static agents, and flame retardants are applied to aid in process control and production efficiency. Compounds utilized range from surfactants to silicones. These process chemicals can be applied at various steps in the manufacturing process, removed in other steps, and essentially reapplied in others. Currently, few automated process control and analysis techniques are available to measure application and removal of these process chemicals.

Heavy use of water-based, wet-processing technology to dye and finish fabrics is typical of the textile industry. Most chemical additives applied during man-made fiber manufacturing must be removed in the fabric preparation or dyeing and finishing stage. This is accomplished through physical washing processes and through the use of surfactants to enhance cleaning. Contaminants found in natural fibers such as cotton and wool also must be removed from finished fabric before it is acceptable to the consumer. Many of these are waxes, trioleins, lanolin, and other "natural" high molecular mass compounds.

Textile chemicals and fiber residues. Determination of fiber contaminants and process lubricants has been accomplished by Soxhlet extraction and application of IR or other spectrophotometric techniques; and recently, analyses of these extracted residues have been performed using LC, GC, and GC/MS.[314,315] SFE/SFC offers a

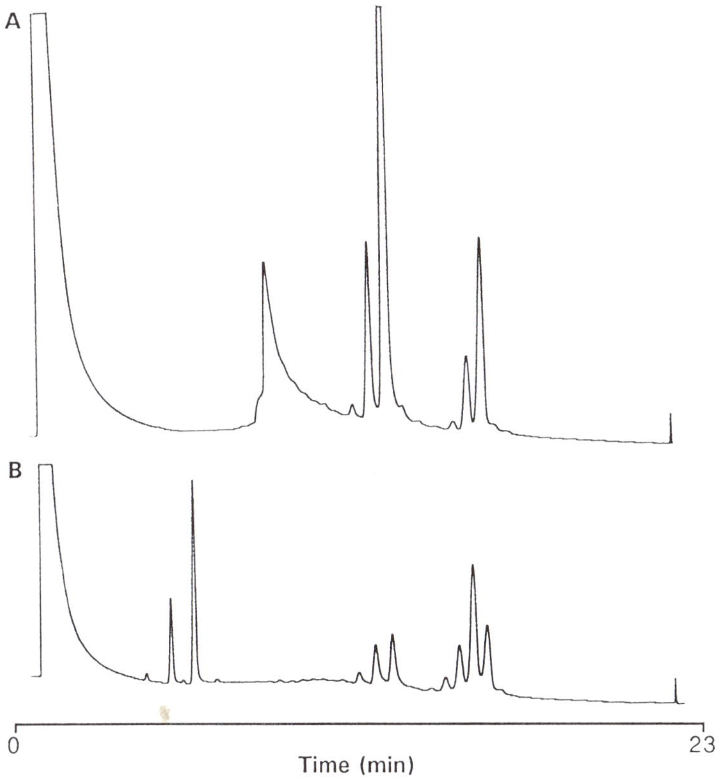

Figure 7.59. SFC chromatograms of (A) food-grade and (B) technical-grade glycerol monostearate. Conditions: 100-mm x 1-mm i.d. packed column, C_8-bonded silica particles (Deltabond); CO_2; 150°C; pressure program from 100 atm to 200 atm at 13.5 atm min^{-1} and to 400 atm at 20 atm min^{-1}; 100 mL min^{-1} flow at 400 atm; 20 μL injected (1% sample conc.); FID (reprinted with permission from Ref. 318).

technique that is useful in the determination and analysis of fiber residues.[316]

Solvents used in the Soxhlet extraction of textile lubricants and finishes such as hexane, 1,1,1-trichloroethane, and Freons are of low polarity. The application of supercritical CO_2 as an extraction solvent with SFE has been relatively simple. A good example of the usefulness of SFE/SFC for textile applications was demonstrated by Yocklovich *et al.*[317] in which SFE/SFC was used to determine the amounts and components of a fiber finish.

Although problems remain with the use of SFC as a routine technique for automated quantitative quality assurance of textile finishing agents on fibers, the technique is useful for determining purity and structural differences in fatty esters and fatty amides such

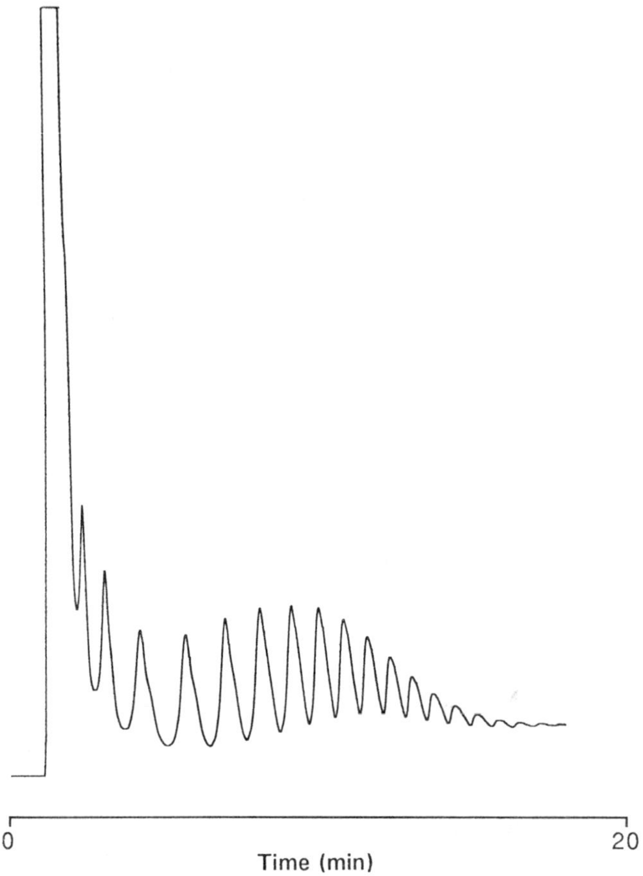

Figure 7.60. SFC chromatogram of alkylphenol ethoxylate (APE) oligomers (EO-3 to EO-17). Conditions: 25-cm x 1-mm i.d. packed column, C_8-bonded silica particles (Deltabond); H_2O saturated CO_2; 120°C; density program from 0.45 g mL^{-1} to 0.75 g mL^{-1} over 75 min; 80 mL min^{-1} flow at 400 atm; 0.1-μL injection (10% sample conc. in CH_2Cl_2); FID (reprinted with permission from Ref. 318).

as glycerol monostearate and stearic diethanolamide (Figure 7.59).[318] Similar information can be obtained for alcohol ethoxylates, alkyl aryl ethoxylates, and other nonionic surfactants.[319] It has also been shown that by using derivatization techniques, it may be possible to separate phosphate esters and sulfated detergents which are too polar for SFE/SFC in their nonderivatized form.[320]

Many fabric residues and the detergents used to remove them in textile wet processing are under scrutiny because of adverse impact on the aquatic environment. SFC can be applied to monitor, for example, the presence of alkyl phenol biodegradation metabolites in wastewaters after concentration and purification (Figure 7.60).[318] SFC has also been useful in identification of alcohol surfactant metabolites.[321]

Wool oils. The use of SFE/SFC for determining amounts of natural wool oils found in raw stock wool is another practical application of SFE/SFC. Because most wool greases and oils have commercial value, selective extraction of the most valuable components in a purified form may be possible by preparative-scale SFE.

UV inhibitors on fibers. UV inhibitors can also be extracted and determined from fabrics using SFE/SFC, similar to extractions from polyethylene films.[322] Extraction of a 10-50 mg sample at 200 atm at 60°C for 30 min has shown good repeatability for fabric treated with nonionic UV inhibitors.

7.15 Metal Chelates and Organometallic Compounds

The analytical capabilities of SFC are more usually applied to organic compounds, even though the first reported SFC separation in 1962 was of nickel porphyrin complexes.[323] Since then, fewer than 15 papers have dealt with the separation of metal chelates or organometallic compounds.

Meanwhile, the detection of metals as their metal chelates by GC has formed the subject of two books, many review articles, and numerous publications. Likewise, LC has been enthusiastically embraced for the separation and determination of metal chelates. Both of these methods have their shortcomings, however. In GC, problems arise from insufficient volatility and thermal instability. The use of high temperatures as a means to increase the volatility of the metal compound often leads to decomposition during the GC run, and the common approach of substituting fluorine atoms for hydrogen atoms in the ligand shell is not available for all compounds. In LC, metal components in the instrumentation may cause severe interferences (ligand exchange, memory effects, and contamination risks), and irreversible adsorption can occur on the stationary phase in the column.

Porphyrins. Separation of metalloporphyrin mixtures is a topic of considerable interest. Not only are complex mixtures of metalloporphyrins found in shale and petroleum, but molecules which are based on porphyrin macrocycles, such as hemoglobin (Fe complex of protoporphyrin IX) and chlorophyll (modified Mg porphyrin), are widespread in biological systems. GC separation of the metalloporphyrins has failed because of their decomposition at the temperatures required for successful elution. Methods based on LC likewise have their limitations, *e.g.*, ligand exchange and adsorption on the stationary phase.

In the first SFC study on metalloporphyrins, nickel etioporphyrin II and nickel mesoporphyrin IX dimethylesters were separated on a column of 33% Carbowax 20M on Chromosorb W, where dichlorodifluoromethane was used as the mobile phase.[323] Later, Cu (or Mg), Ni, VO, and Sn etioporphyrins II were separated with the same mobile phase.[324] The Sn chelate, which was insoluble under the experimental conditions, was sharply separated from the other chelates. An epoxy resin (Epon 1001) was used as a selective liquid phase in a packed column. The study was subsequently extended to include etioporphyrin II chelates of other metals.[325] In general, porphyrin chelates of the transition metals are very stable. Mg(II), TiO, VO, Mn(II), Co(II), Ni(II), Pd(II), Pt(II), Ag(II), and Zn(II) etioporphyrins could be eluted from both packed columns used in the study (Epon 1001 and XE-60), although not without broad peaks and long retention times.

A recent paper reported the separation of mixtures of nickel and vanadyl porphyrins, and of two samples from nature (an extract of ordinary green grass and Maya asphaltene) with 20% methanol-modified CO_2, 4.6-mm i.d. packed columns, and UV detection.[326] The C_{18} stationary phase proved to be more selective for the separation of porphyritic derivatives than were silica and phenyl stationary phases. Nickel and vanadyl porphyrin chelates were successfully separated by both sub- (50°C) and supercritical (100°C) fluid chromatography at constant pressure (~270 atm). The results of this study demonstrated the feasibility of applying SFC to the examination of metalloporphyrin mixtures after careful optimization of conditions.

Other metal chelates and ferrocenes. Metal ß-diketonate and *N,N*-dialkyldithiocarbamate chelates are the most commonly investigated metal compounds in both GC and LC. The ease with which these chelates are formed and the abundant information available with respect to solubility, extractability, and other properties have found for them a wide range of applications in analytical chemistry.

ß-Diketonates are also the most frequently studied chelates in SFC. Only packed columns have been used in their separations, and modifiers (usually methanol in CO_2) have been needed to decrease the retention times and column interactions with components. Because of modifiers, UV detectors have been used in the studies.

In early studies dealing with the separation of metal acetylacetonates and thenoyltrifluoroacetonates by SFC, dichlorodifluoromethane was used as the mobile phase.[327,328] The successful elution of about 65 ß-diketonate chelates has been published in the literature.[327-331] Resolution by SFC on packed columns is usually not very good and peaks are broad, but a mixture of three Cr(III) ß-diketonates has nevertheless been separated with excellent resolution (Figure 7.61).[330] The tailing of peaks so often seen in SFC chromatograms of these compounds could be due to adsorption on the stationary phase or to decomposition during the SFC run.

Separation of some metal salicylaldoximates [Ni(II), Cu(II), and Zn(II)] has been achieved on a packed column with dichlorodifluoromethane as mobile phase.[327]

The 8-hydroxyquinolates (oxinates) are another widely studied group of chelates in analytical chemistry. Five platinoid metal oxinates (Pt, Ru, Ir, Rn, and Os) have been separated on a normal-phase (Si 60) packed column with CO_2/methanol (30 mol%) as mobile phase.[329,332]

Dialkyldithiocarbamates have been separated as diethyl derivatives on a packed column,[329] and as more spherical diisobutyl derivatives on an open tubular column (Figure 7.62).[333] Figure 7.62 shows the excellent resolution of the latter chelates compared with the former ones. Because no modifier was required with the open tubular column, the sensitive FID could be used. Moreover, elution temperatures in open tubular column SFC are at least 100-130°C lower than in open tubular column GC, so that the Fe(III) and As(III) diisobutyl chelates could be separated as well.[333] In GC, these chelates decompose.[334]

Equal amounts of ferrocene, acetylferrocene, and 1,1-diacetylferrocene dissolved in $CHCl_3$ have been separated on four different packed columns (silica, ODS, phenyl, and PRP-1). Methanol-modified CO_2 was needed to remove the substituted ferrocenes from the columns in a reasonable time.[330]

Use of the inductively coupled plasma atomic emission spectrometer as an element selective detector. The packed

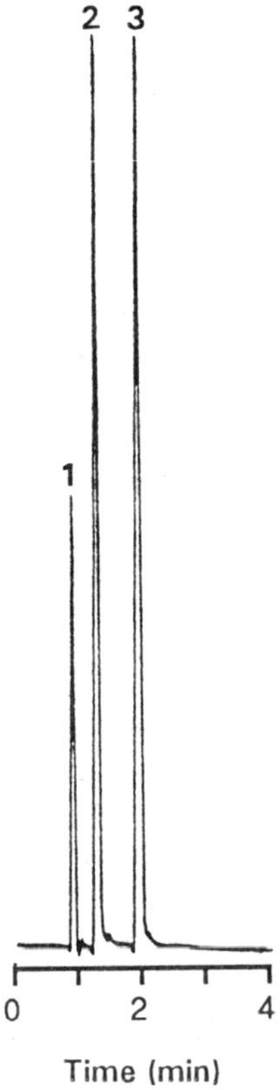

Figure 7.61. SFC chromatogram of Cr(III) β-diketonate chelates. Conditions: 25-cm x 4.6-mm i.d. packed column, 4-μm phenyl-bonded silica particles; 20% methanol in CO_2; 99°C; 257 atm; UV (280 nm). Peak identifications: (1) *tris*-(hexafluoroacetylacetonato)chromium(III), (2) tris(acetylacetonato)-chromium(III), (3) *tris*(1-phenyl-1,3-butanedionato)chromium(III) (reprinted with permission from Ref. 330).

columns used in almost all SFC studies of metal chelates and organometallic compounds required the addition of a modifier to the most commonly used supercritical fluid, CO_2, to increase solubility and, what may be more important, to decrease column activity. The limited success of flame-based detectors with SFC modifiers, and the need at the same time for highly sensitive detection of metals (element selective detectors), has created interest in detection based on plasmas.

Both inductively coupled plasma (ICP)[335,336] and microwave-induced plasma (MIP)[337] atomic emission spectrometers have been successfully employed as element selective detectors in the SFC separation of ferrocene compounds. Also, a hyphenated technique

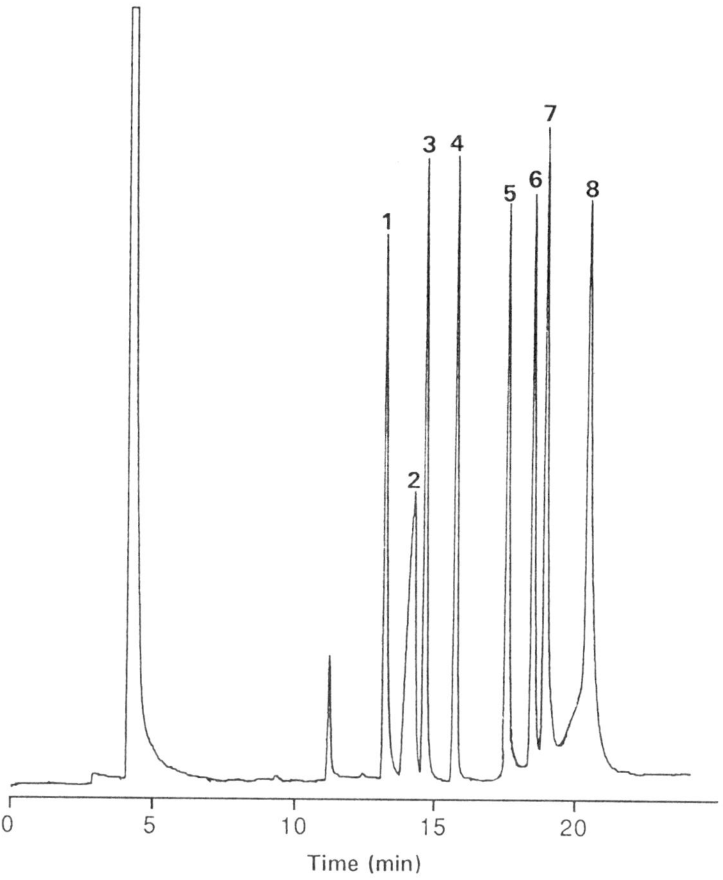

Figure 7.62. SFC chromatogram of dialkyldithiocarbamates. Conditions: 10-m x 50-μm i.d. open tubular column, poly(50% phenyl)methylsiloxane (d_f = 0.25 μm); CO_2; 120°C; density program from 0.31 g mL^{-1} to 0.705 g mL^{-1} at 0.018 g mL^{-1} min^{-1}; FID. Peak identifications: (1) Zn(II), (2) Cd(II), (3) Cu(II), (4) Ni(II), (5) Pb(II), (6) Co(III), (7) Fe(III), (8) As(III) DIBDTC chelates (reprinted with permission from Ref. 333).

involving SFC, a UV multichannel detector, and an ICP has been developed to evaluate the stability of some acetylacetonate complexes under supercritical conditions.[338] A chelating agent was added to the methanol/CO_2 mobile phase to prevent the decomposition of labile chelates, but only Al(III) and Cu(II) complexes were stable enough to be detected by both ICP and UV detectors.

Packed microbore columns of 1-mm i.d. were used in these studies, and the SFC was interfaced to the ICP and MIP detectors with a restrictor that was directly inserted into the plasma torch. An SFC/ICP chromatogram of a mixture of ferrocene, acetylferrocene,

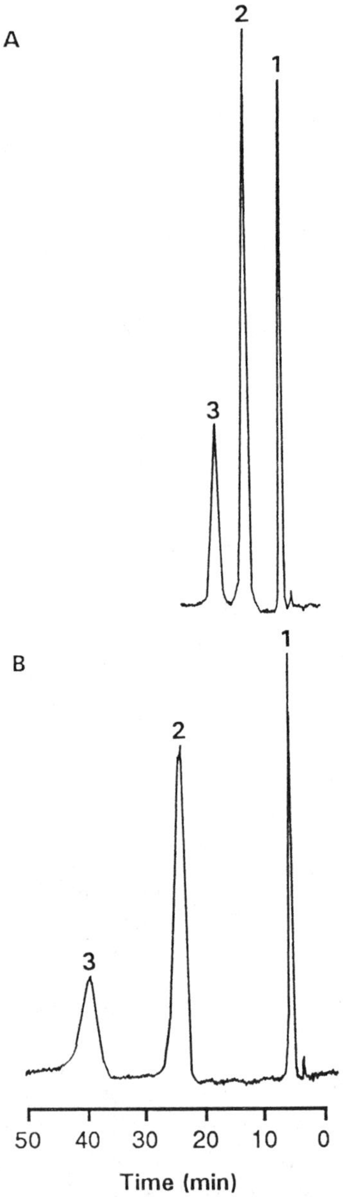

Figure 7.63. SFC chromatograms of a synthetic mixture of ferrocenes with (A) 2.0 mol% methanol in CO_2 and (B) neat CO_2 as mobile phases. Conditions: 25-cm x 1-mm i.d. packed column, diol bonded silica particles (Shimpak); 55°C; 160 atm; 0.5 μL injected. Peak identifications: (1) ferrocene, (2) acetylferrocene, (3) benzylferrocene, (reprinted with permission from Ref. 336).

and benzylferrocene is shown in Figure 7.63.[336] The SFC/ICP and SFC/MIP systems both allow the use of modifiers in CO_2 (SFC/ICP, methanol or acetonitrile; SFC/MIP, methanol) without the performance of the plasma system being degraded.

It has been suggested that in SFC/ICP, the formation of molecular species such as CN could cause the Ca(I) emission to decrease with increasing amounts of CO_2.[339] This hypothesis was confirmed not only for Ca(I), but for Cr(I) and Co(I) in SFC/MIP, indicating that carbon species have a direct effect on plasma energy and thus on analyte signals.[337]

7.16 Enantiomers

Highly specific chiral receptors play important roles in metabolism rates, protein binding and transport,[340] reaction mechanisms,[341] clearance rates,[342] and persistence in the environment. Although the different enantiomers have been shown to have different biological activities, drugs, food additives, and agrichemicals have long been marketed as racemic mixtures. The requirement for chiral purity in the pharmaceutical industry is becoming more strict, which places greater demands on the analysis of reagents and products of asymmetric synethesis, fluids containing enantiomeric drugs for pharmacokinetic studies, and environmental samples for which the determination of the fate of specific enantiomers in synthesis, biological systems, and the environment is important. As the requirements for chiral purity in the pharmaceutical industry become more strict, the means of separation of enantiomers becomes more important.

Methods of achieving chiral separation include the conversion of enantiomers into diastereomers by the formation of diastereomeric salts, derivatization, or by the formation of transient diastereomeric complexes. Transient diastereomeric complexes are readily formed by using a chiral stationary phase (CSP) in chromatography. Chiral stationary phases have been widely applied in GC and LC and several advantages have been reported for the use of CSPs in SFC.

Advantages of SFC for chiral separations. SFC plays an important role in the separation of enantiomers for several reasons: (a) SFC requires relatively low temperatures compared to GC, (b) derivatization is not always necessary, (c) separations by SFC are faster and have higher resolution than can be achieved in LC, (d) open tubular column SFC is more compatible with flow sensitive

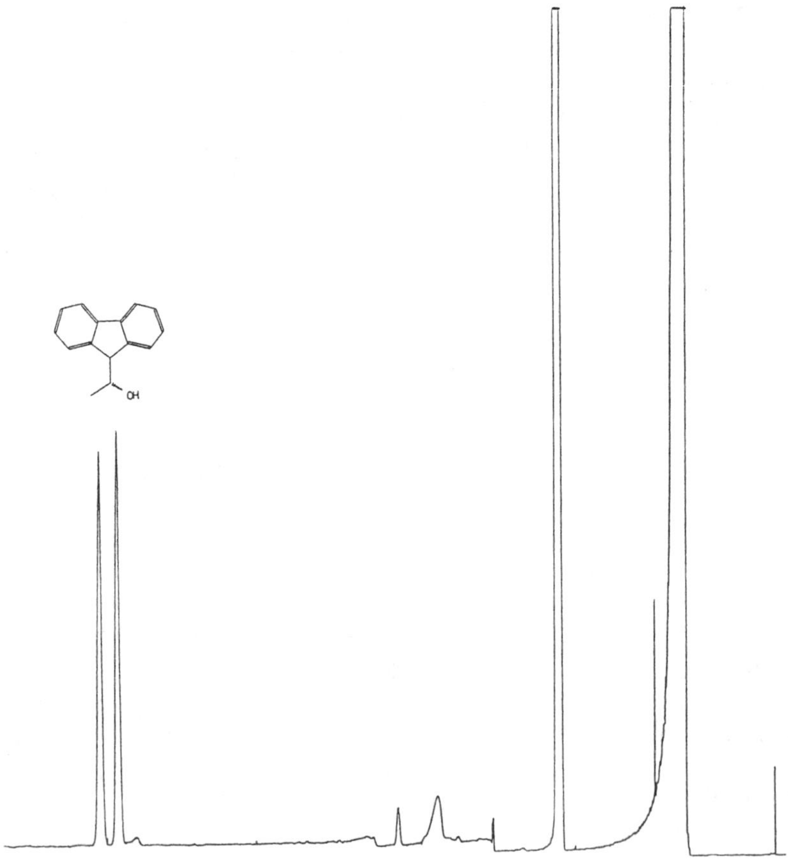

Figure 7.64. SFC chromatogram of diastereomeric derivatives of methyl-9H-fluorenyl-9-methanol. Conditions: 10-m x 50-μm i.d. open tubular column, poly(30% biphenyl)methylsiloxane stationary phase (d_f = 0.25 μm); CO_2 at 100°C; density program from 0.2 g mL^{-1} to 0.75 g mL^{-1} at 0.01 g mL^{-1} min^{-1}; FID (reprinted with permission from Ref. 343).

detectors like the mass spectrometer than conventional LC, and (e) the cavity effects of inclusion-type CSPs are more selective due to the properties of supercritical fluids compared to mobile phases for LC.

Enantioseparations using CSPs in GC are generally temperature limited. Many chiral solutes and their derivatives are thermally unstable; therefore, some solutes cannot be separated by GC. Figure 7.64 represents a case where GC analysis fails. Complete separation of the two diastereomers of the (S)-trolox methyl ether derivatives of racemic methyl-9H-fluorenyl-9-methanol was obtained.[343] At the elevated temperatures required for GC analysis of these enantiomers, this derivative undergoes extensive degradation (presumably β-elimination of trolox methyl ether).

Temperature affects the CSPs in two ways: thermal degradation and racemization. Until the advent of the polysiloxane-type CSPs, most phases had high volatilities (melting points of 80-100°C), and column bleed due to thermal degradation produced high background in the detector. The components of interest had to be derivatized in order to enhance their volatilities for analysis by GC.

Racemization of the chiral center in the stationary phase can also be a limitation. One of the most thermally stable CSPs, a polysiloxane, Chirasil-Val, racemizes at 230°C after 24 h.[344] An N-naphthylethyl benzamide polysiloxane CSP has been developed which shows no loss of efficiency or selectivity after a 10 h conditioning at 280°C.[345] Using this stationary phase in SFC, increased selectivity (over that of GC) for some derivatized amino acids was achieved due to the lower analysis temperature. The enhanced selectivity in SFC more than compensated for the lower efficiency due to the lower diffusion in the supercritical fluid.

SFC has the advantage that high temperatures are not required to achieve successful density or pressure programmed separations. The problems of solute instability at high temperatures, and CSP volatility, thermal stability, and racemization, that plague temperature programmed GC separations are of little consequence in low temperature SFC separations.

Several preferred derivatives of chiral compounds have been established for GC and LC. Chiral compounds are often derivatized to enhance their volatility for GC separations, and to make them detectable by the UV or fluorescence detectors commonly used in LC. The main disadvantages of derivatization include (a) the additional derivatization step increases the overall sample preparation time; (b) a complex mixture of enantiomers is often not readily derivatized; and (c) quantitative analysis of enantiomers can be compromised due to differences in reaction rates and racemization during the derivatization process. One advantage of chiral separations by SFC is that compounds of low volatility can often be eluted without derivatization. Thus, all of the above mentioned problems associated with derivatizations are eliminated. Where derivatization is required, lower analysis temperatures may be used in SFC than in GC, and resolution is often enhanced.

Other advantages of SFC chiral separations include increased speed of analysis and lower volumetric flow rates compared to conventional chiral LC separations. Theoretically, the low viscosity of supercritical CO_2 should generate 5-20 times faster solute diffusion compared to common LC mobile phases. Consequently, chiral SFC analysis times should be increased by at least a factor of 5 without sacrificing separation efficiency. Open tubular column SFC has the additional advantage of low flow rates compared to packed column LC. Interfacing an SFC system with flow sensitive detectors, such as the mass spectrometer, is more straight-forward than for LC.

Röder *et al.*[346] and Macaudiére *et al.*[347] have pointed out that comparison of stereoselectivities of LC and SFC is difficult because of the different operating conditions. These conditions include temperature, mobile phase, and column type (*i.e.*, packed *vs* open tubular). Mourier *et al.*[348] compared the same packed column under SFC and LC conditions. They reported that optimized subcritical and supercritical CO_2 conditions and polar modifiers generated higher resolution per unit time than could be achieved by classical LC. Selectivity was found to be dependent on the polar modifier and temperature. Figure 7.65 compares the analysis times achieved using both techniques in which resolution was held constant. It is obvious that SFC is much faster than LC.

Chiral stationary phases. There are several characteristics that are desirable in a chiral stationary phase. CSPs should be universal so that the separation of a wide variety of compounds is possible. The coated columns should have high efficiency and selectivity for these compounds. CSPs should be effective over a wide temperature range. Not only should the CSP be thermally stable, but racemization should not occur at the operating temperatures used in SFC (60-200°C). An ideal CSP should also be stable in a wide variety of mobile phases. Obviously, a compromise in these requirements has to be made when optimizing a specific enantiomeric separation.

Analogous to the TZ value introduced by Kaiser,[349] Aichholz *et al.*[350] have proposed that the chiral resolution value (cR) can be used to combine enantioselective and purely chromatographic separation potentials into one value. The chiral resolution is given by

$$cR = \frac{t_{r,2} - t_{r,1}}{W_{h,2} - W_{h,1}} - 1 \qquad (7.1)$$

where t_r is the retention time of the solute, W_h is the peak width at half height, and the subscripts 1 and 2 refer to any two well separated chiral homologs (R and S) in a chromatogram. This value is useful because it shows the number of baseline resolved peaks between two enantiomers under programmed conditions.

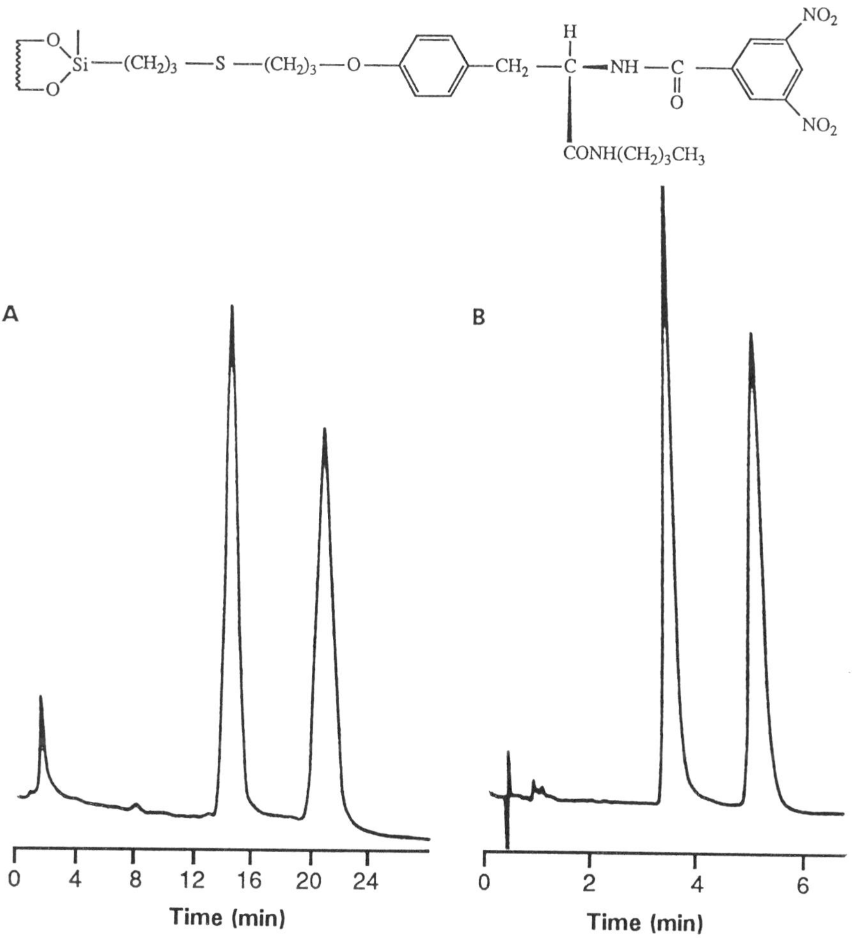

Figure 7.65. Comparison of the analysis times of oxazepam enantiomers at constant resolution using (A) LC and (B) subFC. Conditions: 15-cm x 4.6-mm i.d. packed column, 5-μm (5)-N-(3,5-dinitrobenzoyl)tyrosine *n*-butamide bonded silica; UV (229 nm). LC conditions: 10:90(v/v) ethanol/hexane, 2 mL min^{-1}; 25°C. SFC conditions: 2:98(w/w) ethanol/CO_2; 25°C (reprinted with permission from Ref. 347).

Wainer *et al.*[351-353] have proposed a method for the classification of LC-CSPs which can be extended to GC and SFC. The CSPs are divided into 5 classes. A phase is classified according to how the solute-CSP diastereomeric complexes are formed, and how the stereochemical differences are expressed during and after the formation of these complexes. Type-I CSPs are based on immobilized chiral selectors, such as amino acids, amines, alcohols, amides, or phenyl rings. These types of functional groups use hydrogen

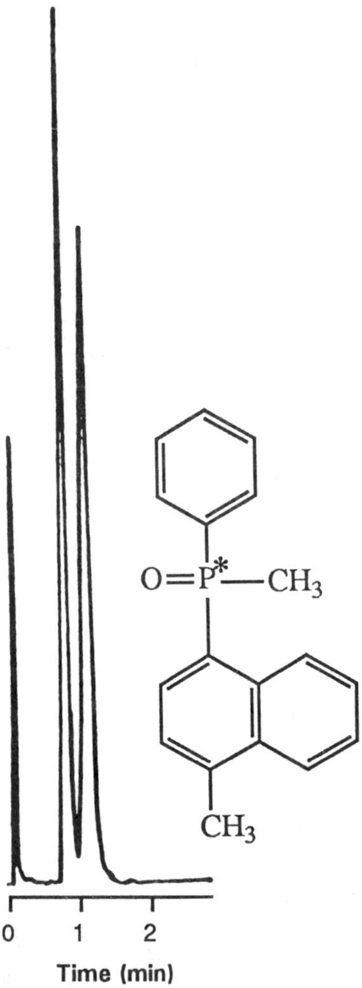

Figure 7.66. SubFC chromatogram of methyl-1-(4-methylnaphthyl)phenyl phosphine oxide. Conditions: 10-cm x 4.6-mm i.d. packed column; 10-μm aminopropyl LiChrosorb NH_2 silica bonded with (R)-N-(3,5-dinitrobenzoyl)-phenylglycine; CO_2/methanol/water at 20°C; 271 atm inlet pressure, 227 atm outlet pressure; UV (280 nm) (reprinted with permission from Ref. 355).

bonding, charge transfer, acid-base, dipole-induced dipole, and/or dipole-dipole interactions to achieve diastereomeric interactions for an enantiomeric separation. The most commonly used Type-I CSPs are the Pirkle-type phases which contain π-donor or π-acceptor functional groups.

The early commercialization of the Pirkle-type phases for LC[354] has made them quite popular among SFC users. The first SFC chromatogram of enantiomers was obtained by using a packed column with a Pirkle-type CSP.[348] Mourier *et al.*[355] later used this phase to study retention and selectivity of methyl-1-(4-methylnaphthyl)phenyl phosphine oxide in supercritical CO_2. The chromatogram in Figure 7.66, obtained under subcritical conditions, shows that a resolution of 3.5 was achieved in 1.5 min using a CO_2/-methanol/H_2O mobile phase. Macaudiére *et al.*[356] used a Pirkle-type phase to resolve amide enantiomers for the comparison of subcritical fluid chromatography (SubFC) and LC. Röder *et al.*[346] were the first to use a Pirkle-type phase in open tubular column SFC. Derivatized amino acid enantiomers were found to have lower enantioselectivities in SFC than in LC under the experimental conditions reported.

Hara *et al.*[357] developed and used an N-formylvaline CSP bonded to silica gel to separate α-amino acid derivatives, and a diamide CSP to separate D- and L-amino acid derivatives.[358] Later, Dobashi *et al.*[359] developed valine-diamide CSPs for the separation of racemic N-4-nitrobenzoylamino acid isopropyl esters in SFC. Gasparrini *et al.*[360] developed a 3,5-dinitrobenzoyl derivative of R,R(-)-1,2-diaminocyclohexane covalently bonded to silica gel and resolved enantiomeric sulfoxides and racemic alprenolol oxazolidin-2-one under subcritical fluid conditions.

Type-II CSPs are phases which include polymers having several chiral centers, such as those described by Ichida *et al.*[361] and Shibata *et al.*[362] Figure 7.67 shows a chiral separation of α-methylene lactam on a cellulose tribenzoate CSP (Chiralcel OB) achieved by Caude and Macaudiére using subcritical CO_2 with an isopropyl alcohol modifier [92:8 (w/w)] at 25°C. Due to low efficiency, the degree of resolution seen in this figure cannot be obtained in LC.

Petersson *et al.*[364] have reported the development of a cyclohexyldiamide polysiloxane copolymer for the resolution of several underivatized cyclic, linear, aromatic, and aromatic diol enantiomers which fits into the Type-II category. The copolymer is made up of copolymerized chiral and achiral monomers resulting in positions in the polymer backbone with a chiral helical twist. Functional groups on the copolymer interact most readily with the enantiomer that fits best into these "grooves." Figure 7.68 shows a successful separation of chiral diols using this CSP.

The Type-III phases include those CSPs which involve the formation of inclusion complexes. The phenylmethacrylate CSPs reported by Okamoto and Hatada[365] and by Blaschke,[366] the cyclodextrin CSPs described by Armstrong and DeMond,[367] and the microcrystalline cellulose triacetate CSP reported by Hesse and Hagel[368] fit into this category.

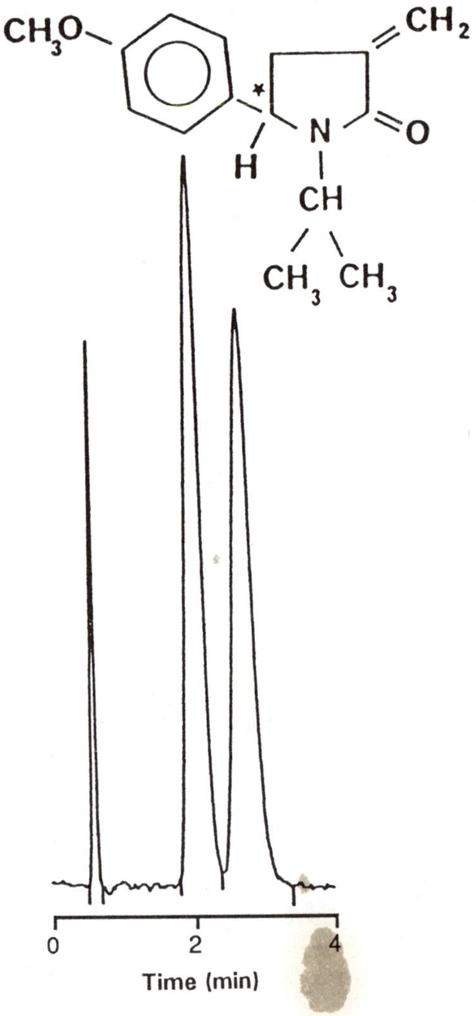

Figure 7.67. SFC chromatogram of racemic α-methylene lactam. Conditions: 25-cm x 4.6-mm i.d. packed column, 10-μm Chiralcel OB; 98:2(w/w) CO_2/isopropanol; 130 atm; 25°C; UV (229 nm) (reprinted with permission from Ref. 363).

Macaudiére *et al.*[369] were the first to report the use of β-cyclodextrin-bonded stationary phases for the resolution of racemic amides and phosphine oxides in packed column subcritical fluid chromatography. Figure 7.69 shows a comparison of the enantiomeric separations of 2-naphthyl and *o*-anisyl phosphine oxides on a β-cyclodextrin-bonded CSP (β-CD) in LC and SubFC. Macaudiére *et al.* found that the free volume of the β-CD cavity is modified by

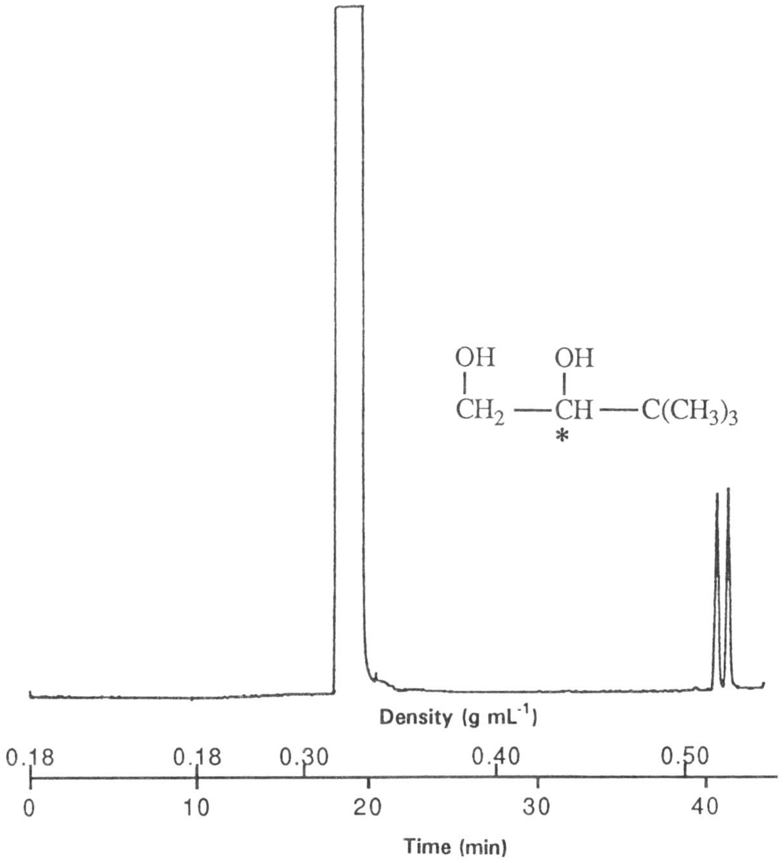

Figure 7.68. SFC chromatogram of racemic 3,3-dimethyl-1,2-butanediol. Conditions: 20-m x 50-μm i.d. open tubular column with a cyclohexyldiamide polysiloxane copolymer stationary phase (d_f = 0.20 μm); CO_2 at 50°C; multilinear density program from 0.18 g mL^{-1} to 0.57 g mL^{-1} over 46 min; FID (reprinted with permission from Ref. 364).

the type of mobile phase. Since the CO_2 molecule (the SubFC mobile phase) is smaller than the hexane molecule (the LC mobile phase), it is more easily displaced from the β-CD cavity. The stereoselectivity (α) for LC was found to be consistently lower than in SubFC. Also, because the 2-naphthylphosphine oxide could form inclusion complexes in SubFC, it was resolved and eluted after the *o*-anisylphosphine oxide. However, the 2-naphthylphosphine oxide eluted more rapidly than the *o*-anisylphosphine oxide and was unresolved using the hexane LC mobile phase. Attempts to resolve these compounds using reversed-phase (water/methanol, 75:25 or 50:50) mixtures in LC also failed. They reported that in normal

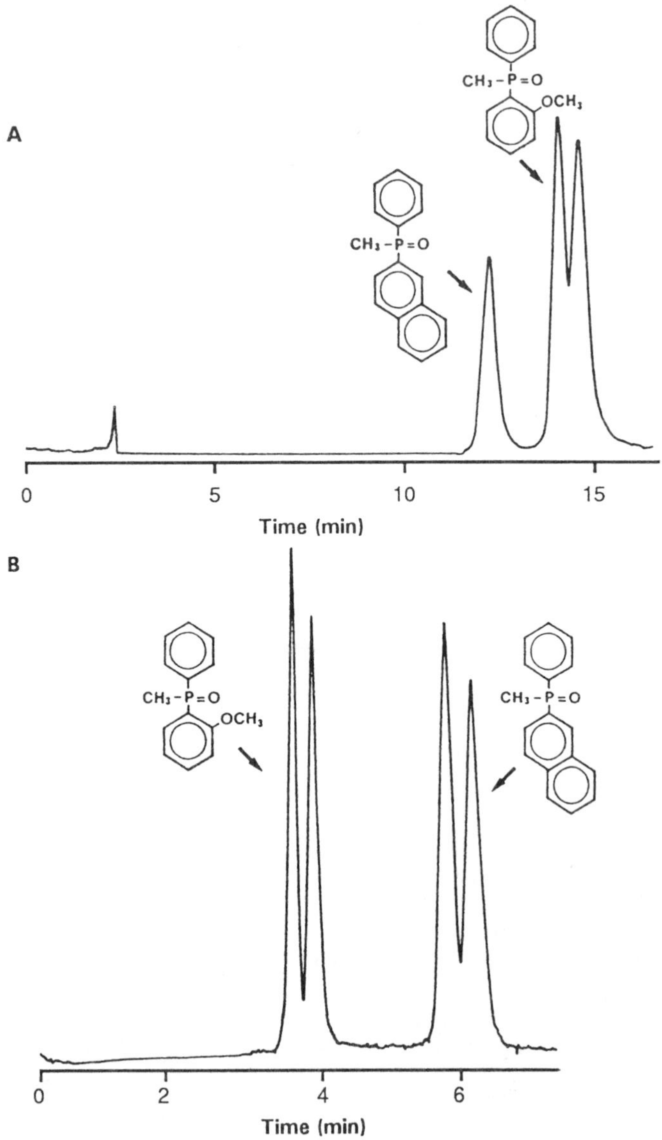

Figure 7.69. SubFC and LC chromatograms comparing the resolution of the racemic 2-naphthyl and *o*-anilyl phosphine oxides. Conditions: 25-cm x 4.6-mm i.d. packed column with a β-cyclodextrin-bonded (Cyclobond I) phase on 5-μm particles; (A) ethanol/hexane LC mobile phase at 1 mL min^{-1}; (B) 8:92(w/w) methanol/CO_2 SubFC mobile phase at 4.5 mL min^{-1}, 25°C, 150 atm; UV (234 nm) (reprinted with permission from Ref. 369).

phase LC, solvents such as *n*-hexane and chloroform are not readily displaced from the cyclodextrin cavity. However, the small size of the CO_2 molecule allowed it to be more readily displaced by the eluting solutes. Later, these authors[370] showed that retention using β-cyclodextrin CSPs was dependent on the nature and content of the modifier, the pH of the mobile phase, and the buffer concentration.

Other classes of CSPs which have not been used in SFC include the Type IV and Type V CSPs. The Type IV CSPs achieve enantiomeric resolution *via* a diastereomeric complex between the solutes of interest, a metal, and an amino acid.[371] Type V CSPs consist of chiral selective protein phases such as bovine serum albumin[372] and α-1-acid glycoprotein.[373]

Mixed mobile phases for chiral SFC separations. One of the difficulties with using a pure CO_2 mobile phase for the resolution of enantiomers is that it is not polar enough to elute many solutes. Thus, polar modifiers have been added to the CO_2. To date, most of the SFC separations of enantiomers were accomplished with the use of modifiers.

References

1. T.A. Norris and M.G. Rawdon, *Anal. Chem.* **56**, 1767 (1984).
2. H.E. Schwartz and R.G. Brownlee, *J. Chromatogr.* **353**, 77 (1986).
3. R.M. Campbell, N.M. Djordjevic, K.E. Markides, and M.L. Lee, *Anal. Chem.* **60**, 356 (1988).
4. W.W. Schulz and M.W. Genowitz, *1990 Pittsburgh Conference on Analytical Chemistry*, New York, NY, March 5-9, 1990, paper No. 133.
5. G.W. Rijnders, *Chem. Ing. Tech.* **42**, 890 (1970).
6. B.W. Wright, H.R. Udseth, E.K. Chess, and R.D. Smith, *J. Chromatogr. Sci.* **26**, 28 (1988).
7. S. Trestianu, G. Zilioli, A. Sironi, C. Sarowalle, F. Munari, M. Galli, G. Gasper, J.M. Colin, and J.L. Jovelin, *J. High Resolut. Chromatogr. Chromatogr. Commun.* **8**, 771 (1985).
8. S. Trestianu, M. Galli, J.L. Jovelin, and G.D. Dupre, in *Proc. Eighth Int. Symp. Cap. Chromatogr.*, P. Sandra, Ed. (Huethig, Heidelberg, 1987), p. 851.
9. H.E. Schwartz, R.G. Brownlee, M.M. Buduzynski, and F. Su, *Anal. Chem.* **59**, 1393 (1987).
10. D.E. Raynie, Ph.D. Dissertation, Brigham Young University, 1990.
11. S. Coulombe and G. Duquette, *Prepr., ACS Div. Fuel Chem.* **33**, 920 (1988).
12. R.E. Jentoft and T.H. Gouw, *Anal. Chem.* **48**, 2195 (1976).
13. T.H. Gouw, R.E. Jentoft, and E.J. Gallegos, *High Pressure Science Tech.* **1**, 583 (1979).

14. W.P. Jackson, R.C. Kong, and M.L. Lee, in *Polynuclear Aromatic Hydrocarbons: Mechanisms, Methods and Metabolism*, M. Cooke and A.J. Dennis, Eds. (Battelle Press, Columbus, 1983), p. 609.
15. H.-C.K. Chang, K.E. Markides, J.S. Bradshaw, and M.L. Lee, *J. Chromatogr. Sci.* **26**, 280 (1988).
16. Z. Juvancz, K.M. Payne, K.E. Markides, and M.L. Lee, *Anal. Chem.* **62**, 1384 (1990).
17. K. Jinno, M. Saito, T. Hondo, and M. Senda, *Chromatographia* **21**, 219 (1986).
18. K. Jinno and S. Niimi, *J. Chromatogr.* **455**, 29 (1988).
19. R. Voyksner, *Research Triangle Institute Hypotenuse*, July-Sept, 12 (1988).
20. B.W. Wright and R.D. Smith, *J. High Resolut. Chromatogr. Chromatogr. Commun.* **8**, 8 (1985).
21. B.E. Richter, *Chrom. Forum*, **1**(6), 52 (1986).
22. B.W. Wright and R.D. Smith, *J. High Resolut. Chromatogr. Chromatogr. Commun.* **9**, 73 (1986).
23. J.R. Wheeler and M.E. McNally, *J. Chromatogr.* **410**, 343 (1987).
24. J.R. Wheeler and M.E. McNally, *Fres. Z. Anal. Chem.* **330**, 237 (1988).
25. J.R. Wheeler and M.E. McNally, *J. Chromatogr. Sci.* **27**, 534 (1989).
26. B.E. Richter, M.R. Andersen, D.E. Knowles, E.R. Campbell, N.L. Porter, L. Nixon, and D.W. Later, in *Supercritical Fluid Extraction and Chromatography: Techniques and Applications, ACS Symp. Ser. 366*, B.A. Charpentier and M.R. Sevenants, Eds. (American Chemical Society, Washington, DC, 1988), p. 179.
27. J.R. Wheeler, *102nd AOAC Annual Int. Meeting and Exposition*, Palm Beach, FL, August 29-September 1, 1988.
28. D.W. Later, D.J. Bornhop, E.D. Lee, J.D. Henion, and R.C. Wieboldt, *LC-GC* **5**, 804 (1987).
29. M. Novotny, *J. High Resolut. Chromatogr. Chromatogr. Commun.* **9**, 137 (1986).
30. M.E. McNally and J.R. Wheeler, *J. Chromatogr.* **435**, 63 (1988).
31. M.E. McNally, J.R. Wheeler, and W.R. Melander, *LC-GC* **6**, 816 (1988).
32. J.E. France and K.J. Voorhees, *J. High Resolut. Chromatogr. Chromatogr. Commun.* **11**, 692 (1988).
33. K. Markides, E.D. Lee, R. Bolick, and M.L. Lee, *Anal. Chem.* **58**, 740 (1986).
34. L.A. Pekay and S.V. Olesik, *Anal. Chem.* **61**, 2616 (1989).
35. S. Kennedy and R.J. Wall, *LC-GC* **6**, 930 (1988).
36. H-C.K. Chang and L.T. Taylor, *J. Chromatogr. Sci.* **28**, 29 (1990).
37. K. Jinno, *Chromatographia* **23**, 55 (1987).
38. C.C. Johnson, J.W. Jordan, L.T. Taylor, and D.W. Vidrine, *Chromatographia* **20**, 717 (1985).
39. K.H. Shafer, S.L. Pentoney, and P.R. Griffiths, *Anal. Chem.* **58**, 58 (1986).
40. Nicolet Instrument Corporation, *Nicolet FTIR Application Note*, AN-8704.
41. R.C. Wieboldt and J.A. Smith, in *Supercritical Fluid Extraction and Chromatography: Techniques and Applications, ACS Symp. Ser. 366*, B.A. Charpentier and M.R. Sevenants, Eds. (American Chemical Society, Washington, DC, 1988), p. 229.
42. R.C. Wieboldt, K.D. Kempfert, D.W. Later, and E.R. Campbell, *J. High Resolut. Chromatogr.* **12**, 106 (1989).

43. M.A. Morrisey and H.H. Hill, Jr., *J. Chromatogr. Sci.* **27**, 529 (1989).
44. H.T. Kalinoski and R.D. Smith, *Anal. Chem.* **60**, 529 (1988).
45. H.T. Kalinoski, H.R. Udseth, B.W. Wright, and R.D. Smith, *J. Chromatogr.* **400**, 307 (1987).
46. E.C. Huang, B.J. Jackson, K.E. Markides, and M.L. Lee, *Anal. Chem.* **60**, 2715 (1988).
47. H.T. Kalinoski, B.W. Wright, and R.D. Smith, *Biomed. Environ. Mass Spectrom.* **13**, 33 (1986).
48. A.J. Berry, D.E. Games, and J.R. Perkins, *J. Chromatogr.* **363**, 147 (1986).
49. B.W. Wright, H.T. Kalinoski, H.R. Udseth, and R.D. Smith, *J. High Resolut. Chromatogr. Chromatogr. Commun.* **9**, 145 (1986).
50. D.A. Laude, Jr., S.L. Pentoney, Jr., P.R. Griffiths, and C.L. Wilkins, *Anal. Chem.* **59**, 2283 (1987).
51. A.J. Berry, D.E. Games, I.C. Mylchreest, J.R. Perkins, and S. Pleasance, *Biomed. Environ. Mass Spectrom.* **15**, 105 (1988).
52. S.B. Hawthorne and D.J. Miller, *Fres. Z. Anal. Chem.* **330**, 235 (1988).
53. M.R. Andersen, *LC-GC* **6**, 566 (1988).
54. B.E. Richter, D.E. Knowles, M.R. Andersen, N.L. Porter, E.R. Campbell, and D.W. Later, *J. High Resolut. Chromatogr. Chromatogr. Commun.* **11**, 29 (1988).
55. M.R. Andersen, *California Department of Food and Agriculture - University of California, Davis: Spring Pesticide Residue Workshop*, Sacramento, CA, February 21-23, 1989.
56. M.E. McNally and J.R. Wheeler, *J. Chromatogr.* **447**, 53 (1988).
57. J.W. King, *J. Chromatogr. Sci.* **27**, 355 (1989).
58. P. Capriel, A. Haisch, and S.U. Khan, *J. Agric. Food Chem.* **34**, 70 (1986).
59. J.R. Wheeler, *California Department of Food and Agriculture - University of California, Davis: Spring Pesticide Residue Workshop*, Sacramento, CA, February 21-23, 1989.
60. J.R. Wheeler, *103rd AOAC Annual Int. Meeting and Exposition*, St. Louis, MO, September 25-28, 1989.
61. M.L. Rowe, *J. Gas Chromatogr.* **4**, 420 (1966).
62. M.L. Rowe, *J. Gas Chromatogr.* **5**, 531 (1967).
63. J.M.F. Douse, *J. Chromatogr.* **445**, 244 (1988).
64. E.C. Huang, in *SFC Applications: 1988 Workshop on Supercritical Fluid Chromatography*, K.E. Markides and M.L. Lee, Comps. (Brigham Young University Press, Provo, UT, 1988).
65. W.H. Griest, C. Guzman, and M. Dekker, *J. Chromatogr.* **467**, 423 (1989).
66. K.M. Payne, B.J. Tarbet, J.S. Bradshaw, K.E. Markides, and M.L. Lee, *Anal. Chem.* **62**, 1379 (1990).
67. C.A. Rouse, Brigham Young University, unpublished results.
68. M. Ashraf-Khorassani and L.T. Taylor, *J. High Resolut. Chromatogr. Chromatogr. Commun.* **12**, 40 (1988).
69. M. Ashraf-Khorassani and L.T. Taylor, *Anal. Chem.* **61**, 145 (1989).
70. S.H.Y. Wong, *Clin. Chem.* **35**, 1293 (1989).
71. C.E. Pippenger, in *Therapeutic Drug Monitoring and Toxicology by Liquid Chromatography*, S.H.Y. Wong, Ed. (Marcel Dekker, New York, 1985), p. 11.
72. C.S. Frings, D.J. Battaglia, and R.M. White, *Clin. Chem.* **35**, 891 (1989).

73. Department of Health and Human Services, Scientific and Technical Guidelines for Federal Drug Testing Programs; Standards for Certification of Laboratories Engaged in Urine Drug Testing for Federal Agencies; Notice of Proposed Guidelines, Fed. Reg. 52:30638-52, 1987.
74. J.B. Crowther, and J.D. Henion, *Anal. Chem.* **57**, 2711 (1985).
75. D.W. Later, B.E. Richter, D.E. Knowles, and M.R. Andersen, *J. Chromatogr. Sci.* **24**, 249 (1986).
76. I. Lurie, in *SFC Applications: 1989 Symposium/Workshop on Supercritical Fluid Chromatography*, K.E. Markides and M.L. Lee, Comps. (Brigham Young University Press, Provo, UT, 1989).
77. R.L. Eatherton, M.A. Morrissey, and H.H. Hill, Jr., *Anal. Chem.* **60**, 2240 (1988).
78. S.H.Y. Wong and S. DellaFera, *J. Liq. Chromatogr.*, in press.
79. B.E. Richter, *J. High Resolut. Chromatogr. Chromatogr. Commun.* **8**, 297 (1985).
80. S.J. Lane, in *SFC Applications: 1989 Symposium/Workshop on Supercritical Fluid Chromatography*, K.E. Markides and M.L. Lee, Comps. (Brigham Young University Press, Provo, UT, 1989).
81. K.E. Markides, S.M. Fields, and M.L. Lee, *J. Chromatogr. Sci.* **24**, 254 (1986).
82. I.J. Koski, B. Janson, K.E. Markides, and M.L. Lee, *J. Biomed. Pharm. Anal.*, submitted.
83. T.A. Berger and J.F. Deye, Hewlett Packard, unpublished data.
84. I.J. Koski, in *SFC Applications: 1989 Symposium/Workshop on Supercritical Fluid Chromatography*, K.E. Markides and M.L. Lee, Comps. (Brigham Young University Press, Provo, UT, 1989).
85. R.M. Smith, *J. Chromatogr.* **505**, 147 (1990).
86. J.C. Kuei, K.E. Markides, and M.L. Lee, *J. High Resolut. Chromatogr. Chromatogr. Commun.* **10**, 257 (1987).
87. P.A. David and M.V. Novotny, *J. Chromatogr.* **461**, 111 (1989).
88. S. Shah, M. Ashraf-Khorassani, and L.T. Taylor, *Chromatographia* **27**, 441 (1989).
89. H.-C.K. Chang, K.E. Markides, J.S. Bradshaw, and M.L. Lee, *J. Microcol. Sep.* **1**, 131 (1989).
90. J.C. Giddings, L.M. Bowman, Jr., and M.N. Myers, *Anal. Chem.* **49**, 243 (1977).
91. L. McLaren, M.N. Myers, and J.C. Giddings, *Science* **159**, 197 (1968).
92. J.C. Giddings, M.N. Myers, L. McLaren, and R.A. Keller, *Science* **162**, 67 (1968).
93. J.C. Giddings, M.N. Myers, and J.W. King, *J. Chromatogr. Sci.* **7**, 276 (1969).
94. F.P. Schmitz and E. Klesper, *Polym. Commun.* **24**, 142 (1983).
95. C. Fujimoto, T. Watanabe, and K. Jinno, *J. Chromatogr. Sci.* **27**, 325 (1989).
96. J.J. Gunderson, M.N. Myers, and J.C. Giddings, *Anal. Chem.* **59**, 23 (1987).
97. S.D. Hawthorne and D.J. Miller, *J. Chromatogr.* **388**, 397 (1987).
98. H.E. Schwartz, J.W. Higgins, and R.G. Brownlee, *LC-GC* **4**, 639 (1986).
99. D. Knowles, in *SFC Applications: 1989 Symposium/Workshop on Supercritical Fluid Chromatography*, K.E. Markides and M.L. Lee, Comps. (Brigham Young University Press, Provo, UT, 1989).

100. C.M. White and R.K. Houck, *J. High Resolut. Chromatogr. Chromatogr. Commun.* **9**, 4 (1986).
101. R.D. Smith and B.W. Wright, in *Microbore Column Chromatography*, F.J. Yang, Ed. (Marcel Dekker, New York, 1989), p. 307.
102. R.E. Jentoft and T.H. Gouw, *J. Polym. Sci.* **7**, 811 (1969).
103. E. Klesper and W. Hartmann, *Eur. Polym. J.* **14**, 77 (1978).
104. D. Leyendecker, D. Leyendecker, F.P. Schmitz, and E. Klesper, *J. High Resolut. Chromatogr. Chromatogr. Commun.* **9**, 525 (1986).
105. A. Nomura, J. Yamada, and K.-I. Tsunoda, *J. Chromatogr.* **448**, 87 (1988).
106. F.P. Schmitz, *Chromatographie mit überkritischen verdichteten Gasen zur Trennung von Oligomeren und Polymeren* (Profil Verlag, München, 1988).
107. Y. Hirata and F. Nakata, *J. Chromatogr.* **295**, 315 (1984).
108. B. Gemmel, Ph.D. Dissertation, Aachen, 1989.
109. M. Taguchi, S. Nagata, K. Matsumoto, S. Tsuga, Y. Hirata, and S. Toda, in *Proc. Int. Symp. Supercr. Fluids,* Vol. 1, M. Perrut, Ed. (Nice, France, 1988), p. 485.
110. F.P. Schmitz, H. Hilgers, and B. Gemmel, *J. Chromatogr.* **371**, 135 (1986).
111. F.P. Schmitz, D. Leyendecker, D. Leyendecker, and B. Gemmel, *J. Chromatogr.* **395**, 111 (1987).
112. F.P. Schmitz and B. Gemmel, in *Progress in HPLC*, Vol. 4, M. Yoshioka, S. Parvez, T. Miyazaki, and H. Parvez, Eds. (VSP, Utrecht, 1989), p. 73.
113. F.P. Schmitz, H. Hilgers, B. Lorenschat, and E. Klesper, *J. Chromatogr.* **346**, 69 (1985).
114. F.P. Schmitz, B. Gemmel, D. Leyendecker, and D. Leyendecker, *J. High Resolut. Chromatogr. Chromatogr. Commun.* **11**, 339 (1988).
115. F.P. Schmitz and H. Hilgers, *Makromol. Chem., Rapid Commun.* **7**, 59 (1986).
116. F.P. Schmitz, H. Hilgers, and E. Klesper, *J. Chromatogr.* **267**, 267 (1983).
117. B. Gemmel, B. Lorenschat, and F.P. Schmitz, *Chromatographia* **27**, 605 (1989).
118. J.D. Pinkston, in *SFC Applications: 1989 Symposium/Workshop on Supercritical Fluid Chromatography*, K.E. Markides and M.L. Lee, Comps. (Brigham Young University Press, Provo, UT, 1989).
119. F.P. Schmitz and B. Gemmel, *Fres. Z. Anal. Chem.* **330**, 216 (1988).
120. T.L. Chester, *J. Chromatogr.* **299**, 424 (1984).
121. R.E. Jentoft and T.H. Gouw, *J. Chromatogr. Sci.* **8**, 138 (1970).
122. J.D. Pinkston, D.J. Bowling, and T.E. Delaney, *J. Chromatogr.* **474**, 97 (1989).
123. H.-C.K. Chang and L.T. Taylor, *J. Chromatogr.*, submitted.
124. P.R. Geissler, *J. Am. Oil Chem. Soc.* **66**, 685 (1989).
125. D. Leyendecker, D. Leyendecker, and F. Höfler, *GIT Fachz. Lab.* **33**, 465 (1989).
126. K.D. Bartle, M.P. Burke, A.A. Clifford, I.L. Davies, J.P. Kithinji, M.W. Raynor, G.F. Shilstone, and A. Williams, *Eur. Chromatogr. News* **2**(5), 12 (1988).
127. Suprex Corporation, Publ. No. 8602-01,02 (1986).
128. R.P. Khosah, in *Modern Supercritical Fluid Chromatography*, C.M. White, Ed. (Hüthig, Heidelberg, 1988), p. 155.
129. H.M. Widmer, in *Modern Supercritical Fluid Chromatography*, C.M. White, Ed. (Hüthig, Heidelberg, 1988), p. 211.

130. A. Giorgetti, N. Pericles, H.M. Widmer, K. Anton, and P. Dätwyler, *J. Chromatogr. Sci.* **27**, 318 (1989).
131. M.L. Lee, B. Xu, E.C. Huang, N.M. Djordjevic, H.-C.K. Chang, and K.E. Markides, *J. Microcol. Sep.* **1**, 7 (1989).
132. D.W. Later, E.R. Campbell, and B.E. Richter, *J. High Resolut. Chromatogr. Chromatogr. Commun.* **11**, 65 (1988).
133. D.E. Knowles, L. Nixon, E.R. Campbell, D.W. Later, and B.E. Richter, *Fres. Z. Anal. Chem.* **330**, 225 (1988).
134. S. Mori, T. Saito, and M. Takeuchi, *J. Chromatogr.* **478**, 181 (1989).
135. L. Schallinger, in *SFC Applications: 1989 Symposium/Workshop on Supercritical Fluid Chromatography*, K.E. Markides and M.L. Lee, Comps. (Brigham Young University Press, Provo, UT, 1989).
136. M.W. Raynor, A.A. Clifford, K.D. Bartle, C. Reyner, A. Williams, and B.W. Cook, *J. Microcol. Sep.* **1**, 101 (1989).
137. Y. Hirata, F. Nakata, and M. Horihata, *J. High Resolut. Chromatogr. Chromatogr. Commun.* **11**, 81 (1988).
138. M.W. Raynor, K.D. Bartle, I.L. Davies, A. Williams, A.A. Clifford, J.M. Chalmers, and B.W. Cook, *Anal. Chem.* **60**, 427 (1988).
139. D.W. Later, B.E. Richter, and M.R. Andersen, *LC-GC* **4**, 992 (1986).
140. M.W. Raynor, I.L. Davies, K.D. Bartle, A.A. Clifford, A. Williams, J.M. Chalmers, and B.W. Cook, in *Analytical Applications of Spectroscopy*, C.S. Creaser and A.M.C. Davies, Eds. (Royal Society of Chemistry, London, 1988), p. 227.
141. F. Höfler, in *InCom 89*, W. Günther and J.P. Matthes, Eds. (GIT Verlag, Weinheim, 1989), p. 275.
142. K. Anton, N. Pericles, S.M. Fields, and H.M. Widmer, *Chromatographia* **26**, 224 (1988).
143. J.A. Nieman and L.B. Rogers, *Sep. Sci.* **10**, 517 (1975).
144. F.J. van Lenten, J.E. Conaway, and L.B. Rogers, *Sep. Sci.* **12**, 1 (1977).
145. T.L. Chester and D.P. Innis, *J. High Resolut. Chromatogr. Chromatogr. Commun.* **9**, 209 (1986).
146. T.L. Chester, L.J. Burkes, T.E. Delaney, D.P. Innis, G.D. Owens, and J.D. Pinkston, in *Supercritical Fluid Extraction and Chromatography: Techniques and Applications, ACS Symp. Ser. 366*, B.A. Charpentier and M.R. Sevevenants, Eds. (American Chemical Society, Washington, DC, 1988), p. 144.
147. V.N. Reinhold, D.M. Sheely, J.C. Kuei, and G.R. Her, *Anal. Chem.* **60**, 2719 (1988).
148. D.R. Gere, *Science* **222**, 253 (1983).
149. H.T. Kalinoski, B.W. Wright, and R.D. Smith, *Biomed. Environ. Mass Spectrom.* **15**, 239 (1988).
150. E.C. Huang, B.J. Jackson, K.E. Markides, and M.L. Lee, *Anal. Chem.* **60**, 2715 (1988).
151. E.C. Huang, B.J. Jackson, K.E. Markides, M.L. Lee, *J. Microcol. Sep.* **2**, 88, (1990).
152. T.A. Berger, J.F. Deye, M. Ashraf-Khorassani, and L.T. Taylor, *J. Chromatogr. Sci.* **27**, 105 (1989).
153. F.P. Schmitz and E. Klesper, *J. Supercrit. Fluids*, **3**, 29 (1990).
154. H.E. Schwartz, P.J. Barthel, S.E. Mohring, T.L. Yates, and H.H. Lauer, *Fres. Z. Anal. Chem.* **330**, 204 (1988).

155. J.L. Perrin and A. Prevot, *Revue Francais des Corps Gras* **35**, 485 (1988).
156. P. Sandra, F. David, F. Munari, G. Mapelli, and S. Trestianu, in *Supercritical Fluid Chromatography*, R.M. Smith, Ed. (Royal Society of Chemistry, London, 1988), p. 137.
157. W.E. Artz and R.M. Sauer, Jr., *American Oil Chemists' Meeting*, Cincinnati, OH, May 3-6, 1989.
158. T.L. Chester and D.P. Innis, *J. High Resolut. Chromatogr. Chromatogr. Commun.* **9**, 178 (1986).
159. T. Gorner and M. Perrut, *LC-GC* **7**, 503 (1989).
160. F.O. Geiser, S.G. Yocklovich, S.M. Lurcott, J.W. Guthrie, and E.J. Levy, *J. Chromatogr.* **459**, 173 (1988).
161. S.B. Hawthorne and D.J. Miller, *J. Chromatogr. Sci.* **27**, 197 (1989).
162. B.E. Berg and T. Greibrokk, *J. High Resolut. Chromatogr.* **12**, 322 (1989).
163. W. Gmür, J.O. Bosset, and E. Plattner, *J. Chromatogr.* **388**, 335 (1987).
164. W. Gmür, J.O. Bosset, and E. Plattner, *Mitt. Gebiete Lebensm. Hyg.* **78**, 21 (1987).
165. A.M.P. Vasconcelos, M. Goncalves, E.J.S. Gomes de Azecedo, M. Nunes da Ponte, and J.J. Chaves das Neves, *J. High Resolut. Chromatogr.* **12**, 244 (1989).
166. G. Schomburg and W. Roeder, *J. High Resolut. Chromatogr.* **12**, 218 (1989).
167. S.M. Fields and M.L. Lee, *J. Chromatogr.* **349**, 305 (1985).
168. Atlas of Chromatograms (SFC15), *J. Chromatogr. Sci.* **25**, 179 (1987).
169. F. Sarlin and J. Higgins, *Analusis* **15**, 25 (1987).
170. J. Doehl, A. Farbrot, T. Greibrokk, and B. Iversen, *J. Chromatogr.* **392**, 175 (1987).
171. T. Greibrokk, B.E. Berg, A.L. Blilie, J. Doehl, A. Farbrot, and E. Lundanes, *J. Chromatogr.* **394**, 429 (1987).
172. G.B. Levy, *Amer. Lab.* **18**, 62 (1987).
173. Atlas of Chromatograms (SFC16), *J. Chromatogr. Sci.* **25**, 179 (1987).
174. C.M. White and R.K. Houck, *J. High Resolut. Chromatogr. Chromatogr. Commun.* **8**, 293 (1985).
175. C.M. White and R.K. Houck, Suprex Technical Bulletin #210, Suprex Corporation, Pittsburgh, PA, 1985.
176. P. Sandra, in *SFC Applications: 1988 Workshop on Supercritical Fluid Chromatography*, M.L. Lee and K.E. Markides, Comps. (Brigham Young University Press, Provo, UT, 1988), p. 123.
177. R. Huopalahti, P. Laakso, J. Saaristo, R. Linko, and H. Kallio, *J. High Resolut. Chromatogr. Chromatogr. Commun.* **11**, 899 (1988).
178. H. Kallio, P. Laakso, R. Huopalahti, and R.R. Linko, *Anal. Chem.* **61**, 698 (1989).
179. M. Proot, P. Sandra, and E. Geeraert, *J. High Resolut. Chromatogr. Chromatogr. Commun.* **9**, 189 (1986).
180. Brownlee Labs, Technical Note 925.
181. T. Greibrokk, *SFC Applications: 1989 Symposium/Workshop on Supercritical Fluid Chromatography*, K.E. Markides and M.L. Lee, Comps. (Brigham Young University Press, Provo, UT, 1989), p. 85.
182. S.J. DeLuca, K.J. Voorhees, T.A. Langworthy, and G.U. Holzer, *J. High Resolut. Chromatogr. Chromatogr. Commun.* **9**, 182 (1986).

183. G.U. Holzer, P.J. Kelly, and W.J. Jones, *J. Microbiol. Methods* **8**, 161 (1988).
184. K. Sakaki, T. Sako, T. Yokochi, T. Sugeta, N. Nakazawa, M. Sako, O. Suzuki, and T. Hakuta, *Yukagaku* **36**, 943 (1987).
185. J.L. Cashaw, R. Seguar, and A. Zlatkis, *J. Chromatogr. Sci.* **8**, 363 (1970).
186. G. Holzer, S. Deluca, and K.J. Voorhees, *J. High Resolut. Chromatogr. Chromatogr. Commun.* **8**, 528 (1985).
187. E.D. Lee and J.D. Henion, *J. High Resolut. Chromatogr. Chromatogr. Commun.* **9**, 172 (1986).
188. K. Matsumoto, S. Tsuge, and Y. Hirata, *Anal. Sci.* **2**, 3 (1987).
189. S.L. Pentoney, Jr., K.H. Shaffer, and P.R. Griffiths, *J. Chromatogr. Sci.* **24**, 230 (1986).
190. J.W. Hellgeth, J.W. Jordan, L.T. Taylor, and M.Ashraf-Khorassani, *J. Chromatogr. Sci.* **24**, 183 (1986).
191. R.C. Wieboldt, G.E. Adams, and D.W. Later, *Anal. Chem.* **60**, 2422 (1988).
192. M. LaFosse, M. Dreux, and L. Morin-Allory, *J. Chromatogr.* **404**, 95 (1987).
193. P. Carroud, D. Thiebaut, M. Caude, R. Rosset, M. LaFosse, and M. Dreux, *J. Chromatogr. Sci.* **25**, 395 (1987).
194. D. Nizery, D. Thiebaut, M. Caude, R. Rosset, M. LaFosse, and M. Dreux, *J. Chromatogr.* **467**, 49 (1989).
195. S. Hoffman and T. Greibrokk, *J. Microcol. Sep.* **1**, 35 (1989).
196. M.W. Ogden and H.M. McNair, *J. High Resolut. Chromatogr. Chromatogr. Commun.* **8**, 816 (1985).
197. E.J. Guthrie and H.E. Schwartz, *J. Chromatogr. Sci.* **24**, 236 (1986).
198. S. Green and W. Bertsch, *J. High Resolut. Chromatogr. Chromatogr. Commun.* **11**, 414 (1988).
199. T.W. Rademacher, R.B. Parekh, and R.A. Dwek, *Ann. Rev. Biochem.* **57**, 785 (1988).
200. S.C. Hubbard and R.J. Ivatt, *Ann. Rev. Biochem.* **50**, 555 (1981).
201. R. Kornfeld and S. Kornfeld, *Ann. Rev. Biochem.* **54**, 631 (1985).
202. M. Novotny, S.R. Springston, P.A. Peaden, J.C. Fjeldsted, and M.L. Lee, *Anal. Chem.* **53**, 407A (1981).
203. J.C. Fjeldsted and M.L. Lee, *Anal. Chem.* **56**, 619A (1984).
204. D.M. Sheeley and V.N. Reinhold, Harvard School of Public Health, unpublished results.
205. IUBAC-IUB Commission on Biochemical Nomenclature **79**, 11 (1977).
206. S. Hakomori, *Ann. Rev. Biochem.* **50**, 733 (1981).
207. D.M. Marcus, *Mol. Immunol.* **21**, 1083 (1984).
208. S. Hakomori, *Sci. Amer.* **254**, 44 (1986).
209. J.C.H. Kuei, G.R. Her, and V.N. Reinhold, *Anal. Biochem.* **172**, 228 (1988).
210. E.T. Rietschel, S. Hase, M.T. King, J. Redmond, and V. Lehmann, in *Microbiology*, D. Schlessinger, Ed. (American Society of Microbiology, Washington, DC, 1977), p. 262.
211. C. Galanos, O. Luderitz, E.T. Rietshel, and O. Wesrphal, *Int. Rev. Biochem.* **14**, 280 (1977).
212. N. Qureshi, K. Takayama, and E. Ribi, *J. Biol. Chem.* **257**, 11808 (1982).
213. N. Qureshi, K. Takayama, D. Heller, and C. Fenselau, *J. Biol. Chem.* **258**, 12947 (1983).
214. C.R.H. Raetz, S. Purcell, M.V. Meyer, N. Qureshi, and K. Takayama, *J. Biol. Chem.* **260**, 16080 (1985).

215. I. Mattsby-Baltzer, P. Gemski, and C.R. Alving, *Rev. Infect. Dis.* **6**, 444 (1984).
216. K. Takayama, N. Qureshi, K. Hyver, J. Honovich, R.J. Cotter, P. Mascagni, and H. Schneider, *J. Biol. Chem.* **261**, 10624 (1986).
217. J. Gmeiner, O. Luderitz, and O. Westphal, *Eur. J. Biochem.* **7**, 370 (1969).
218. N. Qureshi, P. Mascagni, E. Ribi, and K. Takayama, *J. Biol. Chem.* **260**, 5271 (1985).
219. R.S. Johnson, G.R. Her, J. Grabarek, J.J. Hawiger, and V.N. Reinhold, *J. Biol. Chem.* **265**, 8108 (1990).
220. H. Schachter, *Biochem. Cell Biol.* **64**, 163 (1986).
221. A. Kobata, in *Biology of Carbohydrates*, Vol. 2, V. Ginsburg and P.W. Robbins, Eds. (John Wiley & Sons, New York, 1984).
222. T. Feizi, *Nature*, **314**, 53 (1985).
223. J.P. Caesar, D.M. Sheeley, and V.N. Reinhold, *Proc. Thirty-seventh ASMS Conf. Mass Spectrom. Allied Topics*, Miami, FL, May 21-26, 1989, p. 971.
224. V.N. Reinhold, D.M. Sheeley, and J.P. Caesar, *198th National Meeting* (American Chemical Society, Miami, FL, 1989).
225. R. Smith, H.T. Kalinoski, and H.R. Udseth, *Mass Spectrom. Rev.* **6**, 445 (1987).
226. D.M. Sheeley and V.N. Reinhold, *J. Chromatogr.* **474**, 83 (1989).
227. J.D. Pinkston, G.D. Owens, D.S. Millington, L.J. Burkes, and T.E. Delaney, *Anal. Chem.* **59**, 962 (1988).
228. L.J. Halverson and G. Stacey, *Microbiol. Rev.* **50**, 193 (1986).
229. K.J. Miller, V.N. Reinhold, A.C. Weissborn, and E.P. Kennedy, *Biochim. Biophys. Acta* **901**, 112 (1987).
230. K.J. Miller, R.S. Gore, R. Johnson, A. Benesi, and V.N. Reinhold, *J. Bacteriol.* **72**, 136 (1990).
231. M.V. Novotny, *Science* **246**, 51 (1989).
232. R.S. Johnson, D.M. Sheeley, and V.N. Reinhold, Harvard School of Public Health, unpublished results.
233. S. Hara, A. Dobashi, K. Konoshita, T. Hondo, M. Saito, and M. Senda, *J. Chromatogr.* **371**, 153 (1986).
234. W. Röder, F.J. Ruffing, G. Schomburg, and W.H. Pirkle, *J. High Resolut. Chromatogr. Chromatogr. Commun.* **10**, 665 (1987).
235. A. Dobashi, Y. Dobashi, T. Ono, S. Hara, M. Saito, S. Higashidate, and Y. Yamauchi, *J. Chromatogr.* **461**, 121 (1989).
236. R.D. Smith, H.R. Udseth, and B.W. Wright, *J. Chromatogr. Sci.* **23**, 192 (1985).
237. J.A.G. Roach, J.A. Sphon, J.A. Easterling, and E.M. Calvey, *Biomed. Environ. Mass Spec.* **18**, 64 (1989).
238. W.F. Bibb, D.E. Karr, and C.W. Moss, *J. Clin. Microbiol.* **15**, 1044 (1982).
239. J.R. Wheeler, M.E. McNally, *Research & Development* **2**, 134 (1989).
240. K. Grolimund, Ciba-Geigy, unpublished results.
241. H.E. Schwartz, in *SFC Applications: 1989 Symposium/Workshop on Supercritical Fluid Chromatography*, K.E. Markides and M.L. Lee, Comps. (Brigham Young University Press, Provo, UT, 1989), p. 87.
242. F. Yang, in *SFC Applications: 1989 Symposium/Workshop on Supercritical Fluid Chromatography*, K.E. Markides and M.L. Lee, Comps. (Brigham Young University Press, Provo, UT, 1989), p. 83.

243. D.R. Gere, T.J. Stark, and T. Tweeten, Application Note AN800-4 (Hewlett-Packard Co., Avondale, PA).
244. A.W. Francis, *J. Phys. Chem.* **58**, 1099 (1954).
245. D.K. Dandge, J.P. Heller, and K.W. Wilson, *Ind. Chem. Prod. Res. Dev.* **24**, 162 (1985).
246. S.M. Fields and K. Grolimund, *J. High Resolut. Chromatogr. Chromatogr. Commun.* **11**, 897 (1988).
247. F. David and P. Sandra. *J. High Resolut. Chromatogr. Chromatogr. Commun.* **11**, 897 (1988).
248. S.M. Fields and K. Grolimund, *J. Chromatogr.* **472**, 197 (1989).
249. J.W. Hellgeth, M.G. Fessehaie, and L.T. Taylor, *Chromatographia* **25**, 172 (1988).
250. J.L. Janicot, M. Caude, and R. Rosset, *J. Chromatogr.* **437**, 351 (1988).
251. T. Alfredson, *Int. Lab.* **11**, 46 (1981).
252. S.M. Fields, H.J. Grether, and K. Grolimund, *J. Chromatogr.* **472**, 175 (1989).
253. European Patent Application EP 263044A2, 1988, 6 Apr., Rhone-Poulenc Chimie.
254. J.H. Phillips and R.J. Robey, *J. Chromatogr.* **465**, 177 (1985).
255. S.S.H. Rizvi, J.A. Daniels, A.L. Benado, and J.A. Zollweg, *Food Tech.* **40**(6), 55 (1986).
256. M. Toivakka, M. Alkio, and T. Harvala, *Kemia-Kemi* **14**, 1078 (1987).
257. S.M. Hwang, *J. Assoc. Off. Anal. Chem.* **68**, 684 (1985).
258. J.N. Thompson, in *Trace Analysis*, Vol. 2, J.F. Lawrence, Ed. (Academic Press, New York, 1982).
259. K. Matsumoto, S. Tsuge, and Y. Hirata, *Chromatographia* **21**, 617 (1986).
260. C.M. White, D.R. Gere, D. Boyer, F. Pacholec, and L.K. Wong, *J. High Resolut. Chromatogr. Chromatogr. Commun.* **11**, 94 (1988).
261. F. Yang, *1989 Symposium/Workshop on Supercritical Fluid Chromatography*, Snowbird, UT, June 13-15, 1989.
262. H. Hondo, M. Saito, and M. Senda, *Bunseki Kagaku* **35**, 316 (1986).
263. Y. Hirata, *1988 Workshop on Supercritical Fluid Chromatography*, Park City, UT, January 15-18, 1988.
264. R. Board, D. McManigill, H. Weaver, and D.R. Gere, *Publication No. 43-5953-1647* (Hewlett Packard Co., Avondale, PA, 1982).
265. D.R. Gere, *Application Note* AN 800-5 (Hewlett Packard Co., Avondale, PA, 1983).
266. M. Saito, Y. Yamauchi, K. Inomuta, and W. Kottkamp, *J. Chromatogr. Sci.* **27**, 79 (1989).
267. R.A. Scanlan, in *Analysis of Food Contaminants*, J. Gilbert, Ed. (Elsevier Applied Scientific Publishers, London, 1984), p. 321.
268. K. Grolimund, *J. Chromatogr. Sci.* **26**, 194 (1988).
269. D.F. Williams, *Chem. Eng. Sci.* **36**, 1769 (1981).
270. K.-W. Quirin and D. Gerard, in *Grenzen, Möglichkeiten und Anwendungen überkritischer Phasen in der Chromatographie* (Saarbrucken, 1988).
271. S.B. Hawthorne, M.S. Krieger, and D.J. Miller, *Anal. Chem.* **60**, 472 (1988).
272. F. Temelli, C.S. Chen, and R.J. Braddock, *Food Tech.* **42**, 145 (1988).
273. E. Stahl and K.-W. Quirin, *Flavex Naturextrakte Reviews*, 1984, p. 189.
274. M.A. Andersen, J.T. Swanson, N.L. Porter, and B.E. Richter, *J. Chromatogr. Sci.* **27**, 371 (1989).

275. P. Morin, H. Pichard, H. Richard, M. Caude, and R. Rosset, *J. Chromatogr.* **464**, 125 (1989).
276. H. Pichard, M. Caude, P. Sassiat, and R. Rosset, *Analusis* **16**, 131 (1988).
277. P. Morin, H. Pichard, M. Caude, H. Richard, and R. Rosset, in *Progress in Terpene Chemistry*, D. Joulain, Ed. (Frontiers, Gif/Yvette, 1986), p. 343.
278. C.A. Rouse, in *SFC Applications: 1988 Workshop on Supercritical Fluid Chromatography*, K.E. Markides and M.L. Lee, Comps. (Brigham Young University Press, Provo, UT, 1988).
279. H. Engelhardt, in *SFC Applications: 1989 Symposium/Workshop on Supercritical Fluid Chromatography*, K.E. Markides and M.L. Lee, Comps. (Brigham Young University Press, Provo, UT, 1989).
280. E. Stahl and D. Gerard, *Z. Lebensm. Unters. Forsch.* **176**, 1 (1983).
281. M.W. Raynor, R. Moulder, I.L. Davies, J.P. Kithinji, J.E. Hall, A.A. Clifford, and K.D. Bartle, in *Proc. Tenth Int. Symp. Cap. Chromatogr.*, P. Sandra and G. Redant, Eds. (Huethig, Heidelberg, 1989), p. 1366.
282. W.M.A. Niessen, U.R. Tjaden, and J. van Der Greef, *J. Chromatogr. Biomed. Appl.* **492**, 167 (1989).
283. J. Balsevich, L.R. Hogge, A.J. Berry, D.E. Games, and I.C. Mylchreest, *J. Nat. Prod.* **51**, 1173 (1988).
284. G. Holzer, L.H. Zalkow, and C.F. Asibal, *J. Chromatogr.* **400**, 317 (1987).
285. J.B. Crowther and J.D. Henion, *Anal. Chem.* **57**, 2711 (1985).
286. T.A. Langworthy, T.G. Tornabene, and G. Holzer, *Zbl. Bakt. Hyg., I. Abt, Orig. C3* **228** (1982)
287. W.J. Jones and G. Holzer, Georgia Institute of Technology, unpublished results.
288. G. Holzer, Georgia Institute of Technology, unpublished results.
289. C.A. Mancuso, P.D. Nicols, and D.C. White, *J. Lipid Res.* **27**, 49 (1986).
290. T.A. Langworthy, G. Holzer, J.G. Zeikus, and T.G. Tornabene, *System. Appl. Microbiol.* **4**, 1 (1983).
291. J.L. Pond, T.A. Langworthy, and G. Holzer, *Science* **231**, 1134 (1986).
292. D. Jones and M.D. Collins, *Biol. Reviews* **45**, 316 (1981).
293. D.R. Gere, *Application Note* AN 800-2 (Hewlett-Packard Co., Avondale, PA, 1983).
294. P. Morin, M. Caude, and R. Rosset, *Analusis* **15**, 117 (1987).
295. H. Engelhardt and A. Gross, *J. High Resolut. Chromatogr. Chromatogr. Commun.* **11**, 38 (1988).
296. S.B. Hawthorne, D.J. Miller, and M.S. Krieger, *Fres. Z. Anal. Chem.* **330**, 211 (1988).
297. J.R. Bamburg, in *Microbial Toxins*, S. Kadis, A. Ciegler, and S.J. Ali, Eds. (Academic Press, New York, 1971), p. 207.
298. W. Niederberger, P. Schaub, and T. Beveridge, *J. Chromatogr.* **183**, 454 (1980).
299. R.D. Smith and H.R. Udseth, *Anal. Chem.* **59**, 13 (1987).
300. J.C. Giddings, L. McLaren, and M.N. Myers, *Science* **159**, 197 (1968).
301. N.M. Frew, C.G. Johnson, and R.H. Bromund, in *Supercritical Fluid Extraction and Chromatography: Technology and Applications, ACS Symp. Ser. 366*, B.A. Charpentier and M.R. Sevenants, Eds. (American Chemical Society, Washington, DC, 1988), p. 208.
302. S.D. Zaugg, S.J. DeLuca, and K.J. Voorhees, Colorado School of Mines, unpublished results.

303. S.M. Fields, K.E. Markides, and M.L. Lee, *J. High Resolut. Chromatogr. Chromatogr. Commun.* **11**, 25 (1988).
304. L.G. Randall, Hewlett Packard Technical Paper No. 120 (Hewlett Packard Co., Avondale, PA, 1983).
305. E.D. Lee, S.H. Hsu, and J.D. Henion, *Anal. Chem.* **60**, 360 (1988).
306. H-C.K. Chang, K.E. Markides, J.S. Bradshaw, and M.L. Lee, *J. Microcol. Sep.* **1**, 131 (1989).
307. J.C. Fjeldsted, R.C. Kong, and M.L. Lee, *J. Chromatogr.* **279**, 449 (1983).
308. L.G. Randall, in *Ultrahigh Resolution Chromatography, ACS Symp. Ser. 250*, S. Ahuja, Ed. (American Chemical Society, Washington, DC, 1984), p. 135.
309. E.D. Morgan, S.J. Murphy, D.E. Games, and I.C. Mylchreest, *J. Chromatogr.* **441**, 165 (1988).
310. M.W. Raynor, J.P. Kithinji, I.K. Barker, K.J. Bartle, and I.D. Wilson, *J. Chromatogr.* **436**, 497 (1988).
311. R.C. Wieboldt, K.D. Kempfert, D.W. Later, and E.R. Campbell, *J. High Resolut. Chromatogr.* **12**, 106 (1989).
312. P.A. Pospisil and M.F. Marcus, Chemical Waste Management, unpublished results.
313. *Kirk-Othmer Concise Encyclopedia of Chemical Technology* (John Wiley & Sons, New York, 1985) p. 1164.
314. *Analytical Methods for a Textile Laboratory*, J.W. Weaver, Ed. AATCC (Research Triangle Park, NC, 1984).
315. AATCC Test Method 94-1987, *Finishes in Textiles: Identification*, 1988 Technical Manual of the AATCC, p. 136.
316. S.B. Moore, in *SFC Applications: 1989 Symposium/Workshop on Supercritical Fluid Chromatography*, K.E. Markides and M.L. Lee, Comps. (Brigham Young University Press, Provo, UT, 1989).
317. S.G. Yocklovich, S.F. Sarner, and E.J. Levy, *Am. Lab.* **21**, 26 (1989).
318. S.B. Moore, G.E. Coleman, J. Madren, and R.A. Diehl, Burlington Research, Inc., unpublished results.
319. *SFC Applications: 1988 Workshop on Supercritical Fluid Chromatography*, M.L. Lee and K.E. Markides, Comps. (Brigham Young University Press, Provo, UT, 1988).
320. *Symposium and Workshop on Supercritical Fluid Chromatography*, Snowbird, UT, June 1989, unpublished discussions.
321. F.I. Onuska and K.A. Terry, *J. High Resolut. Chromatogr. Chromatogr. Commun.* **11**, 874 (1988).
322. SFE-SFC of Polymer Additives in Polyethylene, *Suprex Focus*, **2**, No. 2, p. 2.
323. E. Klesper, A.H. Corwin, and D.A. Turner, *J. Org. Chem.* **27**, 700 (1962).
324. N.M. Karayannis and A.H. Corwin, *Anal. Biochem.* **26**, 34 (1968).
325. N.M. Karayannis and A.H. Corwin, *J. Chromatogr.* **47**, 247 (1970).
326. M. Ashraf-Khorassani and L.T. Taylor, *J. Chromatogr. Sci.* **27**, 329 (1989).
327. N.M. Karayannis, A.H. Corwin, E.W. Baker, E. Klesper, and J.A. Walter, *Anal. Chem.* **40**, 1736 (1968).
328. N.M. Karayannis and A.H. Corwin, *J. Chromatogr. Sci.* **8**, 251 (1970).
329. F. Bickmann and B. Wenclawiak, *Z. Anal. Chem.* **320**, 261 (1985).
330. M. Ashraf-Khorassani, J.W. Hellgeth, and L.T. Taylor, *Anal. Chem.* **59**, 2077 (1987).

331. K.R. Jahn and B.W. Wenclawiak, *Z. Anal. Chem.* **330**, 243 (1988).
332. B. Wenclawiak and F. Bickmann, *Z. Anal. Chem.* **319**, 305 (1984).
333. P. Manninen and M.-L. Riekkola, University of Helsinki, unpublished results.
334. M.-L. Riekkola, *Mikrochim. Acta* **2**, 381 (1983).
335. C. Fujimoto, H. Yoshida, and K. Jinno, *J. Chromatogr.* **411**, 213 (1987).
336. C. Fujimoto, H. Yoshida, and K. Jinno, *J. Microcol. Sep.* **1**, 19 (1989).
337. C.B. Motley, M. Ashraf-Khorassani, and G.L. Long, *Appl. Spectrosc.* **43**, 737 (1989).
338. K. Jinno, H. Mae, and C. Fujimoto, in *Proc. Tenth Int. Symp. Cap. Chromatogr.*, P. Sandra and G. Redant, Eds. (Huethig, Heidelberg, 1989), p. 771.
339. J.W. Olesik and S.V. Olesik, *Anal. Chem.* **59**, 796 (1987).
340. M. Simonyi, I. Fitos, and J. Vizy, *TIPS* **7**, 112 (1986).
341. E.P.A. Lehmann, *TIPS* **7**, 281 (1986).
342. D.E. Drayer, in *Drug Stereochemistry: Analytical Methods and Pharmacology*, I.W. Wainer and D.E. Drayer, Eds. (Marcel Dekker, New York, 1988), p. 213.
343. W. Walther, Hoffmann-La Roche, unpublished results.
344. H. Frank and G.J. Nicholson, *J. Chromatogr.* **146**, 197 (1978).
345. J.S. Bradshaw, S.K. Aggarwal, C.A. Rouse, B.J. Tarbet, K.E. Markides, and M.L. Lee, *J. Chromatogr.* **405**, 167 (1987).
346. W. Röder, F-J. Ruffing, G. Schomburg, and W.H. Pirkle, *J. High Resolut. Chromatogr. Chromatogr. Commun.* **10**, 665 (1987).
347. P. Macaudiére, M. Caude, R. Rosset, and A. Tambuté, *J. Chromatogr. Sci.* **27**, 383 (1989).
348. P.A. Mourier, E. Eliot, M.H. Caude, R.H. Rosset, and A.G. Tambuté, *Anal. Chem.* **57**, 2819 (1985).
349. R.E. Kaiser, *Chromatographia* **9**, 337 (1976); **10**, 455 (1977).
350. R. Aichholz, U. Bölz, and P. Fischer, *J. High Resolut. Chromatogr. Chromatogr. Commun.* **13**, 234 (1990).
351. I.W. Wainer, *Tr. Anal. Chem.* **6**, 125 (1987).
352. I.W. Wainer and M.C. Alembik, in *Chromatographic Chiral Separations*, M. Zief and L.J. Crane, Eds. (Marcel Dekker, New York, 1988), p. 357.
353. I.W. Wainer, R.M. Stiffin, and Y-Q. Chu, in *Chiral Separations*, D. Stevenson and I.D. Wilson, Eds. (Plenum Press, New York, 1988), p. 13.
354. W.H. Pirkle, C.W. House, and J.M. Finn, *J. Chromatogr.* **192**, 143 (1980).
355. P. Mourier, P. Sassiat, M. Caude, and R. Rosset, *J. Chromatogr.* **353**, 61 (1986).
356. P. Macaudiére, A. Tambuté, M. Caude, R. Rosset, M.A. Alembik, and I.W. Wainer, *J. Chromatogr.* **371**, 177 (1986).
357. S. Hara, A. Dobashi, T. Hondo, M. Saito, and M. Senda, *J. High Resolut. Chromatogr. Chromatogr. Commun.* **9**, 249 (1986).
358. S. Hara, A. Dobashi, K. Kinoshita, T. Hondo, M. Saito, and M. Senda, *J. Chromatogr.* **371**, 153 (1986).
359. A. Dobashi, Y. Doabshi, T. Ono, S. Hara, M. Saito, S. Higashidate, and Y. Yamauchi, *J. Chromatogr.* **461**, 121 (1989).
360. F. Gasparrini, D. Misiti, and C. Villlani, *J. High Resolut. Chromatogr. Chromatogr. Commun.* **13**, 182 (1990).

361. A. Ichida, T. Shabata, I. Okamoto, Y. Yudi, H. Namikashi, and Y. Toga, *Chromatographia* **19**, 280 (1984).
362. T. Shabata, I. Okamato, and K. Ishii, *J. Liq. Chromatogr.* **9**, 313 (1986).
363. M. Caude, in *SFC Applications: 1988 Workshop on Supercritical Fluid Chromatography*, K.E. Markides and M.L. Lee, Comps. (Brigham Young University Press, Provo, UT, 1988).
364. P. Petersson, D. Johnson, M. Eguchi, K.E. Markides, B. Rossiter, J.S. Bradshaw, and M.L. Lee, unpublished results.
365. Y. Okamoto and K. Hatada, *J. Liq. Chromatogr.* **9**, 369 (1986).
366. G. Blashke, *J. Liq. Chromatogr.* **9**, 341 (1986).
367. D.W. Armstrong and W. DeMond, *J. Chromatogr. Sci.* **22**, 411 (1984).
368. G. Hesse and R. Hagel, *Chromatographia* **9**, 62 (1976).
369. P. Macaudiére, M. Caude, R. Rosset, and A. Tambuté, *J. Chromatogr.* **405**, 135 (1987).
370. P. Macaudiére, M. Caude, R. Rosset, and A. Tambuté, *J. Chromatogr.* **450**, 255 (1988).
371. W. Linder and C. Petterson, in *Liquid Chromatography in Pharmaceutical Development: An Introduction*, W. Wainer, Ed. (Aster Publishing Corp., Springfield, OR, 1985), p. 63.
372. S. Allenmark, B. Bomgren, and H. Boren, *J. Chromatogr.* **264**, 63 (1983).
373. J. Hermansson, *J. Chromatogr.* **269**, 71 (1983).

Index